DRUG DISCOVERY AND DEVELOPMENT

DRUG DISCOVERY AND DEVELOPMENT

Volume 1: Drug Discovery

Edited by

MUKUND S. CHORGHADE

A JOHN WILEY & SONS, INC., PUBLICATION

"Epothilone" cover art by Doug Scard
www.sputniknewmedia.com

Published by John Wiley & Sons, Inc., Hoboken, New Jersey.
Published simultaneously in Canada.

For general information on our other products and services or for technical support, please contact our Customer Care Department within the United States at (800) 762-2974, outside the United States at (317) 572-3993 or fax (317) 572-4002.

Wiley also publishes its books in a variety of electronic formats. Some content that appears in print may not be available in electronic formats. For more information about Wiley products, visit our web site at www.wiley.com.

Library of Congress Cataloging-in-Publication Data:

Drug discovery and development/edited by Mukund S. Chorghade.

p. cm.
Includes bibliographical references and index.
ISBN-13: 978-0-471-39848-6
ISBN-10: 0-471-39848-9 (cloth : v. 1)
1. Drug development. I. Chorghade, Mukund S. (Mukund Shankar)
[DNLM: 1. Drug Design. 2. Chemistry, Pharmaceutical–methods.
3. Drug Evaluation, Preclinical–methods. QV 744 D79334 2006]
RM301.25C488 2006
615'.19–dc22

2005021297

Printed in the United States of America
10 9 8 7 6 5 4 3 2 1

CONTENTS

CONTRIBUTORS

John W. Babich, Molecular Insight Pharmaceuticals, Inc., 160 Second Street, Cambridge, MA 02142, USA

Pradeep K. Dhal, Genzyme Corporation, 153 Second Avenue, Waltham, MA 02451, USA

Sham Diwanay, Department of Microbiology, Abasaheb Garware College, Pune 411004, India

Apurba Dutta, Department of Medicinal Chemistry, Malott Hall, 1251 Wescoe Hall Drive, Kansas University, Lawrence, KS 66045-7582, USA

William C. Eckelman, Molecular Tracer, LLC, Bethesda, MD 20814, USA

Paul W. Erhardt, Center for Drug Design and Development, The University of Toledo College of Pharmacy, 2801 West Bancroft Street, Toledo, OH 43606-3390, USA

Paul L. Feldman, GlaxoSmithKline Research and Development, Research Triangle Park, NC 27709, USA

János Fischer, Gedeon Richter Ltd., H-1475 Budapest 10, Hungary

C. Robin Ganellin, University College London, Department of Chemistry, Christopher Ingold Laboratories, 20 Gordon Street, London WC1H 0AJ, UK

Manish Gautam, Bioprospecting Laboratory, Interdisciplinary School of Health Sciences, University of Pune, Pune 411007, India

Anikó Gere, Gedeon Richter Ltd., H-1475 Budapest 10, Hungary

Karl Grozinger, Boehringer-Ingelheim Pharmaceuticals, 900 Ridgebury Road, Ridgefield, CT 06877-0368, USA

Karl Hargrave, Boehringer-Ingelheim Pharmaceuticals, 900 Ridgebury Road, Ridgefield, CT 06877-0368 , USA

S. Randall Holmes-Farley, Genzyme Corporation, 153 Second Avenue, Waltham, MA 02451, USA

Ian Hughes, GlaxoSmithKline Pharmaceuticals, New Frontiers Science Park (North), Third Avenue, Harlow, Essex CM19 5AW, UK

Chad C. Huval, Genzyme Corporation, 153 Second Avenue, Waltham, MA 02451, USA

Susan Dana Jones, BioProcess Technology Consultants, Inc., Acton, MA 01720, USA

Hwa-Ok Kim, CreaGen Biosciences, Inc., 25-K Olympia Avenue, Woburn, MA 01801, USA

Bruce E. Maryanoff, Johnson & Johnson Pharmaceutical Research and Development, Spring House, PA 19477-0776, USA

Lester A. Mitscher, Department of Medicinal Chemistry, 4010 Malott Hall, 1251 Wescoe Hall Drive, Kansas University, Lawrence KS 66045-7582, USA

Richard J. Pariza, Cedarburg Pharmaceuticals, 870 Badger Circle, Grafton, WI 53024, USA

Bhushan Patwardhan, Bioprospecting Laboratory, Interdisciplinary School of Health Sciences, University of Pune, Pune 411007, India

Norton P. Peet, CreaGen Biosciences, Inc., 25-K Olympia Avenue, Woburn, MA 01801, USA

John Proudfoot, Boehringer-Ingelheim Pharmaceuticals, 900 Ridgebury Road, Ridgefield, CT 06877-0368, USA

Peter G. Warren, Independent Biotechnology Consultant, Lexington, MA 02421, USA

Camille G. Wermuth, Prestwick Chemical, Inc., Boulevard Gonthier d'Andernach, 67400 Illkirch, France

PREFACE

The pharmaceutical sector has traditionally been a vibrant, innovation-driven, and highly successful component of industry at large. In recent years, a confluence of spectacular advances in chemistry, molecular biology, genomics, and chemical technology and the cognate fields of spectroscopy, chromatography, and crystallography have led to the discovery and development of numerous novel therapeutic agents for the treatment of a wide spectrum of diseases. To facilitate this process, there has been a significant and noticeable effort aimed at improving the integration of discovery technologies, chemical outsourcing for route selection and delivery of active pharmaceutical ingredients, drug product formulations, clinical trials, and refined deployment of information technologies. Multidisciplinary and multifunctional teams focusing on lead generation and optimization have replaced the traditional, specialized research groups. To develop a drug from conception to commercialization, the biotechnology and biopharmaceutical industries (which have been highly entrepreneurial) have reached out and established global strategic partnerships with numerous companies.

Currently, there is no single book in the market that provides an overview of strategies, tactics, milestones, and benchmarks in the entire sequence of operations involved in discovering a drug and delivering it to the armamentarium of clinicians and medical practitioners. A large number of advanced texts dealing exclusively with medicinal chemistry have been published; process chemistry has not received the attention it deserves (the journal *Organic Process Research and Development* is a useful and overdue step in this direction). Strategic in licensing, virtual company interactions and related topics have hitherto not been chronicled in books on drug discovery. There is usually a great gulf between the medicinal and process chemists in industry; neither has the opportunity to delve into the disparate literature of the other. This book is designed to bridge this gap and provide greater understanding of the target areas.

Conversely, the book is not designed to be a treatise or an encyclopedia. Its scope precludes complete coverage of any defined area. Ideally, it is envisioned to be an advanced-level monograph with appeal to active researchers and investigators in the entire gamut of operations comprising the drug discovery and development process. This two-volume text will be useful to a broad community of academic and industrial chemists. An overview of several recent developments is presented; this will make it valuable as a reference primer. The topics and the extent to which they are summarized are based on decisions by the editor and authors. Each contributor has achieved international distinction in the relevant fields.

The introductory chapter in the first volume, by Dr. Richard Pariza, delineates all the essential elements that comprise the development process, from the initial conception of a program to the successful marketing of a new drug. A time line for making critical decisions, conducting pivotal studies, and the approximate duration of different activities is described. The time line helps to put the entire developmental process into perspective for the reader and serves as a conceptual index that unifies all the contributions. Dr. Pariza elaborates on these concepts by describing some fascinating aspects of the work done on commercially successful analogs of erythromycin.

Professor Paul Erhardt describes the competition in the pharmaceutical industry to be "first to the market" in a chosen therapeutic area and the strategies currently being pursued. These include research in combinatorial chemistry, collaboration with biopharmaceutical and "virtual companies," and strategies in the licensing of drug candidates, among others. Increasingly, the large pharmaceutical corporations have turned to the establishment of strategic links with small biotechnology and biopharmaceutical companies for in-licensing of drug candidates and enhancement of drug portfolios. The author takes a futuristic look at what medicinal chemistry is expected to be in the new millennium. Dr. Erhardt is chairman of the Division of Chemistry and Human Health of the International Union of Pure and Applied Chemistry; his insights gleaned from expertise and experience constitute a valuable lesson.

Professor Lester Mitscher, an internationally renowned academician and expert, and Professor Apurba Dutta take us through the next critical phase of the drug discovery process: detailed studies of the absorption, metabolism, and excretion of potential drug candidates. Such studies are of pivotal importance in determining the suitability of a new compound for further clinical evaluation. His chapter on contemporary drug discovery presents a broad overview of the successive steps in the progression of a drug from mind to marketplace.

Combinatorial chemistry has played a highly visible role in the drug discovery effort in several companies; numerous new companies have been set up to partner established companies in the discovery of new molecular entities. The strategic focus in this field is continually shifting; Dr. Ian Hughes reviews the state of the art with selected examples from his own research at GlaxoSmithKline. This is followed by an excellent exposition by Drs. Norton Peet and Hwa-Ok Kim regarding efficient design and development of parallel solution-phase synthesis. Specific examples of lead identification and optimization are presented.

Dr. János Fischer and Dr. Anikó Gere delve into the important area of the timing of analog research in medicinal chemistry. This work is a remarkable synthesis of knowledge of drugs and their functional congeners and has formed the basis of a major IUPAC project. Professor Camille Wermuth presents fascinating examples of specific new drugs being derived via the functionalization of old drugs. This approach uses the old drugs as new scaffolds and derives benefit from new molecules already having a propensity to be "druglike." Professor Wermuth has worked at the academia–industry interface for collaboration in drug discovery.

Drs. Susan Dana Jones and Peter Warren focus on the impact of proteomics on the discovery of drugs: newer methods for efficient, economical, and safer production, and the development of novel targets and assays for the application of traditional medicinal chemistry methods. A brief survey of novel therapeutic concepts such as gene therapy, antisense, transgenic animals, and pharmacogenomics that have opened new vistas in drug development are surveyed. The authors have familiarized readers with several newer biology-based technologies. Next, Professor Paul Erhardt introduces the concept of using drug metabolism databases during the drug discovery and development process.

Professor C. Robin Ganellin exemplifies the discovery of Tagamet using classical structure–activity relationships and modeling of pharmacophore receptors. This drug was the first "billion-dollar drug." The research work by Sir James Black and Robin Ganellin has long been considered to be a tour de force in modern medicinal chemistry.

The art and science of medicinal chemistry is exemplified and epitomized clearly in the next few chapters. The exponents of the art are highly distinguished and prolific industrial researchers whose work spans the gamut of the therapeutic spectrum. Dr. Bruce Maryanoff brilliantly summarizes research into the discovery of potent nonpeptide vasopressin receptor antagonists. The work is a great tribute to the perseverance and persistence of researchers. Valuable insights are presented into the discovery process: A key idea is followed through despite initial adversity. Dr. Paul Feldman presents an informative case study on the discovery of Ultiva (remifentanil). This is an ultrashort-acting analgesic used as an adjunct to anesthesia. Dr. Paul Feldman introduces the rationale for its discovery and discusses how remifentanil fits into the anesthesia drug regimen. The desire to discover an ultrashort-acting analgesic, the group's medicinal chemistry efforts, and the structure–activity relationships are discussed. The divergent syntheses of analogs and the final process route are described. Finally, the clinical trial data and clinical uses are incorporated in the chapter to give a complete picture of Ultiva. Drs. Karl Grozinger, John Proudfoot, and Karl Hargrave discuss the discovery and development of nevirapine. This drug was a key ingredient in our efforts to combat AIDS, and the success of the researchers is an object lesson in creativity and how various skills were brought to the forefront of research.

Drs. John Babich and William Eckelman present insights into the applications of nuclear imaging in drug discovery and development; the work is technologically complex and involves radiopharmaceuticals. An increasing number of biopharmaceutical companies are involved in this activity; readers will find this to be a new and exciting domain of expertise.

Drs. Pradeep Dhal, Chad Huval, and Randal Holmes-Farley take the reader into a new and somewhat unexplored area of polymer therapeutics. The exciting idea of using a polymer as an active pharmaceutical ingredient was introduced in the 1990s and led to the discovery of drugs such as Renagel and Welchol. A large-molecular-weight polymer when used as a drug manifests its action in the gastrointestinal tract by adsorbing and removing unwanted analytes. The drug is not systemically absorbed in the blood and therefore does not generate any hazardous metabolites or lead to any toxic effects. It is also unnecessary to do long-term toxicity tests. This leads to a significant acceleration of the time required to introduce a drug to meet unmet medical needs.

Professor Bhushan Patwardhan and his collaborators demonstrate the utility of botanical immunomodulators and chemoprotectants in cancer therapy. Much of this work has its genesis in the Indian medicine systems of ayurveda; this turns pharmacology "on its head." It starts with plant extracts that have been used extensively in medicine in Asia and identifies the active ingredients from a complex mixture of ingredients. There is considerable scientific debate and discussion about whether the active moieties exhibit their pharmacological action in tandem or singly.

A detailed introduction to the second volume will be presented in its preface; given here are glimpses of what is to come to whet the reader's appetite. Drs. G. N. Qazi and S. Taneja provide a unique perspective on the therapeutic action of bioactive molecules in medicinal plants. Their group has several years of experience in prospecting natural products in plants and following up with the isolation, characterization, and structure elucidation of natural products.

Professor Steven Ley and his collaborators at Cambridge University enlighten readers is to how natural products have served as inspiration for the discovery of new high-throughput chemical synthesis tools. A salient feature of this masterpiece is the creative use of polymer-supported reagents.

Drs. Braj and Vidya Lohray elaborate on the role of insulin sensitizers in emerging therapeutics. A noteworthy feature of this work is that it was done entirely in India and represents a fast-growing trend: the discovery of new chemical entities in that country.

Drs. Raymond McCague and Ian Lennon at Dowpharma next discuss the criteria for industrial readiness of chiral catalysis technology for the synthesis of pharmaceuticals. They exemplify how and why stereoselective reactions are invented for pharmaceutical researchers: The methodology is applicable in both the discovery and development phases of a drug in making analogs rapidly and by scalable transformations. Dr. Mukund Chorghade then introduces readers to the field of process chemistry: the quest for the elucidation of novel, cost-effective, and scalable routes for production of active pharmaceutical ingredients. The medicinal chemistry routes used in the past have often involved the use of cryogenic reactions, unstable intermediates, and hazardous or expensive reagents. A case study of the development of a process for an antiepileptic drug is presented; readers will also see how problems in the isolation, structure elucidation, and synthesis of metabolites were circumvented.

Drs. Mukund K. Gurjar, J. S. Yadav, G. V. M. Sharma, P. Radha Krishna, C.V. Ramana, Yatendra Kumar, Braj and Vidya Lohray, and Bipin Pandey have each made seminal contributions to process chemistry. They have invented commercial processes for key pharmaceuticals that have resulted in significant economies in cost and minimization of waste, and have engineered "green chemistry" and the development of eco-friendly processes. These scholars describe their work in the next few chapters with case studies of specific compounds. The work is an eloquent testimony to the collaboration and cooperation inherent in the strategic triad of academics institutions government, and industry. The work is applicable to the synthesis of both agricultural and fine chemicals.

Over the last few years, an increasing number of pharmaceutical and biopharmaceutical companies have resorted to outsourcing activities in chiral synthesis, process development, and manufacturing. Dr. Peter Pollack demonstrates this strategy, provides useful pointers about the do's and don'ts, and beautifully elaborates the risks and rewards inherent in outsourcing in the pharmaceutical industry.

Dr. Shrikant Kulkarni exemplifies solving regulatory problems via thorough investigations of processes and processing parameters. Dr. Peter Pollack delineates the fascinating impact of specialty chemicals on drug discovery and development, providing further illustration of the power and utility of outsourcing in drug manufacture.

Chemical engineering plays a central and pivotal role in scale-up operations. Dr. Andrei Zlota discusses chemical process scale-up tools, mixing calculations, statistical design of experiments, and automated laboratory reactors.

Dr. Richard Wife explains how some novel initiatives will lead to rescue of "lost chemistry and molecules," how the net will make research results accessible to the entire chemical world, and how information sharing will lead to better and more efficient research. Thought-provoking and novel studies aimed at predicting compound stability are presented.

In the concluding chapter, Dr. Colin Scott describes some general principles and practices in drug development. A brief review is presented of the history of the requirements for clinical studies leading to the registration of a drug prior to being marketed. This is followed by a discussion of ethical issues related to clinical studies, the phases of drug development, and clinical trial design features. The support operations necessary for the initiation of clinical trials and optimization of results are described. Finally, a global development plan, accelerated development opportunities, international regulatory procedures, and postmarketing requirements are summarized.

There are few courses in academic chemistry departments that deal with drug discovery and development. Graduating students typically have scant exposure to the fascinating world of industrial chemistry. I am confident that the material will excite students interested in careers in the pharmaceutical industry. A salient feature of the book is the inclusion of several case studies that exemplify and epitomize the concepts detailed in each chapter. An instructor interested in developing a course in pharmaceutical chemistry will find the book useful as a teaching text for a one-semester course.

Dr. Raghunath A. Mashelkar, Director General of the Council of Scientific and Industrial Research, has stated: "Rapid paradigm shifts that are taking place in the world as it moves from superpower bipolarity to multipolarity, as industrial capitalism gives way to green capitalism and digital capitalism, as information technology creates netizens out of citizens, as the nations move from 'independence' to 'interdependence,' as national boundaries become notional, and as the concept of global citizenship gets evolved, will see a world full of new paradigms and new paradoxes; there is no doubt that the rapid advance of science and technology will directly fuel many of these. The global pharmaceutical and, in particular, the contract R&D organizations have seen a dramatic change in their capabilities and sophistication. International pharmaceutical companies should now be ideally poised to seek collaborations to bring innovative drugs to the consumers at an affordable price."

Finally, I wish to thank my wife, Veena, my son, Rajeev, and my parents for their encouragement, emotional support, understanding, and love. They have helped immeasurably during this endeavor.

MUKUND S. CHORGHADE

1

FROM PATENT TO PRESCRIPTION: PAVING THE PERILOUS PATH TO PROFIT

Richard J. Pariza
Cedarburg Pharmaceuticals
Grafton, Wisconsin

1.1 INTRODUCTION

A research director at a major pharmaceutical firm used to tell the new scientists in his company that there was no nobler career than to discover and develop a drug that would help alleviate human suffering or cure a deadly disease without causing serious side effects. Many others have doubtless said the same, and added that the complexity of this adventure can be compared to landing people on the moon and getting them home safely to Earth. Notice that safety is paramount in both endeavors. Although we must at first do no harm, our drugs must also do some good. Ethical drug companies spend millions of dollars studying new drugs over many years to determine both safety and efficacy, in order to legitimately promote new chemical entities and formulations to physicians, and more recently directly to the public. Even with enormous research expenditures and careful regulatory scrutiny, safety issues with blockbuster drugs are frequently in the news. Patients do not all respond adequately to existing drugs or even drug classes, and new agents are regularly needed to fight infections caused by microorganisms that become resistant to available antibiotics. So how do we get started along this path to better and safer drugs?

First, a target must be identified. This is a medical and marketing exercise, where a problem is recognized that could be treated with a pharmaceutical drug that fits into a company's portfolio. It is necessary to assure that adequate financial and human resources will be available for this daunting task. Once the commitment is established, teams of scientists must determine how a chemical could possibly be used to help patients. After all, pharmaceuticals are chemicals, and pharmaceutical companies sell chemicals.

Drug Discovery and Development, Volume 1: Drug Discovery, Edited by Mukund S. Chorghade

Biochemists, molecular biologists, physicians, pharmacologists, and others team up with synthetic chemists to determine a strategy to attack a disease. Often, these scientists are in what might be considered a virtual team: not in the same company, not on the same continent, nor even working in the same decade. By following the medicinal literature carefully over many years, often in fields seemingly unrelated to their own, scientists can gain insight into possible treatments and apply their own unique talents to come up with a new drug. There is an enormous amount of information available online, on the World Wide Web and various scientific databases, and modern search engines make it easy to find both obvious and obscure relationships. A small well-equipped startup company with the right mix of desire and talent can make breathtaking strides only dreamed of a few decades ago. They need to understand biology and chemistry, law and economics. To do so, they must seek the wisdom from the past that often made success achievable even without these modern tools. Wisdom translates knowledge into understanding.

Very sophisticated approaches are often envisaged that involve inhibiting complex enzyme pathways, preventing invading microorganisms or invasive cancer cells from multiplying, replacing natural hormones that are lacking in the body, or a host of other possible ways to treat medical conditions. Chemists are involved in every phase, from planning to execution of the research, from the laboratory to the clinic. The resulting product sold will be a chemical, a pure chemical, or a well-defined mixture, often a single enantiomer. It must be stable enough to ship to pharmacies and consumers, who will store it, dispense it, and use it. It must be safe to handle and have unambiguous safety and a predictable side-effect profile once administered. These days especially, it must be cost-effective, offering worthy advantages over cheaper generic drugs, often helping a patient avoid an expensive hospital stay and getting him or her back to work sooner. There is always competition to deal with, so the patent literature must be studied carefully, and risks must sometimes be taken when working in areas where other companies may have also begun research, because earlier priority dates may already have been secured. As you will see below, you may be sowing the seeds for a future partnership by doing research in a crowded field.

The chemical that will become the drug substance or API (active pharmaceutical ingredient) will often be chosen by a process of screening thousands of contending structures, with various attributes evaluated at each stage. Any structural insights that scientists have in the early stages can help enormously to abbreviate this development. Rules of thumb regarding stability, solubility, and toxicity are ubiquitous, and the successful team will know these well. ADME (adsorption, distribution, metabolism, and excretion) concepts must be studied and applied to the drug candidates and their biochemical targets.

Modern approaches that can gain real advantages often involve computer-assisted modeling of potential drug molecules and the sites of their activity. If an x-ray structure of a target enzyme is known, especially with an inhibitor molecule firmly docked, computer modeling can be used to determine what other drug candidates may also bind strongly with that site. NMR techniques are also used to screen and assess the interactions of hosts and potential drugs. With this flood of new technologies only now becoming available to medicinal chemists, it is amazing indeed that so many powerful wonder drugs were discovered and developed in the antediluvian days of the recent past.

At first the cost of producing samples for early testing may not be a major factor, but it will become more and more important as larger quantities are needed for testing and progress is made toward clinical trials and commercialization. It is also essential that chemists and engineers use the most cost-effective syntheses and modern approaches as early as possible along the drug development time line so that when scale-up issues arise, as they always do, the best options are available to solve problems quickly. In 2001, the top 16 pharmaceutical companies spent $90 billion to manufacture their products.[1] Manufacturing costs have become more than twice the cost of R&D and nearly as much as marketing and administrative costs. This is due partially to the enormous regulatory and quality issues, which can lock inefficiencies into a manufacturing process very early in the filing strategy. Detailed process information, equipment specifications, testing protocols, and storage and stability programs must all be put in place long before clinical studies on a new drug are completed and reviewed by the Food and Drug Administration (FDA). This is caused by concerns that any process changes may lead to new impurities or higher levels of extant impurities, or may make a product that will decompose more quickly and lose potency or develop harmful by-products. Companies must choose between delaying a filing, which could allow competitors to move ahead or could lose precious patent life, or must submit a filing with a less than ideal manufacturing process. Because of the enormous profit incentives to get a drug onto the market quickly, the latter is often the course chosen. The drugs discussed in this chapter each sell at least $1 million to $3 million *per day*, so any delays requested to investigate new chemical processes, even for a few weeks, may be considered too costly. This often plays into the hands of the generic companies, which can start refining the manufacturing processes years before the innovator's patents expire. Chemistry is always on the critical path.

A glance into the past may convince the reader that human ingenuity, recognizing the essential features of a problem, and applying Occam's razor[2] can often lead to success. Such cleverness may sometimes be rewarded with a dash of serendipity as well. Genius transforms understanding into beauty.

1.2 A SIMPLE SOLUTION TO A COMPLEX PROBLEM

Erythromycin was introduced into the clinic in 1952, and although it was a useful antibiotic with an excellent safety profile, allowing its use even in children and pregnant women, blood levels were erratic and there were often annoying side effects, such as nausea, upset stomach, and diarrhea. In fact, in 1984, the director of antibiotic sales for Abbott Laboratories announced in a meeting with scientists that if they could come up with "[a compound identical to] erythromycin, but without the belly ache," he could triple the sales. Newer formulations of erythromycin were tried but had only limited success in reducing this relatively benign but market-limiting side effect.

It was recognized very early on that acid instability in the digestive track could be a major cause of these problems. Although the mechanism of the acid-catalyzed degradation was explored in a one-page publication by Abbott chemists in 1971,[3] the acid degradation of erythromycin is not as simple as first envisaged. It was known that erythromycin A (**IA**) formed enol ether (**IIA**), as did erythromycin B (**IB**). Due to the −OH group on C-12 of **IIA**, a further reaction can take place to form anhydroerythromycin A (**III**), a spiroketal that is nearly devoid of antimicrobial activity. A paper published in 1986[4] corroborated this idea, and a more detailed kinetic study in 1989[5] suggested that there is equilibrium between **IA** and **IIA**. This equilibrium was confirmed by the very simple deuterium labeling study shown in Scheme 1.1.[6] Work continues on this intriguing system.[7]

Erythromycins A and B (**IA** and **IB**) were treated with anhydrous CH_3CO_2D to form **IIA** and **IIB**, with the −OH's exchanged for −OD's. When **IIB** was treated further with CH_3CO_2D in D_2O, a more acidic medium, erythromycin B was recovered, with deuterium

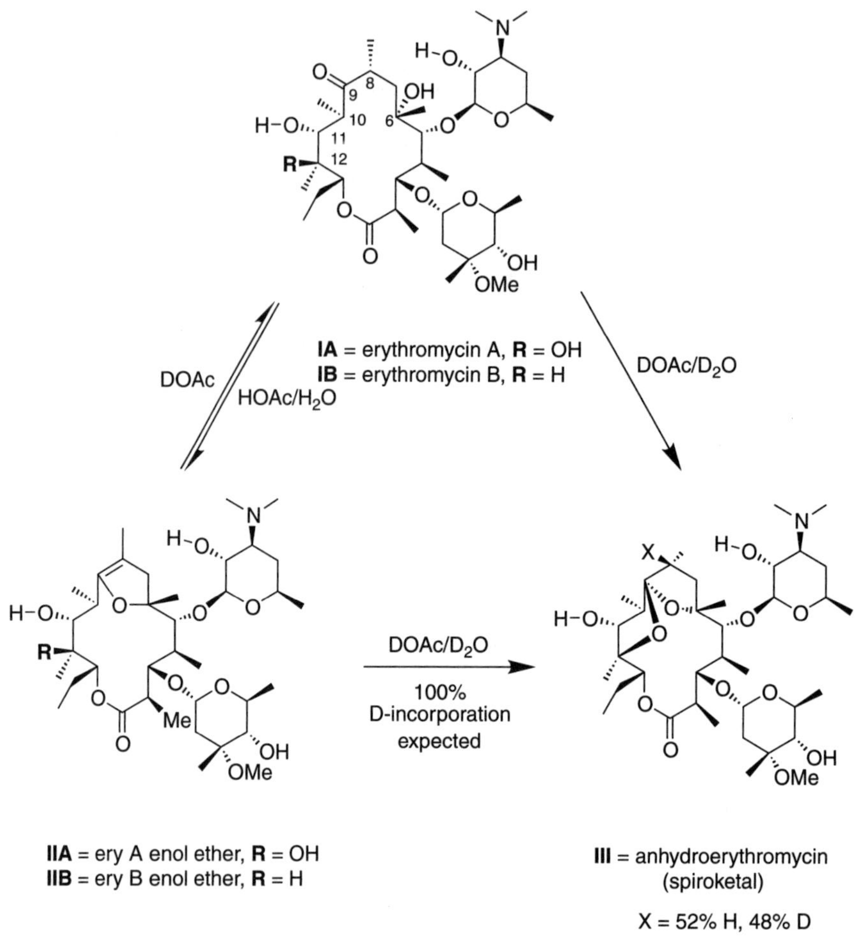

IA = erythromycin A, **R** = OH
IB = erythromycin B, **R** = H

DOAc

$HOAc/H_2O$

$DOAc/D_2O$

$DOAc/D_2O$

100%
D-incorporation
expected

IIA = ery A enol ether, **R** = OH
IIB = ery B enol ether, **R** = H

III = anhydroerythromycin
(spiroketal)

X = 52% H, 48% D

Scheme 1.1

incorporation and some epimerization[8] at C-8. It was known that under similar protic conditions, **IIA** would convert to a single epimer of **III**.[3] However, when, after exchanging the −OH's for −OD's, **IA** was treated directly with CH_3CO_2D in D_2O, within a few minutes the anhydroerythromycin A (**III**) that was formed contained about 50% deuterium at C-8, as analyzed by ^{13}C-NMR. No deuterium was detected at C-10. Furthermore, when naturally labeled (**III**) was treated similarly with CH_3CO_2D in D_2O, deuterium was slowly incorporated at C-8.

Physiologically active compounds often have emergent properties that are due to the unique spatial arrangement and interactions of their functional groups. For example, the macrolactone (macrolide) ring appears to have a hydrophobic and a hydrophilic side in its low-energy conformations, perhaps accounting for the amphiphilic nature of the molecule, with the OH at C-6 sticking out on the hydrophilic side. IR spectra of erythromycin A indicate that there is one OH that is not involved in a hydrogen bond. The x-ray structures as well as molecular modeling show that the OH on carbon 6 is the only one in the molecule not involved in an internal hydrogen bond with a neighboring polar functionality (see Scheme 1.2).

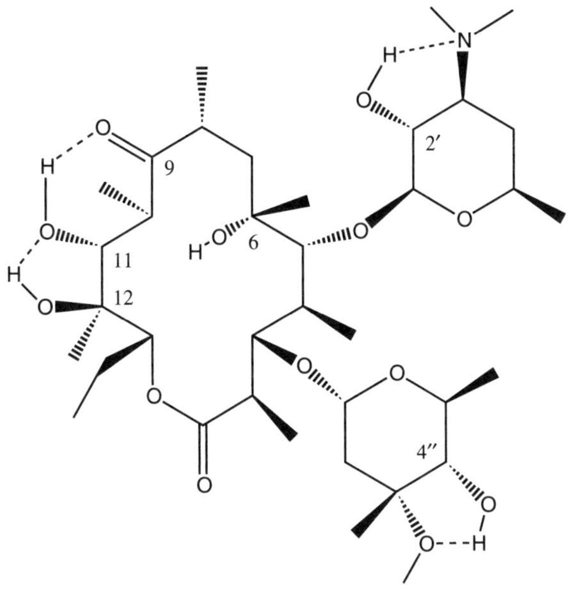

Scheme 1.2 Internal hydrogen bonds of erythromycin A.

Scheme 1.3 shows that the three secondary hydroxyl groups in erythromycin A (2′, 4″, and 11) can readily be differentiated chemically. The most reactive −OH group is on the desosamine sugar moiety by virtue of the 3′-dimethylamino group acting as an intramolecular catalyst. Thus, erythromycin A can easily be converted to its 2′-acetate (**IV**) in dichloromethane by reacting with acetyl chloride and sodium bicarbonate as base, or acetic anhydride and triethylamine, rendering the amino group nearly two p*K* units less basic, due to the neighboring group interaction.[9] Further reaction when DMAP is present leads to acetylation on the cladinose sugar at the 4″-hydroxyl group, and the hydroxyl group at C-11 can be acetylated only after heating. The lone pairs of electrons on the oxygen at C-4″ are more readily accessible to the reagents than those at C-11, which is involved in a hydrogen bond. However, if **IV** is treated with strongly basic conditions capable of fully deprotonating an −OH on the molecule, such as sodium hexamethyldisilazide in THF at −78°C, and acetic anhydride is added, the −OH at C-11 is preferentially acetylated over the −OH at C-4″. This can be understood by stabilization of the C-11 alkoxide by the neighboring proton on the C-12 hydroxyl, while an alkoxide at 4″ is relatively less stable. Interestingly, compound **V** was shown to be a 12,9-hemiacetal by NMR, and in the presence of water–deuterium oxide mixtures undergoes slow hydrogen–deuterium exchange of the proton at the hydroxyl at C-6 on the NMR time scale. A similar 12,9-hemiacetal was reported to be a major contributor to the equilibrium structures of erythromycin A (**IA**) itself.[10]

Although many scientists were studying these issues, a small group of scientists working at Taisho Pharmaceuticals in Japan read Abbott's brief 1971 publication in *Experientia*[3] and thought of a very simple solution to the acid instability of erythromycin. In a meeting with Abbott executives in the 1985, the lead Taisho chemist, Dr. Yoshiaki Watanabe, in somewhat broken English, thanked Abbott for this one-page revelation. The head of the Abbott delegation then mumbled to his scientists, on his side of the table, something to the effect that they were never going to publish another (expletive deleted) paper. Thank

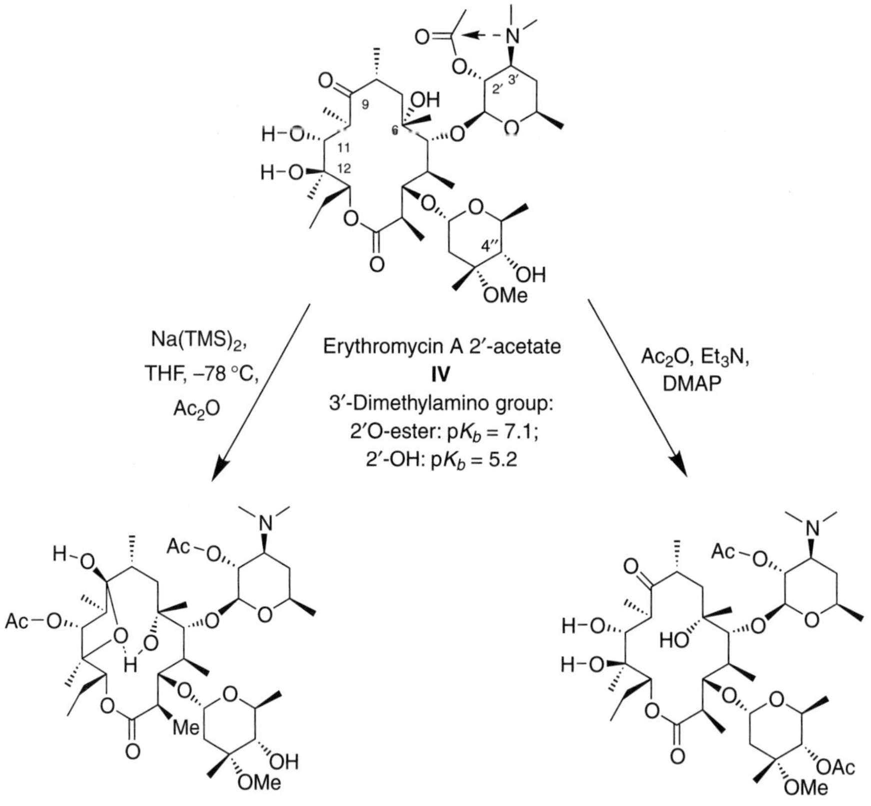

Erythromycin A 2′-acetate
IV
3′-Dimethylamino group:
2′O-ester: pK_b = 7.1;
2′-OH: pK_b = 5.2

Na(TMS)$_2$, THF, −78 °C, Ac_2O

Ac_2O, Et_3N, DMAP

Erythromycin 2′,11-diacetate-12,9-hemiacetal **(V)**

Erythromycin 2′,4″-diacetate **(VI)**

Scheme 1.3

goodness he was joking! A partnership was soon born, bridging time and space, and has flourished. Both clarithromycin and azithromycin, discussed below, have achieved annual sales around $1 billion.

By the amazingly simple idea of blocking the hydroxyl group on carbon 6 with a methyl group, these Japanese chemists were able to prevent formation of the enol ether (**IIA**) or anhydroerythromycin A (**III**). The compound they first made in 1980,[11] now sold as clarithromycin **VII** (Scheme 1.4), not only has superior acid stability, but produces less stomach irritation, a major drawback to the widespread use of erythromycin itself. This serendipitous result may be due at least in part to the inability of clarithromycin to form the enol ether (**II**), which was later shown in animal studies to increase gastrointestinal motility to a much greater extent than the parent structure. This effect was seen even after intravenous administration, so it is not simply the result of contact of the drug with the stomach. Abbott had developed a gastrointestinal motility assay to screen new drug candidates. Pressure transducers were attached along the outside of the GI tract in an anesthetized beagle dog, and peristaltic contractions were recorded after administering an erythromycin analog. Clarithromycin not only demonstrated a

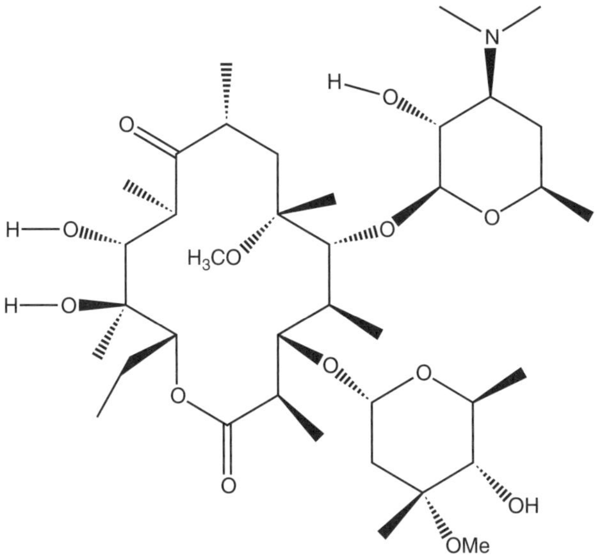

Scheme 1.4 Clarithromycin (VII).

reduction in the recorded contractions relative to erythromycin, but fewer belly aches were reported in the clinic.[12] Clarithromycin has also improved absorption from the GI tract and enhanced blood levels, coupled with lower intrinsic minimum inhibitory concentrations (MICs) against important pathogens. Thus, this second-generation semisynthetic macrolide is a better antibiotic overall than the direct fermentation product from which it is made.

The original synthesis[11] of **VII** involved protecting both the 2′-OH and the 3′-dimethylamino functions on the desosamine sugar with benzyloxycarbonyl groups (Z-groups). This was a method that had been used by many others and results in the loss of one of the methyl groups on the nitrogen. This step was then followed by simple and somewhat selective methylation of the 6-OH with methyl iodide. The Z-groups were then removed by hydrogenolysis, and a methyl group was put back on the nitrogen by reductive amination with formaldehyde. However, as mentioned above, the mere presence of an ester functionality on the 2′-OH, such as an acetate, renders the neighboring nitrogen group much less basic and much less nucleophilic. Therefore, the first process used to prepare larger quantities of **VII** was simply to methylate **IV**, erythromycin 2′-acetate, a compound that is much more easily prepared and subsequently deprotected. This chemistry is shown in Scheme 1.5, where each structure is purposely drawn with a different convention taken from contemporaneous literature, to illustrate how information, even accurate information, does not always lead to clarity![13] A highly crystalline product resulted from the methylation of **IV**, albeit in low yield, which could be purified sufficiently for early studies. In these early studies large supplies of drug were more important than the efficiency of the manufacturing process. The only significant impurity was the 6,11-dimethylated compound, similar to what was seen with the more onerous Z-group protection–deprotection scheme. Dissolving **IV** in methanol, and allowing the methanolysis to take place at room temperature overnight, quantitatively removed the acetyl group. It is interesting to realize

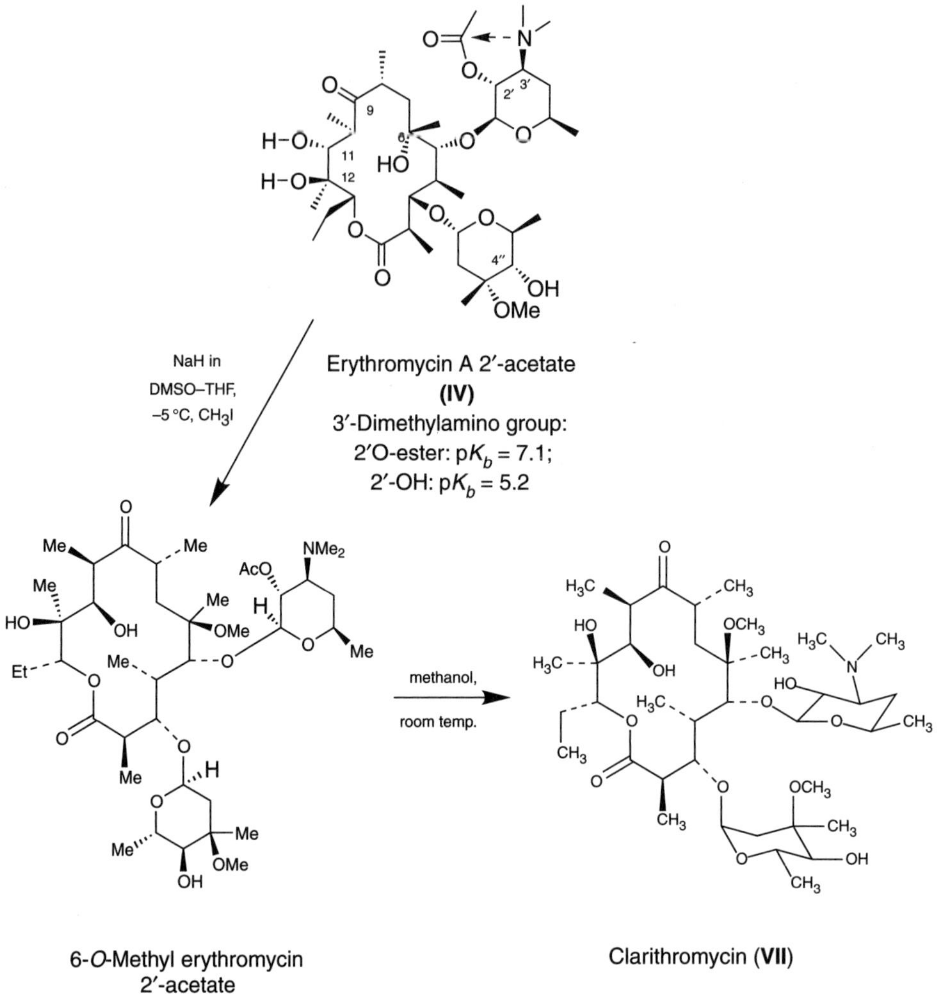

NaH in DMSO–THF, –5 °C, CH_3I

Erythromycin A 2′-acetate
(IV)
3′-Dimethylamino group:
2′O-ester: pK_b = 7.1;
2′-OH: pK_b = 5.2

methanol, room temp.

6-*O*-Methyl erythromycin 2′-acetate

Clarithromycin (**VII**)

Scheme 1.5

how close many scientists were to this novel second-generation macrolide when they were working with the simple 2′-esters of erythromycin many years earlier. This is something to keep in mind when working with a readily available derivative of an active compound: What new chemistry can you do with it?

1.3 AN INTRIGUING PATENT PROBLEM

A different solution to the acid instability and erratic blood-level problems of erythromycin was found with another analog. Ironically, this new compound has such low blood levels that at first it seemed to some researchers that infectious disease physicians would not trust it. The old paradigm was that an antibiotic had to exhibit blood concentrations above the MIC for the particular strain of bacteria causing the illness. In fact, the infectious organism is often compartmentalized in particular tissues such as the tonsils and the prostate gland. An antibiotic that can penetrate and sustain therapeutic levels in those diseased tissues would actually be more useful than one that was largely in the blood serum. This concept is also true of cancer chemotherapy agents, which need to accumulate in the tumor cells rather than in the bloodstream or healthy tissue. Scientists at Sour Pliva in Zagreb, in what was then Yugoslavia,[14] and at Pfizer in Groton, Connecticut,[15] were able, almost simultaneously and concurrently to solve the tissue penetration problems and acid instability issues by cleverly adding an additional basic nitrogen atom in a Beckman rearrangement process followed by reduction. *Almost* simultaneously—and the resulting blockbuster drug, azithromycin (**VIII**), is the subject of *two* U.S. composition of matter patents! Although Pliva filed their patent more than a year earlier, Pfizer's patent was issued seven months sooner. It turns out that the two companies drew their new structures and named their compounds using quite different conventions. They even numbered the macrocyclic ring differently than the classical structures shown in Schemes 1.1, 1.3, and 1.4. Due to this confusion, the U.S. Patent and Trademark Office thought the groups were claiming two distinctly different compounds. Pliva had pioneered work with the ring-expanded macrolides,[16] and since they filed their patent first, Pfizer had to negotiate the rights to a compound that it had discovered and patented independently (Scheme 1.6 and Table1.1). In today's competitive world, the Patent Cooperation Treaty requires publication of patents 18 months after filing, or earlier claimed priority date: for example, from a provisional patent application. Had this process been in place in the early 1980s, it would have allowed Pfizer scientists to see the Pliva application much sooner (the Pliva application would have been published four months after the Pfizer filing), and the Pfizer experts would doubtless have realized that the compounds they both claimed were identical.

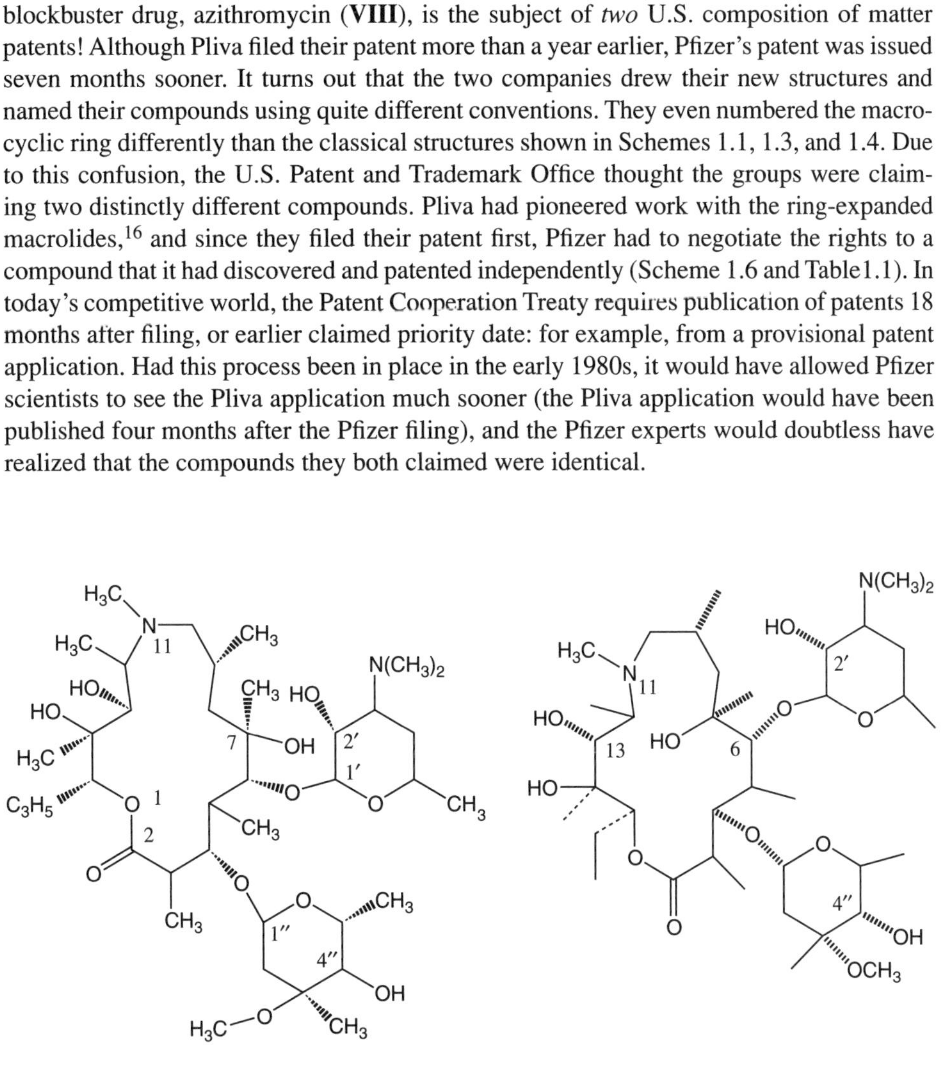

11-Methyl-11-aza-4-*O*-cladinosyl-6-*O*-desosaminyl-15-ethyl-7,13,14-trihydroxy-3,5,7,9,12,14-hexamethyl oxacyclopentadecane-2-one

(*a*)

N-Methyl-11-aza-10-deoxo-10-dihydroerythromycin A

(*b*)

Scheme 1.6 (*a*) U.S. patent 4,517,359, May 14, 1985; (*b*) U.S. patent 4,474,768, October 2, 1984.

TABLE 1.1 Comparison of Paths to Patents

Patent	Filed	18 months[a]	Issued	Extended
Sour Pliva, U.S. 4,517,359	9/22/81	3/22/83	5/14/85 (43.5 months)	5/20/93 (42.2 months)[b]
Pfizer U.S. 4,474,768	11/15/82	5/15/84	10/2/84 (23 months)	

[a]Under current regulations the patents would have been published 18 months after filing.

[b]1,267 days (3.5 years).

The first generic formulations of azithromycin were approved by the FDA on November 14, 2005 for two companies, Teva Pharmaceuticals and Sandoz, to sell this blockbuster drug for a wide variety of indications, permitting Pfizer and Pliva almost exactly a 20.5-year head start, due to the extension of 3.5 years granted May 20, 1993 for the Pliva patent. At that time, patents expired 17 years after issuing rather than 20 years from the date of filing, as is now the case (unless extensions are granted).

Note that in Scheme 1.7 azithromycin is drawn in two additional ways, as shown currently in *SciFinder*[17] (clockwise numbering) and the *Merck Index*[18] (counterclockwise numbering), reversing the order of the sugars. The *Physicians' Desk Reference*[19] draws azithromycin in a fashion related, but not identical, to the Pfizer patent (Scheme 1.6*b*), which depicts the stereochemistry of C-6 as S when in fact it is R. Clarithromycin and erythromycin are drawn in a format similar to that used in *SciFinder*, except inverted (so the numbering runs traditionally counterclockwise); there are some ridiculously long bonds, so the drawings don't overlap; and all the necessary centers are reversed to maintain the correct stereochemistry. It is highly unlikely that anyone has ever looked at structures presented in that fashion and gained any useful insights. How could anyone be expected to see the crucial interaction of the C-6 OH group with the carbonyl at C-9 in such a rendition? [Compare erythromycin (R = H) in Scheme 1.8*b* with Scheme 1.1 and 1.2.] The enormous confusion caused by following such diverse conventions when drawing and naming significant compounds has restricted an understanding of the literature to the few experts who take the time to become familiar with the structures and conventions. It is fortunate that modern desktop computer programs can recognize instantly that these structures are equivalent, calculate empirical formulas, assign stereochemistry, and even predict NMR spectra.[20] The use of such powerful computer routines, *CAS Registry* numbers, and other modern library tools can save time for the expert and be invaluable to the uninitiated. It is hoped that the confusion over structures such as these will become a relic of the past.

1.4 ANOTHER STRUCTURAL INSIGHT

Recently, it has been widely reported that the new class of wonder drugs called COX-2 inhibitors exhibit serious cardiovascular side effects, and several of these drugs have been withdrawn from the marketplace. Meanwhile, another class of blockbuster drugs, the statins, may not only be safe and effective in their intended role of lowering cholesterol, but may have a plethora of other potentially valuable properties. Cancer, Alzheimer's disease, diabetes, osteoporosis, high blood pressure, multiple sclerosis, and macular degeneration are

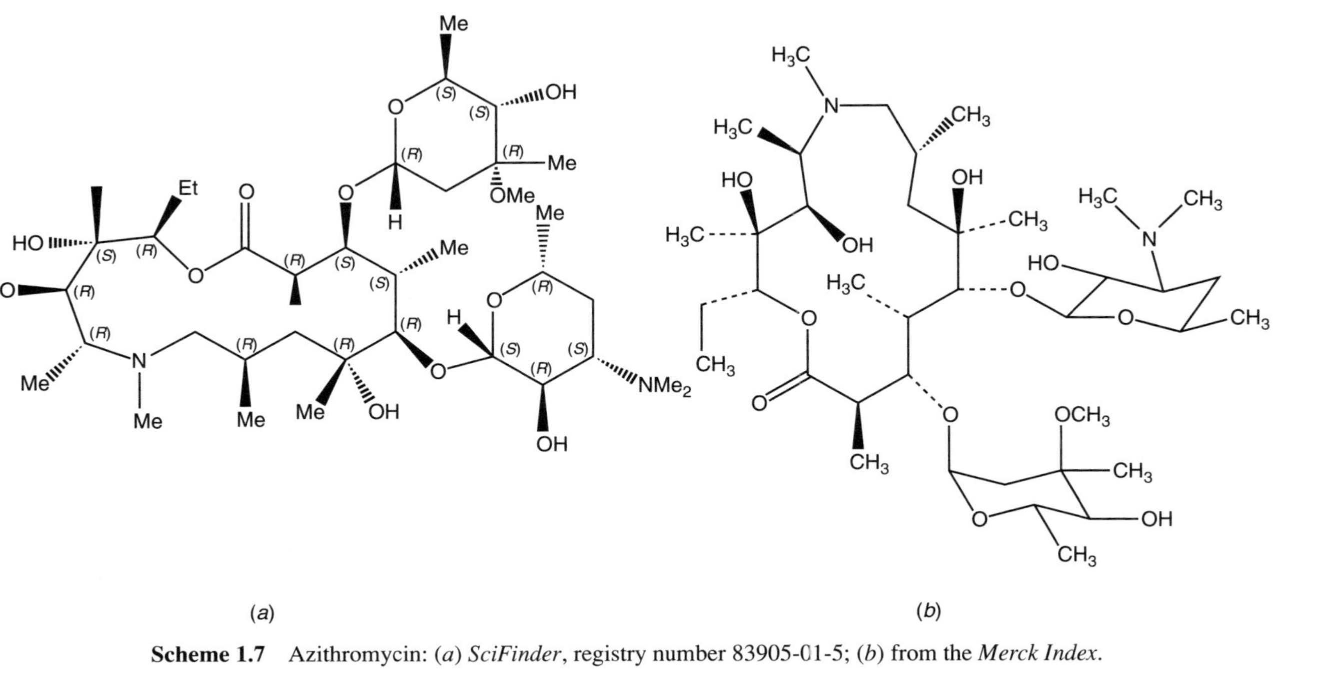

Scheme 1.7 Azithromycin: (*a*) *SciFinder*, registry number 83905-01-5; (*b*) from the *Merck Index*.

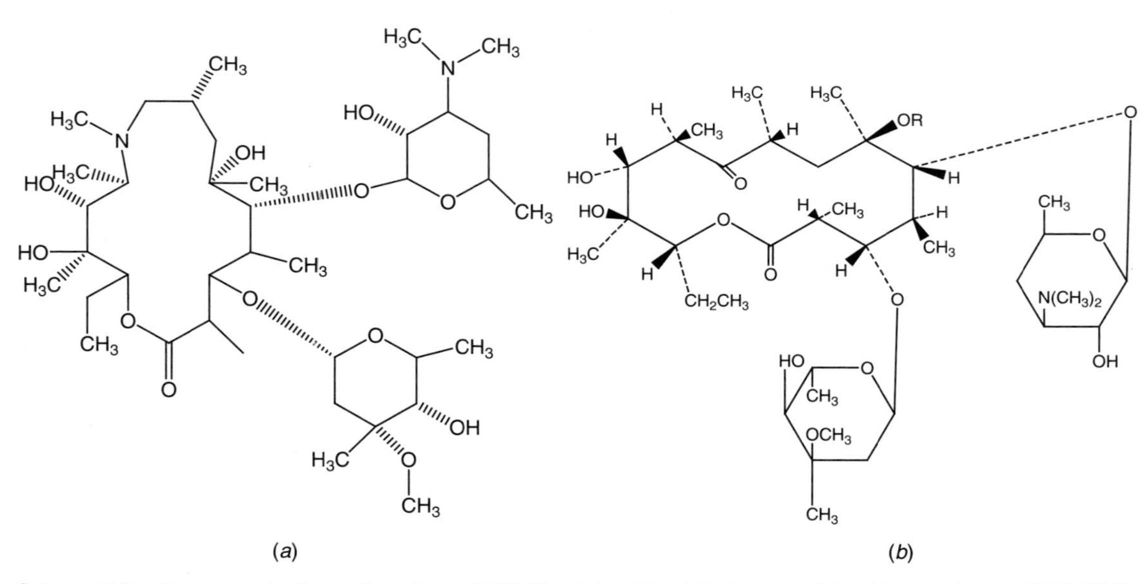

Scheme 1.8 Structures similar to those in the 2002 *Physicians' Desk References*; (*a*) azithromycin, pp. 2739, 2743, 2748; (*b*) clarithromycin, R = CH_3, pp. 403, and erythromycin, R = H, pp. 454, 456.

among the diseases that the statins may ameliorate.[21] It has often been said that drugs are discovered in the clinic. In this sense the clinic consists of the patients using these drugs in the general population. Observational studies on the millions of people taking these drugs revealed the problems of COX-2 inhibitors and the additional potential indications for the statins. Another example of this in a much smaller population can serve as an illustration.

In the late 1960s, Pfizer filed patents on an α_1-adrenergic blocker that came to be known as prazosin.[22] The structure is shown in Scheme 1.9, next to a very similar compound, terazosin, patented in 1977 by Abbott.[23] Both of these compounds are effective in lowering blood pressure and have beneficial effects on the plasma lipid profile. Dr. Marty Winn, a chemist at Abbott, looked at a drawing of the structure of prazosin, and realizing that its failings included problematic intravenous formulation and short duration of action, thought that a similar molecule with higher water solubility might be more effective. He knew that furan is only sparingly soluble in water, whereas tetrahydrofuran (THF) is completely miscible with water. He concluded correctly that simply saturating the furan ring in prazosin might lead to a much more soluble compound. In fact, his first samples were made by direct hydrogenation of prazosin, leading, of course, to a racemate. This was not a problem since in those days the FDA did not require compounds to be pure single enantiomers. In the 1990s, Abbott considered making a new compound as the single enantiomer, a chiral switch, but did not pursue the issue. It turns out that the base form of terazosin is 25 times more water soluble than prazosin, and its elimination half-life is about three times greater, permitting once-daily administration of the new drug. The difference for the corresponding hydrochloride salts is even more dramatic. The terazosin salt is over 500 times more soluble than the corresponding prazosin salt!

Since terazosin was projected to be a relatively low volume (about a ton per year) high-potency (10 mg/day) drug, the cost of manufacturing was not deemed a big issue. Note that one patient would take 3.65 g per year, so 1 metric ton of API is enough to treat 274,000 patients per year. At a price of $1.00 per day, the annual sales would be $100 million. In the late 1980s a clever process chemist[24] in the pharmaceutical division suggested ways to streamline the process and save as much as $500,000 per year. Management decided not to pursue the new chemistry because it was estimated that it would cost over $2 million to run several successful manufacturing scale batches, place the API on stability studies, manufacture tablets, put the tablets on stability for a year, file all the data with the FDA, and wait for approval, before being able to switch over to the new process. The FDA would have to be convinced that the new API made tablets that were identical to those made by the old process, and the review process could take many months. Little did Abbott realize that the market for terazosin was about to increase dramatically, so the new process, despite the costs of implementing it, could have saved them many millions of dollars in manufacturing the drug over the long term.

Soon after the introduction of terazosin into the marketplace as an antihypertensive, it was noticed anecdotally that men with symptomatic benign prostatic hyperplasia (BHP) who were given the drug to treat high blood pressure began reporting relief of their urethral pressure and bladder outlet problems. Sales of terazosin increased slowly as word got out of this promising new treatment for BPH, as it was evidently being prescribed for off-label use. Physicians have the authority to prescribe drugs for conditions other than those promoted by the pharmaceutical companies, so this is quite common: for example, with anti-cancer drugs. However, Abbott had to conduct costly clinical studies and get FDA approval to advertise and market the drug for this new use. Once the new indication was approved by

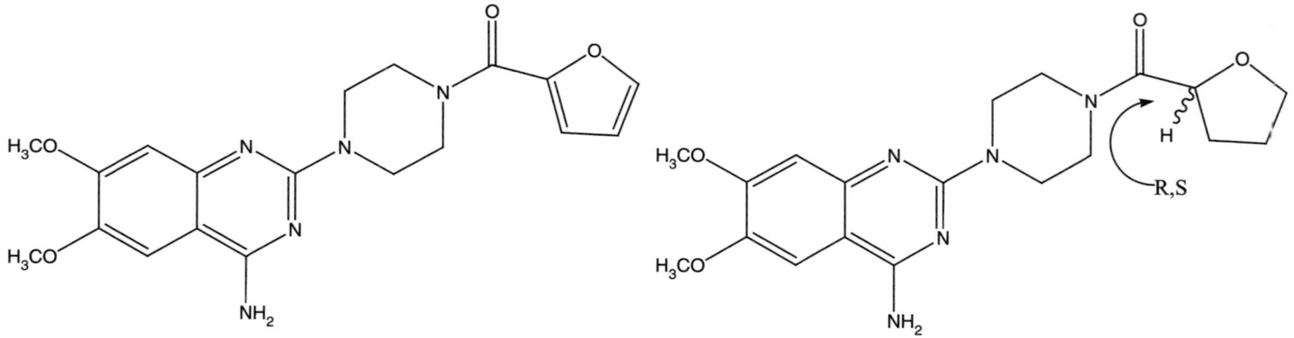

(*a*) (*b*)

Scheme 1.9 (*a*) Prazosin; solubility of the hydrochloride salt in water (pH ca. 3.5) at ambient temperature (mg/mL): 1.4. (*b*) terazosin; solubility in water at 25°C (mg/mL): 29.7-hydrochloride salt (mg/mL): 761.2 (544 times more soluble!).

the FDA for detailing, an unexciting drug that had been third or fourth tier for hypertension, selling much less than $100 million per year, became a major seller at about $500 million per year worldwide. Terazosin was becoming a very profitable drug just as its patents began to run out. It quickly became a very attractive target for generic drug manufacturers. Abbott was able to make deals with the generic competitors to keep them off the market temporarily, extending its very profitable franchise for about four years after the patent ran out. They paid several companies millions of dollars *per month* to keep them off the market with their cheaper generic version of terazosin. However, faced with an antitrust investigation in 1999, they canceled such arrangements with potential competitors.[25] Abbott's sales of terazosin fell 70% in the next year alone, to $141 million.[26]

The insight of a chemist who looked at a structural drawing, and spotted the Achilles' heel of the compound represented, and so elegantly corrected it is astounding. A clearly drawn chemical structure can reveal the beauty of subtle emergent properties of the functional groups to the astute imagination of a skilled scientist. The fact that this increase in solubility makes terazosin a superior drug for BPH makes a mark on the positive side of the ledger of unintended consequences. Serendipity can convert beauty into profits!

We should learn never to ignore the clues that any piece of the puzzle is revealing: by itself or as part of the emerging picture. Each fact that builds toward further understanding can be exploited, but since our fellow scientists, competitors, or future partners may be far from us in time and space, our discoveries must be published with clarity. Abbott's brief paper in the 1970s paved the way for its future Japanese partner to solve a long-standing problem, and Pfizer wisely joined with Pliva when it found success in a research area that the Yugoslavians had pioneered. Despite the dazzling advances in the tools of modern science that you will see in the following chapters, there is no substitute for the wisdom of good common sense, along with painstaking attention to basic details and the occasional flashes of genius that reveal the true beauty of nature. If we only learn from mistakes, then, after we have made all the mistakes possible, we will finally do things right. Discovering wisdom from the past is a much more efficient process!

REFERENCES

1. Leila Abboud and Scott Hensley, *Wall Street Journal*, September 3, 2003, p. A1.
2. The English philosopher William of Occam (1300–1349) propounded Occam's razor: "Entities are not to be multiplied more than necessary." This is especially appropriate when trying to keep microorganisms or cancer cells from dividing.
3. Paul Kurath, P. Hal Jones, Richard S. Egan, and Thomas J. Perun, *Experientia*, **27** (4), 362 (1971).
4. Paul J. Atkins, Tristan O. Herbert, and Norbert B. Jones, *International Journal of Pharmaceutics*, **30**(2–3), 199–207 (1986).
5. T. Cachet, R. Hauchecorne, J. Hoogmartens, G. Van den Mooter, and C. Vinckier, *International Journal of Pharmaceutics*, **55**(1), 59–65 (1989).
6. Richard J. Pariza, and Leslie A. Freiberg, *Pure and Applied Chemistry*, **66**(10–11), 2365–2358 (1994).
7. Yong-Hak Kim, Thomas M. Heinze, Richard Beger, Jairaj V. Pothuluri, and Carl E. Cerniglia, *International Journal of Pharmaceutics*, **271**(1–2), 63–76 (2004).
8. Under these conditions, erythromycin B epimerizes to about a 1 : 1 mixture of epimers at C-8 in a day or two.

9. The *Merck Index* entry for erythromycin (Monograph 03714) gives the pK_{a1} value as 8.8; for erythromycin propionate (Monograph 03719) the pK_a value is 6.9. Merck & Co., Inc., Whitehouse Station, NJ, 2001–2005. $pK_b = 14 - pK_a$.
10. J. Barber, J. I. Gyi, G. A. Morris, D. A. Pye, and J. K. Sutherland, *Journal of the Chemical Society, Chemical Communications*, 1040, (1990).
11. Yoshiaki Watanabe, Shigeo Morimoto, and Sadafumi Omura, Novel erythromycin compounds, U.S. patent 4,331,803, May 25, 1982; filed May 19, 1981. *Merck Index*, Monograph 02362, Merck & Co., Inc., Whitehouse Station, NJ, 2001–2005.
12. Although it is imperative that animal models be validated in the clinic, demonstrating that the responses in the animal model translate to the human condition, the real proof is often found only after large populations have taken the drug.
13. During a plenary lecture on his monumental total synthesis of erythromycin at the 1977 ACS National Organic Symposium in Morgantown, West Virginia, R. B. Woodward had a cartoon on a slide depicting a tombstone with words to the effect: "Cahn–Ingold–Prelog R.I.P." He was annoyed that each time he changed the protecting groups on his intermediates the R, S nomenclature changed, so it appeared that stereogenic centers had been inverted when they had not. The problem of conveniently conveying data by use of various nomenclature conventions has plagued even Nobel laureates in this field! Thank goodness for computers.
14. Gabrijela Kobrehel and Slobodan Djokic, 11-Methyl-11-aza-4-*O*-cladinosyl-6-*O*-desosaminyl-15-ethyl-7,13,14-trihydroxy-3,5,7,9,12,14-hexamethyloxacyclopentadecane-2-one and derivatives thereof, U.S. patent 4,517,359, May 14, 1985; filed September 22, 1981.
15. Gene M. Bright, *N*-Methyl-11-aza-10-deoxo-10-dihydroerythromycin A, intermediates thereof, U.S. patent 4,474,768, October 2, 1984; filed November 15, 1982.
16. Gabrijela Kobrehel, Gordana Radobolja, Zrinka Tamburasev, and Slobodan Djokic, 11-Aza-10-deozo-10-dihydroerythromycin A and derivatives thereof as well as a process for their preparation, U.S. patent 4,328,334, May 4, 1982, filed March 28, 1980; cited in ref. 15.
17. *SciFinder*, registry number 83905-01-5, American Chemical Society, Washington, DC, 2005.
18. *Merck Index*, Monograph 00917, Merck & Co., Inc., Whitehouse Station, NJ, 2001–2005.
19. *2002 Physicians' Desk Reference*, 56th ed. Medical Economics Company, Inc., Montvale, NJ. Since the data in the PDR are identical to the package inserts from the drug companies, it is surprising that Pfizer would approve an incorrect rendition of its own drug.
20. Programs such as *ChemDraw Ultra 9.0* from CambridgeSoft Corporation, Cambridge, MA.
21. Ronald Kotulak, *Chicago Tribune*, December 25, 2005, p. 1.
22. Hans-Jurgen E. Hess, 2,4,6,7-Tetrasubstituted quinazolines, U.S. patent 3,511,836, May 12, 1970; filed December 13, 1967.
23. Martin Winn, Jaroslav Kyncl, Daniel Ambrose Dunnigan, and Peter Hadley Jones, Antihypertensive agents, U.S. patent 4,026,894, May 31, 1977; filled October 14, 1975.
24. Bruce W. Horrom, who started at Abbott in 1946 and became the longest-serving Abbott employee ever when he retired with more than 50 years of service. He came up with this new process idea after having been at Abbott for over 40 years.
25. *New York Times* (National Edition), July 23, 2000, p. 1.
26. *Pharmaceutical Business News Incorporating Biotechnology Business News*, February 7, 2001, p. 11.

2

MEDICINAL CHEMISTRY IN THE NEW MILLENNIUM: A GLANCE INTO THE FUTURE

PAUL W. ERHARDT
The University of Toledo College of Pharmacy
Toledo, Ohio

> Who of us would not be glad to lift the veil behind which the future lies hidden; to cast a glance at the next advances of our science and at the secrets of its development during future centuries?
>
> —David Hilbert in 1900, as quoted recently by R. Breslow[1]

2.1 INTRODUCTION

Given the highly interdisciplinary nature of medicinal chemistry and the potential for its deployment across a myriad of future life science research activities, this review seeks to highlight only those possibilities that stand out upon taking a broad purview of the field's most prominent trends. From this vantage point, however, at least a glance will have then been cast toward some of the more noticeable of the exciting opportunities seemingly in store as medicinal chemistry moves forward into the new millennium.[1,2] An overview of the document's several sections is provided below. It lists what topics are covered as well as those that are not covered and indicates the reasoning behind these choices. The overview also describes the consistent tone that was sought while attempting to elucidate the numerous technologies that necessarily become encompassed by the variously highlighted activities.

While initially contemplating how medicinal chemistry might continue to evolve as both a basic and an applied science, it became apparent that it would first be useful to consider

This chapter has been used by permission of the International Union of Pure and Applied Chemistry (IUPAC) with only minor modification from its original form as published in *Pure and Applied Chemistry* **74**, 703–785 (2002).

Drug Discovery and Development, Volume 1: Drug Discovery, Edited by Mukund S. Chorghade

where medicinal chemistry has been and how it has come to be what it is today. Thus, toward quickly establishing a context from which the future might be better appreciated, and perhaps even seen already to be repeating itself among a new set of players and technologies, in Section 2.2 a short discourse about medicinal chemistry's emergence as a formalized discipline, its early developments, and its present status is provided by considering how medicinal chemistry has been practiced across jumps of about 25-year increments. This section does not include a chronological list of medicinal chemistry's many contributions, nor does it highlight the many accomplishments of its noted investigators. Both of the latter can be found elsewhere as part of more traditional historical treatments.[3]

From the backdrop provided in Section 2.2, medicinal chemistry's near- and longer-term futures are considered in Section 2.3 relative to several of today's trends that are already having a major impact on the drug discovery process. A working definition for medicinal chemistry is recited at the opening of this section so that medicinal chemistry's immediate and future roles can be ascertained more clearly. Section 2.3 also sets the stage for a later consideration of where several drug development topics may be headed in the near and longer terms.

Gene therapy, vaccines, and biotech-derived therapeutic agents are not discussed in Section 2.4, which addresses medicinal chemistry's continued pursuit of efficacy. The aforementioned topics reside primarily within the domains of other disciplines. Readers are, however, encouraged to consult other reviews offered for these areas in order to appreciate how their advances are sure to have a dramatic impact on future life science research and its interface with medicinal chemistry (e.g., refs. 4 to 6, respectively). Alternatively, because assessing molecular conformation is such an integral part of practicing medicinal chemistry along any venue, several aspects of this key topic are considered in Section 2.5. In particular, the handling of chemical structures in database settings (e.g., chemoinformatics) is discussed in detail.

Several drug development topics are regarded as critical factors that will have a pivotal influence on medicinal chemistry's continuing evolution in the new millennium. Each of these topics is addressed briefly in Section 2.6. These key topics include assuring absorption, directing distribution, controlling metabolism, assisting excretion, and avoiding toxicity [i.e., the traditional absorption, distribution, metabolism, excretion, and toxicity (ADMET) studies that previously have been undertaken by pharmaceutical companies during the secondary stages of preclinical drug development]. As an important extension of the ADMET discussions, nutraceuticals considered in parallel with pharmacological synergy are also addressed in Section 2.6.

Issues pertaining to medicinal chemistry's future roles in pharmaceutical intellectual property (IP) and to trends associated with process chemistry are raised in Section 2.7. With today's highly publicized emphasis on genomics and proteomics, at least an abbreviated discourse about process chemistry is included at this juncture so that this fundamental aspect of medicinal chemistry's link with synthetic chemistry remains appreciated. Thus, toward providing just some of such coverage, the unmet need for large-scale stereoselective synthetic methodologies is discussed briefly in Section 2.7. As part of this discussion, an example is cited that concludes by conveying the medicinal chemistry logic that was encompassed as a critical component of the example's investigations.

Although it is beyond the scope of this review to discuss the impact that progress in each of several analytical methods is likely to have on medicinal chemistry,[7] x-ray diffraction has been selected to provide a representative discussion in Section 2.8. As is often acknowledged by researchers from various disciplines, "science moves forward according to what it can measure," and presently, there appear to be numerous promising advances among various analytical techniques that can be used to study drug–receptor interactions. For example, readers are encouraged to seek other reviews in order to appreciate the potential impact that anticipated developments in nuclear magnetic resonance (NMR inclusive of LC-NMR and high-flow-through techniques), mass spectrometry (MS inclusive of LC-MS and LC-MS/MS), microcalorimetry, and surface plasmon resonance may have on medicinal chemistry's future (e.g., refs. 7 to 10, 11 to 13, 14 and 15, and 16 and 17, respectively).

In Section 2.9, practical implications that stem from some of the earlier discussions are revisited in what also serves as an overall summary for this document. After restating medicinal chemistry's anticipated roles in future life science research, concerns pertaining to the training of medicinal chemists, inventorship, and the interplay of patent trends and future research within the context of the IP arena have all been reserved for comment in the concluding summary.

The document's running dialogue has been developed from future possibilities suggested by the current medicinal chemistry and drug discovery literature, as well as from general observations afforded while consulting in both the private and public sectors. Descriptions of specific research projects have been interspersed throughout so that real case examples, along with their chemical structures, could be conveyed explicitly. A concerted effort has been made to keep hype to a minimum. Alternatively, jargon has been used whenever it was thought that such terms portray the mind sets that were important for a given period, or because a particular term or phrase appears to be taking on an enduring significance. Some of the more technical of these terms are listed in Table 2.1 along with a short definition in each case. Since several acronyms have been used for repeating phrases, an alphabetical listing of all acronyms and their definitions is provided in Table 2.2 to assist readers as they move deeper into the document. Numerous references to secondary scientific/primary news journals have been cited because these journals are doing an excellent job of both reporting the most recent trends and forecasting the potential future. In several cases, a single citation has been used to list many of the informational Web sites that can often be found for a given topic.

Topics are considered into the future only to about one-half the distance that has been summarized for medicinal chemistry's past: namely, for about 75 years, with the first 25 being regarded as near term, and the next 50 being regarded as long term. The speculation that has necessarily been interjected throughout the review has been done with the thought that one of the goals for this type of chapter is to prompt the broadest contemplation possible about the future directions that medicinal chemistry might take. Finally, as medicinal chemistry is shown to move forward in time, it has been considered as both a distinct pure science discipline and, equally important, as a key interdisciplinary applied science collaborator seeking to mingle with what should certainly prove to be an extremely dynamic and exciting environment within the life sciences arena of the new millennium.

2.2 PRACTICE OF MEDICINAL CHEMISTRY

2.2.1 Emergence as a Formalized Discipline

Medicinal chemistry's roots can be found in the fertile mix of ancient folk *medicine* and early natural product *chemistry*: hence its name. As appreciation for the links between chemical structure and observed biological activity grew, medicinal chemistry began to emerge about 150 years ago as a distinct discipline intending to explore these relationships via chemical modification and structural mimicry of nature's materials, particularly with

TABLE 2.1 Selected Terms and Abbreviated Definitions

Term	Definition
Ab initio calculations	Quantum chemical calculations that use exact equations to account for the complete electronic structure of each atom in a molecule.[a]
AMBER	Molecular mechanics program commonly used for calculations on proteins and nucleic acids.[a]
Basis set	Set of mathematical functions used in molecular orbital calculations (e.g., 3-21G*, 6-31G*, or B3LYP when used in ab initio calculations refer to the type of mathematical function that was deployed).[a]
Bioinformatics	Application of computer science and technology to address problems in biology and medicine.[b]
Bioisostere	Broadly similar atoms or groups of atoms in terms of physiochemical or topological properties that can be used as replacements in a biologically active compound to create a new structure that retains all selected features of the parent compound's biological properties.[c]
Chemoinformatics	Handling of chemical structure and chemical properties in database settings; when related to biological properties, this field becomes a subdivision of bioinformatics.
Combinatorial chemistry	Synthesis, purification, and analysis of large sets of compounds wherein sets of building blocks are combined at one or more steps during their preparation.[c]
Combinatorial library	Set of compounds prepared by using combinatorial chemistry.[c]
Comparative molecular field analysis (CoMFA)	A 3D-QSAR method that uses statistical correlation techniques for the analysis of the qualitative relationships between the biological activity of a set of compounds with a specified alignment of their 3D electronic and steric properties. Additional parameters such as hydrophobicity and hydrogen bonding can also be incorporated into the analysis.[c]
Druglike	Structural motifs and/or physicochemical profiles that can be associated with providing an overall ADME behavior that is conducive to effective use in humans via the oral route.
Electrostatic potential	Physical property equal to the electrostatic energy between the static charge distribution of an atomic or molecular system and a positive unit point charge. Used in 3D-QSAR, molecular similarity assessment, and docking studies.[a]
e-Research	Research asset management via in-house computer networks and across the World Wide Web.
Extended Hückel calculations	Low-level semiempirical molecular orbital calculations.[a]
Gaussian programs	Type of mathematical function used during ab initio calculations to describe molecular orbitals. Numbers refer to year of program updates.
Genomics	Used herein as the study of the chromosomal and extrachromosomal genes in humans, in particular their complete sequential characterization.
High-throughput screening (HTS)	The level of this activity is moving rapidly from 96- to 384-well microplates.[436]
In silico	Tasks undertaken via computer (e.g., virtual screening).
Isosteres	Molecules or ions of similar size containing the same number of atoms and valence electrons.[c]

TABLE 2.1 (*Continued*)

Term	Definition
Ligand-based drug design	Design of new structural arrangements based on their resemblance to at least a portion of another compound that displays a desired property or biological activity.
Metabophores	Structural features residing in substrates that prompt their specific metabolic conversions.[22]
MNDO calculations	Semiempirical molecular orbital calculations that use a modified neglect of diatomic (differential) overlap approximation.[a]
Molecular mechanics	Calculation of molecular conformational geometries and energies using a combination of empirical force fields.[a]
Molecular orbital calculations	Quantum chemical calculations based on the Schrödinger equation, which can be subdivided into semiempirical (approximated) and ab initio (nonapproximated) methods.[a]
Multivalent ligands	Molecular displays that provide more than one set of binding motifs for interaction with more than one area on one or more biological surfaces.
Nutraceuticals	Dietary supplements purported to have beneficial therapeutic or disease preventive properties (e.g., herbal medicines).[d]
Pharmacogenetics	Study of genetic-based differences in drug response.[106]
Pharmacophores	Structural features needed to activate or inhibit specific receptors or enzyme active sites.
Privileged structures	Molecular frameworks able to provide ligands for diverse receptors.[264,265]
Prodrug	Compound that must undergo biotransformation (e.g., metabolism) before exhibiting its pharmacological effects.[c]
Proteomics	Study of protein structure and function.[438]
Quantum mechanics	Molecular property calculations based on the Schrödinger equation that take into account the interactions between electrons in the molecule.[a]
Semiempirical calculations	Molecular orbital calculations using various degrees of approximation and using only valence electrons.[a]
Single nucleotide polymorphisms (SNPs)	Used herein as single-base alterations in the human genome that occur in specified percentages of distinct portions of the population.
Soft drug	Compound that has been programmed to be biodegraded (e.g., metabolized) to predictable, nontoxic, and inactive metabolites after having achieved its therapeutic role.[c]
Structure-based drug design	Design of new structural arrangements for use as drugs based on protein structural information obtained from x-ray crystallography or NMR spectroscopy.[65,474]
Toxicophores	Structural features that elicit specific toxicities.[481]
Transportophores	Structural features that prompt a specific compound's transport.
Ultrahigh-throughput screening (UHTS)	The level of this activity has moved rapidly from 384- to 1536-well microplates with peak throughput rates of over 100,000 compounds per day. It is estimated that this field will soon be closing in on producing 1200 data points per minute.[478]
Xenobiotic	Compound that is foreign to a given organism.[c]

[a]Adapted from definitions provided by the IUPAC.[432]

[b]Various definitions can be found[433]; this particular definition has been cited because it appears to be used most commonly by the U.S. National Institutes of Health (NIH).[434]

[c]Adapted from definitions provided by IUPAC.[435]

[d]Adapted from definitions provided by the U.S. Office of Dietary Supplements (ODS).[437]

TABLE 2.2 Acronyms and Designations

Acronym	Designation
ADME	Absorption, distribution, metabolism, and excretion
ADMET	ADME and toxicity
CADD	Computer-assisted drug design
CAS	American Chemical Society's Chemical Abstract Services
CCD	Charge-coupled device (as in x-ray area detectors)
CoMFA	Comparative molecular field analysis
CPT	Camptothecin
CYP	Cytochrome P450 metabolizing enzyme[439]
FDA	U.S. Food and Drug Administration
GI	Gastrointestinal tract
HPLC	High-performance liquid chromatography
HTS	High-throughput screening
IND	Investigational new drug[a]
IP	Intellectual property (e.g., trade secrets and patents)
IT	Information technology
KDD	Knowledge discovery in databases[440]
LC	Liquid chromatography
Log P	Log value of a compound's *n*-octonal/H_2O partition coefficient
MDR	Multidrug resistance
MS	Mass spectrometry
NBE	New biological entity,[441] taken to mean that such a compound has potential diagnostic, therapeutic, or prophylactic value
NCE	New chemical entity, taken to mean that such a compound has potential diagnostic, therapeutic, or prophylactic value
NDA	New drug application[a]
NMR	Nuclear magnetic resonance spectroscopy
PAC	Paclitaxel
Pgp	P-glycoprotein pump associated with the ABC class of membrane transporter systems
PK	Pharmacokinetic
QSAR	Quantitative structure–activity relationships; note that the *Q* designation also becomes applicable to all of the other SXR possibilities
SAbR	Structure–absorption relationships
SAR	Structure–activity relationships wherein *activity* is equated herein with therapeutic efficacy-related elements from either an agonist or an antagonist type of interaction
SDM	Site-directed mutagenesis
SDR	Structure–distribution relationships
SER	Structure–excretion relationships
SMR	Structure–metabolism relationships
SNP	Single nucleotide polymorphism within a genome
SPR	Surface plasmon resonance spectroscopy
STR	Structure–toxicity relationships
STrR	Structure–transporter relationships
SXR	Generic representation for simultaneous SAR, SAbR, SDR, SER, SMR, STR, and STrR
UHTS	Ultrahigh-throughput screening
2D	Two-dimensional structure representation
3D	Three-dimensional structure representation

[a]Phrases and acronyms commonly used in the U.S. drug regulatory process.

an eye toward enhancing the efficacy of substances thought to be of therapeutic value.[19] In the United States, medicinal chemistry became formalized as a graduate-level discipline about 75 years ago within the academic framework of pharmacy education. From this setting, overviews of medicinal chemistry's subject matter have been offered to undergraduate pharmacy students for many years.[20,21] Understanding structure–activity relationships (SARs) at the level of inherent physical organic properties (i.e., lipophilic, electronic, and steric parameters) coupled with consideration of molecular conformation soon became the hallmark of medicinal chemistry research. Furthermore, it follows that because these fundamental principles could be useful during the design of new drugs, applications toward drug design became the principal domain for a still young, basic science discipline. Perhaps somewhat prematurely, medicinal chemistry's drug design role became especially important within the private sector, where its practice quickly took root and grew rampantly across the rich fields being staked out within the acres of patents and intellectual property that were of particular interest to the industry.

2.2.2 Early Developments

As a more comprehensive appreciation for the links between observed activity and pharmacological mechanisms began to develop about 50 years ago and then also proceeded to grow rapidly in biochemical sophistication, medicinal chemistry, in turn, entered into what can now be considered to be an adolescent phase. Confidently instilled with a new understanding of what was happening at the biomolecular level, the ensuing period was characterized by the high hope of being able to design new drugs independently in a rational (i.e., ab initio) manner rather than by relying solely on nature's templates and guidance for such. Although this adolescent "heyday of rational drug design"[22] should certainly be credited with having spurred significant advances in the methods that can be deployed for considering molecular conformation, the rate of actually delivering clinically useful therapeutic entities having new chemical structures within the private sector was not significantly improved for most pharmacological targets unless the latter's relevant biomolecules also happened to lend themselves to rigorous analysis (e.g., obtaining an x-ray diffraction pattern for a crystallized enzyme's active site with or without a bound ligand). One of the major reasons that rational drug design fell short of its promise was because without experimental data like that afforded by x-ray views of a drug's target site, medicinal chemistry's hypothetical SAR models often reflected speculative notions that were typically far easier to conceive than were the actual syntheses of the molecular probes needed to assess a given model's associated hypotheses. Thus, with only a small number of clinical success stories to relay, medicinal chemistry's "preconceived notions about what a new drug ought to look like" began to take on negative rather than positive connotations, particularly when being "hand-waved" within the context of a private-sector drug discovery program (e.g., ref. 23). Furthermore, from a practical point of view, the pharmaceutical industry, by and large, soon concluded that it was more advantageous to employ synthetic organic chemists and have them learn some pharmacology than to employ formally trained medicinal chemists and have them rectify any shortcomings in synthetic chemistry that they might have due to their exposure to a broader range of nonchemical subject matter during graduate school. Indeed, given the propensity for like-to-hire-like, the vast majority of today's investigators who practice medicinal chemistry within big pharma, and probably most within the smaller-company segment of the pharmaceutical industry as well, have academic backgrounds from organic chemistry rather than from formalized programs of medicinal chemistry. Although this particular glimpse is important to appreciate as a historical note in that it provides useful insights for the review's later discourse about the formalized training of future medicinal chemists, further references to medicinal chemistry throughout the remainder of this chapter intend to imply medicinal chemistry's practice as a discipline regardless of how a given investigator may have become trained to do so. Importantly, no matter how its practitioners were being derived, and even though it was still very much under-the-gun within the pharmaceutical industry, medicinal chemistry did continue to thrive quite nicely during this period and certainly moved forward significantly as a recognized discipline within all sectors.

Arriving at the next historical segment, however, one finds that medicinal chemistry's inability to accelerate the discovery of new chemical entities (NCEs) by using rational drug design became greatly exacerbated when the biotechnology rainfall began to hover over the field of drug discovery just somewhat less than about 25 years ago.[24] With this development, not only did the number of interesting biological targets begin to rise rapidly, but also, the ability to assay many of these targets in a high-throughput manner suddenly prompted the screening of huge numbers of compounds in very quick time frames. Ultimately, the need to satisfy high-throughput screening's (HTS's)[25,26] immense appetite for compounds was addressed not by either natural product or synthetic medicinal chemistry but by further developments within what had quickly become a flood-level[27] continuing downpour of biotechnology-related breakthroughs. Starting as gene cassette-directed peptide libraries and quickly moving into solid-phase randomly generated peptide and nucleotide libraries,[28–32] this novel technology soon spread across other disciplines, eventually spawning the new field of small-molecule combinatorial chemistry. Today, using equipment and platforms available from a variety of suppliers (e.g., Table 2.3), huge libraries of compounds can readily be produced in either a random or a directed manner.[33–40] Once coupled with HTS, these paired technologies have prompted what has now come to be regarded as a new paradigm for the discovery of NCEs across the entire big-pharma segment of the pharmaceutical enterprise.[41–43] Figures 2.1 and 2.2 provide a comparison of the old (classical) and new drug discovery paradigms, respectively.

TABLE 2.3 Suppliers of Combinatorial Chemistry Systems[a] and Compound Library Trends[b]

Vendor	Web Address
Advanced Chem Tech	Peptide.com
Argonaut Technologies	Argotech.com
Bohdan Automation	Bohdan.com
Charybdis Technologies	Charybtech.com
Chiron Technologies	Chirontechnologies.com
Gilson	Gilson.com
PE Biosystems	Pebiosystems.com
Robbics Scientific	Robsci.com
Tecan	Tecan-us.com
Zymark	Zymark.com

[a]As largely reported in 1999 in an article by Brown.[36]

[b]The number of reported libraries with or without disclosure of their biological activities has grown from only a few that were produced primarily by the private sector in the early 1990s to nearly 1000 in 1999, with nearly half of the latest total now being contributed by the academic sector.[40] Over this same period, the percentage of libraries directed toward unbiased discovery has gone from about 60% to 20%, while that for targeted/optimization of biased structural systems has risen from about 40% to 80%.[40]

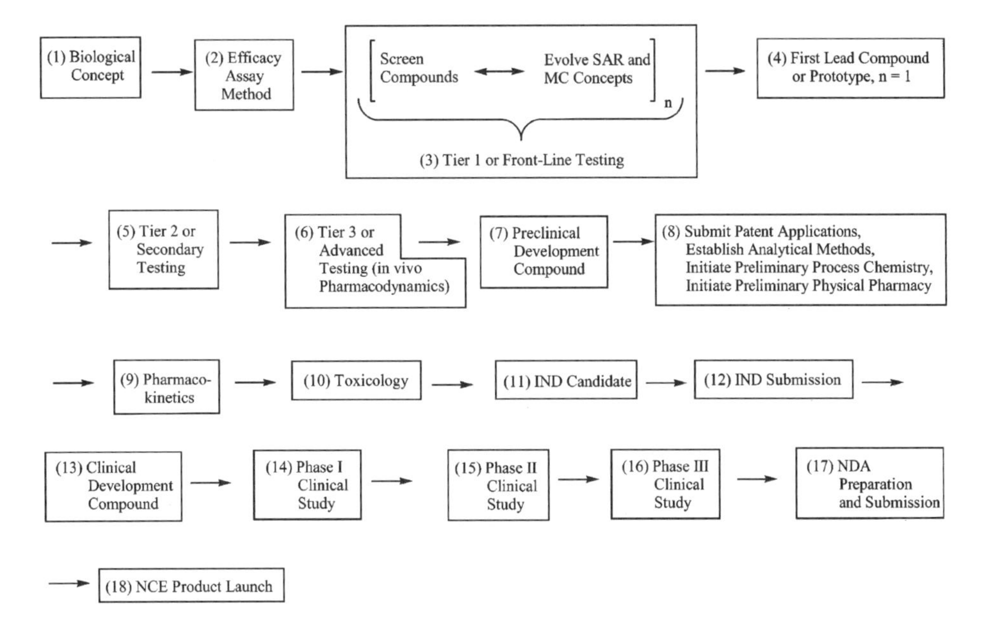

Figure 2.1 Classical drug discovery and development paradigm. This model portrays interactions with U.S. regulatory agencies and uses terms related to those interactions for steps 11 to 17. All of the other terms typify generic phrases that have commonly been used by the international pharmaceutical community. Although some of the noted activities can be conducted in parallel or in an overlapping manner, the stepwise sequential nature of this paradigm's overall process is striking. Furthermore, whenever a progressing compound fails to meet criteria set at an advanced step, the process returns or draws again from step 3 for another reiteration. Numerous reiterations eventually identify a compound that is able to traverse the entire process. A successful passage through the entire process to produce just one product compound has been estimated to require about 15 years at a total cost of about $500 million. While the largest share of these time and cost requirements occur during the later steps, the identification of a promising preclinical development compound, step 7, can be estimated to take about four years from the time of initiating a therapeutic concept. Step 1 is typically associated with some type of physiological or pharmacological notion that intends to amplify or attenuate a specific biological mechanism so as to return some pathophysiology to an overall state of homeostasis. Step 2 typically involves one or two biochemical-level assay(s) for the interaction of compounds intending to amplify or attenuate the concept-related mechanism. As discussed in the text, steps 3 and 4 reflect key contributions from medicinal chemistry and typically use all sources of available information to provide for compound efficacy hits (e.g., everything from natural product surveys to rational approaches based on x-ray diffractions of the biological target). Step 5 generally involves larger in vitro models (e.g., tissue level rather than biochemical level) for efficacy and efficacy-related selectivity. Step 6 generally involves in vivo testing and utilizes a pharmacodynamic (observable pharmacologic effect) approach toward compound availability and duration of action. Step 7 typically derives from a formal review conducted by an interdisciplinary team upon examination of a formalized compilation of all data obtained to that point. Step 8 specifies parallel activities that are typically initiated at this juncture by distinct disciplines within a given organization. Step 9 begins more refined pharmacokinetic evaluations by utilizing analytical methods for the drug itself to address in vivo availability and duration of action. Step 10 represents short-term (e.g., two-week) dose-ranging studies to identify toxic markers initially within one or more small animal populations. Expanded toxicology studies typically progress while overlapping with steps 11 to 14. Steps 11 to 13 represent formalized reviews undertaken by both the sponsoring organization and the U.S. Food and Drug Administration (FDA). Step 14 is typically a dose-ranging study conducted in healthy humans. Steps 15 and 16 reflect efficacy testing in sick patients, possible drug interactions, and so on. Step 17 again reflects formalized reviews undertaken by both the sponsoring organization and the FDA. The FDA's fast-track review of this information is now being said to have been reduced to an average of about 18 months. It is estimated that it costs a company about $150,000 for each day that a compound spends in development. Finally, step 18 represents the delivery of an NCE to the marketplace. (From refs. 41 and 477 to 482.)

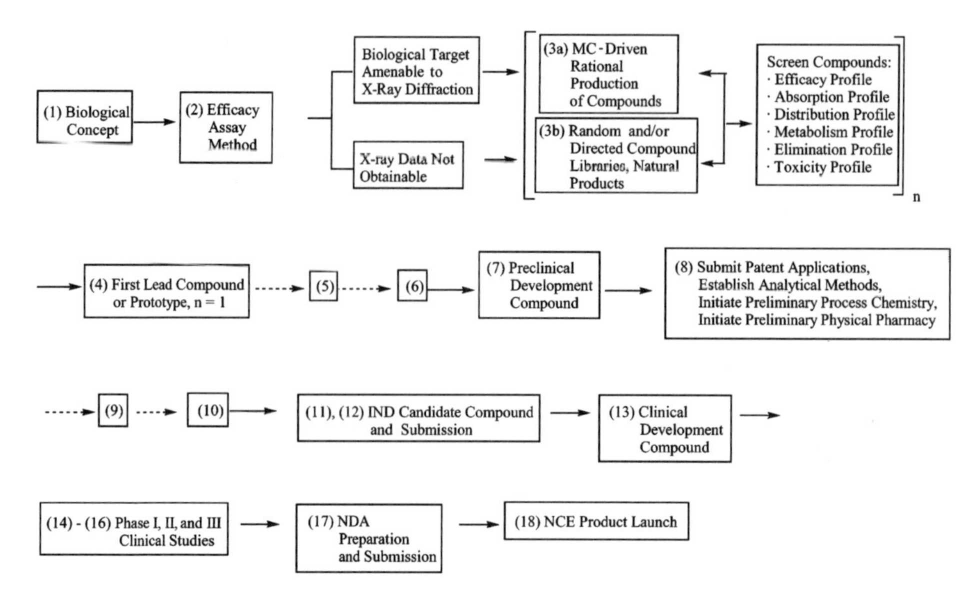

Figure 2.2 New drug discovery and development paradigm. This model portrays interactions with U.S. regulatory agencies and uses terms related to those interactions for steps 11 to 17. All of the other terms typify generic phrases that are commonly used by the international pharmaceutical community. The battery of profiling included in step 3 represents a striking contrast to the classical drug discovery paradigm (Fig. 2.1). Furthermore, all of these screens are/will be of a high-throughput nature such that huge numbers of compounds can be tested simultaneously in extremely short time periods. As the predictive value of the resulting profiles improve, selected compounds will have higher and higher propensities to proceed successfully through steps 5, 6, 9, and 10 (Fig. 2.1) to the point that these assays may become more of a confirmatory nature or may even be able to be omitted completely (hence their dashed lines here).[483] The efficiency of traversing the various clinical testing steps 14 to 16 successfully will also be improved, but their complete removal from the overall process is highly unlikely. After the initial investments to upgrade step 3 and enough time has passed to allow for the generation of knowledge from step 3's raw data results (see text for discussion), the overall time frame and cost for a single NCE to traverse the new paradigm should be considerably improved from the estimates provided in Fig. 2.1. Step 1 is likely to be associated with some type of genomic and/or proteomic derived notion that intends to amplify or attenuate a specific biological mechanism so as to return some pathophysiology to an overall state of homeostasis. Step 2 will be a high-throughput assay derived from using molecular biology and bioengineering techniques. In step 3a, MC = medicinal chemistry. Because step 3a exploits actual pictures of what type of structural arrangements are needed to interact with the biological targets, this approach toward identifying new compound hits will continue to operate with high efficiency. However, because of this same efficiency, the biological targets that lend themselves to such experimental depiction (by affording crystals suitable for x-ray diffraction) are likely to be quickly depleted very early into the new millennium. Step 3b represents various combinatorial chemistry-derived libraries, natural product collections, and elicited natural product libraries (see text for discussion of this topic). Steps 4 to 18 are similar to the descriptions noted in Fig. 2.1. (From refs. 41 and 477 to 482.)

2.2.3 Present Status

Interestingly, the marriage of HTS with combinatorial chemistry has led to a situation where identifying initial lead compounds is no longer considered to be a bottleneck for the overall process of drug discovery and development. Indeed, many of the programs within pharmaceutical companies are presently considered to be suffering from "compound overload,"[44] with far too many initial leads to follow up effectively. ADMET assessments are now regarded as the new bottleneck, along with the traditionally sluggish clinical and regulatory steps. This situation, in turn, has prompted an emphasis to move ADMET-related parameters into more of an HTS format undertaken at earlier decision points. Thus, even though efficacy-related HTS and combinatorial chemistry reflect very significant incorporations of new methodologies, *from a strategic point of view the most striking feature of the new drug discovery and development paradigm shown in Fig. 2.2, compared to the classical approach depicted in Fig. 2.1, actually becomes the trend to place ADMET-related assays closer to the beginning of the overall process* by also deploying HTS methods. Clearly, with the plethora of biologically based therapeutic concepts continuing to rise even further and the identification of lead compounds now being much quicker because of the HTS–combinatorial chemistry approach, more efficient handling of ADMET-related concerns represents one of the most significant challenges now facing drug discovery and development. Because of its importance, this challenge is likely to be resolved within the near term of the new millennium. Medicinal chemistry's critical role during this further evolution of the present drug discovery paradigm is highlighted in subsequent sections.

2.2.4 Examples Involving Site-Directed Mutagenesis

While the drug discovery process has been influenced by biotechnology in numerous ways (e.g., Tables 2.4 and 2.5), one development deserves to be especially noted as this brief account of medicinal chemistry's history and present status is brought to a close. This development is already having a major impact directly on the process of uncovering SARs relevant to small-molecule drug design.[45] The method involves site-directed mutagenesis (SDM). SDM applies to systematic point mutations directed toward specific sites on genes that translate to proteins associated with enzyme-active sites or drug receptor systems such that the targeted changes can be used to study SARs while holding one or more active site/receptor ligands constant during the assessment (Fig. 2.3). Numerous investigators are now utilizing this reverse SAR approach to explore both enzyme and receptor ligand interactions. Three examples are provided below wherein the site-directed mutagenesis studies have been further coupled with one or more of the latest analytical chemistry techniques such as microcalorimetry, as well as with sophisticated computational approaches. The first pair of examples involve the active sites of some isolable enzymes; the third example involves a nonisolable membrane-bound receptor complex.

Slama et al. have conducted studies involving a phosphatidylinositol-4-phosphate phosphatase (Scheme 2.1) designated as Sac1p.[46] With this enzyme's peptide sequence in hand, transformation of a bacterial host with an appropriate gene copy plasmid has allowed at least one point-mutated protein to be examined per month in a functional biochemical assay. While this particular example happens to represent a system that lends itself to over expression coupled with a functional protein product that lends itself to ready isolation, it clearly demonstrates that site-directed mutagenesis should no longer be considered to be lengthy and tedious compared to classical SAR studies undertaken by rational synthetic modifications of an enzyme's substrates. Slama et al.[47] are now studying poly(ADP-ribose)glycohydrolase or PARG (Scheme 2.1) to more definitively assign a tyrosine (i.e., ^{796}Tyr) to a key role within this enzyme's active site. Previously, this particular residue has only been able to be implicated as being important by using classical inhibitor and

TABLE 2.4 Impact of Biotechnology on Small-Molecule Drug Discovery and Development[a]

Activity	Impact
Genomics and proteomics	Plethora of new and better defined mechanisms to pursue as therapeutic targets.
High-throughput efficacy assays	Screen huge numbers of therapeutic candidates in short time frames using low compound quantities.
High-throughput ADMET assays	An evolving development: once validated and coupled with efficacy assays, should eventually allow for selection or drug design/ synthesis of clinical candidate compounds rather than lead compounds that still require considerable preclinical testing and additional chemical tailoring.
Peptide and oligonucleotide compound libraries	Provide huge numbers of compounds for screening (spawned the field of *combinatorial chemistry* as now applied to small organic compounds); can be used as SAR probes and, pending further developments in formulation and delivery, may also become useful as drug candidates.
Site-directed mutagenesis	Allows "reverse" structure–activity relationship explorations.
Transgenic species	Novel in vivo models of pathophysiology that allow *pharmacological proof of principle* in animal models that mimic the human situation; and animal models modified to have human metabolism genes so as to provide more accurate PK data and risk assessment.[442]
Peptide version of pharmacological prototype	Developed to the IND phase as an intravenous agent can allow for *clinical proof of principle* or concept.
Pharmacogenetics	An evolving development: should soon refine clinical studies, market indications/ contraindications, and allow for subgrouping of populations to optimize therapeutic regimens; eventually, should allow for classification of prophylactic treatment subgroups.

[a]This list is not intended to highlight the numerous activities associated with the development of specific "biotech" or large-molecule therapeutics (e.g., see Table 2.5). The arrangement of activities follows the order conveyed in Figs. 2.1 and 2.2 rather than being alphabetical.

photoaffinity labeling studies.[48,49] In this case, one of the PARG assays deploys microcalorimetry measurement of the binding energies for a designated series of ligands that is then held constant as it is surveyed across the various mutant enzymes.[50] Similarly, Messer and Peseckis et al. have collaborated in several site-directed mutagenesis studies involving M1 muscarinic receptors.[51–53] These topographical mapping studies began as a follow-up

TABLE 2.5 Examples of Approved[a] Biotechnology-Based Drugs[b]

Name[c]	Company	Clinical Use
Tissue plasminogen activators (tPAs)		
Activase	Genentech	Dissolution of clots associated with myocardial infarction, pulmonary embolism, and stroke
Retevase	Boeringher-Mannheim; Centocor	
Clotting factors		
BeneFIX	Genetics Inst.	Treatment of various hemophilias
KoGENate	Bayer	
Recombinate	Baxter/Genetics Inst.	
Ceredase-glucocerebrosidase		
Cerezyme	Genzyme	Gaucher's disease
DNAse		
Pulmozyme	Genentech	Cystic fibrosis
Insulins		
Humalog	Eli Lilly	
Humulin[d]	Eli Lilly	
Novolin	Novo Nordisk	Insulin-sensitive diabetes
Novolin L	Novo Nordisk	
Novolin R	Novo Nordisk	
Erythropoiten-related growth factors		
Epogen	Amgen	Anemia associated with renal failure, chemotherapy, surgery, or loss of blood
Procrit	Ortho Biotech	
Growth hormone		
Biotropin	Bio-Tech. Gen.	Growth hormone deficiency in children* and for use in adults and in Turner's syndrome**; AIDS-related wasting and for pediatric HIV patients
Genetropin	Pharmacia & Upjohn	
Humatrope	Eli Lilly	
Neutropin*	Genentech	
Norditropin	Novo Nordisk	
Protropin	Genentech	
Saizen	Serono Labs	
Serostim**	Serono Labs	
Growth hormone-releasing factor		
Sermorelin	Serono Labs	Evaluation and use in pediatric growth hormone deficiency
Platelet-derived growth hormone		
Regranex	Ortho-McNeil	Lower-extremity ulcers in diabetic neuropathy
Fertility hormones		
Gonal-F Serono	Female infertility	

(Continued)

TABLE 2.5 (*Continued*)

Name[c]	Company	Clinical Use
Interferon-α		
Alferon N	Interferon Sci.	Genital warts*; chronic hepatitis C infections**; several antiviral and anticancer indications
Infergen*	Amgen	
Intron**	Schering-Plough	
Roferon-A**	Hoffman-LaRoche	
Interferon-β		
Avonex	Biogen	Multiple sclerosis
Betaseron	Berlex	
Interferon-γ		
Actimmune	Genentech	Chronic graulomatous disease
Granulocyte-colony and macrophage-stimulating factor		
Leukine	Immunex	Treatment of blood disorders during chemotherapy and blood cell/marrow transplants
Neupogen	Amgen	
Interleukins		
Neumega	Genetics Inst.	Thrombocytopenia subsequent to chemotherapy*; certain cancers
Proleukin*	Chiron	
Monoclonal antibodies		
CEA-scan	Immunomedics	Cancer diagnostic
Herceptin	Genetech	Breast cancer
MyoScint	Centacor	Cardia imaging agent
Neumega	Genetics Inst.	Prevention of chemotherapy-induced thrombocytopenia
OncoScint	Cytogen	Cancer diagnostic
Orthoclone	Ortho Biotech	Transplant rejection
Prostascint	Cytogen	Cancer diagnostic
ReoPro	Lilly	Prevention of blood clots during angioplasty
Rituxan	Genctech; IDEC	Non-Hodgkin's lymphoma
Simulect	Novartis	Transplant rejection
Synagis	Med Immune	Respiratory syncytial virus disease
Vezluma	Boechringer-Ingelheim/NesRx	Cancer diagnostic
Zenapax	Hoffman-LaRoche; Protein Design Labs	Transplant rejection
Antisense nucleotides		
Vitravene	ISIS/Ciba	Retinitis

[a]Prior to mid-year 1999 and as largely reported in the three-part series published in 1999 by Hudson and Black.[6] In comparison, a more recent survey reveals that about 375 additional agents were already in, or were nearing entry into, the clinical testing pipeline.[443]

[b]These compounds have also been referred to as *new biological entities* (NBEs) by analogy to *new chemical entities* (NCEs).[441] In general, NBEs use, recreate, or improve on proteins and other biomolecular polymers produced in the body to counter disease.[443]

[c]Asterisks in this column relate to those in the "clinical use" column.

[d]First recombinant-DNA-produced therapeutic protein to enter the market.[6]

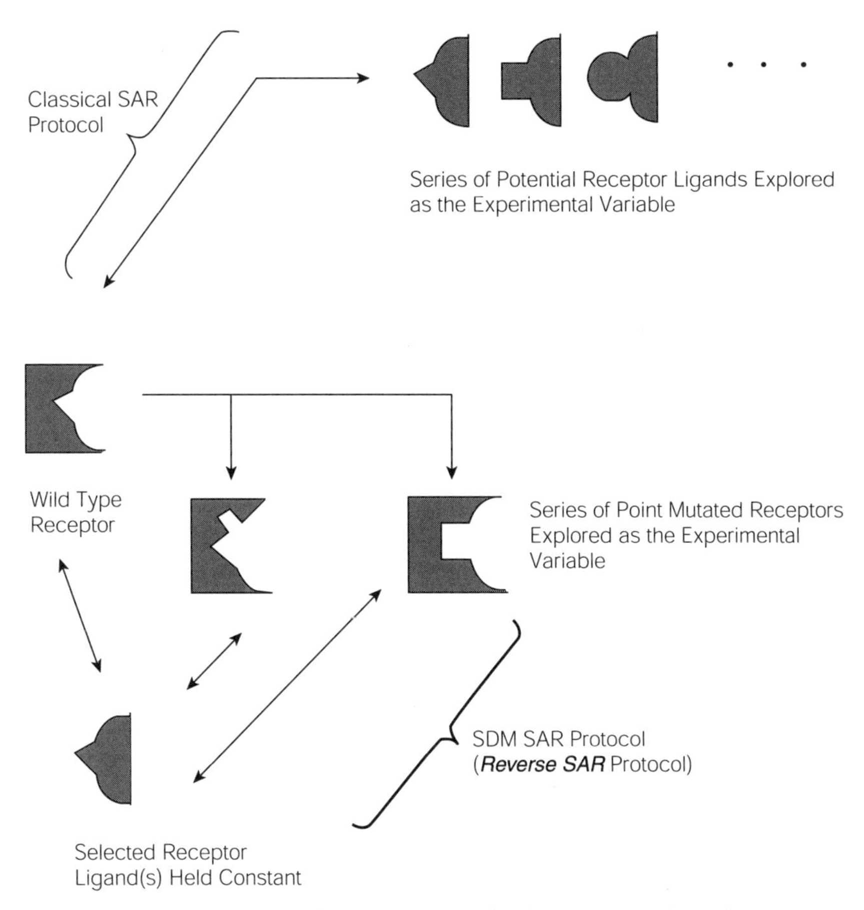

Figure 2.3 Classical versus site-directed mutagenesis (SDM) exploration of structure–activity relationships (SARs).

to better define the receptor interactions exhibited by CDD-0102 (Scheme 2.2), a selective M1 agonist that is undergoing preclinical development for the potential treatment of Alzheimer's disease.[54–56] Interestingly, these investigator's more recent efforts have additionally turned toward addressing questions about the multivalent nature of M1 muscarinic receptors.[57–60] Thus, in this case the site-directed mutagenesis studies are helping to resolve fundamental issues about the nature of these receptors, as well as helping to implicate amino acid residues involved in the binding of agonists for the purpose of identifying drug structural themes having enhanced selectivity for certain of the receptor subtypes. In all of the M1-related studies, the site-directed mutagenesis results are being coupled to computational studies directed toward mapping the receptor's key topological features.[61]

2.2.5 Latest Trends

As has been noted by others, the deployment of these new approaches to studying SARs represents an "exciting"[62] development that has clearly had an invigorating effect on the practice of medicinal chemistry in many areas. Alternatively, that the SAR hallmark and

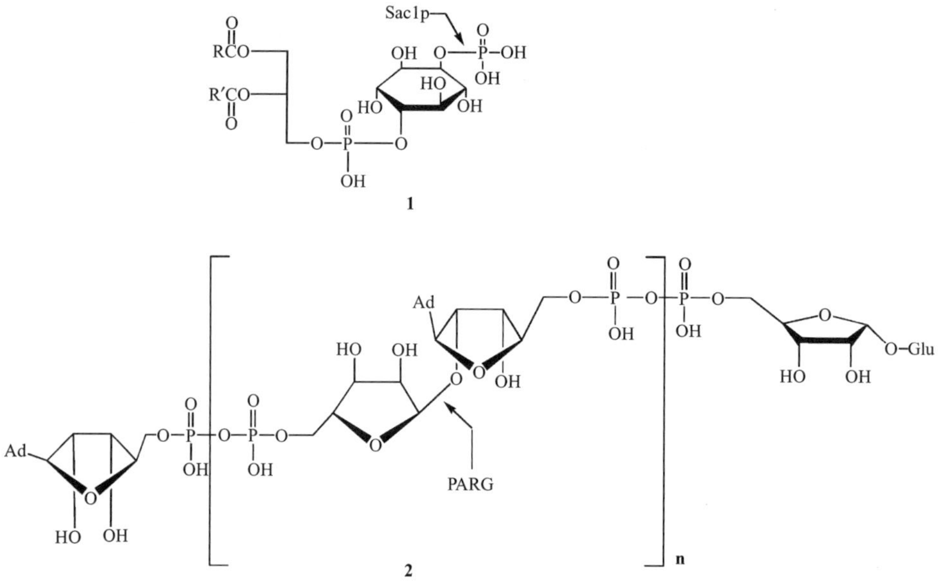

Scheme 2.1 Cleavages effected by Sac1p and PARG. **1**, Phosphatidylinositol-4-phosphate, wherein R and R′ are fatty acid side chains; **2**, poly(ADP-ribose), wherein Ad is adenine and Glu is glucose. In both cases, arrows indicate bonds undergoing enzymatic hydrolyses. (From refs. 46 to 49.)

drug design intellectual domains of medicinal chemistry, along with chemically servicing the experimental approaches toward identifying novel, lead structures, have all been overrun by technologies initially derived from other disciplines, has also had somewhat of an unnerving impact on medicinal chemistry. This is because it has previously been the nearly exclusive deliverance of these roles from a chemical orientation that has served to distinguish medicinal chemistry as a distinct discipline. Thus, as a field, medicinal chemistry has had to mature quickly from its troubled adolescence only to find itself in the middle of an identity crisis. This crisis has been occurring among both its private-sector practitioners

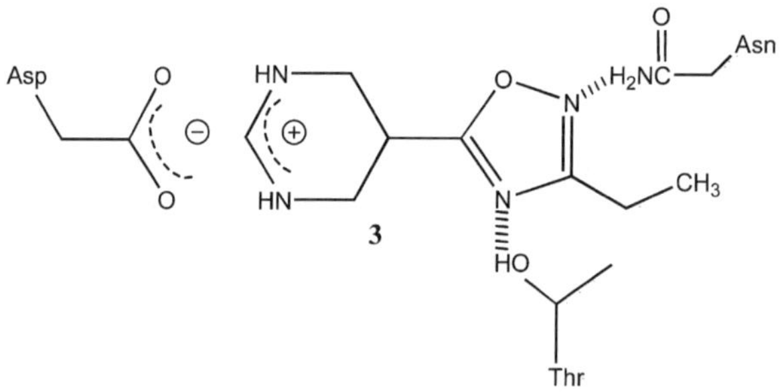

Scheme 2.2 Suspected interactions of CDD-0102 with muscarinic M1 receptors. **3**, 5-(3-Ethyl-1,2,4-oxa-diazol-5-yl)-1,4,5,6-tetrahydropyrimidine (CDD-0102) shown as its protonated species; Asn, asparagine 382; Asp, aspartic acid 105; Thr, threonine 192. Dashed lines signify suspected hydrogen bonds. (From refs. 51 to 53).

and its academicians. Indeed, today's trend within the public-sector funding arena, with a major emphasis on genomics and proteomics, is causing some academic medicinal chemistry and chemistry investigators to turn their intellectual pursuits further and further toward molecular biology.[63] Similarly, the undergraduate instruction of pharmacy students, which has for so long represented medicinal chemistry's academic bread and butter, has shifted its emphasis away from the basic sciences toward more of a clinically oriented curriculum. A final development that has contributed to medicinal chemistry's identity crisis is the fact that the new drug discovery paradigm supplants medicinal chemistry's long-standing position wherein its practitioners have typically been regarded as the primary inventors of the composition of matter specifications associated with NCE patent applications. This last development, along with the overall IP arena of the future, is addressed in Section 2.9.

Although as alluded to earlier and as now practiced in tandem with SDM studies, classical medicinal chemistry rationale is still being relied on heavily to effectively identify and fine-tune lead compounds for systems whose biomolecules have lent themselves to x-ray diffraction and/or NMR analyses (i.e., structure-based drug design),[64,65] medicinal chemists of the new millennium must be prepared to face the possibility that the present complement of these more amenable pharmacological targets is likely to become quickly exploited, perhaps eventually even exhausted. Such a scenario, in turn, suggests that this last stronghold for today's practice of rational medicinal chemistry could also be lost as a bastion against what could then potentially become an even more serious identity crisis in the future.

Present trends thus leave us with the following paradigm for the immediate future of small-molecule drug discovery (Fig. 2.2):

1. Genomics and proteomics will continue to uncover numerous new pharmacological targets, to the extent that choosing the most appropriate and validating such targets among the many therapeutic possibilities will also continue to rise as a growing challenge in itself.
2. Biotechnology will, in turn, continue to respond by generating quick ligand-identification assays for all new targets chosen to be pursued: namely, by deploying HTS protocols.[66]
3. Targets that lend themselves to x-ray diffraction and structure-based drug design are likely to be quickly exploited.
4. Ligands for HTS will be supplied by existing and new combinatorial-derived compound libraries as well as from wild[67] and biotechnology-elicited natural sources.
5. Assessment of ADMET parameters, presently considered to be the bottleneck for the overall drug discovery process (Table 2.6), will continue to move toward HTS modes that can be placed at earlier and earlier positions within the decision trees utilized to select lead compounds for further development as drugs.

It is important to emphasize at this juncture that *in order to place confidence in the predictive value of ADMET HTS surveys, these particular screens must become validated relative to actual clinical-related outcomes*. For the moment, however, this situation is best likened to a deep, dark chasm that the rapidly evolving ADMET HTS surveys still need to traverse if they are ultimately to become successful. Nevertheless, because of its importance, it is proposed that not only will this type of validation be forthcoming within the near term, but as the new millennium then continues into the more distant future,

TABLE 2.6 Assessment of Drug Discovery and Development Bottlenecks

Activity	Estimated Time Frame	Percentage Successfully Traversing Associated Criteria
Biological conception	A plethora of genomic/proteomic characterizations presently lies waiting to be exploited; this situation is expected to prevail well into the new millennium.	The challenge lies in prioritizing which of the numerous mechanisms might be best to pursue (see next entry).
Proof of therapeutic principle	Ultimately requires reaching phase II clinical testing; BIOTECH-derived humanized and/or transgenic disease state models may be able to be substituted at an earlier point, depending on the confidence associated with their validation.	Generally high, although there are some distinct therapeutic categories that continue to have low success rates or lack definitive validation, such as the attempted treatments of septic shock or the pursuit of endothelin modulators.
Identification of lead compound based on efficacy screen	Using HTS, thousands of compounds can be tested in a matter of days or less (10 to 100 times more with UHTS); companies are beginning to have more lead compounds than they can move forward in any given program.	One compound out of 5000 from random libraries/one out of 10 from directed libraries; despite the low efficiency, this is not regarded as a bottleneck because HTS can be done so quickly; much higher percentages can be obtained during ligand- and structure-based drug design, but synthesis is then correspondingly slower.
Progression to preclinical development compound[a]	Approximately two years.	About one out of 50 wherein all can be examined during the time frame indicated.
Progression to clinical development compound[a]	Approximately two years.	About one out of 10 wherein all can be examined during the time frame indicated.
Phase I clinical study[b]	Approximately one year.[c]	About one out of two.
Phase II clinical study[b]	Approximately two years.[c]	About one out of two.
Phase III clinical study	Approximately three years.[c]	About one out of 1.5 and often with modified labeling details.
Product launch	Approximately two years (NDA submission/approval).	About one out of 1.5.[d]

[a]Presently regarded as the bottleneck for the overall process. These are the points where ADMET properties have historically been assessed. Approximately 40% are rejected owing to poor pharmacokinetics and about another 20% because they show toxicity in animals. In the new drug discovery paradigm (Fig. 2.2), the ADMET assessments are being moved to an earlier point in the overall process and are being conducted in an HTS mode. However, in most cases, validation of the new methods relative to clinical success still needs to be accomplished.

[b]Although these studies may not be able to be accomplished any quicker, they may be able to be done more efficiently (e.g., smaller numbers and focused phenotypes within selected patient populations) and with a greater success rate based on making the same improvements in the ADMET assessment area as noted in footnote a.

[c]Timing includes generation and submission of formal reports.

[d]About one of five compounds entering into clinical trials becomes approved. The overall process to obtain one marketed drug takes about 12 to 15 years at a cost of about $500 million.

Source: Refs. 444 to 449.

ADMET-related parameters derived from HTS will become even further manipulated for their potential to allow for synergistic relationships within the overall course of a given therapeutic or prophylactic treatment. These intriguing future possibilities for exploiting ADMET-related parameters in a proactive and synergistic manner rather than in just a negative filtering mode, along with the likely move toward prophylactic medicines, are discussed further in subsequent sections.

2.3 EVOLVING DRUG DISCOVERY AND DEVELOPMENT PROCESS

2.3.1 Working Definition for Medicinal Chemistry

If the targets that are readily amenable to x-ray diffraction and structure-based drug design do become exhausted with time, the only title role presently highlighted for medicinal chemistry within the new paradigm of drug discovery will also disappear. To determine if medicinal chemistry will still be operative under such a circumstance, let us first review medicinal chemistry's present definition. Perhaps further attesting to medicinal chemistry's present identity crisis, however, is the fact that a purview of several of today's major medicinal chemistry texts reveals that although one can find topic-related versions of such within the context of various other discussions, even the textbooks seem reluctant to provide an explicitly stated general definition for medicinal chemistry (e.g., refs. 68 to 73). Turning, instead, to this document's earlier consideration of medicinal chemistry's history, and for the moment disengaging ourselves from any biases that might be interjected by overreacting to the continuing flood of biotechnology-related trends, it does become possible to devise a general, working definition for medicinal chemistry that can be used to address its present-day identity crisis while also serving in a search for any key roles that medicinal chemistry ought to be playing now and into the future of life science research.

As a working definition for this review, let us simply state that *medicinal chemistry uses physical organic principles to understand the interaction of small molecular displays with the biological realm.* Physical organic principles encompass overall conformational considerations, chemical properties, and molecular electrostatic potentials, as well as distinctly localized stereochemical, hydrophilic, electronic, and steric parameters. Understanding such interactions can provide fundamental, basic knowledge that is general as well as compound-specific in its applications directed toward either enhancing the overall profile of a certain molecular display or designing an NCE (e.g., by effecting small molecule-driven perturbations of discrete biological processes or of overall biological pathways for the purpose of eliciting a specified therapeutic endpoint). Small molecular displays should be thought of in terms of low-molecular-weight compounds (e.g., usually less than 1 kg) that are typically of a xenobiotic origin and thus not in terms of biotechnology-derived polymers. While the latter are being addressed aggressively by other fields, it should additionally be noted that a consideration of specific details associated with the interaction of small molecular portions of more complex biomolecular systems still falls within the purview of medicinal chemistry's stated focus. Alternatively, the biological realm should be thought of very broadly, so as to encompass the complete span of new ADMET-related systems as well as the traditional span of biological surfaces that might be exploited for some type of efficacious interaction. The technologies that can be deployed as tools to study these interactions at medicinal chemistry's fundamental level of understanding are, by intent, dissociated from medicinal chemistry's definition. Presently, such tools include

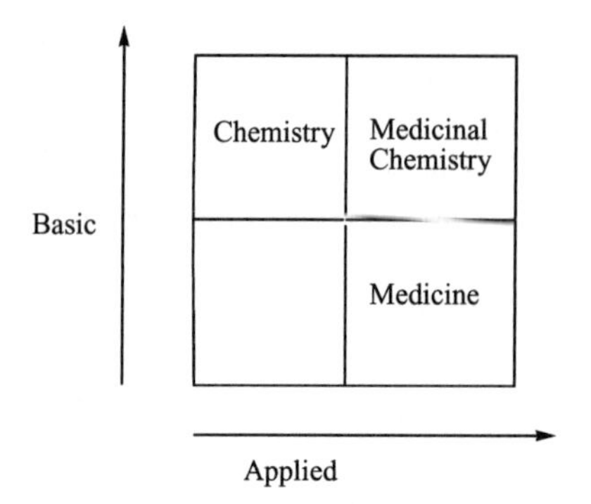

Figure 2.4 Nonlinear relationship of medicinal chemistry to basic and applied research. Surveys suggest that chemical pharmaceutical companies spend about 9%, 37%, and 54% of their research dollars on basic, applied, and developmental aspects of research, respectively.[484] (Adapted from a figure provided by Cotton[75] as part of his summary and commentary about a book entitled *Pasteur's Quadrant*.[74])

biotechnology-related methods such as site-directed mutagenesis, combinatorial chemistry methods provided that the latter are coupled with knowledge-generating structural databases (later discussion), and long-standing synthetic chemistry manipulations that can be conducted in a systematic manner on either one of the interacting species in order to explore SARs. In the future, medicinal chemistry should be equally prepared to exploit whatever new tools and techniques that become available to allow it to proceed more efficiently along the lines of its technology-independent definition. Finally, it should be appreciated that this definition merges both the basic and applied natures of medicinal chemistry's scientific activities into a key mix of endeavors for which a new research paradigm (Fig. 2.4) has also been proposed recently as being a significant trend,[74] even if potentially "dangerous" in that it could compromise the longer-term pursuit of fundamental knowledge by bringing applied science decision criteria into funding programs that have previously supported pure, basic science.[75]

2.3.2 Immediate- and Long-Term Roles for Medicinal Chemistry

Importantly, no matter what pace targets amenable to x-ray may continue to allow structure-based drug design to be pursued in the new millennium, applying the aforementioned definition across the present drug discovery paradigm reveals that there is an even more vital activity with which medicinal chemistry needs to become more involved. Indeed, as HTS results are amassed into mountains not just for efficacy data but for each of the ADMET parameters as well, it should ultimately become medicinal chemistry's role, by definition, to attempt to understand and codify these awesome, crisscrossing ranges if such data are to be merged and used either to select or design the most promising preclinical development compounds. For example, while medicinal chemistry's principles and logic may not be needed to identify hits or leads from a single HTS efficacy parameter survey across a library of potential ligands, or perhaps not even needed for two or three of such consecutive surveys involving a few additional ADMET-related parameters, it is extremely doubtful that the same series of compounds identified from within an initial library as a hit subset, or as further generated within a directed library based on the initial hit subset, will be able to sustain themselves as the most preferred leads upon continued HTS parameter surveys in the future when the latter become ramped up to their full potential. Furthermore, this should still be the case even if the downstream selection criteria become more and more relaxed through any of such progressions that ultimately strive to merge even a preliminary HTS-derived ADMET portfolio in conjunction with one or more selective efficacy HTS parameter surveys. In other words, identification of the optimal end product (i.e., the best preclinical candidate compound) is unlikely to be derivable from an experimental process that does not represent a "knowledge"[76,77] generating system that also allows for rational-based assessments and adjustments, or even complete revamping, to be interspersed at several points along the way. In the end, today's move toward "focused libraries"[78,79] and "smarter," presorted relational databases may thus represent a lot more than just the often-touted desire to "be more efficient."[80] Indeed, this may be the only way for the new drug discovery paradigm, now in its own adolescent phase, eventually to work as it continues to mature and to take on more ADMET-related considerations in an HTS format. In this regard it can now be emphasized that *the common denominator required to correlate the HTS data from one pharmacological setting to that of another ultimately resides in the precise chemical structure language that medicinal chemistry has been evolving since its emergence as a distinct field* (i.e., SAR defined in terms of physical organic properties displayed in three-dimensional space). This, in turn, suggests that within the immediate future of the new millennium *it should be medicinal chemistry that rises to become the central interpreter and distinct facilitator that will eventually allow the entire new drug discovery paradigm to become successful.*

This central role for medicinal chemistry may become even more critical longer term (i.e., for the next 50 to 75 years of the new millennium). Speculating that the new drug discovery paradigm will indeed mature within the next 25 years into a synergistic merger of efficacy and thorough ADMET HTS systems that allows for an effective multiparameter survey to be conducted at the onset of the discovery process, the accompanying validated predictive data that will have been generated over this initial period should be statistically adequate to actually realize today's dream of virtual or in silico screening[81–86] through virtual compound and virtual informational libraries (not just for identifying potential efficacy leads to synthesize, as is already being attempted, but across the entire preclinical portion of the new paradigm wherein the best overall preclinical candidate compound is selected with high precession for synthesis at the outset of a new therapeutic program) (Fig. 2.5). However, this futuristic prediction again depends on the entire maturation process being able to proceed in a knowledge-generating manner. Again central to the latter is medicinal chemistry as the common denominator. For example, with time it can be expected that just as various pharmacophores and toxicophores have already been identified for various portions of the biological realm associated with efficacious or toxic endpoints, respectively, specific molecular properties and structural features will become associated with each of the ADME behaviors. Indeed, work toward such characterizations is already progressing in all of these areas (Table 2.7). Understanding the pharmacophores, metabophores,[22] toxicophores, and so on, in terms of subtle differences in molecular electrostatic potentials (from which medicinal chemistry's physical organic properties of interest are derived), as well as in terms of simple chemical structural patterns, will eventually allow for identifying optimal composites of all of these parameters across virtual compound libraries as long as the latter databases have also been constructed in terms of both accurate three-dimensional molecular electrostatic potentials and gross structural properties.

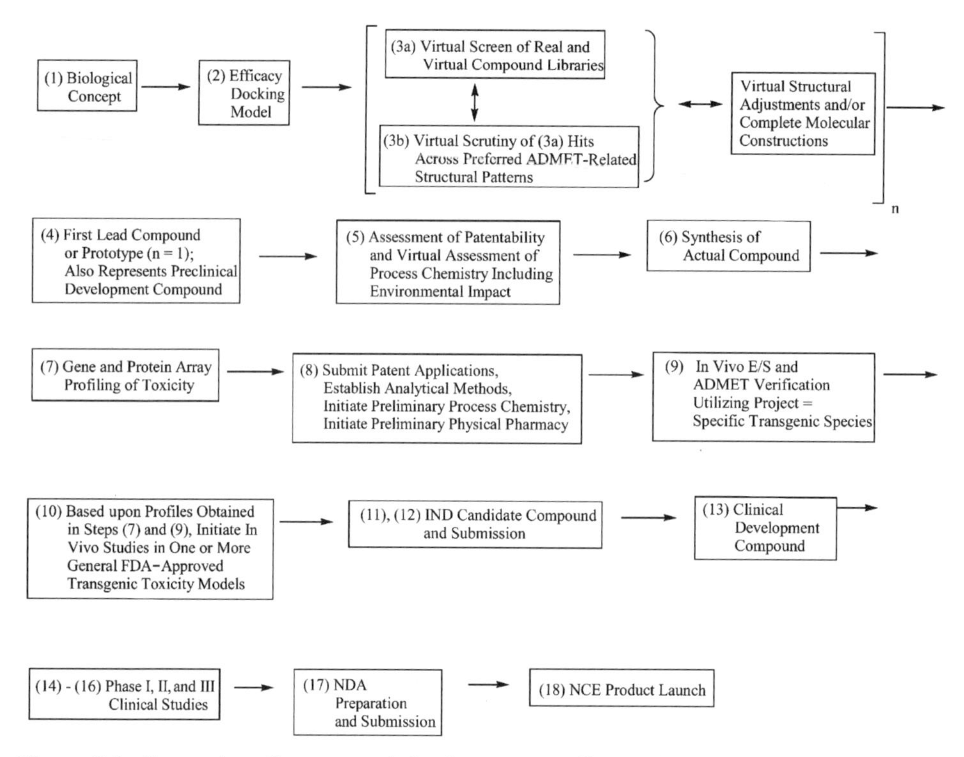

Figure 2.5 Future drug discovery and development paradigm. This model portrays interactions with U.S. regulatory agencies and uses terms related to those interactions for steps 11 to 17. *Future* implies about 50 to 75 years into the new millennium. The most striking feature of this paradigm compared to Figs. 2.1 and 2.2 is the considerable number of decisions that will be made from virtual constructs rather than from experimental results. Confidence in the virtual decisions will be directly proportional to the level of knowledge that is learned from the huge amounts of drug screening data being amassed during the first 25 to 50 years of the new millennium, coupled with the overall ability to predict clinical outcomes. Step 1 is likely to be associated with some type of genomic and/or proteomic derived notion that intends to amplify or attenuate one or more specific biological mechanisms so as either to return some pathophysiology to an overall state of homeostasis or to modify some system in a manner that prevents or provides prophylaxis toward an otherwise anticipated pathophysiological development. A growing emphasis of treatments will be directed toward prevention. Step 2 may be based on an actual x-ray diffraction version of the biological target or on a computationally constructed version derived from similar known systems that have been catalogued for such extrapolations. In either case, docking studies will be conducted in a virtual mode. Steps 3, 4, and 5 will be conducted in a virtual mode. Steps 6 and 7 represent the first lab-based activities. After submission of patents, it is proposed that in vivo testing involving steps 9 and 10 will be able to take advantage of project-specific and FDA-approved generic toxicity model transgenic species. Steps 11 to 18 are similar to those in Figs. 2.1 and 2.2, except that the likelihood for a compound to fall short of the desired criteria will be significantly reduced. Subject inclusion/exclusion criteria will also be much more refined based on advances in the field of pharmacogenetics.

It should be clear that in order to play this key role effectively, the medicinal chemist of the new millennium (Fig. 2.6) will have to remain well versed in physical organic principles and conformational considerations while becoming even more adept at applying them within the contexts of each of the ADMET areas, as has previously been done during

TABLE 2.7 Examples of Efforts to Establish ADME-Related Structural Patterns[a]

ADME Area	Parameters
Absorption	Physicochemical (e.g., log P[312,450]) Functional groups (e.g., rule of five[311])
Distribution	Two-directional flows across Caco-2 cell monolayers[451]
Metabolism	2D structure-metabolism databases[452] (e.g., Metabol Expert,[453] META,[454] Metabolite,[455] and Synopsis Metabolism Database[456]) 3D models (e.g., CYP1A,[457] CYP2A,[458] CYP2B,[459] CYP2C,[560] CYP2D,[461,462] and CYP3A[463]
Excretion (half-lives, etc.)	Structural patterns across species[464] Physicochemical (e.g., log P[465,466])

[a]It is anticipated that the ADME informational area will soon become heavily inundated as pharmaceutical companies begin to share their rapidly accumulating data through publications in the public domain (analogous to what has happened historically for receptor/enzyme active site efficacy information but with the latter having previously had to transpire in a much slower fashion without the arrival of today's HTS methodologies).

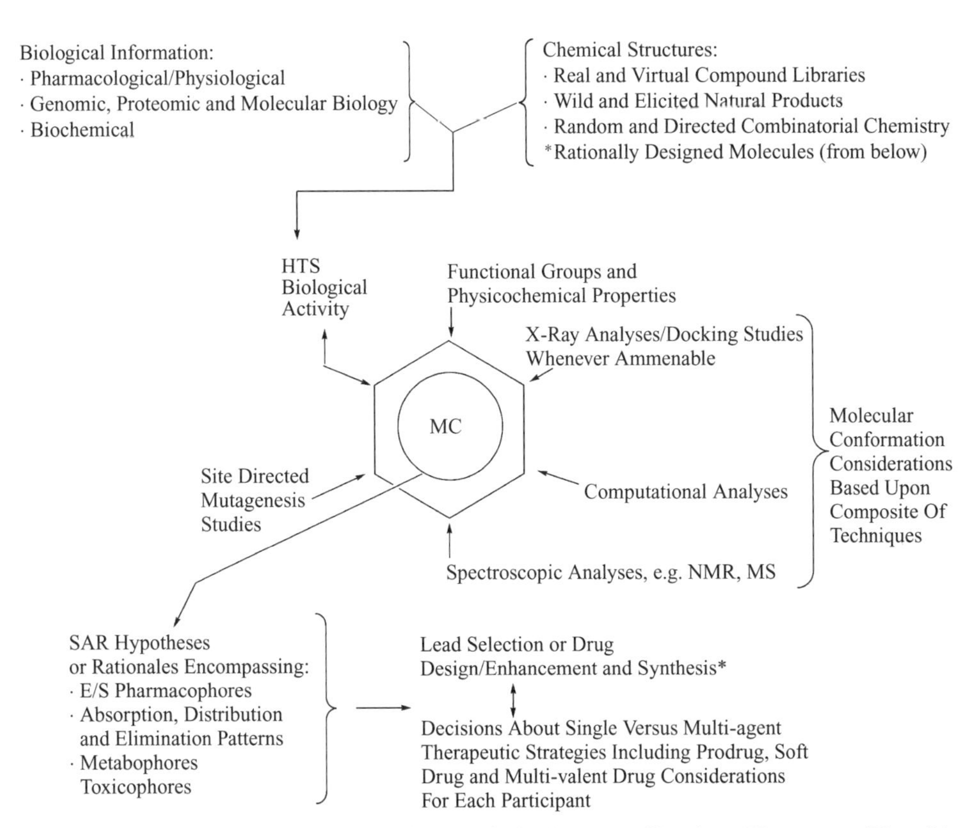

Figure 2.6 Practice of medicinal chemistry (MC) in the new millennium. The most striking differences from the long-standing practice of MC are (1) data reduction of huge amounts of rapidly derived HTS biological results, (2) greater emphasis on multitechnique chemical structure considerations, and most important, (3) the simultaneous attention given to all of the ADMET-related parameters along with efficacy and efficacy-related selectivity (E/S) during lead compound selection and further design or enhancement, coupled with an expanding knowledge base that offers the possibility for achieving synergistic benefits by taking advantage of various combinations of multiagent, prodrug, soft drug, and/or multivalent drug strategies.

medicinal chemistry's pursuit of distinct efficacy-related biochemical scenarios. Because they will continue to represent part of the basic thrust of medicinal chemistry, the pursuit of efficacy into the new millennium and the assessment of molecular conformation are each considered further in the next two sections relative to medicinal chemistry's working definition. Similarly, because the nature of the ADMET area's maturation is regarded as being pivotal toward shaping medicinal chemistry's evolution in the new millennium, each ADMET area is covered within a separate, fifth section. Although space limitations dictate that these topics can be highlighted only in an abbreviated manner, a somewhat longer and more technically oriented discourse is provided for the molecular conformation section because it is envisioned that this area will constitute the alphabet for the universal language that medicinal chemistry will help to elaborate in order to bridge and ultimately unite all of the other areas.

2.4 PURSUING EFFICACY

From an experimental point of view, medicinal chemistry's pursuit of the efficacy displayed by small molecules in a direct manner probably won't change much in the near term beyond what has already been proposed in terms of logical extensions from some of the latest trends. For example, even the intriguing directions that are encompassed by *chemical genomic*[87] *strategies*, where in some cases a small molecule may provide efficacy in an indirect manner through its interactions with biochemically modified genomic and proteomic signaling systems,[88–90] still reduce to being able to exploit the same fundamental principles already cited within medicinal chemistry's definition so as to achieve the specific interaction with the modified biological surface, which then serves as a mediator toward efficacy.

In terms of targets, bioinformatics will certainly associate pathophysiology and individual variation with useful genomic and proteomic information so as to maintain the plethora of traditional and novel pharmacological targets well into the millennium.[91–98] Web-based public efforts[99] and commercial databases[100] coupled with desktop programs[101] are already being positioned to make such information readily available to everyone. The ability of infectious microorganisms and viruses to evolve into resistant forms at a pace at least equal to our ability to produce NCE chemotherapeutic agents can also be counted on to continually provide new targets.[102,103] Alternatively, longer term into the new millennium, gene therapy will hopefully have eradicated many of today's targets that derive strictly from hereditary, gene-based abnormalities.[4] Somewhat along these same lines, the ongoing characterization of single nucleotide polymorphisms (SNPs) and the further pursuit of individually tailored therapies, as presently being promoted by the new field of pharmacogenetics, may also contribute toward some new targets.[104–111] Web-based public and commercial databases in the pharmacogenetic[112] and specific SNP areas,[113] along with commercial sources for the latter's experimental technologies,[114] are also making this type of information generally available. In the near term, the field of pharmacogenetics is likely to have its first major impacts on refining clinical studies for late-stage preclinical candidates and on developing improved indications, contraindications, and dosage regimens for marketed drugs and for compounds undergoing clinical study. Longer term, however, pharmacogenetics should become instrumental in shaping the overall nature and subtleties of the efficacy targets that are pursued rather than the numbers of new targets that relate to novel mechanisms that might be deployed. For example, although one can speculate that the present trend to pursue curative rather than palliative treatments will continue for at least the short term, it is likely that in the longer term a growing emphasis will be placed on preventative rather than on either palliative or curative treatments.[115,116] Toward this end, pharmacogenetics should become of central importance, owing to its potential to divide recipient populations into distinct treatment subgroups based on their predisposition profiles coupled with their general ADMET-related drug handling profiles, wherein both sets of criteria may eventually become accessible before or shortly after birth using gene-based assays and to a lesser extent, administration of diagnostic probe molecules.[117,118] Depending on the variability of an individual's environmental exposures, it can be imagined that in the future, pharmacogenetic profiling will be done at routinely scheduled intervals through the entire course of one's life. Regardless of the number and nature of future pharmacological targets, advances in biotechnology can be expected to continue to flood the overall life sciences arena and to get even better at deriving the required HTS assays. Alternatively, HTS microengineering,[119] it would seem, may have to level off at about 9600-well (or well-less) tests per plate or, perhaps, move to other platforms involving chip or bead technologies.[120]

2.4.1 Gathering Positive, Neutral, and Negative SARs During HTS

As mentioned, HTS efficacy hits per se can certainly be pursued without the aid of medicinal chemistry. Indeed, one can imagine that with one or more compound libraries already in hand from an automated synthesis,[121] and the areas of robotics[122–124] and laboratory information management or LIMS[125] also continuing to evolve rapidly, HTS in the brute force mode may be able to proceed without any significant human intervention, let alone without the need for an interdisciplinary group of investigators from a variety of disciplines. However, as has been emphasized, if the new drug discovery paradigm is ultimately to become successful, this type of screening will need to be accompanied by structure-associated knowledge generation and assessment, with the latter being conducted using the rationale and logic that can only be interjected by human intervention. Furthermore, even though it could run the risk of placing medicinal chemistry into a fall-guy position somewhat analogous to its earlier adolescent phase, it should also be noted that according to the working definition cited herein, as soon as any knowledge assessments become at all sophisticated in terms of molecular structure and biological properties, they quickly fall right into the middle of the domain of medicinal chemistry and its distinct area of expertise.

In this regard, it becomes important to briefly review some aspects of SARs that would be worthwhile to include within the database assemblies that are currently being drawn up to handle the mountains of data already arriving from today's HTS programs. One can predict that once ADMET profiling by HTS is validated in the future, it will become extremely valuable for a knowledge-generating paradigm to be able to discern not just the most active compounds within an efficacy database and to be able to compare their structural patterns to those in another database, but to also be able to flag the regions on compounds that can be altered with little effect on the desired biological activity as well as those areas that are intolerant toward structural modification. The neutral areas, in particular, represent ideal points for seamless merging of one set of a database's hits with that of another regardless of the degree of pattern overlap, or for further chemical manipulation of a hit so as to adjust it to the structural requirements defined by another data set that may be so distant in structural similarity space that attempted overlap or pattern recognition routines are otherwise futile. The regions that are intolerant of modification represent areas to be avoided during knowledge-based tailoring of an efficacy lead. Alternatively, the intolerant regions represent areas that can be exploited when attempting to negate a particular action (e.g., metabolism or toxicity). An actual example of utilizing both neutral and negative SARs to advantage is provided below to further illustrate how these types of data sets might also be deployed simultaneously by future medicinal chemists (albeit with significantly stepped-up complexity) as more and more parameters become added to the process of early lead identification/optimization.

2.4.2 Example Involving Multidrug Resistance of Anticancer Agents

The investigations to be exemplified in this case have been directed toward studying the SAR associated with biological transporter systems with the hope of establishing a database of transportophore relationships that might be generally applicable toward enhancing the selection and/or development of efficacy leads from any other type of data set. To provide immediate relevance to this long-term project, it has initially focused on the P-glycoprotein pump (Pgp)[126,127] that is associated, in part,[128,129] with the development of multidrug resistance (MDR)[130,131] during cancer chemotherapy. Pgp is a 170-kD transmembrane glycoprotein that serves as an energy-dependent unidirectional efflux pump having broad substrate specificity. In humans it is encoded by the MDR gene, MDR1, whose classical phenotype is characterized by a reduced ability to accumulate drugs intracellularly, and thus the deleterious impact of Pgp activity on cancer chemotherapy.[132–136] By way of practical example, the cytotoxicity of paclitaxel (PAC) is decreased by nearly three orders of magnitude when breast cancer cell lines become subject to MDR, largely via a Pgp mechanism.[137] In order to explore a series of probes that will systematically span a specified range of physicochemical properties when coupled to the PAC framework, one needs first to identify a region on PAC that is tolerant toward such modification in terms of PAC's inherent efficacy (i.e., where changes are known not to significantly alter PAC's cytotoxicity toward nonresistant breast cancer cells). Scheme 2.3 provides a summary of the accumulated SAR data obtained from the PAC-related review literature,[138–142] wherein it becomes clear that several positions along the northern edge represent neutral areas that can lend themselves toward such an exploration.

Inhibitors of Pgp have already been identified by several different investigators, and these types of compounds belong to a class of agents referred to as *chemosensitizer drugs* for which there are a variety of mechanisms.[143,144] Although Pgp inhibitors can be coadministered with a cytotoxic agent in order to negate MDR toward the latter when studied in cell culture, to date these types of chemosensitizers have not fared well clinically.[145] One of the reasons that the inhibitors have not fared well is that they must compete with the accompanying cytotoxic agent for access to the Pgp MDR receptors. Thus, it can be imagined that if an SAR can be identified that is unfavorable for binding with Pgp MDR receptors, and furthermore, that if such a negative transportophore could be incorporated onto the original cytotoxic agent in a neutral position, the cytotoxic agent might itself avoid MDR or at the very least become better equipped to do so in the presence of a coadministered Pgp MDR inhibitor (Fig. 2.7).[146] Toward this end, initial studies are being directed toward exploring the possibility that it may be feasible to identify negative SAR that is undesirable to the Pgp system within the specific chemical context

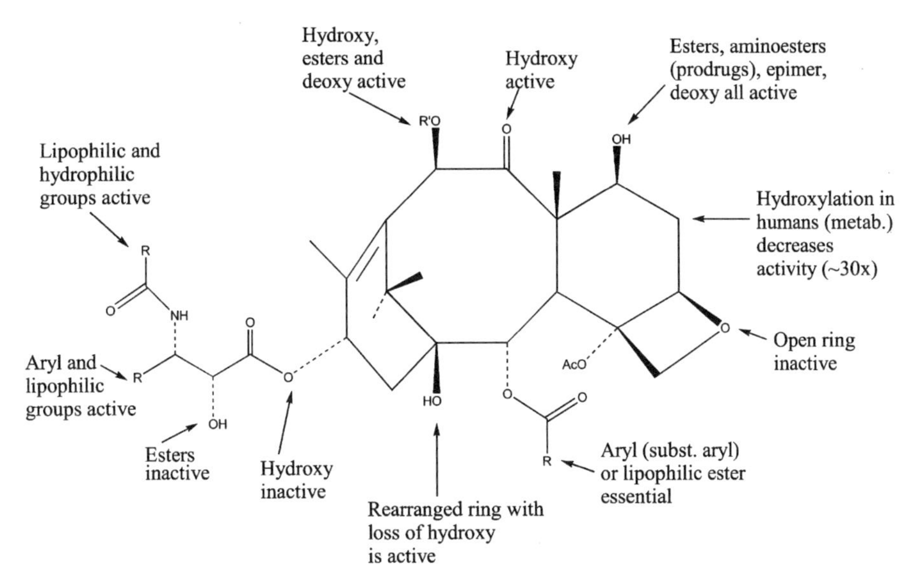

Scheme 2.3 Overall SAR profile for paclitaxel-related compounds. This summary represents a consolidation of SAR information contained in several review articles.[138–142] Note the tolerance for structural modification along the northern hemisphere of the taxane ring system. Paclitaxel has R = ∅ and R′ = CH_3CO.

of PAC by manipulating the latter at neutral positions that do not significantly affect PAC's inherent efficacy. To ascertain the generality toward potentially being able to place such a negative transportophore onto other established chemotherapeutic agents and onto lead compounds being contemplated for preclinical development, an identical series of negative SAR probes is being examined within the context of a completely different molecular scaffold: that of the camptothecin (CPT) family of natural products for which topotecan represents a clinically useful anticancer drug.[147] CPT, accompanied by a summary account of its SAR-related literature,[138,147–170] is depicted in Scheme 2.4, wherein it becomes clear that the 7- and 9-positions represent the key neutral areas in CPT that might be manipulated analogously to those in PAC. Since these two compounds have very different molecular templates and owe their cytotoxicities to two distinctly different mechanisms (i.e., PAC largely, but not exclusively[171–173] to overstabilization of microtubules[174,175] and CPT largely, but not exclusively,[176] to "poisoning" of topoisomerase I[177]), and because topotecan, a clinically deployed CPT analog (Scheme 2.4), is at the lower end of the spectrum in terms of being subject to Pgp-related MDR (it loses about one order of magnitude from its initial potency[137]), taken together these two molecules represent an excellent pair to examine the generality of the transportophore-related SAR findings. Other molecular scaffold systems and biological testing models can also be imagined so as to extend such Pgp investigations into the areas of drug absorption, uptake into hepatic tissue (drug metabolism), and passage across the blood–brain barrier (e.g., for either enhancing or attenuating drug penetration into the CNS). Additional transporters within the ABC class can be explored systematically using a similar approach.

Although this particular example reflects a rational SAR strategy, the same types of informational endpoints can certainly be achieved via a coupled HTS/combinatorial

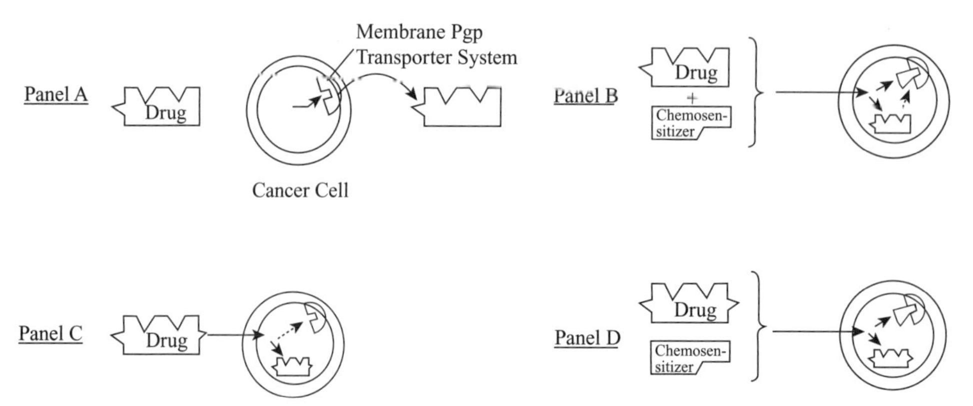

Figure 2.7 Diminishing the Pgp transporter's role in multidrug resistance (MDR). Panel A: Extrusion of chemotherapeutic agents (drug) by cancer cells upon overexpression of the Pgp transporter system is one mechanism associated with MDR. Panel B: Coadministration of a chemotherapeutic agent and a chemosensitizing agent wherein the latter preferentially interacts with Pgp, thus diminishing the extrusion of the desired drug. The chemosensitizer may block the pump competitively (wherein it becomes extruded) or noncompetitively. Panel C: Administration of a modified chemotherapeutic agent (drug) that retains its desired efficacy-related properties but has a diminished affinity for the Pgp system. Panel D: Coadministration of a modified chemotherapeutic agent and a chemosensitizing agent such that the latter has a better chance of interacting with the Pgp system relative to the desired chemotherapeutic agent.

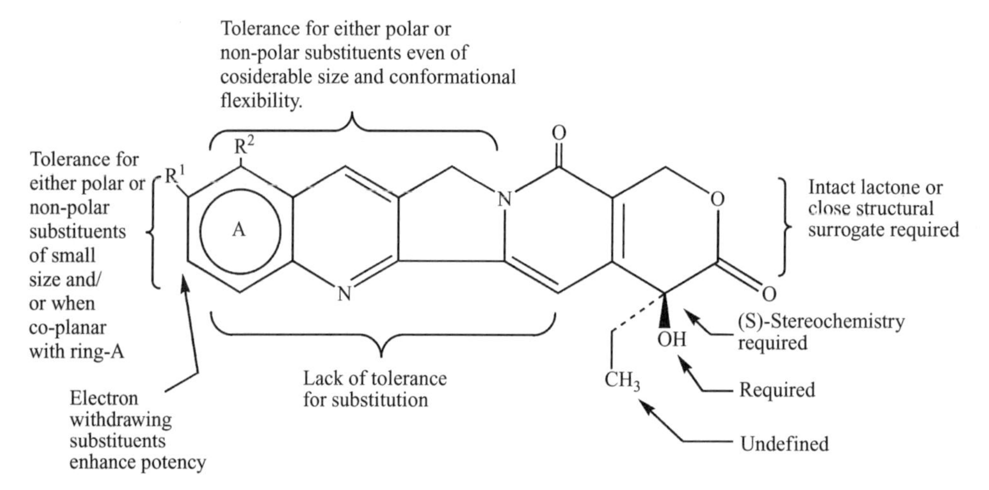

Scheme 2.4 Overall SAR profile for camptothecin-related compounds. This summary represents a consolidation of SAR information contained in several primary references.[138,147–170] Note the tolerance for structural modification along the northern hemisphere of the overall molecule, especially when approaching the western edge. Camptothecin, the parent natural product, has $R^1 = R^2 = H$. Topotecan, used clinically, has $R^1 = OH$ and $R^2 = CH_2N(CH_3)_2$.

chemistry approach, provided that the chemical structure components within the resulting databases are initially constructed with architectures flexible enough to allow for such queries. Likewise, while this example reflects a simple query between two different biological behaviors, the same types of queries can be conducted across multiple databases for multiple parameters. That the next 25 years of medicinal chemistry will involve a considerable amount of making sense out of such multiple parameter correlations based on experimentally derived data is quite clear. That the next 50 to 75 years might then be able to be fruitfully spent in more of a virtual correlations mode is certainly more speculative but is, at least, probably reasonable provided that we can build our knowledge base and fundamental understanding of how the various parameters, as assessed in isolation according to the HTS scenarios described above, interact simultaneously within the entire system.

2.4.3 Compound Libraries: Example of Working with Nature to Enhance Molecular Diversity

Before closing this section, two additional trends need to be mentioned. The first involves the likelihood that industry will even more heavily embrace site-directed mutagenesis as an additional component of its efforts to identify lead compounds. Such modification can be contemplated during the initial development of an HTS assay and then used to contribute to the definition of an overall pharmacophore as the latter is probed via various compound testing paradigms. The importance of gaining a thorough appreciation for the overall pharmacophore rather than for just identifying distinct lead structures is discussed in later sections. The second trend in the efficacy arena that is also likely to become very important in the future can be illustrated by an example from a different research program. Instead of focusing on ADMET-related parameters during rational drug design, this program involves the chemical or library side of the new drug discovery paradigm. In particular, this example seeks to enhance molecular diversity[178,179] along phytochemical structural themes that have shown activity of either a toxic[180] or a promising nature during initial efficacy screening such that having related compound libraries would be highly desirable. As described below, this particular example has a certain appeal in its ultimate practicality since it seeks initially to produce directed molecular diversity within common plants indigenous to the midwestern United States. This possibility is being explored by exposing plants simultaneously to both an elicitor (botany's designation for an inducing agent) and selected biochemical feedstocks. For example, the biochemical pathways leading to the anticancer phytoalexins from soybeans shown in Scheme 2.5 may be able to be elicited by soybean cyst nematode infections to produce a more diverse family of active principles when grown in environments containing biochemically biased nutrients.[181,182] Toward this end, it has been established that the statistical reproducibility of HPLC-derived phytochemical constituent fingerprints from soybean controls is adequate to discern real fluctuations in these types of natural products.[183] Work is now progressing toward ascertaining the differences that result upon exposures of soybeans to various stimuli and feedstocks. Interesting results will be followed-up by studying the genetic control of the involved pathways. In this regard it should be noted that an opposite approach that leads to similar *combinatorial biosynthesis*[184] endpoints is also being undertaken by various other groups, particularly with an interest toward the production of proteins and peptide families from plant systems.[185] In those studies, directed modification of the genetic regions controlling one or more established phytochemical pathways is first effected, and then these types of biotechnology

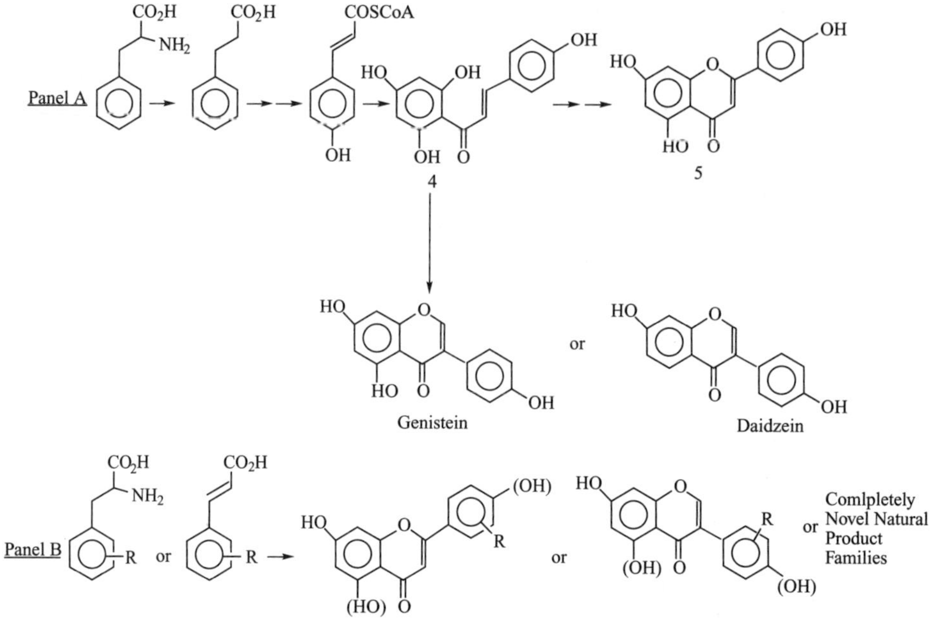

Scheme 2.5 Elicitation of directed and novel natural product families from soybean. Panel A: Normal biochemical pathways leading to the flavones (**5**) to the key anticancer isoflavones genistein and diadzein, wherein phenylalanine is first converted to cinnamic acid, *p*-coumaroyl CoA, and finally to key intermediate **4**, naringenin chalcone. Panel B: Feeding unnatural starting materials such as the aryl-substituted phenylalanine and cinnamic acid derivatives shown as **6** and **7**, respectively, under circumstances where this pathway is also being elicited by external stimuli (e.g., soybean cyst nematode infections), could be expected to produce new flavone derivatives (**8**), isoflavonic derivatives (**9**), or completely novel natural product families. This diagram of the phytochemical pathway leading to flavones and isoflavones represents a composite of several references.[485–487] An array of inexpensive analogs related to **6** and **7** is available from commercial sources.

interventions are followed up by characterization of the altered biocombinatorial expression products.

2.5 ASSESSING AND HANDLING MOLECULAR CONFORMATION

2.5.1 Chemoinformatics

Given the exponential proliferation of technical data and our increasing ability to disseminate it rapidly through a vast maze of electronic networks, it is no wonder that new systems capable of "managing and integrating information"[186] are regarded among "the most important of the emerging technologies for future growth and economic development across the globe."[186,187] That information technology (IT), in turn, is now receiving high priority in all sectors is quite clear,[188–195] particularly with regard to systems directed toward integrating bioinformatic-related information as promoted via the World Wide Web.[196]

Medicinal chemistry's contributions toward this sweeping assessment of the future importance of IT reside primarily in the area of handling chemical structures and chemical information, a specialized exercise complicated enough to merit its own designation as a new field, that of chemoinformatics.[197–200] In this regard, the increasing use of databases to link chemical structures with biological properties has already been alluded to in terms of both real experimental data sets and virtual compilations. Although serious strides are being taken in this area, however, there is a significant need for improvement in the handling of chemical structures beyond what is suggested for the immediate future by what now appears to be occurring within today's database assemblies. For example, that "better correlations are sometimes obtained by using two-dimensional displays of a database's chemical structures than by using three-dimensional displays" only testifies to the fact that we are still not doing a very good job at developing the latter.[201] How medicinal chemistry must step up and rise to the challenges already posed by this situation in order to fulfill the key roles described for its near- and longer-term future is addressed in the next several paragraphs of this section.

Assessment of molecular conformation, particularly with regard to database-housed structures, represents a critical aspect of chemoinformatics. While new proteins of interest can be addressed reasonably well by examining long-standing databases such as the Protein Data Bank[202] and other Web-based resources[203] for either explicit or similar structural motifs and then deploying x-ray (pending a suitable crystal), NMR, and molecular dynamic/simulation computational studies[204–209] as appropriate, the handling of small molecules and of highly flexible molecular systems in general remains controversial.[210] As alluded to above, the only clear consensus is that treatments of small molecules for use within database collections "have, to date, been extremely inadequate."[211] Certainly, a variety of automated three-dimensional (3D) chemical structure drawing programs are available that can start from simple 2D representations by using Dreiding molecular mechanics or other, user-friendly automated molecular mechanics-based algorithms, as well as when data are expressed by a connection table or linear string.[212] Some programs are able to derive 3D structure "from more than 20 different types of import formats."[213] Furthermore, several of these programs can be directly integrated with the latest versions of more sophisticated quantum mechanics packages, such as Gaussian 98, MOPAC (with MNDO/d), and extended Hückel.[200,212] Thus, electronic handling of chemical structures, and to a certain extent comparing them, in 3D formats has already become reasonably well worked out.[197–200,212–215] Table 2.8 provides a list of some of the 3D molecular modeling products that have become available during the 1990s.

Nevertheless, a fundamental problem remains: how the 3D structure is derived initially in terms of its chemical correctness based on what assumptions might have been made during the process. Further, there are still challenges associated with how readily 3D structure information can be linked with other, nonchemical types of informational fields. As has been pointed out by others, the reason that such mingling of data fields often does not afford good fits is because "each was initially designed to optimize some aspect of its own process and the data relationships and structures are not consistent."[197] At this point, inexpensive Web-based tools that can integrate chemical structure data with other types of information from a variety of sources, including genomic data, have already begun to emerge.[216] This trend will continue to pick up into the new millennium, and because of its overall importance for bioinformatics, user-friendly solutions are likely to arrive early on.

TABLE 2.8 Three-Dimensional Molecular Modeling Packages That Became Available During the 1990s

Package	Company	Platform	Description
Low-end sophistication			
Nano Vision	ACS Software	Mac	Simple, effective tool for viewing and rotating structures, especially large molecules and proteins
Ball & Stick	Cherwell Scientific	Mac	Model building and visualization; analysis of bond distances, angles
MOBY	Springer-Verlag	IBM (DOS)	Model building and visualization; classical and quantum mechanical computations; large molecules and proteins; PDB files
Nemesis	Oxford Molecular	IMB (Windows), Mac	Quick model building and high-quality visualization; geometry optimization (energy minimization)
CSC Chem. 3D/Chem 3D Plus	Cambridge Scientific	Mac	Easy-to-use building and visualization; geometry optimization; integrated 2D program and word processing
Alchemy III	Tripos Assoc.	IBM (DOS, Windows), Mac	Quick model building; energy minimization; basic calculations; easy integration to high-end systems
PC Model	Serena Software	IBM (DOS), Mac	Low cost with sophisticated calculations; platform flexibility
Midrange sophistication			
CAChe	Tektronic	Mac	Sophisticated computation tools; distributed processing
HyperChem	Auto Desk	IBM (Windows), Silicon Graphics	Easy-to-use array of computation tools (classical and semiempirical quantum mechanics)
Lab Vision	Tripos Assoc.	IBM (RISC-6000), Silicon Graphics, DEC, VAX	Sophisticated but practical modeling for research
High-end sophistication			
SYBL	Tripos Assoc.	IBM (RISC-6000), Silicon Graphics, DEC VAX, Sun 4, Convex	Integrated computation tools for sophisticated structure determination and analysis; database management
CERIUS	Molecular Simulations	Silicon Graphics, IBM (RISC-6000), Stardent Titan	Suite of high-performance tools for building and simulating properties

Source: Ref. 214.

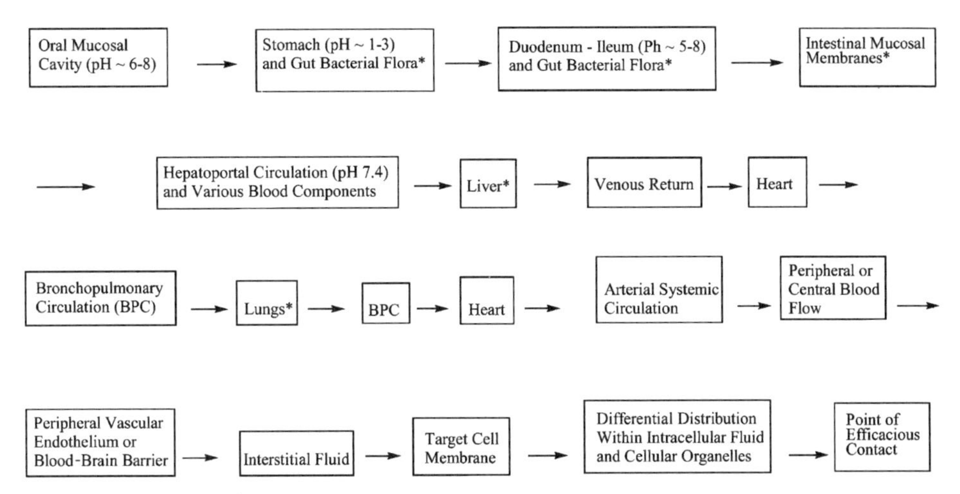

Figure 2.8 Random walk taken by an oral drug on route to its point of efficacious contact within a human target cell. This continuum of interactions between a drug and various biological surfaces within the human biological realm is typically divided into categories associated with ADMET and efficacy. Biological milieu marked with an asterisk represent compartments having particularly high metabolic capabilities. Blood is notably high in esterase capability. In the future, medicinal chemists will utilize knowledge about ADMET-related SARs to more effectively identify the best drug leads and to further enhance the therapeutic profiles of selected compounds. (The phrase "random walk" is taken from ref. 217.)

2.5.2 Obtaining Chemically Correct 3D Structures

Unfortunately, the quick assignment of chemically correct 3D structures may not be readily solvable. Recalling from the first sections of this chapter, medicinal chemistry has been concerning itself with this task for quite some time. Medicinal chemistry's interest in chemical structure is further complicated, however, by the additional need to understand how a given drug molecule's conformational family behaves during its interactions with each of the biological environments of interest. For example, as a drug embarks on its "random walk"[217] through the biological realm (Fig. 2.8), the ensuing series of interactions have unique effects on each other's conformations[218,219] at each step of the journey and not with just the step that finally consummates the drug's encounter and meaningful relationship that is struck with its desired receptor/active site.

To track such behavior in a comprehensive manner, it becomes necessary to consider a drug's multiple conformational behaviors by engaging as many different types of conformational assessment technologies as possible, while initially taking an approach that is unbiased by any knowledge that may be available from a specific interacting environment. For example, the three common approaches depicted in Fig. 2.9 include (1) x-ray (itself prone to bias from solid-state interactions within the crystal lattice); (2) solution spectroscopic methods (i.e., NMR), which can often be done in both polar and nonpolar media (this technique, however, being more limited by the amount of descriptive data that it can generate); and (3) computational approaches that can be done with various levels of solvation and heightened energy content (limited, however, by the assumptions and approximations that need to be taken in order to simplify the mathematical rigor so as to allow solutions to be derived in practical computational time periods). Analogous to the simple, drawing program starting points, programs are also available for converting x-ray

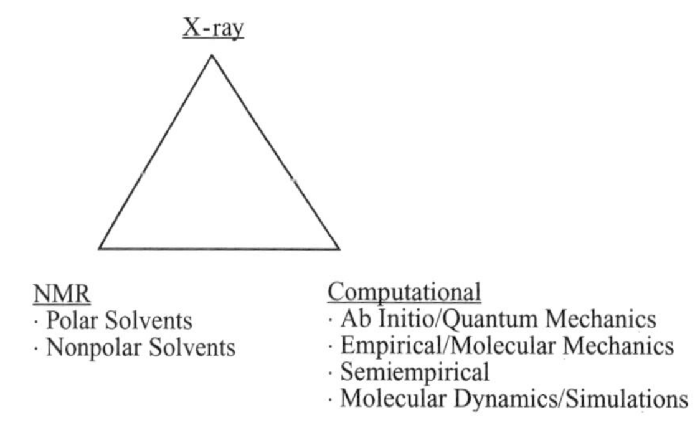

Figure 2.9 Techniques employed to assess conformational detail. X-ray diffraction requires a suitable crystal, and its results are subject to solid-state interactions. Computational paradigms are most accurate when done at the highest levels of calculation, but these types of calculations become computer-time intensive. NMR requires that the molecule be soluble in the chosen solvent and that adequate compound supplies be available. Mass spectrometry is also becoming an important tool for larger molecules, although it provides smaller amounts of descriptive data. A composite of all approaches provides for the best possible assessment of molecular conformation.

and NMR data into 3D structures (e.g., ref. 220). While such a three-pronged approach is not new,[221] it is emphasized herein because today's medicinal chemistry literature suggests that some investigators still fall into the single technique trap from which further extrapolations of data are then sometimes made with great conviction. This may be because it is often difficult to obtain an acceptable crystal for x-ray analysis, have adequate solubility for high-field conformational analysis by NMR, or perhaps, to become aligned with appropriate computational expertise and computing power. At any rate, practical advances in all three of these areas can be expected to alleviate such implementation-related shortcomings so that the medicinal chemist of the near-term future will be more readily able to consider structures from at least a three-pronged starting point either independently or through collaboration with other specialists and experts dedicated to each of these areas. A real example that serves to further illustrate how entry structures might be handled and matured using a computational approach is provided below. This problem pertains to the consideration of structures to be placed in a human drug metabolism database.[221]

2.5.3 Influence of Biological Environments: Example Involving Drug Metabolism

This example is also cited in Chapter 9 of this text. Structures are initially considered as closed-shell molecules in their electronic and vibrational ground states, with protonated and unprotonated forms, as appropriate, also being entered. If a structure possesses tautomeric options or if there is evidence for the involvement of internal hydrogen bonding, the tautomeric forms and the hydrogen-bonded forms are considered additionally from the onset. Determination of 3D structure is carried out in two steps. Preliminary geometry optimization is affected by using a molecular mechanics method. For example, in this case the gas-phase structure is determined by applying the MacroModel 6.5 modeling package running on a Silicon Graphics Indigo 2 workstation with modified (and extended) AMBER parameters also being applied from this package. Multiconformational assessment using systematic rotations about several predefined chemical bonds with selected rotational angles is then conducted to define the low-energy conformers and conformationally flexible regions for each starting structure. In the second step, the initial family of entry structures are subjected to ab initio geometry optimizations, which in this case use a Gaussian 98 package running on a T90 machine housed in a state-level supercomputer center resource. Depending on the size of the molecule, 3-21G* or 6-31G* basis sets[223] are used for conformational and tautomeric assessments. Density functional theory using the B3LYP functional[224] is applied for the consideration of exchange correlation energy while keeping the required computer time at reasonable levels. The highest-level structure determination is performed at the B3LYP/6-31G* level. To ascertain the local energy minimum character of an optimized structure, vibrational frequency analysis is carried out using the harmonic oscillator approximation. Determination of vibrational frequencies also allows for obtaining thermal corrections to the energy calculated at 0 K. Free energies are then calculated at 310 K (human body temperature). The latter values become particularly important for cases where structural (conformational or tautomeric) equilibria occur.

From the calculated relative free energies, the gas-phase equilibrium constant and the composition of the equilibrium mixture can be directly determined. Although these values may not be relevant in an aqueous environment or in the blood compartment, the calculated conformational distribution is relevant for nonpolar environments such as may be encountered when a drug passively traverses membranes or enters the cavity of a nonhydrated receptor/enzyme active site just prior to binding. Repetition of this computational scheme from biased starting structures based on actual knowledge of the interacting biological systems or from x-ray or NMR studies (particularly when the latter have been conducted in polar media), followed by studies of how the various sets of information become interchanged and how they behave when further raised in energy, complete the chemical conformational analysis for each structure being adopted into the human drug metabolism database.

As mentioned earlier, after taking an unbiased structural starting point, medicinal chemistry needs especially to consider structures (and the energies thereof) by ascertaining what their relevant conformations might be during interactions within various biological milieus. It can be imagined that at least within the immediate future, a useful range of such media to be considered will include aqueous solutions at acidic and neutral pH, namely at ca. 2 (stomach) and 7.4 (physiological), respectively; one or more lipophilic settings, such as might be encountered during passive transport through membranes; and finally, specific biological receptors and/or enzyme active site settings that are of particular interest. Importantly, with time this list can then be expected to grow further so as to also include several distinct environmental models deemed to be representative for interaction with various transportophore relationships; several distinct environmental models deemed to be relevant for interaction with specific metabophore relationships such as within the active site of a specific cytochrome P450 metabolizing enzyme; and finally, several distinct environmental models deemed to be relevant for interaction with specific toxicophore relationships. It should also be appreciated that the interaction of even just one ligand within just one of the various biological settings could still involve a wide range of conformational relationships wherein the biological surface may also exist as an equilibrium mixture of various conformational family members. If x-ray, NMR, and so on, can be deployed further to assess any one or combination of these types of interactions, a composite approach that deploys as many as possible of these techniques will again represent the most ideal way to approach future conformational considerations within the variously biased settings. Advances toward experimentally studying the nature of complexes where compounds are docked into real and model biological environments are proceeding rapidly in all of these areas, with MS[11–13] and microcalorimetry[14,15] now adding themselves alongside x-ray (later discussion) and NMR[225,226] as extremely useful experimental techniques for the study of such SARs. Besides the experimental approaches, computational schemes will probably always be deployed because they can provide the relative energies associated with all of the different species. Furthermore, computational methods can be used to derive energy paths to get from the first set of unbiased structures to a second set of environmentally accommodated conformations in both aqueous media and at biological surfaces. Importantly, these paths and their energy differences can then be compared along with a direct comparison of the structures themselves, while attempting to uncover and define correlations between chemical structure and some other informational field within or between various databases.

Finally, it should be noted that by using computational paradigms, these same types of comparisons (i.e., among and between distinct families of conformationally related members) can also be made for additional sets of conformational family members that become accessible at appropriately increased energy levels (i.e., at one or more 5 kcal/mol increments of energy) to thus address the beneficial losses of energy that might be obtained during favorable binding with receptors or active sites.[227] These types of altered conformations can also become candidates for structural comparisons between databases. The latter represents another important refinement that could become utilized as part of SAR queries that will need to be undertaken across the new efficacy and ADMET-related parameters of the future. With time, each structural family might ultimately be addressed by treating the 3D displays in terms of coordinate point schemes or graph theory matrices.[228] This is because these types of methods lend themselves to the latest thoughts pertaining to utilizing intentionally fuzzy coordinates[229,230] [e.g., $x \pm x'$, $y \pm y'$, and $z \pm z'$ (rather than just x, y, and z plots)] for each atomic point within a molecular matrix wherein the specified variations might be intelligently derived from the composite of aforementioned computational and experimental approaches. Alternatively, the fuzzy strategy might become better deployed during the searching routines, or perhaps both knowledgeably fuzzy data entry and knowledgeably fuzzy data searching engines handled, in turn, by fuzzy hardware,[231] will ultimately best identify the correlations that are being sought in any given search paradigm of the future. It should be noted, however, that for the fuzzy types of structural treatments, queries will be most effective when the database has become large enough statistically to rid itself of the additional noise that such fuzziness will initially create. An ongoing example that serves to demonstrate the value of considering the dynamic energy relationships associated with molecular trajectories as well as the more static conformational displays for a particular molecular interaction of interest is provided below.

2.5.4 Dynamic Energy Relationships: Example Involving a Small Ring System

As a different aspect of the aforementioned MDR-related anticancer chemotherapeutic program, an effort has been directed toward replacing the complex scaffold of PAC (Scheme 2.3) with a very simple molecular format that still displays PAC's key pharmacophoric groups in the appropriate 3D orientations purported to be preferable for activity.[232–234] Toward this end, initial interest involved defining the role of the β-acetoxyoxetane system, particularly when the latter is adjacent to planar structural motifs. Since such systems are rather unique among natural products[235] as well as across the synthetic literature,[236–238] it first became necessary to study their formation within model systems relevant for this project. 2-Phenylglycerol was synthesized[239] and deployed as a model to study the molecule's conformation by x-ray, NMR, and computational techniques as a prelude to affecting its cyclization.[240] The energy differences that result as the molecule is reoriented so as to be lined up for the cyclization were also calculated. Finally, once properly oriented in 3D space, the energy required to actually traverse the S_N2 reaction trajectory between the 1 and 3 positions was calculated (one of which positions utilizes its oxygen substituent for the attack while the other relinquishes its oxygen as part of a leaving group). The synthesis of 2-phenylglycerol and the pathway and energies associated with the intermediate species and cyclization process to form the oxetane are summarized in Scheme 2.6.

Not surprisingly, given the strained-ring nature of this system, the energy needed to effect the ring closure from the lowest of three closely related local minima conformations belonging to a family common to the independent x-ray and computationally derived starting points was about 28 kcal/mol. What becomes interesting, however, is that within this particular system, nearly half of this energy requirement results from the need to disrupt hydrogen bonds in order to reorient the molecule initially into a conformation

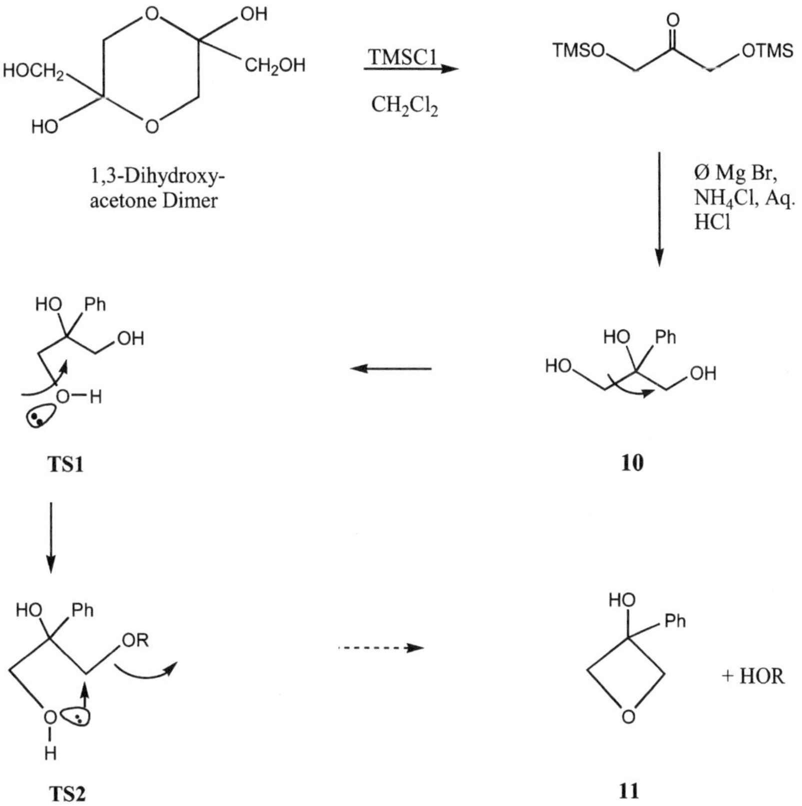

HOCH2 HO O O OH CH2OH
1,3-Dihydroxy-acetone Dimer
TMSCl
CH_2Cl_2
TMSO O OTMS
Ø Mg Br, NH_4Cl, Aq. HCl
HO Ph OH O–H
TS1
HO HO Ph OH
10
HO Ph OR O H
TS2
HO Ph O
11
+ HOR

Scheme 2.6 Synthesis of 2-phenylglycerol and investigation of its conversion to 3-hydroxy-3-phenyloxetane. 2-Phenylglycerol (**10**) is depicted so as to convey the lowest-energy structure of the three close local minima observed during ab initio calculations performed at the HF/6-31G* level. TS1 and TS2 represent transition conformers obtained after the indicated bond rotations, while **11** represents the desired 3,3-disubstituted oxetane sytem. The respective relative energies in kcal/mole for **10**, TS1, and TS2, along with the product oxetane (**11**), are as follows: 0.00, 13.6, 12.4, and 28.2. (From refs. 239, 240, and 488.)

appropriate for the reaction. Despite the resulting strain that becomes placed upon the overall system's bond angles, the actual movement of the relevant atoms along the reaction trajectory (Scheme 2.6, dashed line) then accounts for only slightly more than one-half of the total reaction energy. Therefore, from a synthetic point of view the results suggest that it should be beneficial to employ a hydrogen bond acceptor solvent that has a high boiling point, the first property assisting in disruption of the hydrogen bonds that need to be broken for conformational reorientation and the second property allowing enough thermal energy to be added conveniently to prompt progression across the reaction trajectory. That such favorable conformational perturbations could indeed be achieved simply by deploying these types of solvents was then confirmed by reexamination of the independent results obtained from our initial NMR studies conducted in polar protic media. Alternatively, from a medicinal chemistry point of view, these results serve as a reminder about the long-standing arguments pertaining to the importance and energetics of drug desolvation prior to receptor–active site interaction and, alternatively, the roles that stoichiometric water molecules can play within such sites. That such concerns will be addressed in a much more deliberate manner in the future by using multidisciplinary approaches similar to the example cited herein seems clear. Indeed, a quick survey of the present medicinal chemistry literature suggests that consideration of the dynamic nature of conformational perturbations associated with efficacious events is already beginning to take hold.[241–246] It is predicted, however, that it will be even more critical in the future to correlate SARs from one database to another according to the dynamic energy differences between the various molecule's conformational family members when several ADMET-related interactions are additionally factored into the overall behavior of a molecule being contemplated for further development. In other words, simple comparisons of static structures, even when rigorously assigned in 3D, will probably not be adequate to address a molecule's behavior across all of the efficacy and ADMET-related biological surfaces that become of interest as part of the molecule's optimization during future new drug design and development paradigms.

What this section points to is that, ultimately, structural databases of the future will probably have several "tiers"[247] of organized chemical and conformational information available which can be distinctly mined according to the specified needs of a directed (biased) searching scheme while still being able to be completely mixed within an overall relational architecture such that undirected (unbiased) *knowledge-generating mining paradigms* can also be undertaken.[248–253] Certainly, simple physicochemical data will need to be included among the parameters for chemical structure storage. Similarly, searching engines will need to allow for discrete substructure queries as well as for assessing overall patterns of similarity and dissimilarity[254–261] across entire electronic surfaces.

2.5.5 Druglike Properties and Privileged Structures

It can be noted that it is probably already feasible to place most clinically used drugs into a structural database that could at least begin to approach the low- to midtier levels of sophistication because considerable portions of such data and detail are probably already available in the literature for each drug even if they are spread across a variety of technical journals. On the other hand, it should also be clear that an alternative strategy will be needed to handle the mountains of research compounds associated with just a single HTS parameter survey. Unfortunately, it appears that some of the large compound surveys being conducted today do not even have systematically treated 2D structural representations. Indeed, while the present status of handling chemical structure and data associated with

HTS is wisely being directed toward *controlling the size of the haystack*,[262] the dire status of handling conformational detail is reflected by attempts that try to grossly *distinguish between druglike and nondruglike molecules*[263] in a 2D manner or, at best, to identify certain "privileged structures"[264,265] while using 3D constructs derived from less than completely rigorous experimental and computational assessments. Furthermore, in certain companies, notions about druglike patterns (or actually the lack thereof) are already being set up as the first screen or in silico filter to be deployed against a given compound library's members while the latter are still on route to an HTS efficacy screen. Unfortunately, this scenario can detract from the definition of an initial efficacy pharmacophore along structural motifs that might, alternatively, be able to take advantage of neutral areas by making straightforward chemical modifications that then serve to avoid the non-druglike features. At present, and probably for much of the near term as well, strategies that use non-druglike parameters to limit the number of compounds that can otherwise contribute toward the definition of a given efficacy-related structural space would appear to be premature. At the very least, such strategies are counter to the need to continue to accumulate greater knowledge in the overall ADMET arena, let alone in the specific handling of 3D chemical structure at this particular time. Finally, when it is appreciated that in most cases the connection of HTS ADMET data with actual clinical outcomes still remains to be validated much more securely, the strategy to deploy notions about non-druglike structural hurdles as decision steps prior to efficacy screening becomes reminiscent of medicinal chemistry's own adolescent phase, wherein medicinal chemistry's efforts to design drugs rationally without the benefit of the additional knowledge afforded by an x-ray of the actual target site ultimately did not enhance either the production of NCEs or the image of medicinal chemistry. A more appropriate strategy toward addressing this area that is knowledge building and, instead, can eventually expect to deploy the evolving ADMET druglike patterns in a proactive manner is discussed further in Section 2.6. With regard to chemical structure, the present situation thus indicates that we have a long way to go toward achieving the aforementioned tiers of conformational treatments when dealing with large databases and applying them toward the process of drug discovery. Nevertheless, because of the importance of chemoinformatics toward understanding, fully appreciating, and ultimately, actually implementing bioinformatics along the practical avenues of new drug discovery, it can be imagined that future structural fields within databases, including those associated with HTS, may be handled according to the following scenario, as summarized from the ongoing discussion in this section and as also conveyed in Fig. 2.10.

2.5.6 Tiered Structural Information and Searching Paradigms

For optimal use in the future, it is suggested that several levels of sophistication will be built into database architectures so that a simple 2D format can be input immediately. Accompanying the simple 2D structure field would be a field for experimentally obtained or calculated physicochemical properties (the latter data also to be upgraded as structures are matured). While this simple starting point would lend itself to some types of rudimentary structure-related searching paradigms, the same compound would then gradually progress by further conformational study through a series of more sophisticated chemical structure displays. As mentioned earlier, x-ray, NMR, and computational approaches toward considering molecular conformation will be deployed for real compounds given that it is also likely that advances in all of these areas will allow them to be applied more readily in each case. Obviously, virtual compound libraries and databases will have to rely solely on

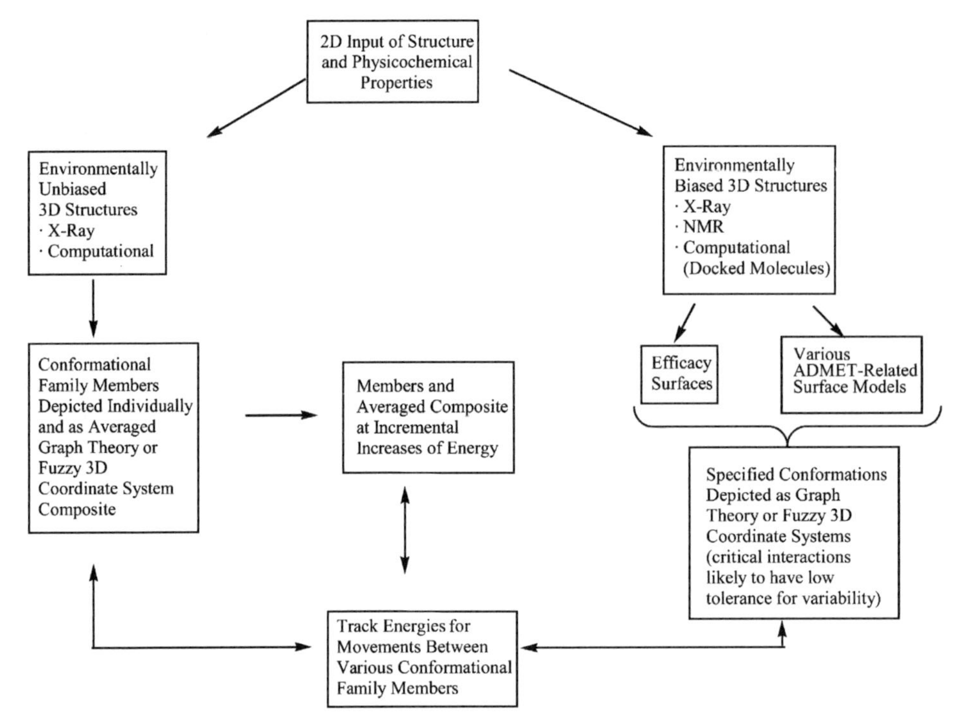

Figure 2.10 Handling chemical structures within databases of the future. This figure depicts the quick entry and gradual maturation of structures. Search engines, in turn, would also provide for a variety of flexible paradigms involving physical properties with both full and partial (sub)structure searching capabilities using pattern overlap/recognition, similarity–dissimilarity, CoMFA, and so on. Structure entry would be initiated by a simple 2D depiction that is gradually matured in conformational sophistication via experimental and computational studies. Note that structures would be evolved in both an unbiased format and in several environmentally biased formats. The highest structural tier would represent tracking or searching the energies required for various conformational movements that members would take when going from one family to another.

computational approaches and on knowledgeable extrapolation from experimental data derivable by analogy to structures within overlapping similarity space. Eventually, structures would be manipulated to a top tier of chemical conformational information. This tier might portray the population ratios within a conformational family for a given structure entry expressed as both distinct member and averaged electrostatic surface potentials, wherein the latter can be expanded further to display their atomic orientations by fuzzy graph theory or fuzzy 3D coordinate systems. Thus, at this point one might speculate that an intelligently fuzzy coordinate system could eventually represent the highest level of development for the 3D *quantitative SAR* (3D-QSAR)[266,267]-based searching paradigms seemingly rising to the forefront of today's trends in the form of comparative molecular field analyses (CoMFA).[268,269] Furthermore, one can imagine that this tier might actually be developed in triplicate for each compound: that is, one informational field for the environmentally unbiased structural entries, another involving several subsets associated with known or suspected interactions with the biological realm, and a third for tracking conformational families when raised by about 5 and 10 kcal/mol in energy. Finally, as the new millennium continues to churn its computational technologies forward, conformational and energetic considerations pertaining to a compound's movement between its various displays, similar to that conveyed by the very simple example provided from the oxetane-related study, can also be expected to be further refined so as to allow future characterization and searching of the dynamic chemical events that occur at the drug–biological interface (e.g., modes and energies of docking trajectories and their associated molecular motions relative to both ligand and receptor/active site). That this top tier is extremely valuable for understanding the interactions of interest to medicinal chemistry is apparent from the large amount of effort already going on today in this area,[241–246] particularly when such studies are able to take advantage of an x-ray-derived starting point.

By the same token, chemical structure search engines of the future will probably be set up so that they can also be undertaken at several tiers of sophistication, the more sophisticated requiring more expert-based enquiries and longer search times for the attempted correlations to be assessed. A reasonable hierarchy for search capability relative to the structural portion of any query might become (1) simple 2D structure with and without physicochemical properties, (2) 3D structure at incremented levels of refinement, (3) 2D and 3D substructures, (4) molecular similarity–dissimilarity indices, (5) fuzzy coordinate matrices, (6) docked systems from either the drug's or the receptor–active site's view at various levels of specifiable precision, and finally, in the more distant future (7) energy paths for a drug's movement across various biological milieu, including the trajectories and molecular motions associated with drug–receptor/active site docking scenarios. Emphasizing informatics flexibility, this type of approach, where data entry can occur rapidly for starting structure displays and then gradually be matured to more sophisticated displays as conformational details are accurately accrued, coupled with the ability to query at different levels of chemical complexity and visual displays[270] at any point during database maturation, should allow for chemically creative database mining strategies to be effected in the new millennium's near term as well as into its more distant future.

2.6 ADMET CONSIDERATIONS

2.6.1 Assuring Absorption

In addition to conducting in vivo bioavailability studies on selected compounds at a later stage of development, early in vitro assessments of structural information that might be useful toward *assuring absorption* after a drug's oral administration have now been going on for several years.[271–273] Somewhat more recently, similar studies also began to be directed toward assessing penetration across the blood–brain barrier (BBB).[274,275] Thus, determination of the pK_a values for ionizable groups, determination of partition coefficients (e.g., using various types of log P calculations and measurements), and measurement of passage across models of biological membranes (e.g., Caco cell lines) represent data that have now been shifted toward HTS experimental and purely computational modes.[276–287] These types of studies can be designated as AHTS (absorption high-throughput screening) (Table 2.2). Since recent results suggest that the biological transporter systems are extremely important factors in this area,[288–290] their study is also becoming part of AHTS (e.g., passage of drugs across Caco cell layers from both directions[291,292]). This trend toward increasing sophistication within AHTS can be expected to continue. That genomics and proteomics will help to identify and initially define absorption-related systems biochemically should be clear.

Alternatively, that beyond establishing the complementary AHTS systems' biotechnology might also be directed toward instilling passageways or specific pores for drugs across the human GI endothelial system[293] is certainly speculative even for the more distant future, as are chemical[294–296] and nanotechnology[116,297–300] approaches to prompting or constructing passageways, respectively. Similarly, that advances in formulation and alternate delivery technologies[301–310] could eventually obviate the need for oral administration is also speculative. Nevertheless, all of these possibilities need to be mentioned because, taken together, they make the point that significant advances in any of the ADMET areas, regardless of their technological source, have the potential to eliminate the need for assaying certain of their presently related parameters, perhaps even returning the initial portion of the present drug discovery paradigm (Fig. 2.2) back to where it originated (i.e., to being concerned primarily with efficacy and selectivity during front-line testing) (Fig. 2.1).

As for deciphering selective efficacy-related SARs, medicinal chemistry's role within the more discernible future is likely to be directed toward making sense out of the AHTS data mountains looming ahead using molecular structure information as the common code, in this instance by relating the latter to structure–absorption relationships, or *SAbRs*. Such efforts might eventually culminate in affording molecular blueprints for affecting absorption-related structural modifications that are correlated with certain structural themes and absorption characteristics for which efficacy hits may be able to be categorized using structural similarity–dissimilarity indices. Notable advances have already been made toward defining useful SAbRs in terms of database and virtual compound profiling (e.g., the *rule of five*).[311] The latter should be recognized as an important first step in this direction that can be expected to continue in a more sophisticated manner in the future (e.g., along the lines of 3D structural considerations relevant to the transporter systems, as well as more refined parameterization of physicochemical properties).[312]

2.6.2 Directing Distribution

The types of studies mentioned above, along with a panel of assays specific for certain depot tissues such as red blood cells, plasma protein binding factors, and adipose tissue,[313–316] will be mobilized toward *directing distribution* of a xenobiotic. Thus, as the handling of chemical structure improves and more sophisticated correlations begin to unfold in the future, AHTS can be thought of as A/DHTS that provides both SAbR and SDR. Simultaneous collection of such data will allow investigators to reflect upon drug absorption and distribution as a continuum of drug events that can effectively be incorporated together at an earlier point of the overall lead decision process. Furthermore, in the case of directing distribution it can be anticipated that genomics and proteomics will become instrumental toward identifying numerous key factors that are overexpressed in various patho-physiological states. For example, cancer cells are already known to overexpress a variety of specified factors.[317–321] Ligands designed to interact with such factors residing on cell surfaces can then be coupled with diagnostic and therapeutic agents so as to be delivered at higher concentrations to these locales. For therapeutics, such strategies can be thought of as placing both an address and a message within a molecular construct[322,323] that may involve an overlap of two small moleculerelated SAR patterns, or perhaps a small molecule conjugated to a bioengineered biomolecule wherein the latter typically serves as the address system. Indeed, the bioconjugate or immunoconjugate strategy has been around for a while[324] and it appears to be benefiting from a renewed interest[325] in that chemotherapeutic "smart bombs"[326] are now being added to our older arsenals of single arrows and combinations of small-molecule "magic bullets."[327] The earlier example involving PAC and CPT (Schemes 2.3 and 2.4) can also be used to further emphasize this theme wherein the chemical knowledge in the area of PAC and CPT protection and coupling reactions can be used to construct compounds that would be directed toward some of the factors that are overexpressed on certain human cancer cells so as to enhance selective toxicity,[328] particularly since there is some precedent in this case that this might be feasible by combining two small molecules. One can imagine that as data are amassed in the future for these types of factors, the most promising ones will be quickly pursued according to both of the aforementioned scenarios, paired small-molecule SARs and small molecule–bioconjugate pairs. Whether undertaken in a rational manner or via the merger of two HTS-generated databases (i.e., one for an efficacious message and one for determining a selective address), these types of pursuits fall into the general category of tailoring a lead. Therefore, it can be expected that the expertise afforded by medicinal chemistry will again be an integral component of such activities. Similarly, as suggested by the earlier PAC/CPT MDR-related example, medicinal chemistry's expertise will also be vital toward exploiting the opposite cases, where it becomes desirable to avoid certain systems (addresses) that become overexpressed as part of a given pathophysiology's resistance mechanisms or because messages delivered to such locales lead to toxicity within a healthy compartment.

Before turning to those parameters that might be considered to be associated with ending a drug's random walk through the biological realm (e.g., metabolism and excretion), it is necessary to discuss a practical limitation to where this overall discourse is leading. Clearly, there will be ceilings for how many molecular adjustments can be stacked into a single compound, no matter how knowledgeable we become about the various ADMET-related structural parameters and how they might be merged so as best to take advantage of molecular overlaps. This will be the case even when prodrug strategies are adopted[329] (Fig. 2.11) wherein certain addresses or messages that have been added to deal with one or more aspects of ADMET, become programmatically jettisoned along the way while simultaneously activating the efficacy payload that is to be delivered only to the desired locale as the final statement.Thus, this situation prompts the prediction that to interact optimally with the entire gamut of efficacy and ADMET-related parameters during a given course of drug therapy, the latter may need to be delivered not as a single agent but as a distinct set of multiple agents wherein each individual component or player makes a specified contribution toward optimizing one or more of the efficacy and ADMET parameters relative to the overall drug team's therapeutic game plan.

2.6.3 Herbal Remedies: Example of Working with Nature to Discover ADMET-Related Synergies

Today's trend to self-administer herbal remedies and preventives, admittedly driven by rampant consumerism in the United States rather than by solid science,[330,331] thus becomes an important topic to be considered at this point in the review. In this regard, the reconnection to medicinal chemistry's historical roots also becomes interesting to note. One of the major, basic science questions about herbals (which do possess validated pharmacological properties) is why their natural forms are sometimes superior to the more purified versions of their active constituents, even when the latter are adjusted to reflect varying concentration ratios thought to coincide with their natural relative abundances. Given the notoriously incomplete analytical characterizations of most herbal products, it should be apparent relative to the present discourse that numerous unidentified, nonefficacious, and otherwise silent constituents within any given herb could have an interaction with one or more of the

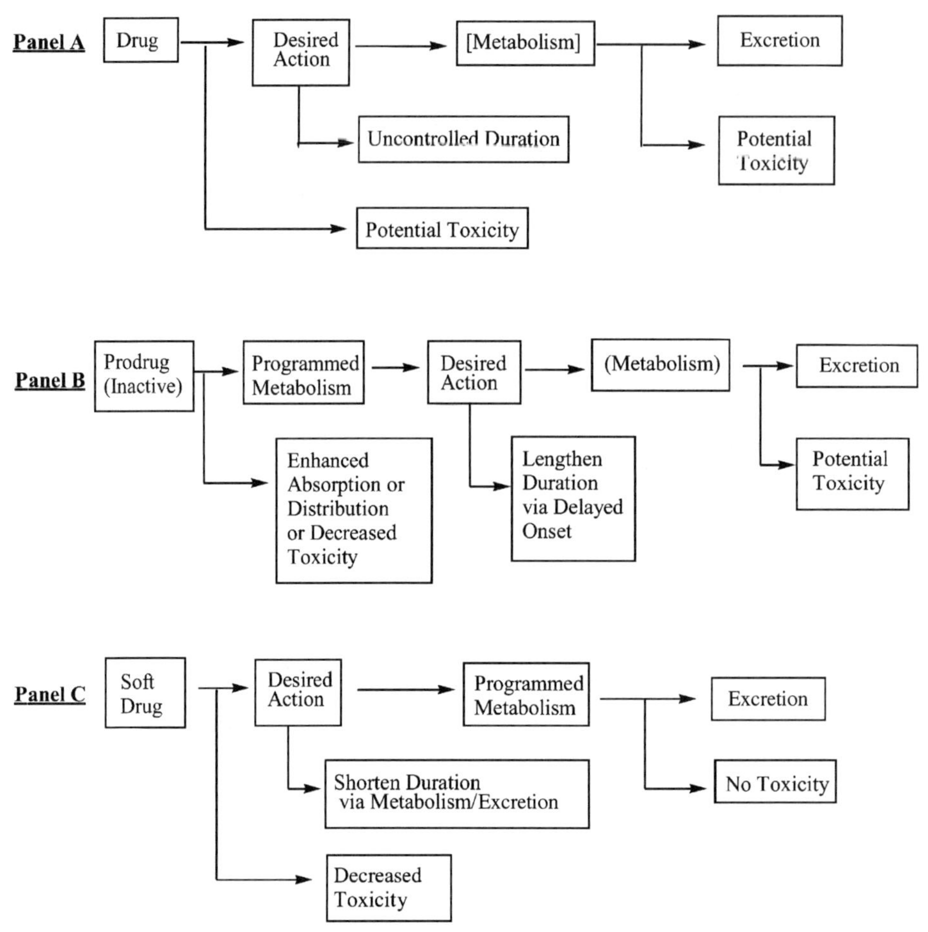

Figure 2.11 Soft drug actions compared to standard drug and prodrug actions. Panel A depicts a generalized version of a standard drug's pattern of observed activities. Panel B depicts how a prodrug approach can be used to modify the entry-side portion of a given drug's overall profile of actions. Panel C depicts how a soft drug approach can be used to modify the elimination-side portion of a given drug's overall profile of action. Both prodrugs and soft drugs can be used to decrease toxicity. (From refs. 329, 352, and 353.)

efficaciously active constituents at any one or more levels of the latter's ADMET steps. When these interactions are favorable, the resulting overall pharmacological profile becomes altered in a seemingly synergistic manner that is obtainable from the more natural forms of the mixture but lost upon purification to matrices containing only the actives.[332] Indeed, there is already some experimental precedent for this scenario relative to efficaciously silent components improving the absorption,[333] enhancing the distribution,[334,335] and favorably altering the metabolism[336] of their active herbal counterparts, as well as more classical synergies involving direct interactions that occur at the sites involved with efficacy.[337] Therapeutic enhancements derived directly from multiple interactions at efficacy sites have been pursued for many years, with multivalent single drug entities reflecting the latest trend in this direction.[338] What will be remarkable is that the new millennium will continue to add the sophistication of the entire ADMET profile into such multi-action-directed considerations.[339–343]

Optimization of the overall pharmacological profile is precisely what is being striven for when selecting and/or chemically tailoring an NCE lead according to either the old or new paradigm of drug discovery. Restating, however, that it may be expecting too much even upon extending the new paradigm into the future as a knowledge-generating process, to obtain complete optimization within a single multiparameterized molecule, perhaps it will be nature that will again lend its hand within the next millennium by revealing some of the modes of ADMET synergy that have long been part of some herbal productions. At the very least, medicinal chemistry should take care not to forget its roots in natural product chemistry as it marches forward with biotechnology just behind genomics and proteomics into the new millennium. For example, efforts can be directed toward uncovering efficacy and ADMET-related synergies that may be present among the constituents of herbs purported to have anticancer or cancer-preventive properties by taking advantage of the common cell culture panels already in place to assess anticancer activity, along with various transporter system interactions via HTS format. However, because anticancer/cancer-preventive synergy could derive from favorable interactions across a wide variety of ADMET processes relative to any combination of one or more efficacy-related endpoints, several mechanism-based assays associated with several key possibilities for efficacy will also need to be deployed as part of such a program. One can only imagine how sophisticated this type of pursuit will become in the future when such highly interdisciplinary efficacy networks are coupled to an even wider network of ADMET parameter experimental protocols.

A more classical approach toward the interactions of multicomponent systems would be to utilize clinical investigations to study the interactions, either positive or negative, that herbals may have with drugs when both are administered to humans. For example, Bachmann and Reese et al.[344,345] have begun to study the interaction of selected herbals with specific markers for several drug metabolism pathways, while V. Mauro and L. Mauro et al.,[346,347] among others, are studying the clinical pharmacokinetic consequences of selected herb–drug administrations, such as ginkgo biloba with digoxin. Importantly, for all of these herb-related studies it becomes imperative that extensive chemical constituent fingerprinting is also undertaken so that the effects observed, particularly those suggestive of synergy, can be correlated with overall chemical composition patterns and not just with the distinct concentrations of preselected components already known to possess established activity.[183]

In contrast to both of the aforementioned types of studies that can be considered to represent systematic examinations of herbal-directed small libraries and specified herb–drug clinical combinations, it becomes interesting to speculate how a truly random brute force approach to identifying synergy might proceed not too far down the road into the new millennium (e.g., as an HTS survey of a huge random compound library in pursuit of optimal pair or even triple compound teams rather than as the pursuit of a single blue-chip drug that can do it all). In this regard, however, it must first be recognized that the present trend to test mixtures of several compounds within a given well does not even begin to address synergy. This is because based on considerable experience with various chemotherapeutic agents,[348] synergy is most likely to be observed at very select ratios within very distinct concentrations of the players involved. In other words, looking at the simplest case of assessing the potential synergy between just two molecules, A and B, requires testing A in the presence of B across a range of molar ratios presented across a range of absolute concentrations. This situation is depicted in Fig. 2.12.[349,350]

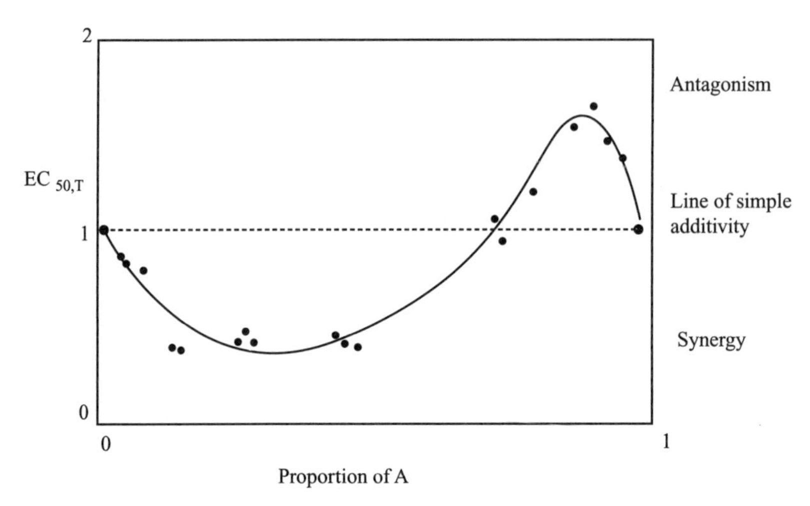

Figure 2.12 Drug interaction plot for two drugs, A and B. $EC_{50,T}$ is the total concentration of the combined drugs which gives 50% of the maximum possible effect. The $EC_{50,T}$ is shown as a function of the fraction of drug A (drug B's fraction is 1 minus the fraction shown). Rescaling of drug concentrations to units of their EC_{50} values allows simple additivity to be set at unity such that deviations below or above this line indicate synergism or antagonism, respectively. The dots are actual experimental results obtained for two anticancer agents, wherein the observed $EC_{50,T}$ values reflect 20 rays of fixed drug fractions as estimated from the data along that ray alone. The fitted curve was generated by the global model for the entire data set and indicates the complicated nature of interaction relationships within even a well-controlled cell culture enviroment. That synergism can be accompanied not only by simple additivity but also by ratio-dependent antagonistic relationships is apparent. (From ref. 348).

2.6.4 Brute Force HTS to Uncover Multicomponent Synergies

Pursuing the brute force approach from a purely mathematical viewpoint and in a minimally elaborated pharmacological format, suppose that the possibility for A plus B synergy relating to just a single efficacy or ADMET-related HTS parameter is examined across a compound library having only 100 members wherein paired combinations are tested at just three relative molar ratios (e.g., A/B at 0.5/1, 1/1, and 1/0.5) at only three total concentrations of both members (e.g., 0.1, 1.0, and 10 μ*M*). Then a total of 44,850 drug tests plus numerous control runs would be required for an $N = 1$ pass through the library.[351] Perhaps because of these rather impressive numbers, brute force HTS in the new millennium will undoubtedly relish such pursuits. Indeed, it strikes this author very surprisingly that nothing along these lines seems yet to have appeared in the literature. At any rate, once the HTS forces do become mobilized in this area, such testing could set up an interesting "John Henry" competition with more directed investigations, such as those that have been elaborated above that seek systematically to identify the specific synergies seemingly present within certain herbals. Ultimately, no matter how the identification of such favorable drug–drug partnering possibilities are uncovered and are able to better deal with the various ADMET parameters of tomorrow, as well as for the classical efficacy relationships of today, they will certainly prove to be invaluable toward alleviating the situation of trying to establish all of the most desired behaviors for a given therapeutic target within the context of a single molecular framework. Furthermore, it can be anticipated that this type of information will become extremely useful when it becomes further elaborated by medicinal chemistry into general structural motifs that have potential synergistic utilities and applications beyond what was initially uncovered by the specific mixtures of defined compounds.

2.6.5 Controlling Metabolism: Example Involving a Soft Drug Strategy

Although aspects of drug metabolism are covered more seriously in other chapters such as Chapter 9, there is one general area pertaining to *controlling metabolism* that falls so specifically into medicinal chemistry's domain of lead tailoring that it merits at least a brief discussion herein. This topic involves exploiting what has come to be called[352,353] *soft drug technology* (Fig. 2.11), where a metabophore is placed within an established drug or lead compound in order to program a specified course of metabolism for the resulting combination. Although nature has provided numerous examples of soft drugs, esmolol (Scheme 2.7) has come to be regarded as the prototypical soft drug that was obtained via rational design.[354] In this case, a methyl propionate was appended to the classical aryloxypropanolamine template associated with β-adrenergic receptor blockade (Scheme 2.7) in order to program the latter's metabolism along the ubiquitous esterase pathways such that the resulting β-blocker would possess an ultrashort duration of action.[355–357] Thus, a methyl 3-arylpropionate system (boldface atoms in Scheme 2.7) represents a useful metabophore already having clinical proof of principle within the molecular context of an aryloxypropanolamine template. This metabophore can be used to program human drug metabolism by esterases. It can be noted that the rational design of esmolol simultaneously drew upon several of medicinal chemistry's basic science principles mentioned thus far: (1) *negative SAR*, wherein it was determined that only lipophilic or, at most, moderately polar groups could be deployed in the aryl portion of the aryloxypropanolamine pharmacophore if activity was to be retained; (2) *electronic physicochemical properties* operative within a biological matrix, wherein it was imagined that while an ester would be permissible in the aryl portion (*neutral SAR*), a carboxylic acid moiety placed in the same aryl portion would become too foreign to be recognized by β-adrenergic receptors upon ionization of the carboxylic acid at physiological pH; (3) general *structure–metabolism relationships* (SMR), wherein it was appreciated that an ester linkage might be relied on to program a

OH O NH

OH O NH
$CH_2CH_2CO_2CH_3$

12 **13**

Scheme 2.7 Esmolol as the prototypical soft drug. Compound **12** represents the classical aryloxypropanolamine pharmacophore associated with blockade of β-adrenergic receptors. Compound **13** is esmolol, a soft drug version of **12** that has been programmed to have an ultrashort duration of action due to hydrolysis of the methyl ester by the ubiquitous esteases. The methyl 3-arylpropionate (bold in **13**) thus represents a useful metabophore[22] for the associated human esterases.[489] (From ref. 354.)

quick metabolism; (4) *steric physicochemical properties*, wherein it was imagined that the metabolic hydrolysis rate could be quickened by extending the initial ester linkages away from the bulky aryl group such that the methyl 3-arylpropionate metabophore was identified; and (5) appreciation for the physiologic *drug excretion structural relationship* (SER), where there is a general propensity to excrete low-molecular-weight acids. These are all fundamental physical organic principles applied in a straightforward manner within very specific contexts of the biological realm. Thus, this example serves four purposes. The first is to emphasize again that beyond activity hits per se, neutral and negative SARs should also be tracked so as to be readily retrievable from the databases associated with a given parameter survey of the future. The second is to emphasize again that medicinal chemistry will need to become an active participant in the merging of various HTS parameter surveys by using chemical structure as a common denominator, especially when such activities become considerably more complicated than in the esmolol case. Third, the esmolol case demonstrates that even when problems can be reduced to what appears to be a rather simple set of factors, it will still be medicinal chemistry's unique desire to systematically characterize the complete pattern of chemical structural relationships that is likely to be called upon to finalize what other disciplines might consider at that point to be rather subtle and mundane details. In other words, who besides a medicinal chemist can be expected to enthusiastically pursue methyl-, ethyl-, propyl-, and so on, relationships either synthetically or by tediously purveying huge databases of the future, just to look for those SXR "Goldilocks" situations[354] that could become relevant toward addressing a problem within another structural setting while attempting to merge the two data sets within a common chemical context? The latter was precisely the case for the esmolol-related metabophore upon comparison of methyl benzoate, methyl α-phenylacetate, methyl 3-phenylpropionate, and methyl 4-phenylbutyrate, wherein the half-lives observed for these systems when incorporated into the molecular context of a β-blocker pharmacophore became about 40, 20, 10, and 60 min, respectively (Fig. 2.13). More current uses of the soft drug technology are

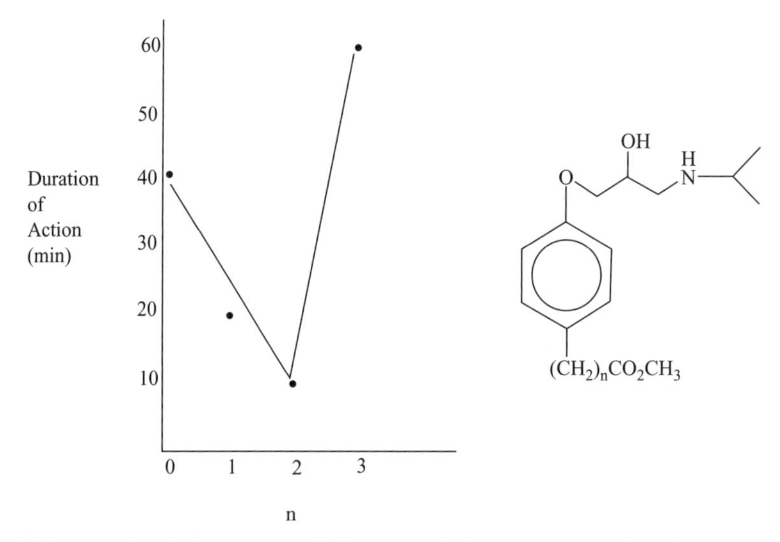

Figure 2.13 Relationship between methylene-extended esters and duration of action within a series of esmolol analogs. The "Goldilocks" nature of the ethylene extension relative to the desired 10-min duration of action is apparent. (From refs. 354 to 357.)

also underway. For example, the esterase capability in newborns was recently compared to that of adults, and it was found that esmolol's half-life in cord blood (baby side) is about twice as long as that in adult blood. Furthermore, individual variation is significantly more pronounced within the newborns.[358,359] These findings, in turn, have prompted an exploration of the generality of deploying the esmolol metabophore within the chemical contexts of several other types of therapeutic agents: namely, those that are commonly used to treat the neonatal population in critical care settings. Thus, this fourth and last aspect of the esmolol example clearly demonstrates the potential impact that such classical medicinal chemistry studies can have on the new field of pharmacogenetics, as the latter is surely to become further evolved within the new millennium. Analogous to the importance of merging SAbRs and SDRs with efficacy and selectivity-related SARs, SMRs and numerous metabophore patterns can be expected to be gradually discerned and put to extensive use by medicinal chemistry in the future either to enhance or to detract from a candidate drug's metabolism.

2.6.6 Optimizing Excretion

Clearly, SAbRs, SDRs, and SMRs can all be used to manipule and optimize the elimination pattern of administered drugs, the esmolol example applying here as well. Analogous to the distribution area, genomics and proteomics can soon be expected to delineate important systems, such as specialized transporters within tissues like the kidney and liver, that are especially responsible for the excretion of xenobiotic drugs and their metabolites. Medicinal chemistry's involvement in uncovering SERs and deriving generally useful structural patterns that might be used to tailor lead compounds or for merging of different types of databases while attempting to select lead compounds again falls into the central theme for medicinal chemistry's future as being elaborated in this review. As for the other areas, more speculative notions in this area can provide some interesting alternatives for this aspect of ADMET. Although SERs are also likely to encompass various endogenous materials and their catabolic fragments, one might still imagine that just like the futuristic examples sited for absorption, in the more distant future, biotechnology, chemical, and nanotechnology approaches might all be used successfully to engineer specific drug excretion passages through selected tissues.

2.6.7 Avoiding Toxicity

It may very well be that the most profound effect that genomics and proteomics are going to have within the ADMET arena will ultimately pertain to *avoiding toxicity*. Indeed, that toxicology has now become a protagonist through its participation in the design or early selection of drug leads already represents a remarkable turnaround from its historical antagonist role as a gatekeeper or policeman standing at an advanced stage of drug development with an eye toward halting the progression of potentially toxic compounds on route to the clinic.[360] Like the field of drug metabolism (Table 2.7), toxicology has been collecting its data within databases for quite some time (Table 2.9). In fact, some of the structural patterns that have come to be associated with distinct toxicities (toxicophores) are probably on much firmer ground than are the metabophore relationships. On the other hand, drug metabolism derives from a finite number of genetic constructs that translate into metabolic activity (albeit notorious for their seeming molecular promiscuities) such that with enough

TABLE 2.9 Toxicology Databases and Related Organizations

Database or Organization	Description
Centers for Health Research [formerly, Chemical Industry Institute of Toxicology (CIIT)]	Industry consortium-sponsored collection/dissemination of toxicology data; also conducts research and training in toxicology[467]
American Chemistry Council Long-Range Research Initiative	Industry consortium-sponsored initiative to advance knowledge about the health, safety, and environmental effects of products and processes[468]
LHASA, Ltd. (UK-based, nonprofit segment)	Facilitates collaborations in which companies share information to establish rules for knowledge bases associated with toxicology[469]
International Toxicology Information Center (ITIC)[a]	Pilot program to share data in order to eventually be able to predict the toxicology of small molecules, thus lessoning the expense of in vitro and in vivo testing[469]
U.S. Environmental Protection Agency (EPA) High-Volume Chemical (HPV) Screening Information Data Set (SIDS)	User-friendly version that will also be submitted to the Organization for Economic Cooperation and Development (OECD) and its tie-in with IUCLID[a][469]
SNP Consortium: nonprofit; makes its information available to public	Addresses phenotypic aspects relative to individual responses to xenobiotics (e.g., metabolic phenotype and toxicity)
Tox Express/Gene Express database offered by Gene Logic (commercial[b])	Offers a gene-expression approach toward toxicity assessment[469]
National Institute for Environmental Health Science (NIEHS)[c]	Compiling a database of results from toxicogenomic studies in order to divide chemicals into various classes of toxicity based on which genes they stimulate or repress[470]
International Program on Chemical Safety/Organization for Economic Cooperation IPCS/OECD	Risk assessment terminology standardization and harmonization[471,472]
MULTICASE (commercial)	Prediction of carcinogenicity and other potential toxicities[473]
MDL Toxicity Database (commercial)	Allows structure-based searches of more than 145,000 (Jan. 2001) toxic chemical substances, drugs, and drug-development compounds[96]
DEREK and STAR (LHASA, commercial segments)	Prediction of toxicity[474]
SciVision's TOXSYS (commercial)	General toxicity database to be developed in collaboration with the U.S. FDA[475]
Phase-1's Molecular Toxicology Platform gene expression microarays (commercial[b])	Allows detection of gene expression changes in many toxicologic pathways[476]

[a]Includes cooperative efforts with the European Union and the European Chemicals Bureau (ECB) in using the International Uniform Chemical Database (IUCLID) and its relationship to high-volume chemicals (HVPs).

[b]This company's product is representative of several such technologies that are also being made available by a variety of other vendors.

[c]Includes cooperative efforts with the U.S. Environmental Protection Agency (EPA) and the Information Division at the National Institute for Occupational Safety and Health (NIOSH).

time the entire set of metabolic options should eventually become well characterized. Toxic endpoints, alternatively, have no such limitation associated with their possible origins. In other words, to show that a drug and its known or anticipated metabolites are completely nontoxic is comparable to trying to prove the null hypothesis, even when a limited concentration range is specified so as to circumvent the situation that everything becomes toxic someplace at high enough concentration. Nevertheless, genomics, proteomics, and biotechnology do, indeed, appear to be producing some promising technologies that can be directed toward this area. For example, array technologies are already becoming available to assess the influence of a drug on enormous numbers of genes and proteins in HTS fashion.[361–368] Once enough standard data of this type are produced by taking known agents up in dose until their toxicity becomes fingerprinted via distinctive patterns of hot spots, array patterns may be used to cross-check against the profiles obtained in the same HTS mode for new lead compounds. Given the quick rate that these important trends are likely to be further developed and eventually validated within the new millennium, medicinal chemistry could certainly become overwhelmed trying to keep up with its complementary role to identify the corresponding STR for each array hot spot.

In the case of toxicity, then, medicinal chemistry will probably need to approach STRs in a different manner [e.g., initially from just the exogenous compound side of the equation for a given toxicity relative to the observed hot spot patterns (unless genomics, proteomics, and biotechnology also quickly step in to define the biochemical nature of the actual endogenous partners that are involved in a given toxic event)]. Taking a chemically oriented starting point, however, should serve reasonably well for at least awhile into the new millennium in that there will likely become a finite number of chemical reactivity patterns that can be associated with toxicity. Medicinal chemistry can be expected to elaborate these reactivities into general STRs and then to use them toward defining the liabilities in new compounds. The notion that there should be a finite number of structurally identifiable toxiclike patterns is analogous to the notion that there should be a restricted number of amenable druglike patterns that reside within structural databases having high degrees of molecular diversity. Indeed, the case for toxicity is certainly on firmer ground at this particular point in time since there will likely be little added to the area of fundamental chemical reactivity in the new millennium as opposed to proteomic's anticipated revelation of numerous new biochemistries that will, in turn, provide numerous new pharmacologic targets wherein many can be expected to have their own distinct pharmacophore (and potentially new druglike patterns). Finally, since the precise locales where the toxicity hot spots may ultimately occur are endless, the latter will perhaps be better addressed by directing a second set of database queries toward the ADME profile and intracellular localization patterns that a given drug may exhibit. In the end, after array technologies are producing useful toxicology-related knowledge, the interplay of all of the ADME parameters with STR should become just as important as they are for efficacy in terms of what type of toxicity may ultimately be observed within the clinic.

2.6.8 Weighting Decision Criteria from Efficacy and ADMET SAR

Figures 2.14 to 2.17 convey how all of the efficacy and ADMET HTS profiling data may eventually be simultaneously deployed toward the design of an optimal preclinical candidate compound. Based on the magnitudes of the various molecular similarities and dissimilarities across each HTS parameter survey of the compound data set relative to the locations of the generalized pharmacophores associated with various parameters, a

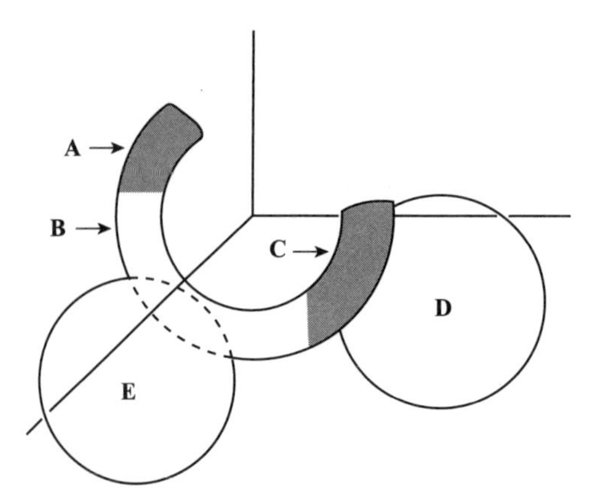

Figure 2.14 3D pattern recognition example pertaining to the simultaneous consideration of efficacy and ADMET-related pharmacophoric parameters during lead selection and drug design. In this example, SXR space has been mapped according to experimental HTS results from a moderately sized, directed compound library wherein an efficacious template was identified. A and C represent requisite pharmacophoric features oriented in space by structural elements B. Lacking distinctive functionality, B, in turn, tends to prolong the elimination half-life and eventually causes toxic concentrations to be produced in compartments not associated with efficacy. D represents the structural space that provides for a desirable absorption and initial distribution profile. E represents a structural space that is subject to rapid metabolism. Thus, in this case a range of suitable functionality defined by A can be utilized in the northwest region of the efficacious pharmacophore while selected functionalities or bioisosteres that reside within the structural space defined by the overlap of C and D should be utilized in the southeastern region. The latter strategy optimizes absorption and initial distribution features while retaining efficacy such that more complicated prodrug scenarios are not necessary. A soft drug version, however, should be contemplated so that the eventual toxicity problems derived from the connecting chain can be circumvented, especially since the SMR portion of the overall map indicates that there is an intrusion of B by structural features E that prompt rapid metabolism and elimination. The latter can thus be readily exploited by incorporating them into the overall molecular construct and then making adjustments or fine-tuning them to a desired metabolic rate by the incremental insertion of steric impediments near the point of metabolic contact.

balance will be sought between options involving any combination of single or coadministered multiple entities wherein each member has been further tailored according to an unchanging multivalent prodrug and/or soft drug strategy once interactions within the biological realm have been initiated. Figure 2.14 attempts to convey some of these possibilities using a hypothetical set of structural space. Alternatively, serving as a real example that couples several of the aforementioned anticancer studies, Fig. 2.15 captures some of the results already obtained from the pursuit of transportophore SARs relative to the paclitaxel template by placing the results within the formalism of the present discussion. Figure 2.16 provides an overall step-by-step strategy for deploying the consideration of ADMET parameters proactively to enhance the process of drug discovery rather than just to play a negative role as a series of filters toward either the entry or the continued progression of a given compound through such a process. Concerns about today's trend toward filtering the entry of compounds have been alluded to in Section 2.5. In particular, it should be noted from Fig. 2.16 that it is a completely defined efficacy-related pharmacophore that serves as the central structural theme to drive the proactive ADMET strategy in that all

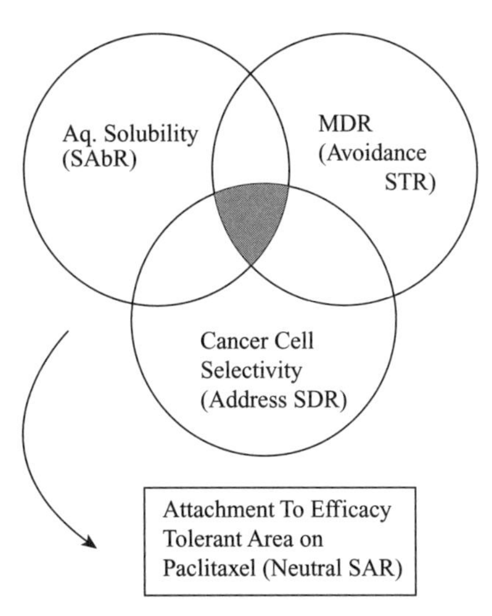

Figure 2.15 Design of optimized drug candidate based on simultaneous consideration of several fields of paclitaxel-related SARs. Studies indicate that there is a distinct region of structural space that is overlapped simultaneously by SARs pertaining to enhanced aqueous solubility, avoidance of multidrug resistance (MDR), and the propensity to associate selectively with cancer cells compared to healthy cells. Coupling of this distinct structural space (depicted as the shaded region) onto an area of paclitaxel that can accommodate structural modification without losing efficacy provides an optimized drug candidate. The specific details of the distinct structural space and the synthetic methods that can be used to couple its useful molecular displays onto paclitaxel are the subject of pending patent applications.[490,491]

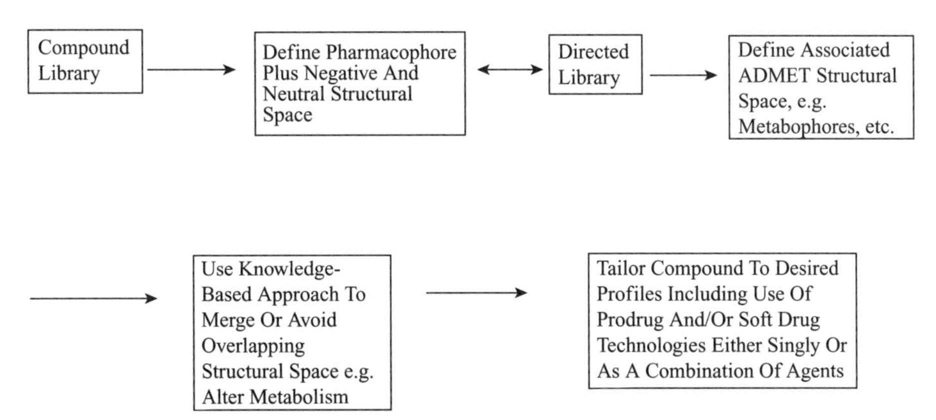

Figure 2.16 Drug design and development strategy. Note that this strategy emphasizes a complete definition of the efficacy-related pharmacophore as the central theme such that it can be merged with simultaneously generated ADMET-related pharmacophores via a knowledge-driven proactive process that will ultimately produce the optimized clinical candidate or combination of agents to be deployed for therapy or prophylaxis. It should be noted that this knowledge-based decision tree contrasts some of the futuristic schemes that have been suggested by others wherein the various ADMET issues are simply used as consecutive or simultaneous filters to eliminate compounds being selected from huge compound databases. As elaborated within the text, it is this author's opinion that by medicinal chemistry input, HTS data of the future will be able to take drug discovery investigations to significantly greater heights of knowledge such that the latter can then be used for proactively assembling the positive type of enhanced property molecular constructs mentioned above. Indeed, if this scenario does not unfold, the overall new and future processes will forever be locked into a negative mode that simply keeps eliminating compounds failing to meet certain criteria placed at each parameter.

other structural modifications/biological enhancements are then conducted according to knowledge-based scenarios generated via previous ADMET profiling experiences with analogous structural space or by immediate HTS ADMET profiling of appropriately directed libraries. Clearly, while nondruglike structural space may thus be used effectively to contribute toward defining the efficacy pharmacophore's electronic surface potentials, such structural components within a library should still be tagged with red flags, indicating that they are also destined to be altered or removed completely by tailoring of the spatial overlaps according to the type of knowledge base that can be afforded by medicinal chemistry. That the various ADMET-related pharmacophores will eventually be evolved so as to be able to be deployed more independently from a given efficacy-related pharmacophore and in a completely in silico manner at any point along the drug design and development flowchart will indeed occur as well. However, the latter will probably be realized only after we are well into the future and probably only after we have accumulated considerable knowledge about how to manipulate their structural patterns for optimal overlap and/or avoidance with a variety of model efficacy platforms wherein all structural space has been accounted for by 3D electrostatic potential maps derived from rigorous experimental and computational considerations of molecular conformation. Finally, drawing from the overall strategy listed in Fig. 2.16, Fig. 2.17 illustrates the specific interplay of all of the aforementioned efficacy, selectivity, and ADMET considerations, along with some other practical drug discovery considerations.

2.7 PROCESS CHEMISTRY CONSIDERATIONS

As shown in Fig. 2.17, practical issues pertaining to intellectual property (IP), such as the structural novelty of biologically interesting compositions of matter, as well as to the latter's synthetic accessibility (relative to process chemistry and manufacturing costs) will also continue to be factored into earlier decision points intending to select the optimal preclinical candidate compounds of the future.[369–371] The changing landscape of IP-related structural novelty is addressed in the final, summary section of this review. Three key issues pertaining to the interplay of medicinal chemistry with process chemistry's responses to current trends are mentioned below.

2.7.1 Cost and Green Chemistry

First, the eventual production cost for a new therapeutic agent is much more important today than it has been in the past. This is because pharmaceutical companies must now garner their profits from a marketplace that has become sensitized about the cost of ethical pharmaceutical agents. The days of simply raising the price of such products in parallel to increasing costs associated with discovering and developing them have been over for quite some time.[372] In this regard, the cost-effectiveness of small-molecule drugs will probably maintain an edge over biotechnology-derived therapeutic agents for at least the near-term portion of the new millennium. The second point to be mentioned pertains to the impact of the green chemistry movement.[373–376] This movement has prompted pharmaceutical companies to ensure that their productions of drugs are friendly toward the environment in terms of all materials and methods that may be deployed in the process. Finally, the U.S. FDA's initiative to have all stereoisomers present within a drug defined both chemically and biologically has prompted industry's pursuit of drugs that either do not contain

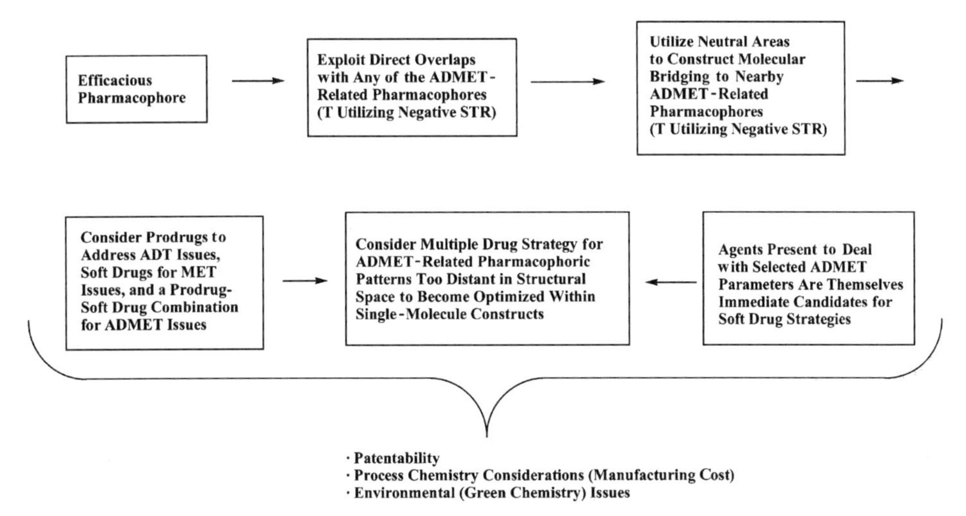

Figure 2.17 Lead selection and drug design decision flowchart based on efficacy- and ADMET-related pharmacophoric parameters. This flowchart has been set up to represent the case where a single-molecule construct having the lowest level of complexity/sophistication is initially sought. However, in the future it is also likely that well-established templates that optimize a certain parameter will be able to be paired with the efficacy-related agent at an early point in the overall design process. For example, a compound that inhibits a transporter system responsible for a given lead's poor passage through the GI endothelium might ideally be coadministered as a soft drug version. In this way, the partner compound would solve the oral absorption problem and then be metabolized and eliminated quickly without doing much of anything else. Structural manipulation of the efficacy construct could then be directed toward enhancing other DMET-related profiles.

asymmetric centers or are enantiomerically pure.[377] This, in turn, has prompted the need for better stereochemically controlled processes during production. Stereocontrol has always represented an extremely interesting area for synthetic chemistry exploration and now for biotechnology-derived chemistry and reagent research as well (e.g., exploitation of enzymes at the chemical manufacturing scale). Considerable progress is being made toward developing such methods on many fronts, including enzymatic[378–380] and microarray technologies.[381] Often, however, the new laboratory techniques do not readily lend themselves to inexpensive scale-up/manufacturing type of green chemistry. Alternative approaches that seek to address this situation in a very practical manner can be exemplified by the following study that intends to exploit simple α-substituted benzylamine systems. The latter are being explored as chiral auxiliary synthetic reagents for the delivery of a nitrogen atom in a stereochemically biased fashion during the synthesis of end-product amines that contain neighboring asymmetry (Scheme 2.8).

2.7.2 Defining Stereochemistry: Example Involving Benzylamine Chiral Auxiliary Synthetic Reagents

Nitrogen systems having α, β, or γ asymmetry represent an extremely common structural motif within drug molecules such that the proposed methods immediately become

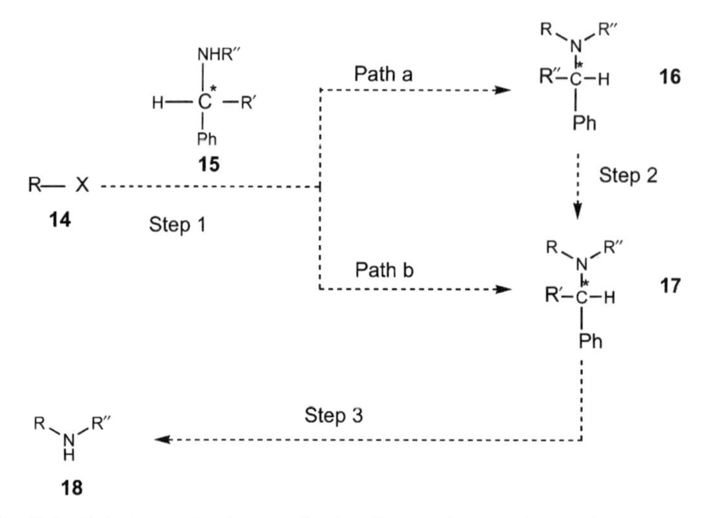

Scheme 2.8 Potential stereoselective synthesis of secondary amines using asymmetric benzylamine-related systems as chiral auxiliary reagents. **14** is a racemic precusor where R has an asymmetric center α, β, or γ to X; **15** is an optically pure chiral (*) auxiliary reagent wherein **R′** is a substituent having selected physical properties such as specified elements of steric bulk; **16** and **17** are diastereomeric tertiary amine intermediates which are mixtures or separated single diastereomers, respectively; and **18** is the desired optically pure secondary amine product. Step 1 represents a variety of N-alkylation or reductive alkylation methods (e.g., X = halide, carbonyl, etc.), step 2 represents a fractional recrystallization or chromatographic separation if necessary, and step 3 represents a catalytic hydrogenolysis reaction. It is reasonable to anticipate that during step 1, racemic **14** could combine with optically pure reagent **15** to provide one or the other of the two possible diastereomers directly (path b). Even when no asymmetric bias is observed during step 1 (path a), the two diastereomers present as intermediate **16** will differ in physical properties such as solubility, chromatographic behavior, boiling point, or melting point. Thus, the desired diastereomer, **17**, may be able to be separated conveniently under selected conditions involving recrystallization or chromatography during the workup (step 2) of the first reaction. Alternatively, it may also be possible to effect an asymmetric cleavage during step 3 if the chiral auxiliary in one or the other of the diastereomeric tertiary amines can be removed preferentially by hydrogenolysis. Also note that step 3 can be delayed so as first to effect other chemical modifications associated with an overall synthesis while the amino functionality is still 3°. (From ref. 389.)

of interest to medicinal chemistry. Given the ease of selectively debenzylating tertiary amines, the methodology should be particularly well suited for the production of asymmetric secondary amines. While a benzyl moiety has previously been deployed as either a common N− or O− protecting group,[382,383] the utilization of benzylamine to deliver a nitrogen while simultaneously controlling the degree of substitution is less common, even though such systems can sometimes be quite effective in this regard, due to the phenyl group's inherent steric properties.[384] The use of chiral α-substituted benzylamine systems has been even more limited, with such systems most typically being deployed as resolving agent counterions for carboxylic acid partners.[385] The rarer deployment of such systems in covalent relationships probably results from the prevailing notion that as the steric environment about the N− atom is increased, it eventually becomes more difficult to effect debenzylation via catalytic hydrogenolysis.[386] Thus, in order to first scope the overall

2 mol 1 mol MeOH or when R = CO_2CH_3 or CONHØ

R = CH_3, CO_2CH_3, CONHØ, CH_2OH, CH_2OCH_3, or $CH_2OC(CH_3)_3$

Scheme 2.9 Attempted diastereoselective opening of a model epoxide using benzylamine-related chiral auxiliary synthetic reagents. (From ref. 391.)

applicability of this type of chemistry, recent studies have systematically examined the relationship between a nitrogen's immediate steric environment and the propensity toward debenzylation within relevant model systems. Surprisingly, these results indicate that steric attenuation of debenzylation is not likely to be problematic.[387,388] Explorations are now proceeding toward assessing the diastereomer bias that may be achievable upon reaction of racemic electrophiles with various members of a readily obtainable family of chiral auxiliary benzylamine synthetic reagents.[389] Scheme 2.9 depicts the initial approach toward an area that targets enantiomerically pure aryloxypropanolamines as an appropriately challenging chemical model that is very relevant to the production of pharmaceuticals.[390,391] In addition to its relation to practical asymmetric pharmaceutical process chemistry applications of the future, the benzylamine example has been cited herein because it further illustrates the important role that physical organic chemistry considerations play within the medicinal chemistry thought process, and vice versa. By analogy to medicinal chemistry terms, the preliminary benzylamine studies have determined that there is an allowable region of potential *structure stereochemical relationship* (SSR) space to explore because of the neutral effects that were observed relative to a co-event that is required for the overall chemical process (i.e., subsequent debenzylation). The neutral space was mapped out by using precisely defined *structure–debenzylation relationship* (SDebR) steric probes. The present benzylamine studies, in turn, are now taking advantage of the neutral space to identify useful synthetic stereophores that can elicit selected asymmetries analogous to the situation of pharmacophores that can elicit selected efficacies. Finally, upon turning all of these analogies around, exactly the same "library" of steric probes can also be used to actually explore the steric tolerance associated with metabolic N-dealkylation (i.e., wherein a cytochrome P450 biological surface and its requisite cofactors then substitute for the inorganic catalytic surface and its hydrogenated atmosphere that were present during the aforementioned hydrogenolysis studies). Importantly, such metabolism data will have the potential to be knowledge-generating in that they can then be used to assist in the prediction of the susceptibility toward metabolic N-dealkylation not just by specific structural pattern recognition, but also by a parameterized and well-defined physicochemical property that is inherently localized in the space immediately residing about the N− atom within any metabolic candidate. The value of this example's endpoint merits reemphasis. In the future, efficacy and ADMET parameters will be defined maximally in terms of their associated pharmacophore's electrostatic potential space and not just in terms of distinct compound hits or leads. Although compound hits and leads can certainly reflect desirable structural prototypes for a given parameter, they can also become restrictive toward a broader conceptualization of the more diverse structural space that can be invoked to better purview the full spectrum of efficacy and ADMET overlaps, as the latter parameters are all being addressed simultaneously. Thus, from a broader, knowledge-derived vantage point, the ideal candidate drugs (candidate drug teams) of the future will ultimately be designed by considering all of the efficacy and ADMET parameter pharmacophores simultaneously while not being restricted to any type of predetermined structural template associated with a hit or lead structure that may have been obtained from any one of them.

2.8 ANALYTICAL CHEMISTRY/X-RAY DIFFRACTION

As mentioned in the introduction, new developments in any of several key analytical techniques can be expected to have a profound impact upon medicinal chemistry's future. X-ray diffraction is taken herein as just one example for these types of possibilities.

2.8.1 Latest Trends

At the forefront of current progress and trends in the field of x-ray technology are efforts to properly derive and readily portray electrostatic potentials.[392] As alluded to earlier, the initial recognition and driving forces associated with the interactions between a xenobiotic and the biological surfaces that it will encounter in vivo are due to a complementary match between the topography of the electrostatic potential of the xenobiotic ligand and that of the biological site (less anything energetically favorable that is given up when the xenobiotic leaves its solvated environment). As discussed in Section 2.5, considerable effort has been expended toward calculating accurate depictions of the electrostatic potentials of molecules theoretically. This has been most beneficial, however, for only very small molecules because extended basis sets are required to obtain accurate results. Calculations with smaller basis sets have been carried out for larger molecules, but as already pointed out, the results are typically unreliable, owing to the nature of the mathematical approximations that will have necessarily been taken. Using larger basis sets for small fragments and then using the fragments as building blocks for larger molecules is also being done. The latter approach will probably become much more prevalent in the new millennium.

Alternatively, it has been possible for several years to obtain experimental electrostatic potential maps from the molecular charge distribution derived from x-ray diffraction data. This approach has had limited appeal, however, because the experiments needed to provide the large amount of quality data that is used as the starting point have themselves been extremely time-consuming, typically taking many weeks even for quite small molecules. Nevertheless, from the studies undertaken to date,[392] there is growing evidence that results for small molecules can indeed be extrapolated to similar fragments in larger molecules,[393] just like what is being suggested by the theoretical approaches that were reiterated above. *The practical implications of potentially using this approach to simultaneously address huge numbers of compounds when they reside along distinct structural themes within compound libraries or databases (e.g., wherein a given scaffold then becomes the common fragment) is worth noting for both the x-ray and computational types of approaches.*

Today, the experimental approach afforded by x-ray diffraction has become much more tractable. With the new generation of x-ray diffractometers using charge-coupled device (CCD) area detectors, the necessary experimental data can be obtained in just a few days, a duration comparable to that currently required for a routine x-ray structure determination.[394,395] Furthermore, with access to x-ray synchrotron beam lines, the time of the experiment may be reduced to a few hours.[396] Shorter experiments, in turn, have allowed development of cooling devices using liquid helium, thus giving access to lower temperatures and improved data.[397] In addition, whereas with a serial diffractometer the length of the experiment scales with the number of atoms in the molecule under study, the size of the molecule is less important when collecting the data using a CCD detector. How large a molecule is tractable in terms of converting these data into electrostatic potentials is still unknown but is likely to be forthcoming within just the near-term future. In this regard, one might speculate that this approach may even be able to handle larger molecules more reliably than will the theoretical approaches elaborated in Section 2.5. From the present situation it is already clear that it is now possible to map the topology of the electrostatic potential for a typical small-molecule therapeutic agent within very reasonable time periods. This means that the electrostatic potentials may be able to be compared for series of molecules having established biological activities so as to produce refined SAR and to provide the most meaningful data possible relative to the x-ray contributed structures within future databases. Not too much further into the new millennium, one might imagine that it will also be feasible to use this approach toward mapping the complementary electrostatic potential of receptor/enzyme active sites at highly improved resolutions,[397–401] especially as promising results have already been obtained for some small proteins.[393] Indeed, certain of the techniques described above for small molecules are already being applied to the analysis of macromolecules.[401]

2.8.2 Examples Involving Dopamine Receptors, c-AMP Phosphodiesterase Enzymes, and the Dynamics of Protein Folding

Since applications of these latest trends in x-ray technology are themselves under investigation, studies like the ones being undertaken by Pinkerton et al.,[402] which intend to validate the utility of deploying the CCD type of cutting-edge diffractometer approaches toward the study of drug design, represent a critical step at this juncture. For these studies, classical SAR that is available within some established systems of therapeutic interest, such as that for renal dopamine receptors[403] and for c-AMP phosphodiesterase active sites,[404] are being reexamined. Long-standing topographical models for these systems (Schemes 2.10 and 2.11, respectively) will certainly be interesting to reevaluate based on refined analyses of the most relevant structural probes that lend themselves to crystallization and extremely accurate x-ray analysis. In this regard, the example depicted by Scheme 2.11 is especially noteworthy because it clearly demonstrates the "broadest conceptualization of a pharmacophore" theme that was emphasized as being extremely important at the close of Section 2.8.1 (i.e., note that it is the electrostatic surface potential of a lead compound's imidazolone system that is further likened to the electronic topography of the cyclic phosphate's trigonal bipyramid transition-state species traversed when cAMP is hydrolyzed by phosphodiesterase).

Assuming that refined x-ray techniques will become commonplace within the first 25 years of the new millennium and that they will be coupled with HTS approaches toward crystallization and actual obtainment of diffraction data,[122,405,406] it becomes interesting

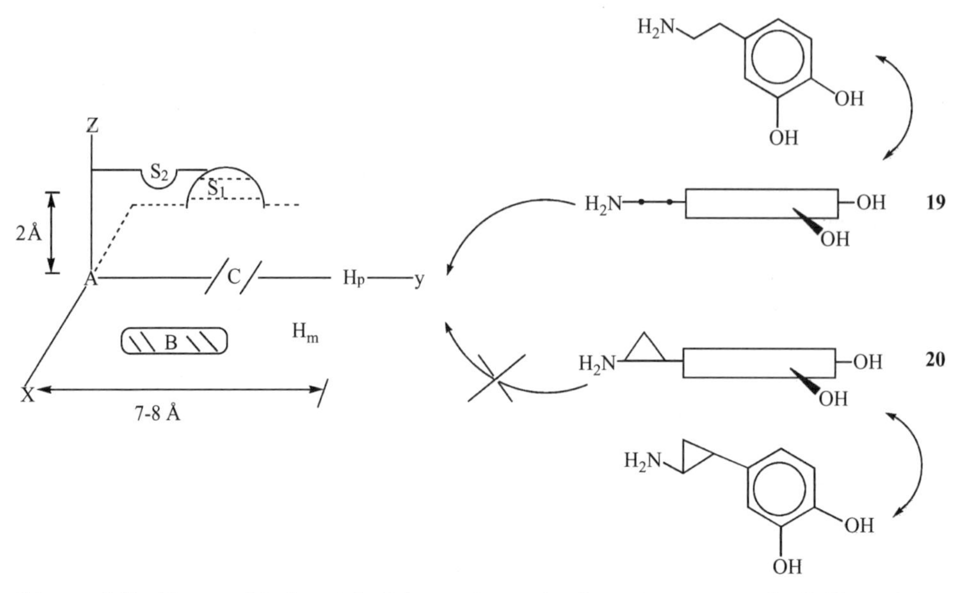

Scheme 2.10 Topographical model of the renal vascular dopamine receptor. A, C, Hp, and Hm reside in a single plane and represent regions that interact with dopamine's (**19**) amine, catechol ring, *p*-hydroxy and *m*-hydroxy, respectively. Region B represents an auxiliary binding site suggested by apomorphine's SAR, whereas S1 and S2 represent steric limitations toward the binding of receptor ligands. Note that while dopamine, **19**, is nicely accommodated by the model, neither enantiomer of the cyclopropyl analog, **20**, can be accommodated because they will collide with either the receptor's planar floor or S2 ceiling. The cyclopropyl analogs were found to be devoid of activity at either peripheral (renal) or central dopamine receptors.[492] (From ref. 403.)

to ponder what might be on the more distant horizon of x-ray-related technologies. Certainly, the ability to derive x-ray diffraction patterns from noncrystalline small and large molecules would allow medicinal chemistry to embark on a rational approach toward drug design immediately upon the obtainment of such "pictures" for every new pharmacological target of the future. In silico screening of real and virtual libraries, having matured considerably at that point as well, would also benefit enormously from such a development and would be expected to be equally interactive with such data in terms of docking virtual compounds into x-ray pictures of enzyme active sites and/or receptors followed by ranking them as potential drug candidates.[407] Further out, but perhaps not too far past the next 75 years of imaginable future, it might be expected that x-ray-type data (even if no longer strictly derived from x-ray's present physics-related principles[408,409]) will be able to be collected at fast enough real-time intervals such that it may become possible to observe molecular interactions and motions within userfriendly videos after the data are processed appropriately. One can imagine that videos of events such as the actual docking (binding or *affinity*[410]) of a drug, the motions that transpire upon an agonist's triggering of a receptor (*intrinsic activity*[410]), or a substrate's alterations upon action by an enzyme, could all become commonplace in the somewhat more distant future. These displays would be similar to the cartoon (i.e., in Section 2.1, "hand-waved") versions that we presently generate using computational or theoretical approaches coupled to molecular modeling packages[411–414] except that the entire process would be based on actual experimental data obtained in a

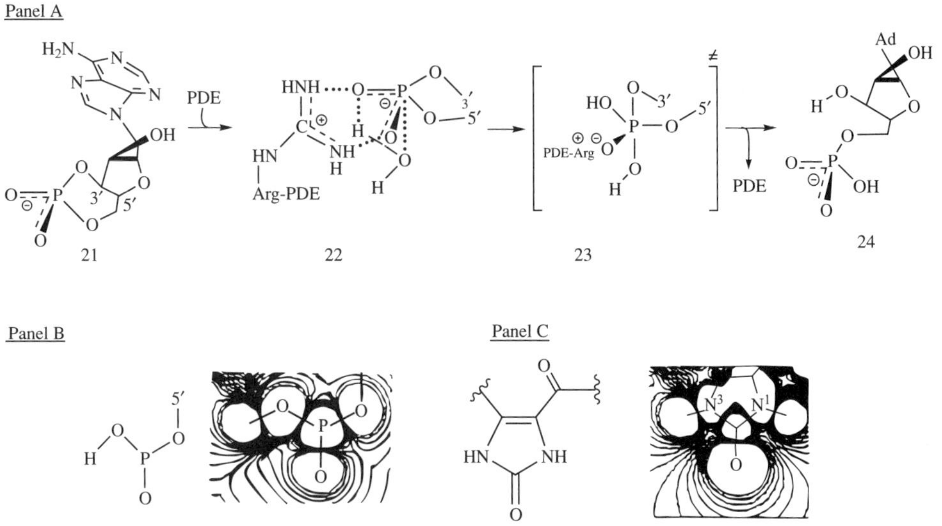

Scheme 2.11 Conformational and electrostatic potential topographies of c-AMP phosphodiesterase III (PDE III) active site ligands. In panel A, compound **21** represents a c-AMP substrate with its adenine (Ad) and ribose moieties in an "anti" relationship. Interaction **22** depicts binding of the phosphate portion using an arginine residue and a water molecule that was initially associated with Mg^{2+} in a stoichiometric relationship. Complex **23** depicts S_N2 attack of phosphorus by H_2O, with formation of a trigonal bipyramid (TBP) transition state (TS). Compound **24** represents 5′-AMP as its inverted product. The electronic charges indicated conserve the net charge overall and across the TS. Panel B represents an overhead view of the atoms in the single plane of the TS, which forms the common base for the two pyramids of the TBP system. Also shown is the proposed electrostatic potential map for the same atoms. Panel C shows a classical PDE III inhibitor ligand and the AM1 derived in-plane molecular electrostatic potential map of its imidazolone ring. Because of the very notable similarity between these two electrostatic potential maps, it has been proposed that these types of compounds, as well as several other heterocycles that have electron-rich heteroatoms in analogous locations, act as TS inhibitors of PDE-III.[404,493,494] (From ref. 404.)

real-time fashion, as the event actually occurred rather than by what we have only been able to so far catch experimental glimpses of at stabilized junctures or what we are able to imagine what such conversions may look like. For example, observing an actual trigonal bipyramid transition state for the phosphorus atom within c-AMP during its hydrolysis by a phosphodiesterase (Scheme 2.11) would certainly be a crowning analytical experimental achievement of the future, especially since today's sophisticated theoretical approaches cannot even provide a good cartoon version for this process due to collapse of such a species to lower energies during minimization. Similarly, observing the methylene portion of cyclopropyl dopamine actually crashing into the proposed molecular ceiling on the renal vasculature's dopamine receptor (Scheme 2.10) would be equally impressive since this bit of older SAR appears to have gradually become lost from the dopamine field, due to lack of timely advances able to clarify its reality. Finally, observing that the conformationally redundant nature of endothelin's 3,11-disulfide bond (Scheme 2.12) actually serves to initiate its post-translational folding paradigm because that bond is easier (in terms of distance) to form than the subsequent 1,15-disulfide link[415] represents just one of many older hypotheses that might be taken from the area of proteomics.

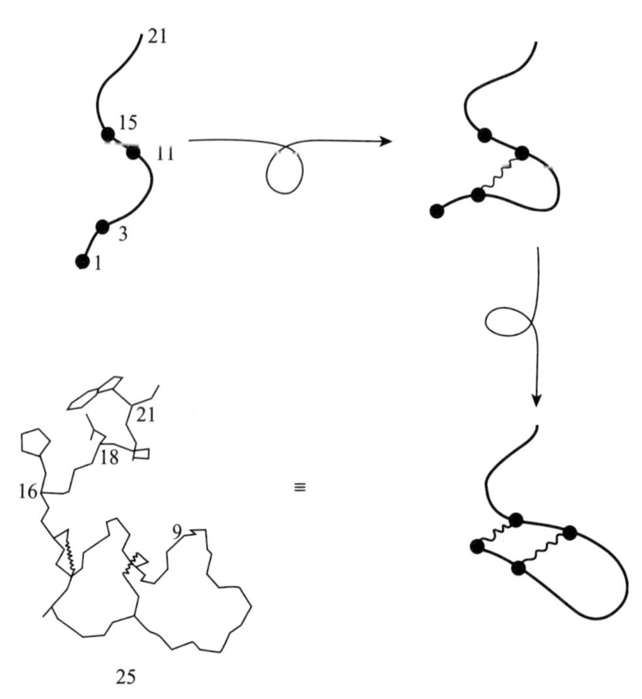

Scheme 2.12 Proposed folding scheme for the endothelins (ETs). Positions 1, 3, 11, and 15 are cysteines. Note that formation of the 3,11-disulfide linkage is proposed to occur first given their closer proximity. Subsequent formation of the 1,15-disulfide can then occur more readily to finally adopt the tightly folded form depicted as **25**. At this juncture, the 3,11-disulfide may then represent a conformational redundancy that is not actually required in the final structure. This sequence of events is equally applicable to big endothelin and to pre-proendothelin. Structure **25** represents a molecular dynamics-derived minimum energy conformation for ET-1 after starting from an entry structure where residues 9 to 16 were initially constrained in an α-helix.[495] Note that the α-helix is retained and that ^{16}His, ^{18}Asp, and ^{21}Trp form a close triad wherein ^{16}His and ^{18}Asp are probably internally hydrogen bonded.[415] (From ref. 415.)

2.9 SUMMARY

2.9.1 General Points

Medicinal chemistry has been defined as a pure science that explores fundamental physical organic principles to understand the interactions of small molecular displays with the biological realm. Regardless of discipline or background, when investigators seek this level of understanding, they are embarking upon basic medicinal chemistry research. Using a variety of input data, medicinal chemistry's applications, in turn, become that of designing or selecting new drug candidates as well as providing molecular blueprints for improving the therapeutic profiles of existing drugs or of new pharmaceutical agent lead structures. Historically, medicinal chemistry has also been heavily involved with generating input data by designing and synthesizing probe molecules that can systematically test the roles being played by the various physical organic principles during a given interaction. More recently, medicinal chemistry has begun to utilize site-directed mutagenesis as an additional tool to understand these same types of interactions. Future developments in store for drug discovery are summarized in Table 2.10. The overall progression conveyed by Table 2.10 can be

TABLE 2.10 Future Events Predicted for Drug Discovery and Development

- Genomics, and especially proteomics, will continue to unveil myriad new biomechanisms applicable to modification for therapeutic gain and applicable to better understanding processes associated with ADMET.
- Biomechanistic systems lending themselves to crystallization followed by x-ray diffraction will be readily exploited by structure-based drug design.
- BIOTECH will continue to evolve HTS/UHTS methodologies to assay therapeutic mechanisms and to assess ADMET properties as a front-line initiative.
- For cases where structure-based drug design is not possible, combinatorial libraries along with both wild and genetically altered or elicited collections of natural materials will provide initial hits, which will then be followed up by directed libraries and by ligand-based drug design.
- Accumulating data in all areas will become well managed via a variety of databases, including, in particular, an evolved level of 3D sophistication within chemoinformatics. The latter will ultimately provide the common language for the integration of information across all databases.
- As ADMET information accumulates from preclinical models, which will indeed become validated relative to the human case, distinct structural motifs will arise that can be used with statistically derived confidence limits in a predictive manner during drug design.
- UHTS of man-made libraries, directed analyses of nature's mixtures, and importantly, rational exploitation of the newly elaborated and fully understood ADMET structural motifs will reveal synergistic combinations that involve more than one compound.
- For each case of new drug design, a summation of the accumulated knowledge from efficacy testing, ADMET, and potential synergistic relationships will guide final tailoring of the clinical candidate drug or multidrug combination. Whereas a reductionist approach toward achieving the most simple single-molecular system possible will remain appealing, any combination of one or more agents wherein each member can independently use prodrug, soft drug, and multivalent strategies will be considered within the context of deploying whatever is envisioned ultimately to work the best for the clinical indication at hand.
- As clinical successes unfold, a gradual move from experimental to virtual screening will occur. Although this is already being done for initial lead finding by docking druglike virtual compound libraries into pockets defined by x-ray, virtual methods should eventually be able to replace the entire preclinical testing paradigm.
- Paliative and curative targets will remain, the latter particularly for combating the ever-evolving populations of microorganisms and viruses, and the former for retaining the quality of life as the human lifespan continues to elongate. However, preventive and prophylactic treatment paradigms will gradually take on more significance and will eventually reside at the highest priority of life sciences research.
- For all treatments, and especially for preventive and prophylactic paradigms, pharmacogenetics will define population subgroups that will then receive treatment protocols optimized for their individuality relative to the particular treatment that is being rendered. Testing to ascertain an individual's pharmacogenetic profile will begin at birth and continue at periodic intervals throughout one's life relative to the optimal deployment of preventive protocols. Such testing will also be conducted immediately prior to any treatment of a detected pathophysiology.
- In the more distant future, the combination of nanotechnologies with bioengineering and biotechnology will allow instillation of devices not only for immediate diagnoses of any deviations from homeostasis, but also of programmable portals for user-friendly drug administration other than via the oral route. Eventually, this same mix of technologies will also allow for instillation of programmable exits for the controlled excretion of drugs with or without the need for metabolism. Such developments will, in turn, cause drug discovery to return primarily to the pursuit of only efficacy and toxicity issues.

(Continued)

TABLE 2.10 *(Continued)*

- Somewhere during gene therapy's complete eradication of defective gene-based disease and bolstering of gene-linked defenses, along with the nanotechnology–bioengineering effort to reconstruct humans relative to improving health, public policies and opinions dealing with what other attributes might be manipulated to enhance the quality of life will need to be clarified by significant input from the nonbasic science disciplines. Thus, the social sciences, humanities, and philosophy fields, along with religion and the lay public at large, should look forward to providing what will soon become desperately needed input into the continued directions that life sciences research is likely to take well before the end of the next century, let alone before the conclusion of the present millennium.

TABLE 2.11 Molecular Attributes of Discovery Libraries, Compound Hits, Lead Structures, and Final Compounds[a]

Discovery libraries

- Size is less important than diversity (with allowance for structurally redundant series, e.g., Me, Et, etc.).
- Diversity is also much more important than druglike properties (e.g., presence of non-druglike members can be extremely useful toward initially probing overall structural space).
- Alternatively, assay likable properties are mandatory [i.e., compound members must be able to be delivered (e.g., solubilized, etc.) according to demands of a given assay].

Compound hits

- Druglike properties are less important than efficacy tolerance [i.e., flexibility for altering structure without altering efficacy (access to neutral regions)].

Initial lead structures

- Individual structures are less important than a detailed description of the pharmacophore in terms of electrostatic surface potentials plus knowledge about structural space that is neutral or intolerable toward modifications relative to the measured biological parameter (e.g., efficacy).
- Nondruglike features that may be present within certain members contributing to the overall pharmacophore should be "red-flagged" but not necessarily ruled out as potential building blocks, while the overall process of merging structural space across all parameters is continued (e.g., efficacy plus ADMET).

Final lead compounds

- Optimal blend of efficacy and druglike properties (nondruglike features now completely removed or adequately modified according to experimentally ascertained criteria that have been validated for their correlation to the clinical response).
- At least one neutral region or prodrug/soft drug option remains such that unanticipated hurdles presenting themselves during further development might still be addressed by additional chemical modification.
- One or more backup compounds having distinctly different molecular scaffolds while still fulfilling the overall ensemble of pharmacophore and druglike patterns.

[a]Notice that while emphasis is placed on defining a given pharmacophore to the maximum possible detail by deemphasizing the use of druglike property parameters as an early filtering mechanism, components of the pharmacophore that are presumed to be undesirable should still be red-flagged as such. ADMET SAR can then be superimposed within the distinct molecular contexts of each identified pharmacophore so as to be deployed more efficiently as a filter and, importantly, in a proactive manner while initial lead structures move toward final lead compounds. This two-step approach will also allow for continued knowledge building within all of the key parameters relative to various therapeutic areas.

thought of as being driven by an ever-increasing accumulation of knowledge within the life science arena. Table 2.11 summarizes some key points relevant to chemical compound categories associated with drug discovery.

2.9.2 Attributes of Drug Discovery Libraries, Compound Hits, and Lead Compounds

Table 2.11 is important because it addresses a noteworthy concern, namely that today's trend to aggressively filter non-druglike compounds out of the initial drug discovery HTS process may work against the accumulation of knowledge that will be vital toward continually improving the overall process. In other words, although it can be argued that a certain efficiency in the production of NCEs might already be obtainable at this juncture by engaging in this type of negative strategy, an overemphasis upon this approach runs the risk of having the drug discovery and development process becoming forever locked into just this point of evolution. Alternatively, Table 2.11 conveys how some of today's trends that pertain to the molecular attributes of discovery libraries, compound hits, and lead compounds, might best be deployed so as to allow the future developments conveyed in Table 2.10 to be derived from a truly solid base of accumulating knowledge.

At the front of the push to move forward remain the trends inspired by biotechnology. HTS has already led to the situation where there are now mountains of in vitro data available for input toward drug-related considerations. Within just the near term of the new millennium, the entire gamut of ADMET parameters can be expected to join efficacy surveys being conducted by HTS. Importantly, during this period the latter's output will have also become validated in terms of predicting clinical correlates. The common link between these databases will be molecular structure as afforded by the probe compounds or compound library members that become deployed during a given assay. Molecular structure can best be appreciated by the precise language that medicinal chemistry has been learning since its formalization as a distinct discipline about 75 years ago. Thus, medicinal chemistry is also obliged to step to the forefront and assist in understanding and translating what the mountains of new data mean so that they might be optimally applied toward the development of new therapeutic agents.

Recognizing that we have not been very effective to date, the appropriate handling of 3D chemical structure within large databases represents a significant challenge that needs to be resolved by a cooperative effort between medicinal chemists, computational chemists, and both bioinformatic and chemoinformatic database experts as quickly as possible. That medicinal chemists have a good appreciation for the biological nature of the data within one mountain versus that of another is an equally challenging interdisciplinary problem that will need to be addressed by cooperative efforts between medicinal chemists and investigators from all of the biochemical- and biological-related sciences. Resolving both of these challenges will eventually allow the in vitro data sets to be intermeshed so as to provide knowledge-generating assemblies that accurately predict the results that are eventually obtained in vivo and, ultimately, within the clinic.

2.9.3 Formalized Instruction of Medicinal Chemistry

Faced with these immediate critical roles for medicinal chemistry within drug discovery research, how should academia be preparing doctoral-level investigators to contribute as medicinal chemists of the new millennium? First and foremost should be to retain medicinal chemistry's emphasis on the physical organic principles that define chemical behavior in any setting. This is fundamental to being a medicinal chemist. Such principles cannot be learned well by relying only on textbook/e-instruction or even by predesigned laboratory outcome exposures. Thus, a laboratory-based thesis project that involves physical organic principles as the underlying variables of its scientific enquiries seems mandated. While several types of chemical problems might be envisioned to provide such a learning experience, the laboratory practices of synthetic and physical organic chemistry represent extremely useful tools to drive home permanently the principles associated with intra- and intermolecular behavior and chemical reactivity. Similarly, with regard to synthesis/compound production, it is also extremely important to learn first how to isolate and characterize pure materials. Combinatorial mixtures and biochemical manipulations that rely on chemical kinetics and the process of natural selection to dictate their concentrations can then be better appreciated if approached at a later point in time. Finally, while a multistep synthesis of a complex natural product can instill fundamental chemical principles, it may be more effective for a budding medicinal chemist to prepare one or more series of probes wherein most of the members in the series are novel in structure but are (seemingly) still readily obtainable via reasonably close literature precedent according to short synthetic sequences (e.g., five or six steps to each template that is to be further derivatized). This positions the student closer to eventually appreciating structural trends and patterns that may reside within databases.

While this solid foundation is being derived from experimental lab work, graduate-level exposures to various other fields and aspects of life science research will, instead, have to rely on available courses, seminars, or independent reading. Merging a student's chemical learning base with a specific biological area being targeted by the student's molecular probes, however, ought to be feasible via actual experimentation without jeopardizing either subject's rigor. In the end, however, continuing postgraduate education is probably the only way that an investigator intending to practice medicinal chemistry will even be able to purview the explosion of information occurring in all of the areas relevant to assuming the roles needed to resolve the aforementioned challenges of the new millennium. That a practitioner may be able to have a head start along this learning path by initially pursuing a formalized medicinal chemistry curriculum rather than an organic chemistry curriculum has recently been suggested by others.[416]

It should be emphasized that the broader exposure to the life sciences is a critical component for a medicinal chemist's continuing education not because medicinal chemists should eventually attempt to pursue such endeavors independently but because these exposures will allow them to interact more intelligently and meaningfully and collaborate with dedicated experts in each of the numerous other fields. Thus, the ability of medicinal chemists to participate in interdisciplinary research while serving as scientific "scholars" during their attempts to integrate knowledge across broad sets of data and scientific fields[417] are key operational behaviors that also need to be instilled early in the overall graduate-level educational process. By their very nature, medicinal chemistry experiments often prompt fundamental questions or hurdles that may be related to a variety of other disciplines. For example, while in pursuit of dopamine receptor ligands having a cyclopropyl template (Scheme 2.10), it became necessary to devise a new chemical method for effecting the Curtius conversion under neutral conditions while preserving benzyl-protecting groups that were located on the catechol moiety.[418] A similar chemical hurdle pertaining to the formation of β-acetoxyoxetane systems relative to the paclitaxel studies was likewise encountered[239,240] as part of an example cited earlier. From the biological side, one of the first questions that a medicinal chemist cannot help but ponder immediately upon entry into the paclitaxel arena[175] is whether or not an endogenous material also exists, perhaps similar to but distinct from the microtubule-associated proteins, which normally interacts with the paclitaxel receptors purported to reside on intact microtubules in a stoichiometric manner.[419] Similar biological questions pertain to the results from probing the SAR associated with the Pgp MDR system. The latter appear to place the investigators in the middle of a "one versus two distinct Pgp binding sites" controversy.[420] In this case, further experimental clarification of this situation may eventually allow these medicinal chemists to pursue site-directed selective Pgp ligands that are less prone to affect normal cells than cancer cells, or to pursue bivalent superligands, well before the details of this controversy have become fully defined and resolved by genomic and proteomic approaches. Thus, *the interplay of the subject matter from various biological disciplines during medicinal chemistry research is as inherent in the broader medicinal chemistry intellectual process and notion of scholarship as is the practical requisite for a solid-based knowledge of fundamental physical organic principles.*

In the future, increasing numbers of formalized short programs pertaining to a given biological area are likely to be offered to practicing medicinal chemists at technical meetings, academic centers, home cites, and via e-instruction.[421] Given the interdisciplinary nature of the problems already at hand, along with the proposition that they will become significantly more complex as we progress further into the new millennium, it is likely that companies that encourage such interdisciplinary types of continuing education will also eventually become the leaders that are able to most effectively implement the new paradigm of drug discovery (i.e., not just toward generating more data faster while working on smaller scale, but toward producing knowledge systems that actually lead to better NCEs at a quicker pace while spending less money).

2.9.4 Intellectual Property Considerations

Before closing this chapter it also becomes appropriate to consider how all of the aforementioned technical and operational possibilities could affect where medicinal chemistry may be headed in terms of pharmaceutical IP.[422,423] Comments in this area will be directed only toward small-molecule compounds and not toward biomolecules, despite the noted turmoil that was initially created in the gene-related arena.[424,425] As indicated earlier, the highly interdisciplinary nature of today's life science research endeavors, coupled with the new paradigm in drug discovery, indicates that the small-molecule composition of matter arena is no longer the exclusive domain of medicinal chemistry. Nevertheless, even though the appropriate list of inventors for any given case that has utilized HTS and combinatorial chemistry could become quite complex, with patience these situations should all be reconcilable. Alternatively, there are some other issues that are also beginning to hit the IP arena for which answers and appropriate operational models may not be as clear. Given that the desirable goal of enhancing world trade has prompted the need to recognize (if not to completely harmonize) patents on a global basis,[426,427] it is likely that the unique position held by the United States with regard to acknowledging notebook entries as the earliest dates of an invention's conception will ultimately give way to the more practical European process that simply acknowledges the first to file. However, this move will further encourage the filing of patents on technologies that are still very immature. For example, casting this possibility within the trends elaborated throughout this review, companies will need to resist the urge to file on complete compound libraries and instead focus on claims that protect a reasonable family of leads for which several members have indeed been identified as being meritorious by both efficacy and selectivity and at least preliminary ADMET HTS (i.e., experimentally ascertained privileged structures). Unfortunately, an even worse scenario has already begun, in that applications appear to be pending and arguments are being directed toward the validity of patenting huge virtual libraries considered to be druglike in their makeup. Emphasizing the notion that an actual reduction to practice is paramount for a patent, this author presently stands in opposition to the attempts to garner protection of virtual libraries. Along this same line, this author feels compelled to further note that the current motion to assign CAS numbers to virtual compounds also represents a step in the wrong direction. Finally, while patent protection of an existing scaffold that has experimentally demonstrated its utility in one or more therapeutic areas is certainly meritorious, in the future, companies will still need to refrain from overelaborating these same scaffolds in an attempt to generate NCEs across several other therapeutic areas based solely on having already secured IP protection within the context of compositions of matter. In other words, force-fitting a given scaffold via its array of appendage options rather than by conducting an HTS survey of other structural systems across the complete profile of selective efficacy and ADMET parameters could easily be taken as a step backward in terms of both the molecular diversity and therapeutic quality that is ultimately being delivered to the marketplace down the road of the new millennium. A broader discourse on business and scientific ethics at this juncture is beyond the scope of this review even though the rapid biotechnology advances are certainly pressing the need for in-depth discussions in these areas and their fundamental ties with philosophy and religion.[428]

2.9.5 Knowledge Versus Diversity Paradox

In this same regard, however, a seeming paradox will be created by the insertion of knowledge systems into the new drug discovery paradigm. Since medicinal chemistry will seek to define SXR in terms of 3D electrostatic potentials that become predictive of preferred ADMET and efficacy and selectivity behaviors so that their various assemblages can lead to privileged drug structural motifs (or to ensembles of privileged structural motifs that are deployed as drug teams), once this process begins to become effective, it will also play against molecular diversity. Although this situation is not nearly as limiting as the situation conveyed in the preceding paragraph and will always be subject to an expansion of diversity based on the uniqueness of the efficacy pharmacophoric components, enhanced ADMET knowledge in particular will indeed work in a direction away from overall diversity. Hopefully, however, the saving factor in this evolution will remain the pursuit of therapeutically preferred arrangements and not the overutilization of a particular motif just because it has been able to garner an exceptionally favorable ADMET profile, perhaps accompanied by a strong patent position as well. Finally, while enhanced knowledge inherently leads to more credible and useful predictions, it is the overextrapolation, extra weight, or zeal that is sometimes placed on a given prediction versus other options, including that of having no prediction, that can become problematic. Thus, even when all of the challenges cited in this review appear to be resolved, the various disciplines caught up in drug discovery, including that of medicinal chemistry, should all remain cognizant of the earlier days of "preconceived notions" while also recalling the old adage that "a little bit of knowledge can sometimes be dangerous" such that when the ideal drugs and drug ensembles of the near-term future are constructed from experimental data, and those of the more distant future from virtual data, the subsequent lab-based preclinical and clinical investigations will still remain open to the possibility that at any point along the way, anything might still be able to happen. Casting this last sentiment in a favorable direction, medicinal chemists of the future, no matter how knowledgeably and guided by wisdom the overall process of drug discovery may seem to have become, should always remain on the alert for serendipity. Toward this end, the following three quotes, already revived by others relative to recent developments important to medicinal chemistry, have been strung together as an apt closing for this chapter. Each is just as relevant today as it was when it was first pronounced.

> We have scarcely as yet read more than the title page and preface of the great volume of nature, and what we do know is nothing in comparison with that which may be yet unfolded and applied.
>
> —Joseph Henry, more than 100 years ago, as quoted recently by Jacobs[2]

> And if we are indeed to go forth and "...see further, then it will be by standing on the shoulders of the giants..." as well as on the shoulders of the many others like us who have gone before and who have thus brought us to where we are now.
>
> —Isaac Newton, as quoted recently by Wedin[429] and as modified in tense so as to be used within the present context

> For "...we shall not cease from exploration. And the end of all of our exploring will be to arrive where we started, and know the place for the first time."
>
> —T. S. Eliot, as recently quoted by the International Human Genome Sequencing Consortium while concluding their report on the analysis of a substantially complete version of the human genome sequence,[430] a historic accomplishment also reported recently in a similar manner by a Celera Genomics–led consortium[431]

ACKNOWLEDGMENTS

The secretarial assistance of Mrs. Pam Hennen is gratefully acknowledged.

REFERENCES AND NOTES

1. R. Breslow. *Chem. Eng. News*, Nov. 20, 5 (2000).
2. M. Jacobs. *Chem. Eng. News*, Dec. 6, 5 (1999).
3. M. Jacobs, A. Newman, J. Ryan (Eds.). *The Pharmaceutical Century*, American Chemical Society, Washington, DC (2000).
4. J. Steiner. *Drug Discov. World* **1**(2), 61–66 (2000); **2**(1), 55–61 (2000/1).
5. H. O. Alpar, K. R. Ward, E. D. Williamson. *STP Pharm. Sci.* **10**(4), 269–278 (2000).
6. R. A. Hudson and C. D. Black. *Biotechnology—The New Dimension in Pharmacy Practice: A Working Pharmacist's Guide*, 2nd ed., ACPE Publications, Columbus, OH (1999).
7. C. M. Henry. *Chem. Eng. News*, July 3, 41–47 (2000).
8. U. Holzgrabe, I. Wawer, B. Diehl (Eds.). *NMR Spectroscopy in Drug Development and Analysis*, Wiley-VCH, New York (1999).
9. W. Maas. *Drug Discov.*, Aug./Sept., 87 (2000).
10. M. Mayer and B. Meyer. *J. Med. Chem.* **43**, 2093–2099 (2000).
11. P. J. Hajduk, S. Boyd, D. Nettesheim, V. Nienaber, J. Severin, R. Smith, D. Davidson, T. Rockway, S. W. Fesik. *J. Med. Chem.* **43**, 3862–3866 (2000).
12. K. Tang, D.-J. Fu, D. Julien, A. Braun, C. R. Cantor, H. Koster. *Proc. Natl. Acad. Sci. USA* **96**, 10016–10020 (1999).

13. R. H. Griffey, S. A. Hofstadler, K. A. Sannes-Lowery, D. J. Ecker, S. T. Crooke. *Proc. Natl. Acad. Sci. USA* **96**, 10129–10133 (1999).
14. G. Karet. *Drug Discov. Develop.*, Oct., 38–40 (2000).
15. M. L. C. Montanari, A. E. Beezer, C. A. Montanari, D. Pilo-Veloso. *J. Med. Chem.* **43**, 3448–3452 (2000).
16. M. L. C. Montanari, A. E. Beezer, C. A. Montanari. *Thermochim. Acta* **328**, 91–97 (1999).
17. L. Stolz. *Annu. Rep. Med. Chem.* **33**, 293–299 (1998).
18. E. Danelian, A. Karlen, R. Karlsson, S. Winiwarter, A. Hansson, S. Lofas, H. Lennernas, M. Hamalainen. *J. Med. Chem.* **43**, 2083–2086 (2000).
19. A. Burger (Ed.). *Medicinal Chemistry*, 3rd ed., Part 1, Wiley-Interscience, New York (1970) (first published in 1951).
20. C. O. Wilson, O. Gisvold, R. F. Doerge (Eds.). *Textbook of Organic Medicinal and Pharmaceutical Chemistry*, 6th ed., Lippincott, Philadelphia (1971) (first published in 1949).
21. J. H. Block, E. B. Roche, T. O. Soine, C. O. Wilson (Eds.). *Inorganic Medicinal and Pharmaceutical Chemistry*, Lee & Febiger, Philadelphia (1974).
22. P. W. Erhardt. Drug metabolism data: past, present and future considerations, in *Drug Metabolism: Databases and High-Throughput Testing During Drug Design and Development*, P. W. Erhardt (Ed.), pp. 2–15, IUPAC/Blackwell, Geneva (1999).
23. P. W. Erhardt. Esmolol, in *Chronicles of Drug Discovery*, D. Lednicer (Ed.), Vol. 3, pp. 191–206, American Chemical Society, Washington, DC (1993).
24. M. C. Venuti. *Annu. Rep. Med. Chem.* **25**, 289–298 (1990).
25. K. R. Oldenburg. *Annu. Rep. Med. Chem.* **33**, 301–311 (1998).
26. D. J. Ausman. *Mod. Drug Discov.*, Jan., 18–23 (2001).
27. M. J. Felton. *Mod. Drug Discov.*, Jan., 24–28 (2001).
28. P. J. Hylands and L. J. Nisbet. *Annu. Rep. Med. Chem.* **26**, 259–269 (1991).
29. W. J. Dower and S. P. Fodor. *Annu. Rep. Med. Chem.* **26**, 271–280 (1991).
30. W. H. Moos, G. D. Green, M. R. Pavia. *Annu. Rep. Med. Chem.* **28**, 315–324 (1993).
31. M. A. Gallop, R. W. Barrett, W. J. Dower, S. Fodor, E. M. Gordon. *J. Med. Chem.* **37**, 1234–1251 (1994).
32. E. M. Gordon, R. W. Barrett, W. J. Dower, S. Fodor, M. A. Gallop. *J. Med. Chem.* **37**, 1385–1401 (1994).
33. N. K. Terrett. *Combinatorial Chemistry*, American Chemical Society/Oxford University Press, New York (1998).
34. G. L. Trainor. *Annu. Rep. Med. Chem.* **34**, 267–286 (1999).
35. J. N. Kyranos and J. C. Hogan, Jr. *Modern Drug Discov.*, July/Aug., 73–82 (1999).
36. R. K. Brown. *Mod. Drug Discov.*, July/Aug., 63–71 (1999).
37. S. Borman. *Chem. Eng. News*, May 15, 53–65 (2000).
38. E. LeProust, J. P. Pellois, P. Yu, H. Zhang, X. Gao. *J. Comb. Chem.* **2**, 349–354 (2000).
39. D. Scharn, H. Wenschuh, U. Reineke, J. Schneider-Mergener, L. Germeroth. *J. Comb. Chem.* **2**, 361–369 (2000).
40. R. E. Dolle. *Mod. Drug Discov.*, Feb., 43–48 (2001) [the complete survey article, which appeared in *J. Comb. Chem.* **2**(5), 383–433 (2000) and includes 290 references, 10 tables, and 26 figures, is also available on the Web at http://pubs.acs.org/hotarticl/jcchff/2000/cc000055x_rev.html].
41. J. Boguslavsky. *Drug Discov. Develop.*, Sept., 71 (1999).
42. G. Karet. *Drug Discov. Develop.*, Mar., 28–32 (2000).
43. D. C. U'Prichard. *Drug Discov. World*, Fall, 12–22 (2000).
44. Personal communications while visiting several large pharmaceutical companies during 2000.

45. (a) C. D. Strader. *J. Med. Chem.* **39**, 1 (1996); (b) P. S. Portoghese. Alfred Burger award in medicinal chemistry. Molecular recognition at opiod receptors: from concepts to molecules, 219th ACS Natl. Meet., Mar. 26–30, Abstract 302 (2000).
46. T. Maehama, G. S. Taylor, J. T. Slama, J. E. Dixon. *Anal. Biochem.* **279**, 248–250 (2000).
47. J. T. Slama, S. Ramsinghani, D. W. Koh, J.-C. Ame, M. K. Jacobson. Identification of the active site of poly(ADP-ribose)glycohydrolase using photoaffinity labeling and site directed mutagenesis. 20th Midwest Enzyme Chem. Conf., Sept. 23, abstract 73 (2000).
48. J. T. Slama, N. Aboul-Ela, D. M. Goli, B. V. Cheesman, A. M. Simmons, M. K. Jacobson. *J. Med. Chem.* **38**, 389–393 (1995).
49. S. Ramsinghani, D. W. Koh, J.-C. Ame, M. Strohm, M. K. Jacobson, J. T. Slama. *Biochemistry* **37**, 7801–7812 (1998).
50. H. Zeng, R. S. Miller, R. A. Flowers II, B. Gong. *J. Am. Chem. Soc.* **122**, 2635–2644 (2000).
51. X.-P. Huang, F. E. Williams, S. M. Peseckis, W. S. Messer, Jr. *J. Pharmacol. Exp. Ther.* **286**, 1129–1139 (1998).
52. X.-P. Huang, F. E. Williams, S. M. Peseckis, W. S. Messer, Jr. *Am. Soc. Pharmacol. Exp. Ther.* **56**, 775–783 (1999).
53. X.-P. Huang, P. I. Nagy, F. E. Williams, S. M. Peseckis, W. S. Messer, Jr. *Br. J. Pharmacol.* **126**, 735–745 (1999).
54. W. S. Messer, Jr., Y. F. Abuh, Y. Liu, S. Periyasamy, D. O. Ngur, M. A. N. Edgar, A. A. El-Assadi, S. Sbeih, P. G. Dunbar, S. Roknich, T. Rho, Z. Fang, B. Ojo, H. Zhang, J. J. Huzl III, P. I. Nagy. *J. Med. Chem.* **40**, 1230–1246 (1997).
55. W. S. Messer, Jr., Y. F. Abuh, K. Ryan, M. A. Shepherd, M. Schroeder, S. Abunada, R. Sehgal, A. A. El-Assadi. *Drug Develop. Res.* **40**, 171–184 (1997).
56. W. S. Messer, Jr., W. G. Rajeswaran, Y. Cao, H.-J. Zhang, A. A. El-Assadi, C. Dockery, J. Liske, J. O'Brien, F. E. Williams, X.-P. Huang, M. E. Wroblewski, P. I. Nagy, S. M. Peseckis. *Pharm. Acta Helv.* **74**, 135–140 (2000).
57. Y. Cao, M. E. Wroblewski, P. I. Nagy, W. S. Messer, Jr. Synthesis and biological evaluation of bis[3-(3-alkoxy-1,2,5-thiadiazol-4-yl)-1,2,3,4-tetrahydropyrid-1-yl]alkane dihydrochlorides as muscarinic agonists, 9th Int. Symp. Subtypes of Muscarinic Receptors, Nov., abstract (2000).
58. W. S. Messer, Jr., W. G. Rajeswaran, Y. Cao, X.-P. Huang, M. E. Wroblewski, P. I. Nagy. Synthesis and biological evaluation of bivalent xanomeline analogs as M1 muscarinic agonists, 9th Int. Symp. Subtypes of Muscarinic Receptors, Nov., abstract (2000).
59. Y. Cao, X.-P. Huang, W. S. Messer, Jr. *Soc. Neurosci.* **26**, abstract, p. 1912 (2000).
60. W. S. Messer, Jr., F. Liu, Y. Cao, F. E. Williams, X.-P. Huang. *Soc. Neurosci.* **26**, abstract, p. 1912 (2000).
61. P. I. Nagy. *Recent Res. Develop. Phys. Chem.* **3**, 1–21 (1999).
62. P. S. Portoghese. *Am. J. Pharm. Educ.* **63**, 342–347 (1999).
63. M. Jacobs. *Chem. Eng. News*, Oct. 30, 43–49 (2000).
64. H. Kubinyi. *Curr. Opin. Drug Discov. Develop.* **1**(1), 4–15 (1998).
65. M. A. Murcko, P. R. Caron, P. S. Charifson. *Annu. Rep. Med. Chem.* **34**, 297–306 (1999).
66. K. Rubenstein. *Drug Market Develop.*, Report 1993, 1–200 (2000).
67. J. Josephson. *Mod. Drug Discov.*, May, 45–50 (2000).
68. C. Hansch. *Comprehensive Medicinal Chemistry*, Wiley, New York (1990).
69. R. Silverman. *The Organic Chemistry of Drug Design and Drug Action*, Academic Press, New York (1992).
70. M. E. Wolff (Ed.). *Burger's Medicinal Chemistry and Drug Discovery*, 5th ed., Vol. 1, Wiley, New York (1994).

71. W. Foye, T. Lemke, D. Williams (Eds.). *Principles of Medicinal Chemistry*, 4th ed., Williams & Wilkins, Baltimore (1995). During the last stages of this manuscript's review, the author was pleased to see that the latest edition (5th) of this text has explicitly incorporated an earlier IUPAC definition for *medicinal chemistry*.
72. A. Gringauz. *Introduction to Medicinal Chemistry: How Drugs Act and Why*, Wiley, New York (1997).
73. H. J. Smith (Ed.). *Smith and Williams' Introduction to the Principles of Drug Design and Action*, 3rd ed., Harwood, Amsterdam (1998).
74. D. E. Stokes. *Pasteur's Quadrant: Basic Science and Technological Innovation*, Brookings Press, Washington, DC (1997).
75. F. A. Cotton. *Chem. Eng. News*, Dec. 4, 5 (2000).
76. T. Flores, D. Garcia, T. McKenzie, J. Pettigrew. In *Innovations in Pharmaceutical Technology*, P. A. Barnacal (Ed.), pp. 18–22, Chancery Media., London (2000).
77. T. Studt. *Drug Discov. Develop.* Mar., 7 (2000).
78. E. Hodgkin and K. Andrews-Cramer. *Modern Drug Discov.*, Jan./Feb., 55–60 (2000).
79. N. Sleep. *Mod. Drug Discov.*, July/Aug., 37–42 (2000).
80. G. Karet. *Drug Discov. Develop.*, Nov./Dec., 39–43 (1999).
81. J. Boguslavsky. *Drug Discov. Develop., Mar.*, 67 (2000).
82. S. Young and J. Li. *Drug Discov. Develop.*, Apr., 34–37 (2000).
83. M. B. Brennan. *Chem. Eng. News*, June 5, 63–73 (2000).
84. E. Estrada, E. Uriarte, A. Montero, M. Teijeira, L. Santana, E. De Clercq. *J. Med. Chem.* **43**, 1975–1985 (2000).
85. H.-J. Boehm, M. Boehringer, D. Bur, H. Gmuender, W. Huber, W. Klaus, D. Kostrewa, H. Kuehne, T. Luebbers, N. Meunier-Keller, F. Mueller. *J. Med. Chem.* **43**, 2664–2674 (2000).
86. C. Bissantz, G. Folkers, D. Rognan. *J. Med. Chem.* **43**, 4759–4767 (2000).
87. S. Borman. *Chem. Eng. News*, Dec. 18, 24–31 (2000).
88. S. Borman. *Chem. Eng. News*, Nov. 15, 55–57 (1993).
89. D. M. Spencer, T. J. Wandess, S. L. Schreiber, G. R. Crabtree. *Science* **262**, 1019–1024 (1993).
90. A. C. Bishop, J. A. Ubersax, D. T. Petsch, D. P. Matheos, N. S. Gray, J. Blethrow, E. Shimizu, J. Z. Tsien, P. G. Schultz, M. D. Rose, J. L. Wood, D. O. Morgan, K. M. Shokat. *Nature* **407**, 395–401 (2000).
91. I. Wickelgren. *Science* **285**, 998–1001 (1999).
92. R. Pettipher. In *Innovations in Pharmaceutical Technology*, P. A. Barnacal (Ed)., pp. 62–67, Chancery Media, London (2000).
93. M. Swindells. In *Innovations in Pharmaceutical Technology*, P. A. Barnacal (Ed.), pp. 69–74, Chancery Media, London (2000).
94. J. Boguslavsky. *Drug Discov. Develop.*, June/July, 11–12 (2000).
95. S. Borman. *Chem. Eng. News*, Sept. 11, 6–7 (2000).
96. R. Willis and M. J. Felton. *Modern Drug Discov.*, Jan., 31–36 (2001).
97. P. Elmer-DeWitt. *Time* **157**(2), 57 (2001).
98. M. D. Lemonick. *Time* **157**(2), 58–69 (2001).
99. National Human Genome Research Institute at www.nhgri.nih.gov; National Center for Biotechnical Information in the NIH at www.ncbi.nlm.nih.gov; Human Genome Project via Gene Bank at www.ncbi.nlm.nih.gov/genbank; and European Bioinformatics Institute at www.ebi.ac.uk; as well as the home pages for the 16 institutions participating in the Human Genome Sequencing Consortium.

100. AxCell Bioscience Corp. at www.axcellbio.com; BIND at http://bioinfo.mshri.on.ca; Celera Genomics at www.celera.com; Curagen Corp. at www.curagen.com; Gene Logic at www.genelogic.com; Geneva Bioinformatics SA at www.genebio.com; Genome Database at www.gdb.org; Genome Therapeutics Corp. at www.genomecorp.com; Hybrigenics SA at www.hybrigenics.com; Incyte Genomics, Inc. at www.incyte.com; Lab On Web at www.labonweb.com; Life Span BioSciences at www.lsbio.com; PRESAGE at http://csb.Stanford.edu; Protein Data Bank at www.rcsb.org/pdb/; Proteome, Inc. at www.proteome.com; Structural Bioinformatics, Inc. at www.strubix.com; and Swiss Institute of Bioinformatics at www.isbsib.ch/.

101. J. Buguslavksy. *Drug Discov. Develop.*, Mar., 52–54 (2000).

102. C. S. W. Koehler. *Mod. Drug Discov.*, Nov./Dec., 75–76 (2000).

103. J. M. Nash. *Time* **157**(2), 90–93 (2001).

104. W. E. Evans and M. V. Relling. *Science* **286**, 487–491 (1999).

105. D. S. Bailey and P. M. Dean. *Annu. Rep. Med. Chem.* **34**, 339–348 (1999).

106. K. F. Lau and H. Sakul. *Annu. Rep. Med. Chem.* **35**, 261–269 (2000).

107. R. Pettipher and R. Holford. *Drug Discov. Develop.*, Jan./Feb., 53–54 (2000).

108. M. P. Murphy. *Pharmacogenomics* **1**(2), 115–123 (2000).

109. M. P. Murphy. *Drug Discov. World* **1** (2), 23–32 (2000).

110. G. Karet, J. Boguslavsky, T. Studt. *Drug Discov. Develop.*, Nov./Dec., 55–514 (2000).

111. J. Boguslavsky. *Drug Discov. Develop.*, Nov./Dec., 521–524 (2000).

112. Curagen at www.curagen.com; Gene Logic at www.genelogic.com; Genome Therapeutics Corp. at www.genomecorp.com; Immunogenetics Database at http://imgt.cnusc.fr:8104/; Signalling Pathway Database at www.gvt.kyushu-u.ac.jpl/spad/.

113. The National Center for Biotechnical Information at www.ncbi.nlm.nih.gov/SNP/; the SNP Consortium at http://snp.cshl.org/; as well as the home pages for the several organizations participating in or supporting the SNP Consortium.

114. Affymetrix, Inc. at www.affymetrix.com; Applera Corp. at www.applera.com; Applied Biosystems at www.pebio.com/ab; Molecular Devices Corp. at www.moldev.com; Orchid BioSciences Inc. at www.orchid.com; QIAGEN Genomics Inc. at www.qiagengenomics.com; Sequenom Inc. at www. sequenom.com; Third Wave Technologies at www.twt.com.

115. J. A. Halperin. *The Standard* (U.S. Pharmacopeia newsletter), Mar./Apr., 2 (2000).

116. C. M. Henry. *Chem. Eng. News*, Oct. 23, 85–100 (2000).

117. L. P. Rivory, H. Qin, S. J. Clarke, J. Eris, G. Duggin, E. Ray, R. J. Trent, J. F. Bishop. *Eur. J. Clin. Pharmacol.* **56**, 395–398 (2000).

118. O. Zelenkova, E. Hadasova, E. Ceskova, M. Vojtiskova, M. Hyksova. *Human Psychopharmacol. Clin. Exp.* **15**, 303–305 (2000).

119. L. Bellavance, J. Burbaum, D. Dunn. In *Innovations in Pharmaceutical Technology*, P. A. Barnacal (Ed.), pp. 12–17, Chancery Media, London (2000).

120. J. Burbaum, C. Dilanni-Carrol, M. Traversari. *Biotech. Int.* **II**, 213–216 (1999).

121. J. Harness. In *Innovations in Pharmaceutical Technology*, P. A. Barnacal (Ed.), pp. 37–45, Chancery Media, London (2000).

122. G. Karet. *Drug Discov. Develop.*, June/July, 34–40 (2000).

123. J. Boguslavsky. *Drug Discov. Develop.*, Aug./Sept., 51–54 (2000).

124. J. Boguslavsky. *Drug Discov. Develop.*, Oct., 54–58 (2000).

125. J. Gibson and G. Karet. *Drug Discov. Develop.*, June/July, 61–64 (2000).

126. J. A. Endicott and V. Ling. *Annu. Rev. Biochem.* **58**, 137–171 (1989).

127. M. M. Gottesman and I. Pastan. *Annu. Rev. Biochem.* **62**, 385–427 (1993).

128. G. L. Scheffer, M. Kool, M. Heijn, M. deHaas, A. Pijnenborg, J. Wijnholds, A. van Helvoort, M. C. de Jong, J. H. Hooijberg, C. Mol, M. van der Linden, J. de Vree, P. van der Valk, R. Elferink, P. Borst, R. J. Scheper. *Cancer Res.* **60**, 5269–5277 (2000).
129. J. D. Allen, R. F. Brinkhuis, L. van Deemter, J. Wijnholds, A. Schinkel. *Cancer Res.* **60**, 5761–5766 (2000).
130. S. Gupta and T. Tsuruo (Eds.). *Multidrug Resistance in Cancer Cells: Molecular, Biochemical, Physiological and Biological Aspects*, Wiley, Chichester, UK (1996).
131. J. A. Moscow, E. Schneider, S. P. Ivy, K. H. Cowan. *Cancer Chemother. Biol. Response Modif.* **17**, 139–177 (1997).
132. M. M. Gottesman, I. Pastan, S. V. P. Ambudkar. *Curr. Opin. Genet. Develop.* **6**, 610–617 (1996).
133. P. Gros, J. Croop, D. Housman. *Cell* **47**, 371–380 (1986).
134. S. P. Cole, G. Bhardwaj, J. H. Gerlach, J. E. Mackie, C. E. Grant, K. C. Almquist, A. J. Stewart, E. U. Kurz, A. M. Duncan, R. G. Deeley. *Science* **258**, 1650–1654 (1992).
135. L. A. Doyle, W. Yang, L. V. Abruzzo, T. Krogmann, Y. Gao, A. K. Rishi, D. D. Ross. *Proc. Natl. Acad. Sci. USA* **95**, 15665–15670 (1998).
136. C. Ramachandran and S. J. Melnick. *Mol. Diagn.* **4**, 81–94 (1999).
137. Values reported in the NIH/NCI database at http://dtp.nci.nih.gov/docs/cancer/searches/standard_mechanism_list.html#topo1. Similar values are obtained in our labs using MCF-7 versus NCI/ADR-RES cell lines (unreported data).
138. M. E. Wall. In *Chronicles of Drug Discovery*, D. Lednicer (Ed.), pp. 327–348, American Chemical Society, Washington, DC (1993).
139. M. Suffness. *Annu. Rep. Med. Chem.* **28**, 305–314 (1993).
140. G. George, S. M. Ali, J. Zygmunt, L. R. Jayasinghe. *Exp. Opin. Ther. Patents* **4**(2), 109–120 (1994).
141. K. C. Nicolaou, W.-M. Dai, R. K. Guy. *Angew. Chem. Int. Ed. Engl.* **33**, 15–44 (1994).
142. G. I. Georg, T. Chen, I. Ojima, D. M. Vyas (Eds.) *Taxane Anticancer Agents: Basic Science and Current Status*, American Chemical Society, Washington, DC (1995).
143. P. Sonneveld and E. Wiemer. *Curr. Opin. Oncol.* **9**, 543–548 (1997).
144. V. Ling. *Cancer Chemother. Pharmacol.* **40**, S3–S8 (1997).
145. G. A. Fisher and B. I. Sikic. *Hematol. Oncol. Clin. North Am.* **9**, 363–382 (1995).
146. J. Sarver, P. Erhardt, W. Klis, J. Byers. Modification of anticancer drugs to avoid multiple drug resistance, SGK 2000 Reaching for the Cure/Making a Difference Mission Conf., Sept. (2000).
147. W. D. Kingsbury, J. C. Boehm, D. R. Jakas, K. G. Holden, S. M. Hecht, G. Gallagher, M. J. Caranfa, F. L. McCabe, L. F. Faucette, R. K. Johnson, R. P. Hertzberg. *J. Med. Chem.* **34**, 98–107 (1991).
148. L. Liu. *Adv. Pharmacol.* **29B**, 1–298 (1994).
149. B. Sinha. *Drugs* **49**(1), 11–19 (1995).
150. P. Pantazis, B. Giovanella, M. Rothenberg (Eds.). *Ann. N.Y. Acad. Sci.* **803**, 1–328 (1996).
151. C. Jaxel, K. Kohn, M. Wani, M. Wall, Y. Pommier. *Cancer Res.* **49**, 1465–1469 (1989).
152. M. Wall, M. Wani, A. Nicholas, G. Manikumar, C. Tele, L. Moore, A. Truesdale, P. Leitner, J. Besterman. *J. Med. Chem.* **36**, 2689–2700 (1993).
153. Y. Pommier. *Semin. Oncol.* **23**, 3–10 (1996).
154. M. Wall and M. Wani. *Ann. N.Y. Acad. Sci.* **803**, 1–12 (1996).
155. D. Uehling, S. Nanthakumar, D. Croom, D. Emerson, P. Leitner, M. Luzzio, G. McIntyre, B. Morton, S. Profeta, J. Sisco, D. Sternbach, W.-Q. Tong, A. Vuong, J. Besterman. *J. Med. Chem.* **38**, 1106–1118 (1995).

156. S. Sawada, S. Matsuoka, K. Nokata, H. Nagata, T. Furuta, T. Yokokura, T. Miyasaka. *Chem. Pharm. Bull.* **39**, 3183–3188 (1991).

157. M. Sugimori, A. Ejima, S. Ohsuki, K. Uoto, I. Mitsui, K. Matsumoto, Y. Kawato, M. Yasuoka, K. Sato, H. Tagawa, H. Terasawa. *J. Med. Chem.* **37**, 3033–3039 (1994).

158. B. Giovanella, J. Stehlin, M. Wall, M. Wani, A. Nicholas, L. Liu, R. Silber, M. Potmesil. *Science* **246**, 1046–1048 (1989).

159. H. Hinz, N. Harris, E. Natelson, B. Giovanella. *Cancer Res.* **54**, 3096–3100 (1994).

160. H. Wang, S. Liu, K. Hwang, G. Taylor, K.-H. Lee. *Bioorg. Med. Chem.* **2**, 1397–1402 (1994).

161. M. Luzzio, J. Besterman, D. Emerson, M. Evans, K. Lackey, P. Leitner, G. McIntyre, B. Morton, P. Myers, M. Peel, J. Sisco, D. Sternbach, W.-Q. Tong, A. Truesdale, D. Uehling, A. Vuong, J. Yates. *J. Med. Chem.* **38**, 395–401 (1995).

162. K. Lackey, J. Besterman, W. Fletcher, P. Leitner, B. Morton, D. Sternbach. *J. Med. Chem.* **38**, 906–911 (1995).

163. K. Lackey, D. Sternbach, D. Croom, D. Emerson, M. Evans, P. Leitner, M. Luzzio, G. McIntyre, A. Vuong, J. Yates, J. Besterman. *J. Med. Chem.* **39**, 713–719 (1996).

164. A. Nicholas, M. Wani, G. Manikumar, M. Wall, K. Kohn, Y. Pommier. *J. Med. Chem.* **33**, 972–978 (1990).

165. R. Crow and D. Crothers. *J. Med. Chem.* **35**, 4160–4164 (1992).

166. C. Jaxel, K. W. Kohn, M. C. Wani, M. E. Wall, Y. Pommier. *Cancer Res.* **49**, 1465–1469 (1989).

167. Y. Pommier, C. Jaxel, C. Heise. Structure–activity relationship of topoisomerase I inhibition by camptothecin derivatives: evidence for the existence of a ternary complex, in *DNA Topoisomerases in Cancer*, M. Potmesil and K. Kohn (Eds.), pp. 121–132, Oxford University Press, New York (1991).

168. K. Lackey, J. Besterman, W. Fletcher, P. Leitner, B. Morton, D. Sternbach. *J. Med. Chem.* **38**, 906–911 (1995).

169. M. Sugimori, A. Ejima, S. Ohsuki, K. Uoto, I. Mitsui, K. Matsumoto, Y. Kawato, M. Yasuoka, K. Sato, H. Tagawa, H. Terasawa. *J. Med. Chem.* **37**, 3033–3039 (1994).

170. S. Carrigan, P. Fox, M. Wall, M. Wani, J. Bowen. *J. Comput.-Aided Mol. Des.* **11**, 71–78 (1997).

171. D. J. Rodi, R. W. Janes, H. J. Sanganee, R. A. Holton, B. A. Wallace, L. Makowski. *J. Mol. Biol.* **285**, 197–203 (1999).

172. C.-P. Yang and S. B. Horwitz. *Cancer Res.* **60**, 5171–5178 (2000).

173. N. Andre, D. Braguer, G. Brasseur, A. Goncalves, D. Lemesle-Meunier, S. Guise, M. A. Jordan, D. Briand. *Cancer Res.* **60**, 5349–5353 (2000).

174. P. B. Schiff, J. Fant, S. B. Horwitz. *Nature* **277**, 665–667 (1979).

175. P. W. Erhardt. *The Taxane J.* **3**, 36–42 (1997).

176. A. Nakashio, N. Fujita, S. Rokudai, S. Sato, T. Tsuruo. *Cancer Res.* **60**, 5303–5309 (2000).

177. S. Froelich-Ammon and N. Osheroff. *J. Biol. Chem.* **270**, 21429–21432 (1995).

178. W. Moos, G. D. Green, M. R. Pavia. *Annu. Rep. Med. Chem.* **28**, 315–324 (1993).

179. D. C. Spellmeyer and P. D. J. Grootenhuis. *Annu. Rep. Med. Chem.* **34**, 287–296 (1999).

180. A. Dorfman. *Time* **157**(2), 97–99 (2001).

181. J.-S. Huang and K. Barker. *Plant Physiol.* **96**, 1302–1307 (1991).

182. J. J. Cheong, R. Alba, F. Cote, J. Enkerli, M. G. Hahn. *Plant Physiol.* **103**, 1173–1182 (1993).

183. J. Faghihi, X. Jiang, R. Vierling, S. Goldman, S. Sharfstein, J. Sarver, P. Erhardt. *J. Chromatogr.* **A915**, 61–74 (2001).

184. S. Borman. *Chem. Eng. News*, Oct. 30, 35–36 (2000).

185. V. Gruber. *Annu. Rep. Med. Chem.* **35**, 357–364 (2000).

186. Interactive video conference: The role of higher education in economic development, Old Dominion University Academic Television Service, Jan. 31, University of Toledo site (1997).

187. P. W. Erhardt. In *Drug Metabolism Databases and High-Throughput Testing During Drug Design and Development*, P. W. Erhardt (Ed.), pp. viii–ix, IUPAC/Blackwell, Geneva (1999).

188. D. Ausman (Ed.). Data, data, everywhere. *Molecular Connection* (MDL's Newsletter) **18**(2), 3 (1999).

189. A. J. Recupero. *Drug Discov. Develop.*, Nov./Dec., 59–62 (1999).

190. T. Agres. *Drug Discov. Develop.*, Jan./Feb., 11–12 (2000).

191. T. Studt. *Drug Discov. Develop.*, Jan./Feb., 68–69 (2000).

192. S. Zarrabian. *Chem. Eng. News*, Feb. 7, 5 (2000).

193. A.M. Thayer. *Chem. Eng. News*, Feb. 7, 19–32 (2000).

194. S. Tye and G. Karet. *Drug Discov. Develop.*, Mar., 35 (2000).

195. W. Schulz. *Chem. Eng. News*, Feb. 5, 25 (2001).

196. DARPA at www.darpa.mil; FAO at www.fao.org/GENINFO/partner/defautl.htm; Gene Bank at www.ncbi.nlm.nih.gov/entrez/nucleotide.html; DDBJ at www.ddbj.nig.ac.jp/; EMBL at www.ebi.ac.uk/ebi_docs/embl_db/ebi/topembl.html; Human Genome Sequencing at www. ncbi.nlm.nih.gov/genome/seq; KEEGG 12.0 at www.genome.ad. jp/kegg/; NASA at www. nasa.gov; National Academy of Science STEP at www4.nas.edu/pd/step.nsf; National Center for Biotechnical Information at www.ncbi.nlm.nih.gov; National Human Genome Research Institute at www. nhgri.nih.gov/index.html; National Science Foundation Mathematics/ Physical Sciences at www.nsf.gov/home/mps /start.htm; Net Genics SYNERGY at www. netgenics.com/SYNERGY.html; Pfam 4.3 at pfam.wustl.edu/; PROSITE at www3.oup.co.uk/ nar/Volume_27/Issue_01/html/gkc073_ gml.html#hdO; Sanger Center at www.sanger.ac.uk/; SRS at www.embl–heidelberg–de/srs51; TIGER Database at www.tigr.org/tdb/tdb.html; UNESCO at www.unesco.org/general/eng./ programmes/science/life/index.htm; University of California–Berkeley at www.Berkeley.edu.

197. F. K. Brown. *Annu. Rep. Med. Chem.* **33**, 375–384 (1998).

198. M. Hann and R. Green. *Curr. Opin. Chem. Biol.* **3**, 379–383 (1999).

199. P. Ertl, W. Miltz, B. Rohde, P. Selzer. *Drug Discov. World* **1** (2), 45–50 (2000).

200. ChemFinder at www.chemfinder.com; Chem Navigator at www.chemnavigator.com; Info Chem GmbH at www.infochem.com; Institute for Scientific Information at www.isinet.com; MDL Information Systems at www.mdli.com; MSI at www.msi.com; Oxford Molecular at www.oxmol.com; Pharmacopeia at www.pharmacopeia.com; QsarIS at www.scivision. com/ qsaris.html; Synopsys Scientific Systems at www.synopsys.co.uk; Tripos at www.tripos.com.

201. Personal communications while attending several recent workshops directed toward expediting drug development.

202. Protein Data Bank at www.rcsb.org/bdb/.

203. Biotech Validation Suite For Protein Structure at http://biotech.embl–heidelberg.de:8400/; BLAST at www.ncbi.nlm.nih.gov/BLAST/; Build Proteins from Scratch at http://csb.stanford. edu/levitt and www.chem.cornell.edu/department/Faculty/Scheraga/ Scheraga.html; CATH at www.biochem.ucl.ac.uk/bsm/cath/; FASTA at www2.ebi.ac.uk/ fasta3/; MODELLER at http://guitar.rockfeller.edu; PDB at www.pdb.bnl.gov/; Pedant at MIPS at http:pedant.mips. biochem.mpg.del; PIR at www.nbrf.georgetown.edu/ pir/searchdb.html; Predict Protein at www.embl–heidelber.de/predictprotein/ predictprotein.html; PSIPRED at http://insulin. brunel.ac.uk/psipred; Rosetta at http://depts..Washington.edu/bakerpg; SAM–98 at www.cse. ucsc.edu/research/compbio; SCOP at http://scop.mrc–lmb.cam.ac.uk/sopl; Structure at www. ncbi.nlm.nih.gov/Structure/ RESEARCH/res.shtml; Swiss–Model at http://expasy.hcuge. ch/swissmod/ SWISSMODEL.html; SWISS–PROT at http://expasy.hcuge.ch/sprot/sprottop. html; Threader at http://insulin.brunel.ac.uk/threader/threader.html.

204. G. E. Schulz and R. H. Schirmer. In *Principles of Protein Structure,* C. R. Cantor (Ed.), pp. 108–130, Springer-Verlag, New York (1999).
205. E. K. Wilson. *Chem. Eng. News*, Sept. 25, 41–44 (2000).
206. L. Holliman. *Mod. Drug Discov.*, Nov./Dec., 41–46 (2000).
207. M. J. Felton. *Mod. Drug Discov.*, Nov./Dec., 49–54 (2000).
208. M. Vieth and D. J. Cummins. *J. Med. Chem.* **43**, 3020–3032 (2000).
209. M. A. Kastenholz, M. Pastor, G. Cruciani, E. Haaksma, T. Fox. *J. Med. Chem.* **43**, 3033–3044 (2000).
210. G. Karet. *Drug Discov. Develop.*, Jan., 28–32 (2001).
211. Personal communications while attending several recent scientific conferences in the area of medicinal chemistry.
212. ChemPen 3D at http://home.ici.net/~hfevans/chempen.htm; CS Chem Draw and CS Chem Draw 3D at www.camsoft.com; Corina at www.mol–net.de.
213. A. Nezlin. *Chem. News Commun.* **9**, 24–25 (1999).
214. M. Endres. *Today's Chem. at Work*, Oct., 30–44 (1992).
215. G. Karet. *Drug Discov. Develop.*, Aug./Sept., 42–48 (2000).
216. R. S. Rogers. *Chem. Eng. News*, Apr. 26, 17–18 (1999).
217. C. Hansch and T. Fujita. *J. Am. Chem. Soc.* **86**, 1616–1626 (1964).
218. D. E. Koshland, Jr. and K. E. Neet. *Annu. Rev. Biochem.* **37**, 359–410 (1968).
219. B. Belleau. *Adv. Drug Res.* **2**, 89–126 (1965).
220. QUANTA (system for determining 3D protein structure from x-ray data) and FELIX (system for determining 3D protein structure from NMR data) available from MSI-Pharmacopeia (www.msi.com).
221. J. H. Krieger. *Chem. Eng. News*, Mar. 27, 26–41 (1995).
222. P. W. Erhardt. Epilogue. In *Drug Metabolism Databases and High-Throughput Testing During Drug Design and Development*, P. W. Erhardt (Ed.), p. 320, IUPAC/Blackwell, Geneva (1999).
223. W. J. Hehre, L. Radom, P. vR. Schleyer, J. A. Pople. *Ab Initio Molecular Orbital Theory*, Wiley, New York (1986).
224. A. D. Becke. *J. Chem. Phys.* **98**, 5648–5652 (1993).
225. A. Medek, P. J. Hajduk, J. Mack, S. W. Fesik. *J. Am. Chem. Soc.* **122**, 1241–1242 (2000).
226. M. Rouhi. *Chem. Eng. News*, Feb. 21, 30–31 (2000).
227. (a) B. C. Oostenbrink, J. W. Pitera, M. van Lipzig, J. H. Meerman, W. F. van Gunsteren. *J. Med. Chem.* **43**, 4594–4605 (2000); (b) N. Wu, Y. Mo, J. Gao, E. F. Pai. *Proc. Natl. Acad. Sci. USA* **97**, 2017–2022 (2000); (c) A. Warshel, M. Strajbl, J. Villa, J. Florian. *Biochemistry* **39**, 14728–14738 (2000).
228. J. Devillers and A. T. Balaban (Eds.). *Topological Indices and Related Descriptors in QSAR and QSPR*, Gordon and Breach, Amsterdam (1999).
229. C. Wrotnowski. *Modern Drug Discov.*, Nov./Dec., 46–55 (1999).
230. C. M. Henry. *Chem. Eng. News*, Nov. 27, 22–26 (2000).
231. E. K. Wilson. *Chem. Eng. News*, Nov. 6, 35–39 (2000).
232. Z. Hu. Synthesis of C-13 side chain analogues related to the anticancer agent paclitaxel, M.S. thesis, University of Toledo (1997).
233. Z. Hu and P. W. Erhardt. *Org. Proc. Res. Develop.* **1**, 387–390 (1997).
234. Z. Hu, M. J. Hardie, P. Burckel, A. A. Pinkerton, P. W. Erhardt. *J. Chem. Crystallog.* **29**(2), 185–191 (1999).
235. A. P. Marchand, Y. Wang, C.-T. Ren, V. Vidyasagar, D. Wang. *Tetrahedron* **52**, 6063–6072 (1996).

236. (a) M. Pomerantz and P. H. Hartman. *Tetrahedron Lett.* 991–993 (1968); (b) N. J. Turro and P. A. Wriede. *J. Am. Chem. Soc.* **92**, 320–329 (1970).
237. F. D. Lewis and N. J. Turro. *J. Am. Chem. Soc.* **92**, 311–320 (1970).
238. L. E. Friedrich and P. Y.-S. Lam. *J. Org. Chem.* **46**, 306–311 (1981).
239. W. A. Klis and P. W. Erhardt. *Synth. Commun.* **30**, 4027–4038 (2000).
240. P. W. Erhardt, W. A. Klis, P. I. Nagy, K. Kirschbaum, N. Wu, A. Martin, A. A. Pinkerton. *J. Chem. Crystallogr.* **30**, 83–90 (2000).
241. H. A. Carlson, K. M. Masukawa, K. Rubins, F. D. Bushman, W. L. Jorgensen, R. D. Lins, J. M. Briggs, J. A. McCammon. *J. Med. Chem.* **43**, 2100–2114 (2000).
242. E. K. Bradley, P. Beroza, J. E. Penzotti, P. D. Grootenhuis, D. C. Spellmeyer, J. L. Miller. *J. Med. Chem.* **43**, 2770–2774 (2000).
243. M. Graffner-Nordberg, J. Marelius, S. Ohlsson, A. Persson, G. Swedberg, P. Andersson, S. E. Andersson, J. Aqvist, A. Hallberg. *J. Med. Chem.* **43**, 3852–3861 (2000).
244. R. Garcia-Nieto, I. Manzanares, C. Cuevas, F. Gago. *J. Med. Chem.* **43**, 4367–4369 (2000).
245. A. Vedani, H. Briem, M. Dobler, H. Dollinger, D. R. McMasters. *J. Med. Chem.* **43**, 4416–4427 (2000).
246. M. L. Lopez-Rodriguez, M. J. Morcillo, M. Fernandez, M. L. Rosada, L. Pardo, K. J. Schaper. *J. Med. Chem.* **44**, 198–207 (2001).
247. B. Ladd. *Mod. Drug Discov.*, Jan./Feb., 46–52 (2000).
248. V. Venkatsubramanian. Computer-aided molecular design using neural networks and genetic algorithms, in *Genetic Algorithms in Molecular Modeling*, J. Devillers (Ed.), Academic Press, New York (1996).
249. A. Globus. *Nanotechnology* **10**(3), 290–299 (1999).
250. T. Studt. *Drug Discov. Develop.*, Aug./Sept., 30–36 (2000).
251. R. Resnick. *Drug Discov. Develop.*, Oct., 51–52 (2000).
252. J. Jaen-Oltra, T. Salabert-Salvador, F. J. Garcia-March, F. Perez-Gimenez, F. Tohmas-Vert. *J. Med. Chem.* **43**, 1143–1148 (2000).
253. Bio Reason at www.bioreason.com; Columbus Molecular Software (Lead Scope) at www.columbus- molecular.com; Daylight Chemical Information Systems at www.daylight.com; IBM (Intelligent Miner) at www.ibm.com; Incyte (Life Tools and Life Prot) at www.incyte.com; Molecular Applications (Gene Mine 3.5.1) at www.mag.com; Molecular Simulations at www.msi.com; Oxford Molecular (DIVA 1.1) at www.oxmol.com; Pangea Systems (Gene World 3.5) at www.pangeasystems.com; Pharsight at www.pharsight.com; SAS Institute (Enterprise Miner) at www.sas.com; SGI (Mine Set 3.0) at www.sgi.com; Spotfire (Spotfire Pro 4.0 and Leads Discover) at www.spotfire.com; SPSS (Clementine) at www.spss.com; Tripos at www.tripos. com.
254. M. A. Johnson, E. Gifford, C.-C. Tsai. In *Concepts and Applications of Molecular Similarity*, M. A. Johnson (Ed.), pp. 289–320, Wiley, New York (1990).
255. E. Gifford, M. Johnson, C.-C. Tsai. *J. Comput.-Aided Mol. Des.* **5**, 303–322 (1991).
256. E. M. Gifford, M. A. Johnson, D. G. Kaiser, C.-C. Tsai. *J. Chem. Inf. Comp. Sci.* **32**, 591–599 (1992).
257. E. M. Gifford, M. A. Johnson, D. G. Kaiser, C.-C. Tsai. *Xenobiotica* **25**, 125–146 (1995).
258. E. M. Gifford. Applications of molecular similarity methods to visualize xenobiotic metabolism structure–reactivity relationships, Ph.D. thesis, University of Toledo (1996).
259. K. M. Andrews and R. D. Cramer. *J. Med. Chem.* **43**, 1723–1740 (2000).
260. T. Borowski, M. Krol, E. Broclawik, T. C. Baranowski, L. Strekowski, M. J. Mokrosz. *J. Med. Chem.* **43**, 1901–1909 (2000).
261. E. A. Wintner and C. C. Moallemi. *J. Med. Chem.* **43**, 1933–2006 (2000).

262. D. B. Boyd. *Modern Drug Discov.*, Nov/Dec., 41–48 (1998).

263. Personal communications while visiting several large pharmaceutical companies and while attending recent scientific conferences in the area of medicinal chemistry.

264. B. E. Evans, K. E. Rittle, M. G. Bock, R. M. DiPardo, R. M. Freidinger, W. L. Whitter, G. F. Lundell, D. F. Veber, P. S. Anderson, R. S. L. Chang, V. J. Lotti, D. J. Cerino, T. B. Chen, P. J. Kling, K. A. Kunkel, J. P. Springer, J. Hirshfield. *J. Med. Chem.* **31**, 2235–2246 (1988).

265. A. A. Patchett and R. P. Nargund. *Annu. Rep. Med. Chem.* **35**, 289–298 (2000).

266. M. Pastor, G. Cruciani, I. McLay, S. Pickett, S. Clementi. *J. Med. Chem.* **43**, 3233–3243 (2000).

267. C. Gnerre, M. Catto, F. Leonetti, P. Weber, P.-A. Carrupt, C. Altomare, A. Carotti, B. Testa. *J. Med. Chem.* **43**, 4747–4758 (2000).

268. R. E. Wilcox, W. H. Huang, M. Y. Brusniak, D. M. Wilcox, R. S. Pearlman, M. M. Teeter, C. J. DuRand, B. L. Wiens, K. A. Neve. *J. Med. Chem.* **43**, 3005–3019 (2000).

269. P. R. N. Jayatilleke, A. C. Nair, R. Zauhar, W. J. Welsh. *J. Med. Chem.* **43**, 4446–4451 (2000).

270. R. Wedin. *Modern Drug Discov.*, Sept./Oct., 39–47 (1999).

271. S. H. Yalkowski and S. C. Valvani. *J. Pharm. Sci.* **69**, 912–922 (1980).

272. R. P. Mason, D. G. Rhodes, L. G. Herbette. *J. Med. Chem.* **34**, 869–877 (1991).

273. L. G. Herbette, D. G. Rhodes, R. P. Mason. *Drug Des. Deliv.* **7**, 75–118 (1991).

274. M. H. Abraham, S. Chada, R. Mitchell. *J. Pharm. Sci.* **83**, 1257–1268 (1994).

275. S. C. Basak, B. D. Gute, L. R. Drewes. *Pharm. Res.* **13**, 775–778 (1996).

276. M. Kansy, F. Senner, K. Gubernator. *J. Med. Chem.* **41**, 1007–1010 (1998).

277. K. Palm, K. Luthman, A.-L. Ungell, G. Strandlund, F. Beigi, P. Lundahl, P. Artursson. *J. Med. Chem.* **41**, 5382–5392 (1998).

278. D. A. Smith and H. Van de Waterbeemd. *Curr. Opin. Chem. Biol.* **3**, 373–378 (1999).

279. B. H. Stewart, Y. Wang, N. Surendran. *Annu. Rep. Med. Chem.* **35**, 299–307 (2000).

280. P. Crivori, G. Cruciani, P.-A. Carrupt, B. Testa. *J. Med. Chem.* **43**, 2204–2216 (2000).

281. F. Yoshida and J. G. Topliss. *J. Med. Chem.* **43**, 2575–2585 (2000).

282. F. Lombardo, M. Y. Shalaeva, K. A. Tupper, F. Gao, M. H. Abraham. *J. Med. Chem.* **43**, 2922–2928 (2000).

283. P. Ertl, B. Rohde, P. Selzer. *J. Med. Chem.* **43**, 3714–3717 (2000).

284. W. J. Egan, K. M. Merz, Jr., J. J. Baldwin. *J. Med. Chem.* **43**, 3867–3877 (2000).

285. A. Chait. *Drug Discov. Develop.*, Apr., 63 (2000).

286. L. Pickering. *Drug Discov. Develop.*, Jan., 34–38 (2001).

287. Absorption Systems' BCS Biowaivers at www.absorption.com; ArQule's Pilot at www.argule.com; Camitro Corp's ADME/Tox at www.camitro.com; Cerep SA's BioPrints at www.cerep.fr; Exon Hit Therapeutics SA's Genetic Makeup/Drug Toxicity at www.exonhit.com; LION Bioscience AG's i-Biology at SRS Databases at www.lionbioscience.com; Quintiles' Early ADMET Screening at qkan.busdev@quintiles.com; Schrodinger, Inc's QikProp at http://www.schrodiner.com; Simulations Plus's GastroPlus at www.simulationsplus.com; Trega Biosciences' In Vitro ADME at www.trega.com.

288. G. L. Amidon, P. I. Lee, E. M. Topp (Eds.). *Transport Processes in Pharmaceutical Systems*, Marcel Dekker, New York (1999).

289. G. Zimmer (Ed.). *Membrane Structure in Disease and Drug Therapy*, Marcel Dekker, New York (2000).

290. J. B. Dressman and H. Lennernas (Eds.). *Oral Drug Absorption: Prediction and Assessment*, Marcel Dekker, New York (2000).

291. E. W. Taylor, J. A. Gibbons, R. A. Braeckman. *Pharm. Res.* **14**, 572–577 (1997).

292. J. A. Gibbons, E. W. Taylor, C. M. Dietz, Z.-P. Luo, H. Luo, R. A. Braeckman. In *Drug Metabolism: Databases and High-Throughput Testing During Drug Design and Development*, P. W. Erhardt (Ed.), pp. 71–78, IUPAC/Blackwell, Geneva (1999).
293. M. Rouhi. *Chem. Eng. News*, Jan. 29, 10 (2001).
294. V. Janout, C. DiGiorgio, S. L. Regen. *J. Am. Chem. Soc.* **122**, 2671–2672 (2000).
295. P. A. Wender, D. J. Mitchell, K. Pattabiraman, E. T. Pelkey, L. Steinman, J. B. Rothbard. *Proc. Natl. Acad. Sci. USA* **97**, 13003–13008 (2000).
296. J. B. Rothbard, S. Garlington, Q. Lin, T. Kirschberg, E. Kreides, P. L. McGrove, P. A. Wendy, P. A. Kharari. *Nature Med.* **6**, 1253–1257 (2000).
297. M. Rouhi. *Chem. Eng. News*, Sept. 4, 43 (2000).
298. R. Dagani. *Chem. Eng. News*, Oct. 16, 27–32 (2000).
299. W. Schulz. *Chem. Eng. News*, Oct. 16, 39–42 (2000).
300. P. Zurer. *Chem. Eng. News*, Mar. 12, 12 (2001).
301. K. Park and R. J. Mrsny (Eds.). *Controlled Drug Delivery*, American Chemical Society, Washington, DC (2000).
302. M. Mort. *Med. Drug Discov.* Apr., 30–34 (2000).
303. C. M. Henry. *Chem. Eng. News*, Sept. 18, 49–65 (2000).
304. S. Bracht. *Innovations Pharm. Technol.* **5**, 92–98 (2000).
305. A. Levy. *Innovations Pharm. Technol.* **5**, 100–109 (2000).
306. J. Southall and C. Ellis. *Innovations Pharm. Technol.* **5**, 110–115 (2000).
307. C. Winnips and M. Keller. *Innovations Pharmaceut. Tech.* **6**, 70–75 (2000).
308. C. Winnips. *Drug Discov. World* **2**(1), 62–66 (2000/2001).
309. K. J. Watkins. *Chem. Eng. News*, Jan. 8, 11–15 (2000).
310. M. Jacoby. *Chem. Eng. News*, Feb. 5, 30–35 (2001).
311. C. A. Lipinski, F. Lombardo, B. W. Dominy, P. J. Feeney. *Adv. Drug Deliv. Rev.* **23**, 3–25 (1997).
312. F. Darvas, K. Valko, G. Dorman, P. W. Erhardt, M. T. D. Cronin, I. Szabo. *Biol. Techniques*, Accepted for publication.
313. A. Frostell-Karlsson, A. Remaeus, H. Roos, K. Andersson, P. Borg, M. Hamalainen, R. Karlsson. *J. Med. Chem.* **43**, 1986–1992 (2000).
314. N. Diaz, D. Suarez, T. L. Sordo, K. M. Merz, Jr. *J. Med. Chem.* **44**, 250–260 (2001).
315. P. J. Hajduk, M. Bures, J. Praestgaard, S. W. Fesik. *J. Med. Chem.* **43**, 3443–3447 (2000).
316. M. P. Czech. *CIIT Activities* (Chemical Industry Institute of Toxicology newsletter), May/June, 2–4 (2000).
317. M. A. Dechantsreiter, E. Planker, B. Matha, E. Lohof, G. Holzemann, A. Jonczyk, S. L. Goodman, H. Kessler. *J. Med. Chem.* **42**, 3033–3040 (1999).
318. R. Pasqualini, E. Koivunen, R. Kain, J. Lahdenranta, M. Sakamoto, A. Stryhn, R. A. Ashmun, L. H. Shapiro, W. Arap, E. Ruoslahti. *Cancer Res.* **60**, 722–727 (2000).
319. A. Eberhard, S. Kahlert, V. Goede, B. Hemmerlein, K. H. Plate, H. G. Augustin. *Cancer Res.* **60**, 1388–1393 (2000).
320. L. Wyder, A. Vitaliti, H. Schneider, L. W. Hebbard, D. R. Moritz, M. Wittmer, M. Ajmo, R. Klemenz. *Cancer Res.* **60**, 4682–4688 (2000).
321. M. Matsuda, S.-I. Nishimura, F. Nakajima, T. Nishimura. *J. Med. Chem.* **44**, 715–724 (2001).
322. T. G. Metzger, M. G. Paterlini, P. S. Portoghese, D. M. Ferguson. *Neurochem. Res.* **21**, 1287–1294 (1996).
323. D. L. Larson, R. M. Jones, S. A. Hjorth, T. W. Schwartz, P. S. Portoghese. *J. Med. Chem.* **43**, 1573–1576 (2000).

324. J. L. Marx. *Science* **216**, 283–285 (1982).

325. P. A. Baeuerle and E. Wolf. *Mod. Drug Discov.* Apr., 37–42 (2000).

326. G. W. Geelhoed. *S. Afr. J. Surg.* **26**, 1–3 (1988).

327. P. Ehrlich. *Lancet*, 445–451 (1913).

328. A. Albert. *Selective Toxicity*, Wiley, New York (1960).

329. (a) T. Higuchi and V. Stella (Eds.). *Prodrugs as Novel Drug Delivery Systems*, American Chemical Society, Washington, DC (1975); (b) H. Bundgaard (Ed.). *Design of Prodrugs*, Elsevier, Amsterdam (1985).

330. F. Wood. *The Blade*, Toledo, OH Jan. 25, 3A (2000).

331. K. Howe. *San Francisco Chronicle*, Jan. 21, 5 (2000).

332. J. A. Barnes. *Inpharma* **1185**, May (1999).

333. W.-M. Keung, O. Lazo, L. Kunze, B. Vallee. *Proc. Natl. Acad. Sci. USA* **93**, 4284–4289 (1996).

334. F. Stermitz, P. Lorenz, J. Tawara L. Zenewicz, K. Lewis. *Proc. Natl. Acad. Sci. USA* **97**, 1433–1438 (2000).

335. N. R. Guz and F. R. Stermitz. *J. Nat. Prod.* **63**, 1140–1145 (2000).

336. W. M. Keung, O. Lazo, L. Kunze, B. L. Vallee. *Proc. Natl. Acad. Sci. USA* **92**, 8990–8993 (1995).

337. S. Verma, E. Salamone, B. Goldin. *Biochem. Biophys. Res. Commun.* **233**, 692–696 (1997).

338. M. Mammen, S. K. Choi, G. M. Whitesides. *Angew. Chem. Int. Ed.* 2754–2794 (1998).

339. L. L. Kiessling, L. E. Strong, J. E. Gestwicki. *Annu. Rep. Med. Chem.* **35**, 321–330 (2000).

340. H. Annoura, K. Nakanishi, T. Toba, N. Takemoto, S. Imajo, A. Miyajima, Y. Tamura-Horikawa, S. Tamura. *J. Med. Chem.* **43**, 3372–3376 (2000).

341. A. Gangjee, J. Yu, J. J. McGuire, V. Cody, N. Galitsky, R. L. Kisliuk, S. F. Queener. *J. Med. Chem.* **43**, 3837–3851 (2000).

342. S. Borman. *Chem. Eng. News*, Oct. 9, 48–53 (2000).

343. S. J. Sucheck, A. L. Wong, K. M. Koeller, D. D. Boehr, K. Draker, P. Sears, G. D. Wright, C.-H. Wong. *J. Am. Chem. Soc.* **122**, 5230–5231 (2000).

344. K. Bachmann and R. Belloto, Jr. *Drugs and Ageing* **15**, 235–250 (1999).

345. K. Malmstrom, J. Schwartz, T. F. Reiss, T. J. Sullivan, J. H. Reese, L. Jauregui, K. Miller, M. Scott, S. Shingo, I. Peszek, P. Larson, D. Ebel, T. L. Hunt, R. Huhn, K. Bachmann. *Am. J. Ther.* **5**, 189–195 (1998).

346. L. S. Mauro, D. A. Kuhl, J. R. Kirchhoff, V. F. Mauro, R. W. Hamilton. *Am. J. Ther.* **8**, 21–25 (2001).

347. V. F. Mauro, L. S. Mauro, J. F. Kleshinski, P. W. Erhardt. *Pharmacotherapy* **21**, 373, abstract (2001).

348. Y. Brun, C. Wrzossek, J. Parsons, H. Slocum, D. White, W. Greco. *Am. Assoc. Cancer Res. Proc.* **41**, 102 (Abstract) (2000).

349. D. White and L. James. *J. Clin. Epidemiol.* **49**, 419–429 (1996).

350. (a) W. Greco, G. Bravo, J. Parsons. *Pharmacol. Rev.* **47**, 331–385 (1995); (b) H. M. Faessel, H. K. Slocum, Y. M. Rustum, W. R. Greco. *Biochem. Pharmacol.* **57**, 567–577 (1999).

351. Paired agent combinations = 100!/(2! 98!) = 4950 agent pairs. Test each pair at 3 concentrations for each agent = 9 combinations per pair. Number of pair tests = 9(4950) = 44,550. Number of single agent tests at 3 concentrations per agent = 3(100) = 300. Total tests for an $N = 1$ pass through the library = 44,550 + 300 = 44,850.

352. N. Bodor. In *Strategy in Drug Research*, J. A. K. Buisman (Ed.), pp. 137–164, Elsevier, Amsterdam (1982).

353. N. Bodor and P. Buchwald. *Med. Res. Rev.* **20**, 58–101 (2000).

354. P. Erhardt. *Chron. Drug Discov.* **3**, 191–206 (1993).

355. P. W. Erhardt, C. M. Woo, R. J. Gorczynski, W. G. Anderson. *J. Med. Chem.* **25**, 1402–1407 (1982).

356. P. W. Erhardt, C. M. Woo, W. G. Anderson, R. J. Gorczynski. *J. Med. Chem.* **25**, 1408–1412 (1982).

357. P. W. Erhardt, C. M. Woo, W. L. Matier, R. J. Gorczynski, W. G. Anderson. *J. Med. Chem.* **26**, 1109–1112 (1983).

358. J. Liao, K. Golding, K. Dubischar, J. Sarver, P. Erhardt, H. Kuch, M. Aouthmany. Determination of blood esterase capability in newborns, 27th Nat. ACS Med. Chem. Symp., June (2000).

359. J. Liao. Development of soft drugs for the neonatal population, M.S. thesis, University of Toledo (2000).

360. P. Erhardt. Using drug metabolism databases as predictive tools during drug design and development: past, present and future status, Soc. Tox. Canada 32nd Annu. Symp., Dec. (1999).

361. J. F. Sina. *Annu. Rep. Med. Chem.* **33**, 283–291 (1998).

362. R. B. Conolly. *CIIT Activities* (Chemical Industry Institute of Toxicology newsletter) **19** (10), 6–8 (1999).

363. M. Jacobs. *Chem. Eng. News*, Mar. 13, 29 (2000).

364. J. Boguslavsky. *Drug Discov. Develop.* June/July, 43–46 (2000).

365. M. Eggers. *Innovations Pharm. Technol.* **6**, 36–44 (2000).

366. J. Boguslavsky. *Drug Discov. Develop.*, Oct., 30–36 (2000).

367. R. E. Neft, S. Farr. *Drug Discov. World* **1**(2), 33–34 (2000).

368. For general toxicology: Cellomics, Inc. at www.cellomics.com; Celera, Inc. at http://www.celera.com; Chemical Industry Institute of Toxicology at http://www.ciit.org; Inpharmatica at www.inpharmatica. co.uk; National Center for Biotechnical Information at http://www.ncbi.nlm.nih.gov/Web/Genebank/index.html and at http://www.ncbi.nlm.nih.gov/Entvez/Genonome/org.html; Rosetta Inpharmatics at www.vii.com; SciVision (ToxSYS) at www.scivision. com. For drug–drug interactions: Georgetown University at http://dml.Georgetown.edu/depts/pharmacology/clinlist.html; FDA at http://www.fda.gov/cder/drug/advisory/stjwort.htm; Public Citizen Health Research Group at http://www.citizen.org/hrg/newsletters/pillnews.htm.

369. R. Kumar. *Innovations Pharm. Technol.* **6**, 120–123 (2000).

370. B. Lewis, M. Renn, K. Manning. *Innovations Pharm. Technol.* **6**, 125–128 (2000).

371. M. McCoy. *Chem. Eng. News*, Mar. 12, 27–42 (2001).

372. Personal communications during the 175th Anniversary USP Convention, Mar. (1995).

373. P. Anastas and T. C. Williamson (Eds.). *Green Chemistry: Frontiers in Benign Chemical Syntheses and Processes*, Oxford University Press, New York (1998).

374. P. Anastas and J. Warner. *Green Chemistry: Theory and Practice*, Oxford University Press, New York (1998).

375. P. T. Anastas, L. G. Heine, T. C. Williamson (Eds.). *Green Chemical Syntheses and Process* (Part 1) and *Green Engineering* (Part 2), Oxford University Press, New York (2000).

376. P. Tundo and P. Anastas (Eds.). *Green Chemistry: Challenging Perspectives*, Oxford University Press, New York (2000).

377. S. C. Stinson. *Chem. Eng. News*, Oct. 23, 55–78 (2000).

378. K. Auclair, A. Sutherland, J. Kennedy, D. J. Witter, J. P. Van den Heever, C. R. Hutchinson, J. C. Vederas. *J. Am. Chem. Soc.* **122**, 11519–11520 (2000).

379. K. Watanabe, H. Oikawa, K. Yagi, S. Ohashi, T. Mie, A. Ichihara, M. Honma. *J. Biochem.* **127**, 467–473 (2000).

380. K. Watanabe, T. Mie, A. Ichihara, H. Oikawa, M. Honma. *J. Biol. Chem.* **275**, 38393–38401 (2000).

381. S. Borman. *Chem. Eng. News*, Jan. 15, 9 (2001).

382. T. W. Greene and P. G. M. Wuts. *Protective Groups in Organic Synthesis*, 3rd ed., Wiley, New York (1999).

383. L. J. Silverberg, S. Kelly, P. Vemishetti, D. H. Vipond, F. S. Gibson, B. Harrison, R. Spector, J. L. Dillon. *Org. Lett.* **2**, 3281–3283 (2000).

384. P. W. Erhardt. *Synth. Commun.* **13**, 103–113 (1983).

385. S. Stinson. *Chem. Eng. News*, Dec. 4, 35–44 (2000).

386. (a) R. Baltzly and P. B. Russell. *J. Am. Chem. Soc.* **75**, 5598–5602 (1953); (b) M. Freifelder (Ed.) *Catalytic Hydrogenation in Organic Synthesis Procedures and Commentary*, Wiley, New York (1978).

387. Y. Ni and P. Erhardt. Synthesis of steric probes to examine the chemical behavior of potential benzylamine-related chiral auxiliary synthetic reagents, ACS Nat. Meet., Apr. (1997).

388. Y. Ni. Synthesis and N-debenzylation of steric probes to define the practical limit for employing potential benzylamine-related chiral auxiliary reagents, M.S. thesis, University of Toledo (1996).

389. P. Erhardt. U.S. patent 5,977,409, 1–25 (1999).

390. C. Zhang and P. Erhardt. Benzylamine-related chiral synthetic reagents, Natl. ACS Med. Chem. Symp., June (2000).

391. C. Zhang. Benzylamine-related chiral synthetic reagents: asymmetric opening of terminal epoxides, M.S. thesis, University of Toledo (1999).

392. P. Coppens (Ed.). *X-Ray Charge Densities and Chemical Bonding,* Oxford University Press, New York, (1997).

393. (a) V. S. Lamzin, R. J. Morris, Z. Dauter, K. S. Wilson, M. M. Teeter. *J. Biol. Chem.* **274**, 20753–20755 (1999); (b) D. Housset, F. Benabicha, V. Pichon-Pesme, C. Jelsch, A. Maierhofer, S. David, J. C. Fontecilla-Camps, C. Lecomte. *ACTA Crystallogr. Sect. D Biol. Crystallogr.* **56**, 151–160 (2000); (c) C. Jelsch, M. M. Teeter, V. Lamzin, V. Pichon-Pesme, R. H. Blessing, C. Lecomte. *Proc. Natl. Acad. Sci. USA* **97**, 3171–3176 (2000); (d) M. V. Fernandez-Serra, J. Junquera, C. Jelsch, D. Lecomte, E. Artacho. *Solid State Commun.* **116**, 395–400 (2000).

394. A. Martin and A. A. Pinkerton. *Acta Crystallogr.* **B54**, 471–478 (1998).

395. A. A. Pinkerton. In *Electron, Spin and Momentum Densities and Chemical Reactivity*, P. G. Mezey and B. Robertson (Eds.), pp. 213–223, Klewer, London (2000).

396. B. B. Iversen, F. K. Larsen, A. A. Pinkerton, A. Martin, A. Darovsky, P. A. Reynolds. *Acta Crystallogr.* **B55**, 363–374 (1999).

397. K. Palczewski, T. Kumasaka, T. Hori, C. A. Behnke, H. Motoshima, B. A. Fox, I. Le Trong, D. C. Teller, T. Okada, R. E. Stenkamp, M. Yamamoto, M. Miyano. *Science* **289**, 739–745 (2000).

398. V. Comello. *Drug Discov. Develop.* Apr., 26–28 (2000).

399. (a) N. Ban, P. Nissen, J. Hansen, P. B. Moore, T. A. Steitz. *Science* **289**, 905–920 (2000). (b) P. Nessen, J. Hansen, N. Ban, P. B. Moore, T. A. Steitz. *Science* **289**, 920–930 (2000).

400. F. Schluenzen, A. Tocilj, R. Zarivach, J. Harms, M. Gluehmann, D. Janell, A. Bashan, H. Bartels, I. Agmon, F. Franceschi, A. Yonath. *Cell* **102**, 615–623 (2000).

401. (a) B. T. Wimberly, D. E. Brodersen, W. M. Clemons, Jr., R. J. Morgan-Warren, A. P. Carter, C. Vonrhein, T. Hartsch, V. Ramakrishnan. *Nature* **407**, 327–339 (2000); (b) B. L. Hanson, A. Martin, J. M. Harp, C. G. Bunick, D. A. Parrish, K. Kirschbaum, G. J. Bunick, A. A. Pinkerton. *J. Appl. Crystallogr.* **32**, 814–820 (1999); (c) B. L. Hanson, J. M. Harp, K. Kirschbaum, D. A. Parrish, D. E. Timm, A. Howard, A. A. Pinkerton, G. J. Bunick. *J. Cryst. Growth.* In press.

402. M. J. Hardie, K. Kirschbaum, A. Martin, A. A. Pinkerton. *J. Appl. Crystallogr.* **31**, 815–817 (1998).

403. P. W. Erhardt. *J. Pharm. Sci.* **69**, 1059–1061 (1980).

404. P. W. Erhardt and Y.-L. Chou. *Life Sci.* **49**, 553–568 (1991).

405. S. Moran and L. Stewart. *Drug Discov. World* **2**(1), 41–46 (2000).

406. Z. P. Wu. Accelerating lead and solid form selection in drug discovery and development using non-ambient x-ray diffraction, New Chemical Technologies Satellite Program, 221st Nat. ACS Meet., Mar. (2001).

407. D. A. Pearlman and P. S. Charifson. *J. Med. Chem.* **44**, 502–511 (2001).

408. R. W. Schoenlein, S. Chattopadhyay, H. H. W. Chong, T. E. Glover, P. A. Heimann, C. V. Shank, A. A. Zholents, M. S. Zolotorev. *Science* **287**, 2237–2240 (2000).

409. E. Wilson. *Chem. Eng. News*, Mar. 27, 7, 8 (2000).

410. E. J. Ariens (Ed.). *Molecular Pharmacology: The Mode of Action of Biologically Active Compounds*, Vol. I, Academic Press, New York (1964).

411. I. Schlichting, J. Berendzen, K. Chu, A. M. Stock, S. A. Maves, D. E. Benson, R. M. Sweet, D. Ringe, G. A. Petsko, S. G. Sligar. *Science* **287**, 1615–1622 (2000).

412. A. M. Rouhi. *Chem. Eng. News*, Mar. 13, 50–51 (2000).

413. E. K. Wilson. *Chem. Eng. News*, Apr. 24, 39–45 (2000).

414. U. Kher. How to design a molecule, *Time*, Jan. 15, 67 (2001).

415. P. W. Erhardt. In *Endothelin*, G. M. Rubanyi (Ed.), pp. 41–57, Oxford University Press, New York (1992).

416. C. R. Ganellin. *Chem. Int.* **23**(2), 43–45 (2001).

417. J. Brent. CLA scholars spin curiosity into discovery, *Univ. Minn. CLA Today* (1998–1999 Annual Report). Fall, 10–13 (1999).

418. P. W. Erhardt. *J. Org. Chem.* **44**, 883 (1979).

419. J. Parness and S. B. Horwitz. *J. Cell Biol.* **91**, 479–587 (1981).

420. A. B. Shapiro and V. Ling. *Eur. J. Biochem.* **250**, 130–137 (1997).

421. For example, respectively: the long-standing Pharmacology for Medicinal Chemistry course often offered by the ACS as a satellite program to their national scientific meetings; the Residential School on MC offered at Drew University (http://www.depts.drew.edu/resmed); Drug Metabolism for Medicinal Chemists short course that I offer on-site; and a long-distance Continuing Learning Program in MC offered by the University of Nottingham (http://www.nottingham.ac.uk/pazjac/sbddpapertext. html).

422. E. G. Wright. *Mod. Drug Discov., Oct.*, 69–70 (2000).

423. E. Ledbetter. *Mod. Drug Discov.*, Apr. and May, 25–28 and 81–84 (2000).

424. M. Jacobs (Ed.). Challenging times for patent office, *Chem. Eng. News*, Apr. 10, 39–44 (2000).

425. M. Jacobs (Ed.). PTO finalizes guidelines for gene patents (http://www.uspto.gov/.) *Chem. Eng. News*, Jan. 15, 33 (2001).

426. D. J. Hanson. *Chem. Eng. News*, Sept. 11, 19–20 (2000).

427. M. D. Kaminski. *Mod. Drug Discov.*, Jan., 36–37 (2001).

428. W. Schulz. *Chem. Eng. News*, Nov. 27, 18–19 (2000).

429. R. Wedin. *Chemistry* (ACS newsletter), Spring, 1, 17–20 (2001).

430. International Human Genome Sequencing Consortium (representing more than 100 contributing authors). Initial sequencing and analysis of the human genome, *Nature* **409**, 860–921 (2001).

431. Human Genome Research Team (over 200 participating authors represented largely from Celera Genomics). The sequence of the human genome, *Science* **291**, 1304–1351 (2001).

432. H. van de Waterbeemd, R. E. Carter, G. Grassy, H. Kubinyi, Y. C. Martin, M. S. Tute, P. Willett. *Annu. Rep. Med. Chem.* **33**, 397–409 (1998).

433. K. J. Watkins. *Chem. Eng. News*, Feb. 19, 29–45 (2001).

434. T. Agres. *Drug Discov. Develop.*, Feb. 9, 10 (2001).
435. C.-G. Wermuth, R. Ganellin, P. Lindberg, L. A. Mitscher. *Annu. Rep. Med. Chem.* **33**, 385–395 (1998).
436. M. Beggs. *Drug Discov. World* **2**(1), 25–30 (2000).
437. http://dietary-supplements.info.nih.gov/.
438. S. Borman. *Chem. Eng. News*, July 31, 31–37 (2000).
439. K. A. Bachmann. *Drug Metabolism: Databases and High-Throughput Testing During Drug Design and Development*, P. W. Erhardt (Ed.), pp. 79–121, IUPAC/Blackwell, Geneva (1999).
440. W. Klosgen and J. Zytko (Eds.). *Handbook of Data Mining and Knowledge Discovery*, Oxford University Press, New York (2000).
441. A. M. Doherty. *Annu. Rep. Med. Chem.* **35**, 331–356 (2000).
442. D. Hanson. *Chem. Eng. News*, June 19, 37 (2000).
443. M. Jacobs (Ed.). *Chem. Eng. News*, Apr. 3, 15 (2000).
444. G. Karet. *Drug Discov. Develop.*, Nov./Dec., 71–74 (1999).
445. M. S. Lesney. *Mod. Drug Discov.*, May, 31–32 (2000).
446. M. B. Brennan. *Chem. Eng. News*, June 5, 63–73 (2000).
447. F. Golden. *Time* **157**(2), 74–76 (2001).
448. A. Thayer. *Chem. Eng. News*, Feb. 5, 18 (2001).
449. Clinical trial information: Royal Coll. Physicians of Edinburgh at www.rcpe.ac.uk/cochrane/intro.html; Pharmaceutical Information Network at htt://pharminfo.com/conference/clintrial/fda_1.html; PhRMA (Assoc. of 100 major pharmaceutical comp.) at www.pharma.org/charts/approval.html; Univ. Pittsburgh IRB manual at www.ofreshs.upmc.edu/irb/preface.htm; Center Watch (a comprehensive venue and listing for clinical trials) at www.centerwatch.com; FDA Ctr. For Drug Evaluation and Res. (CDER) at www.fda.gov/cder; eOrange Book (Approved drug products with therapeutic equivalence evaluations) at www.fda.gov/cder/ob.; FDA CDER Handbook at www.fda.gov/cder/handbook/index.htm; FDA CDER Fact Book 1997 at www.fda.gov/ cder/reports/cderfact.pdf; FDA Consumer Magazine at www.fda.gov/fdac/fdacindex.html; and FDA Guidelines on content and format for INDs at www.fda.gov/cder/guidance/clin2.pdf.
450. M. Kansy, F. Senner, K. Gubernator. *J. Med. Chem.* **41**, 1007–1010 (1998).
451. C. J. Cole and W. P. Pfund. *Mod. Drug Discov.*, Oct., 73–74 (2000).
452. T. T. Wilbury. *Drug Metabolism: Databases and High-Throughput Testing During Drug Design and Development*, P. W. Erhardt (Ed.), pp. 208–222, IUPAC/Blackwell, Geneva (1999).
453. F. Darvas, S. Marokhazi, P. Kormos, G. Kulkarni, H. Kalasz, A. Papp. In *Drug Metabolism: Databases and High-Throughput Testing During Drug Design and Development*, P. W. Erhardt (Ed.), pp. 237–270, IUPAC/Blackwell, Geneva (1999).
454. G. Klopman, M. Tu. In *Drug Metabolism: Databases and High-Throughput Testing During Drug Design and Development*, P. W. Erhardt (Ed.), pp. 271–276 IUPAC/Blackwell, Geneva (1999).
455. R.W. Snyder and G. Grethe. In *Drug Metabolism: Databases and High-Throughput Testing During Drug Design and Development*, P. W. Erhardt (Ed.), pp 277–280, IUPAC/Blackwell, Geneva (1999).
456. J. Hayward. In *Drug Metabolism: Databases and High-Throughput Testing During Drug Design and Development*, P. W. Erhardt (Ed.), pp. 281–288, IUPAC/Blackwell, Geneva (1999).
457. D. F. Lewis and B. G. Lake. *Xenobiotica* **26**, 723–753 (1996).
458. D. F. Lewis and B. G. Lake. *Xenobiotica* **25**, 585–598 (1995).
459. D. F. Lewis, C. Ioannides, D. V. Parke. *Biochem. Pharm.* **50**, 619–625 (1995).
460. T. Walle. *Methods Enzymol.* **272**, 145–151 (1996).

461. M. J. de Grout, G. J. Buloo, K. T. Hansen, N. P. Vermeulen. *Drug Metab. Dispos.* **23**, 667–669 (1995).

462. M. J. de Groot, N. P. Vermeulen, J. D. Kramer, F. A. van Acker, G. M. Donne-Op den Kelder. *Chem. Res. Toxicol.* **9**, 1079–1091 (1996).

463. D. F. Lewis, P. J. Eddershaw, P. S. Goldfarb, M. H. Tarbit. *Xenobiotica* **26**, 1067–1086 (1996).

464. K. W. DiBiasio. *Univ. Micro. Int.* (University of California–Davis), 1–5 (1989).

465. K. Bachmann, D. Pardoe, D. White. *Environ. Health Perspect.* **104**, 400–407 (1996).

466. J. G. Sarver, D. White, P. Erhardt, K. Bachmann. *Environ. Health Perspect.* **105**, 1204–1209 (1997).

467. B. J. Kuypu (Ed.). *CIIT Activities* (Chemical Industry Institute of Toxicology newsletter) **20**(11–12), 1–2 (2000).

468. C. J. Henry and J. S. Bus. *CIIT Activities* (Chemical Industry Institute of Toxicology newsletter) **20**(7), 1–5 (2000).

469. M. J. Felton. *Mod. Drug Discov.*, Oct., 81–83 (2000).

470. C. Hogue. *Chem. Eng. News*, Mar. 19, 33–34 (2001).

471. P. Lewalle. *Terminol. Standard. Harmon.* **11**(1–4), 1–28 (1999).

472. J. H. Duffus. *Chem. Int.* **23**(2), 34–39 (2001).

473. G. Klopman. *Quant. Struct.–Act. Relat.* **11**, 176–184 (1992).

474. D. Sanderson, C. Earnshaw, P. Judson. *Hum. Exp. Toxicol.* **10**, 261–273 (1991).

475. S. C. Gad. *SciVision Update* (SciVision newsletter), Jan., 1–2 (2001).

476. R. E. Neft and S. Farr. *Drug Discov. World*, Feb., 33–34 (2000).

477. W. Cabri and R. Di Fabio (Eds.). *From Bench to Market*, Oxford University Press, New York (2000).

478. K. Knapman. *Modern Drug Discov.*, June, 75–77 (2000).

479. S. Hume. *Innovations Pharm. Technol.* **6**, 108–111 (2000).

480. J. Court and N. Oughton. *Innovations Pharm. Technol.* **6**, 113–117 (2000).

481. M.D. Lemonick. *Time*, Jan. 15, 58–67 (2001).

482. M. Zall. *Mod. Drug Discov.*, Mar., 37–42 (2001).

483. M. B. Brennan. *Chem. Eng. News*, Feb. 14, 81–84 (2000).

484. M. S. Reisch. *Chem. Eng. News*, Feb. 12, 15–17 (2001).

485. T. Geissman and D. Crout (Eds.). *Organic Chemistry of Secondary Plant Metabolism*, Freeman, San Francisco (1969).

486. R. Herbert (Ed.). *The Biosynthesis of Secondary Metabolites*, Chapman & Hall, New York (1989).

487. T. Murashige and F. Skoog. *Physiol. Plant* **15**, 473–497 (1962).

488. P. Erhardt and W. Klis. U.S. patent allowed.

489. P. Erhardt. U.S. Patent pending.

490. P. Erhardt, J. Sarver, W. Klis. U.S. patent pending.

491. P. Erhardt, W. Klis, J. Sarver. U.S. patent pending.

492. P. W. Erhardt, R. J. Gorczynski, W. G. Anderson. *J. Med. Chem.* **22**, 907–911 (1979).

493. P. W. Erhardt, A. A. Hagedorn III, M. Sabio. *Mol. Pharmacol.* **33**, 1–13 (1988).

494. P. W. Erhardt. Second generation phosphodiesterase inhibitors: Structure–activity relationships and receptor models, in *Cyclic Nucleotide Phosphodiesterases: Structure, Regulation and Drug Action*, J. Beavo and M. Housley (Eds.), Wiley, New York (1990).

495. J. C. Hemple, W. A. Ghoul, J.-M. Wurtz, A. T. Hagler. In *Peptides: Chemistry, Structure and Biology*, J. E. River and G. R. Marshall (Eds.), pp. 279–284, ESCOM, Leiden, The Netherlands (1990).

3

CONTEMPORARY DRUG DISCOVERY

LESTER A. MITSCHER AND APURBA DUTTA
Kansas University
Lawrence, Kansas

3.1 INTRODUCTION

It is unlikely to come as a startling revelation that the challenges faced by chemists involved in drug seeking are significantly different from those faced by chemists occupied in drug development even though they both work toward the same ultimate objective. The reasons for this are, perhaps, not fully realized by those occupying either one of the two camps. Primarily, medicinal chemists find their route to success, with all of its subthemes, significantly constrained, but the final product is not, whereas process chemists find their final product highly constrained, but the route to it is significantly variable. In this essay we attempt to illustrate the challenges facing the medicinal chemist that lead to this aphorism.

At its core, every branch of chemistry involves achieving an understanding of the relationship between chemical structure and molecular properties. At its zenith, this involves the ability to design a new substance intended to fill a perceived need and, following preparation and evaluation, getting it right. To design and construct a molecule to possess, say, a particular color is complex enough, but it is dramatically simpler than the problems involved in designing and constructing a novel therapeutic agent. Many more characteristics must be introduced into a molecule before it can become a drug, and many of those are often found to be in opposition, so that one must make a series of appropriate structural compromises.

3.1.1 Getting Started

The drug-seeking chemist seeks a molecule with a sufficient number of favorable properties to justify the very major investment in time and treasure required to bring the drug before the public and return a profit. The fundamental complexity of the process stems from the Byzantine complexity of the human body and its processes. Whereas enormous

Drug Discovery and Development, Volume 1: Drug Discovery, Edited by Mukund S. Chorghade

strides have been made in unraveling these in the last half-century, our understanding of these is still in a comparatively primitive state. The average time required between conception and introduction of a drug approaches 12 years and the cost in 2002 (factoring in the failed initiatives) lies somewhere between \$300 million and \$800 million. Only then can it begin to amortize the cost of its discovery and earn a return on investment. These breath-catching costs stem in large part from the high failure rate and significant constraints put upon the prices that society will accept. It also clearly explains why no firm will make such an investment unless a proprietary patent position can be achieved and the disease target is sufficiently prevalent among persons capable of paying for their treatment.

At the outset of the project, the specific identity of the compound sought is almost always obscure. Indeed, finding a suitable starting molecule is often the most challenging feature of the search. As the project proceeds and the manifold barriers to success are progressively overcome by creative analoging, the final structure comes more and more into focus and the molecular options become fewer. During the bulk of this time, the particular chemistry involved in the route is not a major concern. It is necessary only that the route be amenable to the reasonably efficient production of the necessary analogs. It is also important in the later stages that the chemistry be suitable for the production of the quantities required for intensive animal studies. The cost of goods is not trivial but is not yet a major consideration. The synthetic steps are usually somewhat optimized, but much room is present for improvement and many of the steps will involve reagents and processes that cannot safely be translated into large-scale production when this becomes a relevant consideration. The contemporary position is that ease of economic synthesis is increasingly being considered at earlier stages, but this is not always possible. In any case, handing the process over from discovery to development at the optimal stage requires judgement.

After the project is assigned to development, the structure of the molecule to be produced is quite clear and will not be changed casually. Hundreds to thousands of chemicals will have been prepared, evaluated, and discarded by this time. The challenge now is to develop a synthetic route to the surviving molecule that is economically attractive, scalable, uses safe chemicals and processes, does not require chromatography, produces benign wastes, and in which each step is optimized. It is not surprising therefore that the discovery route is rarely the development route and that success in both stages requires a high level of expertise considering the constraints imposed.

3.2 CHARACTERISTICS OF A SUITABLE LEAD SUBSTANCE

There is a partially resolved debate among medicinal chemists as to which property is more important at the outset of a drug-seeking campaign: potency and selectivity or suitable druglike characteristics. Ultimately, as shown in Fig.3.1, both are required for success. The

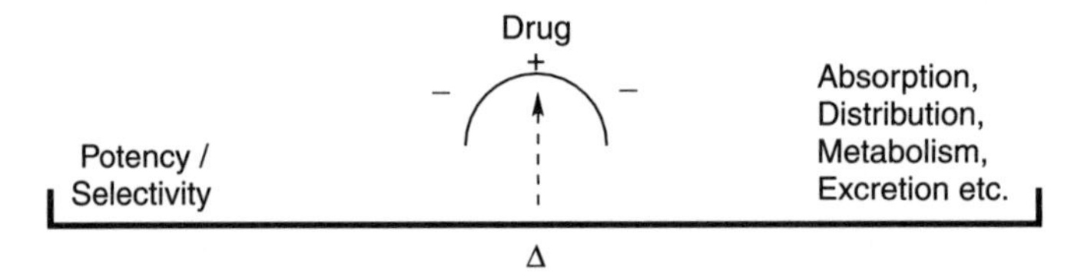

Figure 3.1 Successful drug seeking requires a satisfactory balance between efficacy and other drug-like characteristics.

real question is which one to start with if one can have only one of the two. Most discovery chemists now display a bias toward druglike characteristics in lead molecules, as many more variables are involved and many more macromolecular fits must be optimized in solving these than are involved with enhancing potency and selectivity.

3.2.1 Potency and Selectivity

A drug and its receptor mostly possess exquisitely demanding mutual molecular compatibilities, leading to a tight host–guest complex possessing signaling properties for cells that neither the drug nor the receptor alone possesses. Similarly, an inhibitor must bind to an enzyme and inhibit its functioning. The difficulty in accomplishing this stems from the large and complex structure of the receptor or the enzyme, whose detailed composition and topography are generally unknown at the outset of the project.

The original drug–receptor theory was that they were both rigid substances, so that the drug needed to be crafted to fit the desired receptor and no other, just as a specific key is required to open a specific lock. This simple picture is now recognized to be only one extreme of a continuum. The opposite of this concept is a flexible drug and a flexible receptor. In this view both interacting molecules must adjust in shape to accommodate each other. This is analogous to a zipper. Following an initial docking interaction, the two components interact progressively and ultimately form a new supermolecule with new properties. All intermediate degrees of flexibility between these extremes are recognized (i.e., rigid drug and flexible receptor, flexible drug and rigid receptor, etc.). An example can help clarify these concepts for those who find these ideas to be new.

Methotrexate is an antitumor drug classified as an antimetabolite in that it inhibits the formation of an essential component of DNA. DNA is the central biochemical from which all other molecules flow. It is made up of four essential monomers—adenine, cytosine, guanine, and thymine—joined together in a linear manner in specific sequences as deoxyribosyl phosphodiesters. Of these, thymine is unique to DNA, not being present in RNA. Instead, it is biosynthesized by addition of a methylene group to deoxyuridine 5′-monophosphate in a reaction catalyzed by thymidylate synthase, an enzyme using tetrahydrofolate as a source of the needed carbon. Cellular growth, repair, and reproduction are impossible without this reaction. One way to prevent cancer cells from growing is to poison them selectively by starving them of thymine by inhibiting this reaction. Methotrexate acts as an anticancer agent in just this way as it competes with the cofactor for binding to the enzyme. It was found following up a natural product lead and its molecular mode of action, not known at the time, is now generally understood.

Dihydrofolate reductase alters its conformation slightly but importantly when it is liganded with its cofactor or its inhibitors. This complicates the rational design of inhibitors, for crafting a fit to the ground state that did not trigger this movement would produce molecules that would probably be ineffective. Fortunately, none of this was known at the time that methotrexate was developed.

Later, numerous x-ray studies with the normal substrate and its inhibitors, including methotrexate, became available so it was possible not only to understand generally how this inhibition works but to design novel agents using this information. From Scheme 3.1, which illustrates schematically mutual interactions between the enzyme and methotrexate as well as dihydrofolic acid, it may be seen that many distinct molecular interactions are involved. Prime among these is the docking interaction between one of the glutamate carboxyls and a conserved arginine of the enzyme. For the inhibitor to be potent and specific, the remaining interactions must be not only electronically but also sterically correct. It is important to note

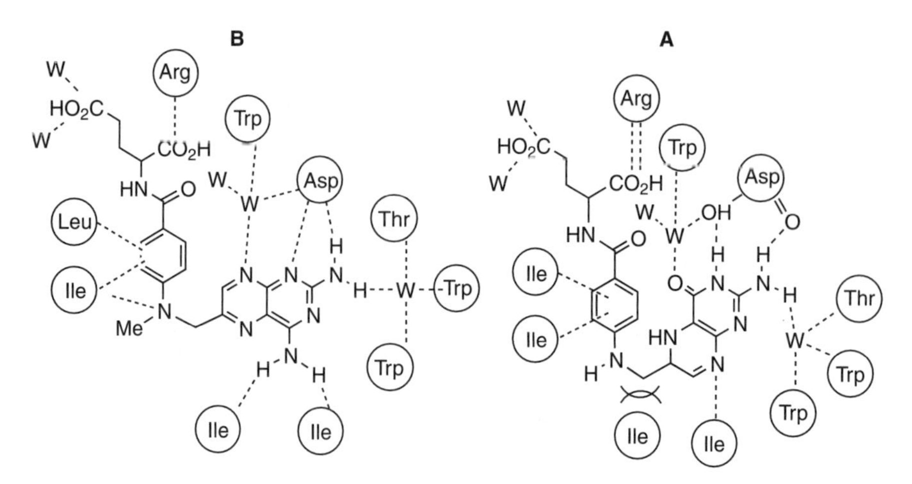

Scheme 3.1 Dihydrofolic acid (A) and methotrexate (B) in the active site of human dihydrofolate reductase. [From Klebe, G. (1994), *J. Mol. Biol.* 237: 212 – 235.]

that the natural substrate, dihydrofolic acid, although structurally similar to methotrexate, actually binds rather differently to the enzyme in that the pteridine rings are rotated with respect to one another! This is not common in drug seeking and could have been a big problem. More commonly, one starts with the structure of the normal substrate and designs inhibitors following the belief that they will bind in the same manner. As is common in enzyme–substrate interactions, the binding site for dihydrofolic acid and for methotrexate is in a groove in the enzyme's structure that is accessible only in specific ways from the exterior of the molecule. One might visualize it as a frankfurter in a bun. Most important, methotrexate bound in the enzyme's active site is not capable of providing the one-carbon unit needed for thymine biosynthesis and also blocks dihydrofolic acid's access.

Having a reliable picture of how a drug interacts with its target is a powerful aid to drug seeking. This particular mode is known as *structure-based drug design*. Fortunately, it is becoming more common. One notes the complexities of the task even with this information in hand. Designing from first principles a successful alternative ligand for a pair that has been optimized for each other evolutionarily is a tall order. The job is even more complex when the alternative ligand binds to the same site but in a different way!

Because of the difficulty of representing complex three-dimensional relationships in two dimensions, for clarity of exposition the steric aspects of the relationship illustrated in Scheme 3.1 have been simplified by drawing the molecules as though they were coplanar. This is not actually the case. The central *para*-aminobenzoate moiety, for example, actually projects upward toward the viewer's eyes.

This example serves as one of many indicating that finding potent and selective ligands is not easy. It is, perhaps, comforting to note that it was originally accomplished empirically without knowing the molecular details presented in Scheme 3.1. It is discomforting to note that a ligand that is successful in vitro all too often fails as a drug because other factors prevent it from reaching its target. Getting a good fit is only part of the job. This reemphasizes the thought conveyed by Fig. 3.1.

The parts of a drug such as methotrexate that are essential for its pharmacological action are those in direct contact with its target macromolecule and are known as the *pharmacophore*. The other functionalities of a pharmaceutical are less essential and are manipulated in order to deal with pharmacokinetic deficiencies to be discussed later. These might be termed *auxophores*.

The point of this exposition is that a successful inhibitor must possess suitable functionality and stereochemistry to fit tightly into the critical region of the receptor but not function in the same manner as the normal ligand. Whereas nature solves the problem through evolutionary processes, a chemist must prepare such an agent through analoging, often without knowing all of the significant details and in a much, much shorter time period. The medicinal chemist must be familiar with the many ways in which successes have been achieved in the past and recognize when these may be applied productively to a present problem and be prepared to devise novel approaches as well.

3.2.2 Structure–Activity Relationships

Molecules are altered progressively, guided by biological data so that the desirability of the substances is dramatically enhanced as one after another of the perceived deficiencies is overcome. The structural and biological scorecard kept during this process is known as a *structure–activity relationship* (SAR). The simplest of these involve lists of structures and the comparative potency associated with each. From such lists the chemist can rapidly sort out which molecular features are helpful and which are not. Multiple SARs involve analogous data associated with serum protein binding, achievable blood levels following oral administration, and so on.

3.2.3 Toxicity

The reactivity of the functional groups in the molecule should not be great. For example, drugs containing α-haloketone moieties or other reactive electrophilic groups react relatively indiscriminately with tissue macromolecules, producing covalently bound products. At best these are sites of loss of molecular identity. Highly reactive molecules often fail to reach their drug targets. At worst, they can result in drug allergy or overt toxicity, even mutagenicity. These untoward effects are rarely tolerable. Stability of the agent is also related to reactivity. The drug must be stable enough to allow for preparation of dosage forms that will survive in active form to reach the patient and be administered.

The importance of lack of toxicity is obvious. *Acute toxicity* is the amount of a given agent that will kill or injure following a single or a few doses. The bigger the separation between the effective dose and the toxic dose (the therapeutic index), the safer the agent will be to use. A minimum separation of 1000-fold or higher is a common target. *Chronic toxicity*, on the other hand, refers to the dose that will produce deleterious effects when the drug is administered for a significant number of doses. The chronically deleterious dose is almost always significantly smaller than the acutely toxic dose.

Certain functional groups are recognized to produce toxicities all too often and are therefore put into drug molecules only with great reservation. These groups are often called *toxicophores*. Flat three-ring-containing aromatic moieties intercalate into DNA and interfere with its function. Usually, this is undesirable. Aromatic nitro groups are reduced to a variety of intermediately oxygenated species such as hydroxylamines and nitroxides that are capable of self-condensation into toxic moieties. Similarly, aniline groups are capable of oxidation into the same sort of undesirable metabolites. Thiols are capable of condensation with sulfhydryl/disulfide-containing tissue constituents, leading to toxic problems. Many other toxicophores are well known to medicinal chemists and are avoided.

Genetic toxicity is the most dangerous kind. DNA is a self-replicating polymer, and a damaged DNA molecule can be copied into multiple altered copies, magnifying and preserving the initial injury. Since damaged DNA, if not repaired promptly and accurately, leads to mutations that can even lead to cancer, there is really no consensually safe dose of a mutagen. Consequently, mutagenicity is a serious liability. Not all mutations are associated with cancer, but almost all cancers involve mutagenicity. If mutagenicity is relatively molecule specific, it can often be engineered out of a structure by appropriate analoging. If it is series related, however, this is a showstopper and another lead series must be sought. Known mutagenic moieties are therefore avoided at the outset.

3.2.4 Changing Appellation of the Best in Series: Analog Attrition

There are some key definitions that are associated with molecules as the hunt progresses. A *hit* is a substance that produces a significant response in a screen designed to reveal promising substances. Its potency and selectivity might be quite low, but it should have a promising structure for further elaboration. High-throughput screens enable one to search rapidly through large collections of molecules (often called *compound libraries*). Usually, high-throughput screens are based on particular biological mechanisms, are cell-free, sample sparing, and have relatively modest information content. A *lead* is a more important molecule most often descended from a hit that produces promising activity in a whole cell, intact organ system, or whole animal, confirming the relevance of the activity found in the initial screen. Much more compound is needed to reveal this, the criteria for acceptance are much higher, and the information content is much higher. The number of compounds that advance from screening substances to hits to leads is an order of magnitude less at each stage. A *candidate drug* is one of the few that has graduated from a lead substance by demonstrating favorable properties in more than one relevant animal species suffering from a model disease and so becomes a serious contender for clinical evaluation. A *clinical candidate* is a molecule that is judged to be worth the major costs involved in evaluation in humans, both healthy and diseased. A *drug* is a molecule that has passed successfully through all of these stages and advances into human use. It is only at this last point that it begins to return a profit to justify the expenditures already made.

Figure 3.2 illustrates schematically that the number of compounds one starts with filters down quickly into a very few survivors. The aim of the medicinal chemist is to make the attrition significantly smaller.

3.3 SOME CRITERIA THAT A HIT MUST SATISFY TO BECOME A DRUG

In addition to satisfying requirements for suitable potency and selectivity, at least 20 additional characteristics are significant. A drug must possess all of these to a favorable degree, but they are almost never equally important. Early identification of the most significant variables hindering further advancement is of prime importance in drug seeking. One can almost always enhance any one of these characteristics through analoging, but this often results in a detriment to at least one of the other characteristics. Thus, one is almost always forced to make compromise choices. For example, if one has outstanding potency, one may be willing or forced to sacrifice some to enhance some other feature that must be improved, such as solubility or absorption. Thus, the guiding principle is "good enough, soon enough." Perfection is rarely achieved. One interesting consequence of this multiplicity of

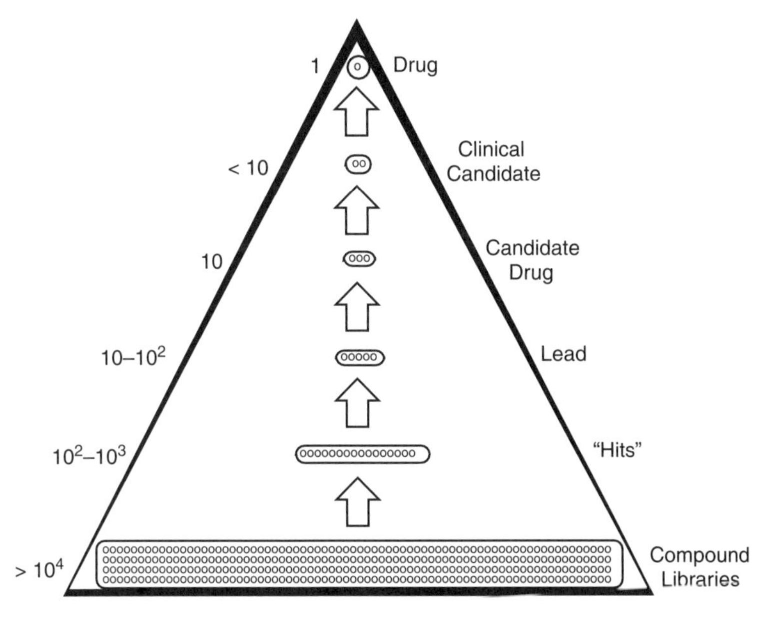

Figure 3.2 The drug discovery pyramid.

choices is that two different groups of investigators starting with the same lead molecule will usually finish with different final drugs.

The steps along the way are usually listed as though they are traversed linearly. All too often, however, the medicinal chemist becomes stymied at a given point and has to back up and bring a different substance forward to make additional progress. Whenever possible, to save time, the problems are identified and attacked in parallel rather than sequentially.

Some of the properties that must be present or introduced into a drug substance are:

Potency	Efficacy	Selectivity
Toxicity	Absorption	Distribution
Metabolism	Excretion	Acute toxicity
Chronic toxicity	Mutagenicity	Stability
Accessibility	Cost	Patentability
Clinical efficacy	Water solubility	Taste
Formulability	Idiosyncratic problem(s)	

3.3.1 Level of Potency

The potency of the substance must be significant in order to identify early hits worth pursuing. Since there are comparatively few copies of a given receptor in a cell, the potency needed for a hit is usually on the order of 100 μmol/mL or less. The potency of a candidate drug is often in the range of 10 nmol/mL or less. This is somewhat related to specificity. It is generally recognized that any effect of a drug other than that for which it is administered represents a side effect and is therefore undesirable. The more potent a drug is, the more

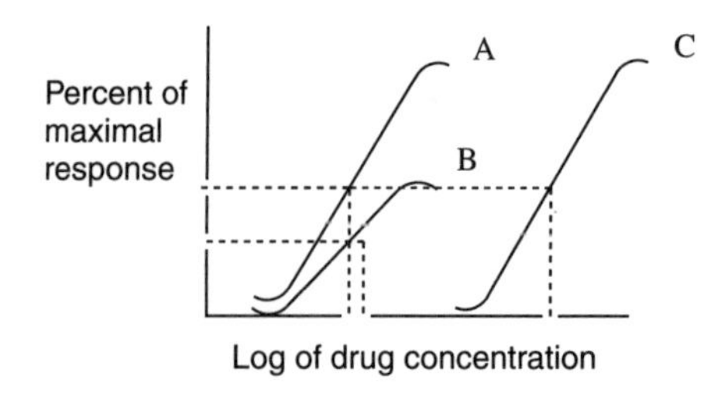

Figure 3.3 A comparison of potency and efficacy for drugs A, B, and C.

tightly and specifically it fits its receptor, and thus it is expected that it will not bind equally well to extraneous receptors and so will have fewer side effects. As with most generalizations, this one has significant exceptions, but it nonetheless colors one's attitude in drug seeking. Thus, potency and selectivity are often considered as closely related.

3.3.2 Comparison of Potency and Efficacy

There is sometimes confusion about the difference between potency and efficacy. *Potency* is the concentration of an agent that must be present to produce a biological result. *Efficacy* is how efficient a given dose is in producing a given result. Two substances possessing equal potency (in this case, producing their half-maximal effect at a similar dose) may produce significantly different efficacies (produce much different total responses, regardless of the dose). In Fig. 3.3, compounds A and B are similary potent and both are much more potent than C. Compounds A and C are equally efficacious, but compound B is significantly less so. Unfortunately, much more than potency, efficacy, and selectivity is required. Otherwise, drug seeking would be a much simpler task than it is in reality.

3.3.3 Druglike Character

Pharmacodynamics is the study of the effect that a drug has on the body (e.g., lowering blood pressure); *pharmacokinetics* is the study of the effect that the body has on the drug (e.g., metabolism). Contemporarily, it is recognized that the pharmacokinetic (PK) properties of a substance are extremely important but are complex to produce at satisfactory levels in a lead substance. Compound collections to be screened are most valuable, in this view, if they contain many molecules that are likely to possess satisfactory pharmacokinetic properties at the outset, that is, are intrinsically druglike. These properties are often referred to by the acronym ADME (absorption, distribution, metabolism, and excretion). Getting these right in a lead series as soon as possible avoids wasting resources on molecules that have little to no chance of becoming drugs.

3.3.4 Efficacy Following Oral Administration

One of the most desired characteristics of a drug is its ability to deliver therapeutically useful concentrations in the body following oral administration. Although this property can often be manipulated favorably by suitable analoging, it is a complex undertaking. The drug substance must first dissolve and escape digestion into polar small pieces. Next, it must pass through the cells lining the gastrointestinal tract and pass into the blood. A number of competing processes are involved, but most drugs do so by passive diffusion. Since the cell membranes are lipoidal in nature, most drugs must be at least partially nonionized at the cell surface for passive diffusion through this barrier. Since the small intestine has about the surface of a tennis court, due to its highly convoluted nature, this is the body chamber where the bulk of food and drug absorption takes place. Since the normal pH of the upper small intestine ranges from 4 to 7, nonpolar neutral compounds and compounds only slightly ionized at this pH range (e.g., weak acids and unprotonated tertiary amines) are absorbed well by passive diffusion. Thus, a drug's pK_a value should usually be less than 5, so that a sufficient percentage is nonionized. In contrast, if it is too lipophilic, it will remain in the lipid bilayer and not pass through to the other side. If it is too water soluble, it will not penetrate at all (see Fig. 3.4). Small water-soluble compounds can enter cells through

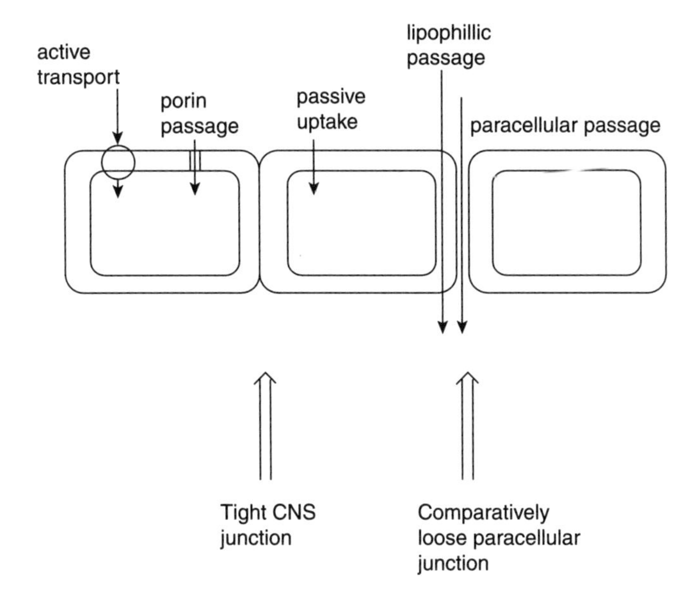

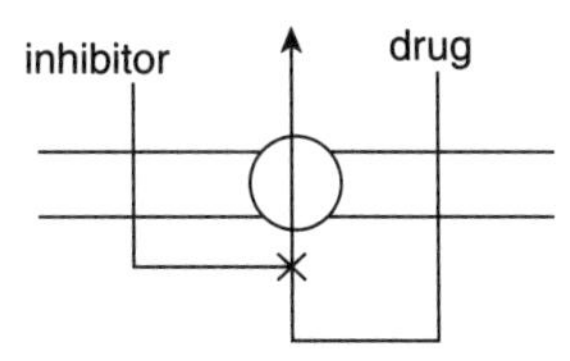

Figure 3.4 Drug uptake and exclusion at cells possessing a lipid bilayer. Passive diffusion requires lack of ionization and dehydration. Small hydrophilic molecules can pass between cells in the periphery but not in the CNS where the capillary cells are more tightly assopciated with their neighbors. More hydrophilic drugs can pass through proteins with a water-lined interior passage (porins) that span the membrane or by active transport using protein receptors. The active transport is very specific in structural requirements and can be in or out.

porins (i.e., membrance-spanning proteins with a central water-lined well) or may even pass between cells (paracellular passage), avoiding their membrance barriers altogether. There are also protein-based selective uptake mechanisms in the membranes providing active transport for some polar materials (e.g., amino acids, glucose), but this does not often work for drugs. Very precise host–guest compatibilities must be present for active transport to work. These active transporters can also work in the opposite direction to expel drugs from cells. These processes prevent many drugs from reaching their cellular targets and play a very significant role in resistance to antibiotic and anticancer chemotherapy.

To avoid excess polarity, molecules not only must not be too ionized but also must be able, reversibly, to shed their adhering water. As a general matter, molecules should have fewer than five hydrogen-bond donating moieties and fewer than 10 hydrogen-bond accepting groups for this to be efficient. More than these and it will be too difficult to desolvate.

Peripheral cells are not tightly bound together, often leaving small spaces between cells through which small hydrophilic molecules (such as ethyl alcohol) can pass. On the other hand, the capillaries of the central nervous system are so tightly packed together that this does not take place. Consequently, drugs intended to act in the central nervous system (CNS) must be more lipophilic than drugs intended to act elsewhere in the body. Conversely, drugs to be excluded from the CNS should be more hydrophilic.

Also, the influence of molecular weight on oral absorption is significant. Within reasonable limits, molecules exceeding about 500 atomic mass units are often poorly absorbed in the intestine. This is largely accounted for by the fact that passive diffusion becomes much slower as molecular weight increases.

The overall shape of the substance is also of significance. Compact, rigid molecules generally are absorbed more efficiently than floppy molecules with significant cross-sectional areas.

The combined influence of all these factors lead to the conclusion that the physicochemical characteristics of a substance are in many ways more important than its specific chemical structure if it is to benefit from passive uptake. The medicinal chemist must learn how to design molecules with favorable absorption characteristics.

This brief exposition avoids mention of ion channels that allow certain ions to penetrate into and escape from cells against concentration gradients. This important topic will have to be covered at another time in another place.

3.3.5 Lipinski Rules for Oral Absorption

The Lipinski rules were clearly enunciated and given scalar dimensions by Christopher Lipinski and are widely known as the *Lipinski rules of five*. They were developed by comparing the properties of a large number of candidate drugs to their relative absorption following oral administration. Although there are many exceptions, particularly for drugs taken up by active transport mechanisms for which the rules do not apply, the Lipinski rules are a very useful. Their popularity stems from their simplicity, their accord with reality, and the fact that they can readily be compreheneded by chemists and applied through inspection of structural formulas in advance of synthesis. The rule of five is recognized as being less useful for hits and leads than for the final drugs from which they were derived. To improve the properties of these classes of agents it is usually necessary to introduce additional functionality. Almost inevitably this results in molecular weight and grease creep. If the molecular weight of a lead compound is too high and/or the molecule is quite lipophilic to start with, it is difficult to produce a final molecule with a good chance of being orally active following further analoging. To accommodate this, chemists now often refer to a rule of threes, instead, for hits and leads.

It is also well known that compact molecules are more efficiently absorbed than those with a big cross section. Flexible molecules with many rotatable bonds are among the bad actors. Daniel Veber has pointed out that molecules with no more than 10 rotatable bonds are more likely to be orally active, and the *Veber rule* is often added to the Lipinski rules in drug seeking.

3.3.6 Injectable Medications

When an illness is acute or a patient is unable to achieve benefit from oral absorption, the drug is injected into the bloodstream or into tissues. For these purposes, significant water solubility is required so that the volume administered does not itself create problems for the patient. In this form of administration, the drug is rapidly available for the needs of the various tissues, and the Lipinski rules are not relevant for these.

3.3.7 Distribution

Distribution refers to the ability of an agent to arrive at the source of pathology in the body in concentrations sufficient to produce a favorable response. Many potentially intervening molecular interactions play a role in this process, as illustrated in Fig. 3.5. Some of these have just been discussed in dealing with oral uptake. However, there are many more not yet mentioned. Ability to deliver an agent to a particular target organ by analoging is not yet highly developed. The first organ that receives a drug-following uptake through the gut wall and passage into the blood is the liver. The liver is a sort of metabolic factory that transforms lipoidal materials into more polar molecules that are more readily transported in the blood and excreted either back into the intestines or distributed throughout the body, followed by passage into the urine via the kidneys. The liver in this sense is seen as a kind of grease trap. These concepts are illustrated schematically in Fig. 3.5.

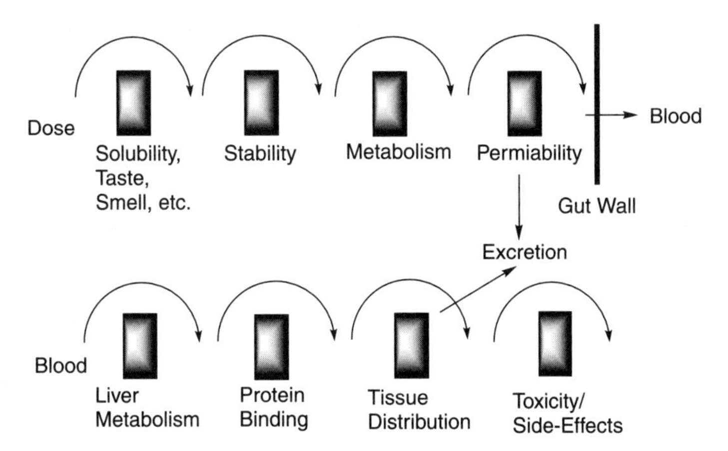

Figure 3.5 Some of the successive hurdles that a lead must overcome before marketing would be possible. In this diagram potency is taken for granted.

3.3.8 Serum Protein Binding

After a drug enters the bloodstream, it is subject to a series of competing equilibriums. The first of these involves binding to serum proteins. One of the major normal physiological functions of serum protein binding is to carry fatty acids safely to the tissues. Otherwise, these surface-active molecules would cause hemolysis of red blood corpuscles, destroying their ability to function in their normal role in carrying gases to and from tissues. Acidic, lipophilic molecules are particularly subject to this phenomenon, although several other drug-binding systems are also found in the blood. The percentage of the drug that is bound to serum proteins is inhibited from entering into the other equilibriums that otherwise would come into play. The capacity of serum protein albumen to bind drugs is rather high compared to the normal doses administered, so it can be a major factor in drug performance. If the drug is highly protein bound and released slowly, it is not readily available to other tissues. This does protect it from liver metabolism but also makes it difficult for it to reach the target tissues where its receptors or enzyme target lies and so interferes with its action. If it is minimally protein bound and readily released, it can freely reach tissue enzymes and receptors, so serum protein binding is less consequential. For these reasons the medicinal chemist avoids making lipophilic acids if possible.

3.3.9 Metabolism

Drug metabolism can take place in most tissues of the body, including the gut wall, but the liver is the main site of metabolism. The liver is the first organ to receive blood-carrying absorbed materials from the intestines. Metabolism is a complex process often strongly affecting the performance of a drug. Metabolism can enhance, result in an indifferent effect, or be deleterious to drug action. In many cases, enzymes alter the structure of administered drugs by converting them to more polar substances by hydrolysis or oxidation. The modified molecule will distribute differently and have a different fit to its receptor(s) than the molecule administered. Metabolism is often benign, but in some cases results in toxic products and must be prevented or avoided. Medicinal chemists have become skilled in designing molecules that are protected from metabolic change or in utilizing metabolism to enhance the desirable properties of a drug. For example, changing a hydrogen atom to a halogen on an aromatic ring is often effective in preventing oxidative hydroxylation at that position. Many other metabolic reactions are recognized. Hydrolysis and oxidation are common metabolic reactions. When metabolism fails to produce sufficient water solubility, conjugation with sulfate or glucuronic acid (an acidic sugar) can be employed to enhance solubility further. When these processes are complete, the drug escapes from the liver and passes either back into the gut or into the general circulation. Many drug molecules, particularly those with sufficient intrinsic water solubility, are not metabolized at all in the liver and pass rapidly into the general circulation without significant change. The medicinal chemist must be able to control or modify metabolic processes.

3.3.10 Distribution

Most tissues are richly vascularized (e.g., lung and heart), whereas others are not (e.g., bone). Thus, even under ideal conditions, organ distribution can be expected to be uneven and to play a role in the action of a drug. The brain deserves special mention in this regard. Only quite lipophilic materials can enter the brain by passive diffusion, and the range of molecules that are taken in actively (e.g., glucose) is quite limited. This special characteristic is referred to as the *blood–brain barrier*. Drugs intended for central nervous system delivery generally must be more lipophilic than drugs designed to perform in other tissues. The lungs, heart, liver, and kidneys receive comparatively significant concentrations of drug administered early. Other organs get their drug concentrations in more dilute form and more slowly.

3.3.11 Excretion

Excretion, primarily in the urine, requires significant water solubility to be efficient. Another important point to consider is that normal urine is a protein-free filtrate. Thus, compounds that are both extensively and tightly serum protein bound are excreted slowly, resulting in long residence times in the body. The ability to adjust the excretion rate of a substance is crucial to ensure that the disease being treated benefits from a sufficient quantity of the agent being present at the receptor for a sufficient period of time.

3.3.12 Patenting

For intellectual property rights and freedom to operate, they must be achievable. The drug industry is a worldwide enterprise, and the costs of a new agent are so high that no drug will be introduced if there are insufficient proprietary rights in all significant global markets to give the investing group a reasonable chance for a satisfying return on their money. Since a patent requires at a minimum both novelty and utility, the agent must be not only chemically unprecedented but also efficacious in an economically significant disease state.

3.3.13 Pharmaceutical Properties

The organoleptic properties of a drug must also be satisfactory. The agent must not have an unattractive color, a disgusting taste, or a pungent odor. The agent must also be compatible with the ingredients normally necessary for formulation in the form of tablets, capsules, injectable solutions, and the like. The agent must not be hygroscopic. It also cannot possess metastable isomeric crystal habits. This last factor can lead to subsequent alterations in behavior due to undesirable changes in crystal form and thus modified solubility properties. In that case, the entire evaluation process must be redone at a huge expense.

3.3.14 Idiosyncratic Problems

There are other problems of a more idiosyncratic nature (i.e., associated with individual projects and not occurring generally) with which a medicinal chemist must be prepared to deal. In every project there are unanticipated difficulties that develop between inception and introduction. The medicinal chemist must be skillful and resourceful enough to provide acceptable solutions for these as the need arises.

3.3.15 Summary

Taken together, it can be seen that the synthetic chemist must be able to modulate structures through thoughtful analoging so as to find final structures with an acceptable collection of additional properties without losing the efficacy sought. The development chemist can readily appreciate that a compound that survives all of this is not discarded lightly. Even if the synthesis in use up to this point is complex, dangerous, and costly, these represent challenges that, fortunately, the development chemist is skilled at resolving without requiring that the clinical candidate be discarded.

3.4 EXAMPLE OF DRUG DEVELOPMENT THAT ILLUSTRATES MANY OF THE AFOREMENTIONED CONSIDERATIONS

Unfortunately, space is not available to give specific examples of all of these problems and of the various solutions that have been devised. Nonetheless, it is likely that a general discussion such as the preceding may not be entirely clear to those who are new to these considerations. The methotrexate example given earlier deals mostly with host–guest relationships rather than ADME and other considerations. Thus, the following is a brief recounting of a different drug-seeking campaign from which the interested reader can see how many of those influences play out in actual practice. The practical result of the investigations recounted briefly in this section is that more than a dozen commercially successful compounds have been introduced into human use, and a few of these are billion-dollar-a-year drugs.

3.4.1 Control of Blood Pressure with Drugs

Despite many decades of intensive research and many notable successes, cardiovascular disease remains one of the major killers of humankind. Unfortunately, the symptoms of hypertension are subtle, and many sufferers are unaware of their problem until significant and irreversible pathology to vital organs has set in, leading to strokes, heart failure, and kidney damage. There are three main mechanisms by which blood pressure becomes pathologically high: increased pumping of blood by the heart, decreasing elasticity of the arteries, and decrease in kidney filtration so that fluids in the body build up.

3.4.2 Historical Background

A century ago the major therapy available for treating high blood pressure comprised diuretics and related materials (e.g., digitalis) that reduced blood volume. Added to this was nitroglycerine. Discovered accidentally as a by-product of explosives manufacture, nitroglycerine alleviates some of the symptoms of angina pectoralis (chest pain associated with narrowing of the arteries that supply the heart with its own blood supply) by relaxing certain blood vessels. Nitroglycerine serves as a source of the gaseous neurohormone nitric oxide. Another useful substance was quinidine, whose action is to treat cardiac arrhythmia (weak and improperly synchronized heart beats) by altering nervous conductivity. Quinidine was discovered as the result of observing a useful side effect in treating patients suffering from malaria.

All of these agents treat symptoms rather than the underlying pathology, so are not curative. Even half a century ago we had hardly progressed much beyond these means, so that the causes and treatment of cardiovascular disorders were still comparatively primitive. Since that time a group of remarkably effective agents have been discovered that have saved a multitude of lives and returned multibillions of dollars to shareholders.

The story below is offered to demonstrate how an incompletely met medical need can be addressed, utilizing knowledge gained painfully from close observation of natural phenomena. In the right hands, this can be translated through intelligence, diligence, creativeness, and luck into a medically useful solution by analoging.

3.4.3 Finding a Starting Point: A Clue from Nature

To do something different than previously existed, how should one begin? In this case the initial clue came from an understanding of a natural phenomenon. The deadly South American pit viper *Bothrops jararaca* has solved in an interesting way the problem that many snakes have. That is, if they kill their prey successfully through the injection of venom, it is important that the prey die quickly so that they cannot run off before dying, as the snake cannot follow far and it does the pit viper little good to kill something that it cannot eat. A profound drop in blood pressure following intoxication leads to fainting, if not outright death. The venom that produces this result solves the snake's dietary problem promptly because it can now catch up with its immobilized victim and consume it.

A few insightful scientists understanding these considerations recognized the potential value of snake venom to control malignant blood pressure if administered carefully to sufferers. To understand how the venom works in reducing blood pressure and how it might be used for deliberate human therapeutic work, it is important to understand a bit of physiology. The entire process is wonderfully complex, and the nonspecialized reader can easily lose the thread of the narrative if it is explicated in complete detail. Thus, the following is an abridged version that covers only the main points.

3.4.4 Renin–Angiotensin–Aldosterone System

Maintenance of blood pressure and composition is vital for survival. The volume and initial contents are provided by intake of water and nutrients into the gestrointestinal tract, from which they pass into the blood. The heart keeps the blood moving. It supplies oxygen for energy generation from burning (oxidizing) food by circulation to the lungs, from which at the same time volatile wastes are removed. The oxygenated blood is returned to the heart and pumped into the blood vessels. These must distend to hold this volume. They posses the property of elastic rebound, which process squeezes the blood onward to the tissues. Backflow is prevented by a one-way valve in the heart. The body clears its wastes in large part by excretion in the urine. The kidneys play an essential role in this and function as a pressure-driven filter, leading to excretion of salt and water. Blood volume and content are regulated by selective reuptake of necessary amounts of these things. Equilibrium between intake and output is important for the system to work property. If the kidney filtration rate drops too low, the heart can beat more strongly, and if it is too high, it can beat less powerfully. The blood vessels can alter their diameter to cooperate with this process such that decreased diameter raises the blood pressure. These considerations are illustrated schematically in Fig. 3.6.

To regulate these activities, a variety of hormones and enzymes play various roles. One of these is the enzyme renin, produced and stored in kidney blood vessel wall, from which it is released as needed when sodium ion concentration or blood pressure falls. A simplified version of this complex sequence of events is given in Fig. 3.7.

Renin, an aspartoyl peptidase, catalyzes the cleavage of four amino acids from the acid end of angiotensinogen by water, so producing angiotensin I. This 10-unit peptide is only weakly active in conrolling blood pressure. A second enzyme, *angiotensin-converting enzyme* (ACE), cuts dipeptide (two amino acids) units from the acid ends of a number of peptides, of which angiotensin I is most relevant to our story. This produces the octapeptide angiotensin II. This octapeptide hormone acts very powerfully to increase blood pressure following binding to specific receptors. Angiotensin II also triggers the formation of another

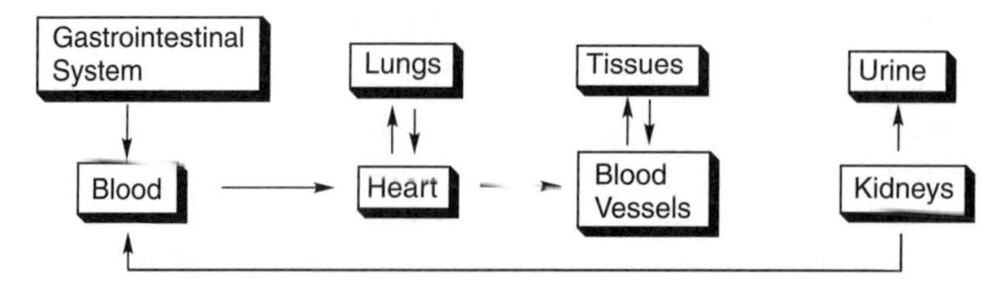

Figure 3.6 Interplay of several organs in the regulation of blood pressure. Blood volume and content is regulated from intake from the GI system and excretion of fluid from the kidneys. The heart keeps it moving and the lungs add oxygen for the generation of energy. The kidneys filter the blood, removing soluble wastes and passing them into the urine. Water, salt, etc., are reabsorbed to prevent excessive losses. The whole system is also regulated by the brain in ways not illustrated here.

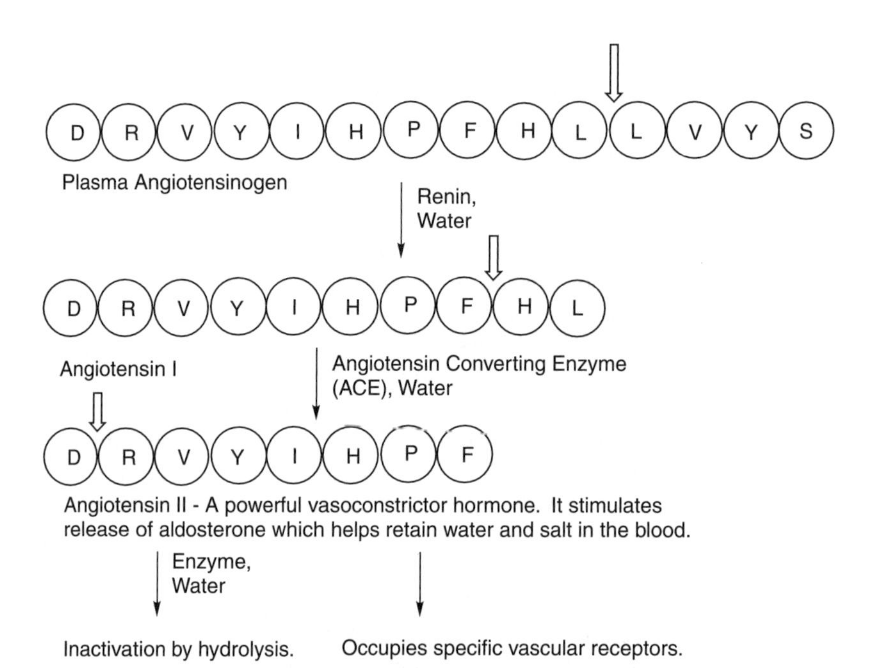

Figure 3.7 A simplified diagram of the renin-angiotensin system for regulating blood pressure. Renin is released by certain kidney cells when the blood pressure or salt concentration fall too low. The circles represent individual amino acids and the letters are the standard one-letter code for the amino acids. The fat arrows indicate where hydrolysis takes place. Angiotensin-converting enzyme (ACE) is released from the lungs. The letters in the circles are individual amino acids joined together by peptide bonds. Bradykinin is another peptide hormone that plays a role. It combats the blood pressure elevating action of angiotensin II. It is also hydrolyzed with the aid of ACE.

contributing hormone, aldosterone. The action of aldosterone is significant in reabsorbtion of sodium ions from kidney tubule filtrates. This salt reabsorption concomitantly also produces significant water reuptake, maintaining blood volume and consequently blood pressure. Fine control of blood pressure in this system is achieved by yet another hormone, bradykinin which antagonizes the action of angiotensin II and so is hypotensive. The action of ACE is nearly equipotent in forming angiotensin II and destroying bradykinin, both actions resulting in blood pressure rise. The action of angiotensin II is essentially terminated by cleavage into a smaller peptide, angiotensin III, by an aminopeptidase (an enzyme that chews off an amino acid from the amino end). Other angiotensin peptides are known, but a discussion of these would needlessly confuse the story.

It may seem strange and cumbersome to involve so many molecules in this process, but it allows for fine control. In sum, angiotensin II and aldosterone combine to increase blood pressure. Bradykinin antagonizes the action of angiotensin II. Thus, it seemed reasonable to presume that inhibiting one of these enzymes would lead to a decrease in blood pressure. In case the reader is wondering why bradykinin was not chosen as a target instead, much experience has shown that it is far easier to inhibit an enzyme than it is to enhance its action. Thus, renin- and angiotensin-converting enzymes appeared to be more tractable targets.

3.4.5 Attempts to Inhibit Renin

Initially, a major effort was made in many laboratories to search for suitable inhibitors of renin, the enzyme that starts the cascade. Despite finding a number of inhibitors with powerful in vitro potency and the discovery of much useful information about the various ways to convert peptides into more druglike molecules, no orally active agent emerged from this work. Among the difficulties that could not be solved satisfactorily, the molecular weight of strong inhibitors always turned out to be too high for satisfactory absorption. Clearly, control of blood pressure through the renin–angiotensin–aldosterone cascade would have to be found elsewhere.

3.4.6 Attempts to Inhibit Angiotensin-Converting Enzyme

Attention was turned to the next enzyme in the system, ACE. It was clear at the outset from a study of snakes that this was a feasible means of reducing blood pressure, so proof of principle was not a problem. By about 1965 it was found that the active principles in the venom of the South American pit viper *Bothrops jararaca* were a mixture of closely related peptides that inhibit angiotensin-converting enzyme. The venom is collected in the gums of the snake and is injected into the victim through hollow teeth (fangs) when the snake strikes because the gums are compressed as a consequence of the bite. This inhibition reduces blood pressure and immobilizes the victim. The question, then, was whether medicinal chemists and physicians could utilize this mechanism to control hypertension through the crafty application of drugs based on the venon. Unfortunately, the active constituents in the venom, being peptides, must be injected in order to work. The snake has no problem with this. Physicians and patients, however, would. As noted, most hypertensive individuals do not perceive unpleasant symptoms from their disease until their disease is far advanced. Consequently, there would be little motivation to take a lifelong series of injections when generally feeling rather fit. Under these circumstances, the treatment would be more apparently troublesome than the disease. Consequently, successful drugs for this indication would have to be taken orally and not have unpleasant side effects or they would not be used.

3.4.7 Peptides Make Poor Orally Active Drugs

Peptides make poor drugs for chronic oral administration, as most of them are digested before absorption from the gastrointestinal tract. An intense effort has been expended to overcome this. In most cases where this has been successful, they have been molecules with comparatively low molecular weight and whose peptide bonds have been altered to make them resistant to digestion. These altered, unnatural analogs are known as *peptidomimetics*, and the hypotensive agents inspired by snake venom peptides are one of the classical examples of how this can be done.

3.4.8 Analoging Studies of Pit Viper–Inspired Peptides

By 1971 several synthetic peptide-based analogs of the pit viper agents had been prepared at Squibb (now Bristol–Myers Squibb) without finding an orally active analog. The main understanding that came out of this preliminary work was that the intensity of the effect could be modulated by changing the specific identity and sequence of amino acids in the inhibitor and that proline was a good amino acid to have in the first position. Persistence ultimately produced dramatic results when the peptide character of the inhibitors was successfully reduced.

3.4.9 Peptidomimetics

The next important clue came from the literature. Converting enzyme was known to be a zinc protease, and the general catalytic mechanism of such enzymes was known. Water itself is much too weak to cleave peptides (or taking a shower would be a deadly act!); it takes centuries to cleave a peptide bond unless the reaction is catalyzed. The zinc atom of the enzyme bonds to a key water residue and greatly activates it so that peptide hydrolysis is rapid, specific, and effective. This is illustrated in Fig. 3.8.

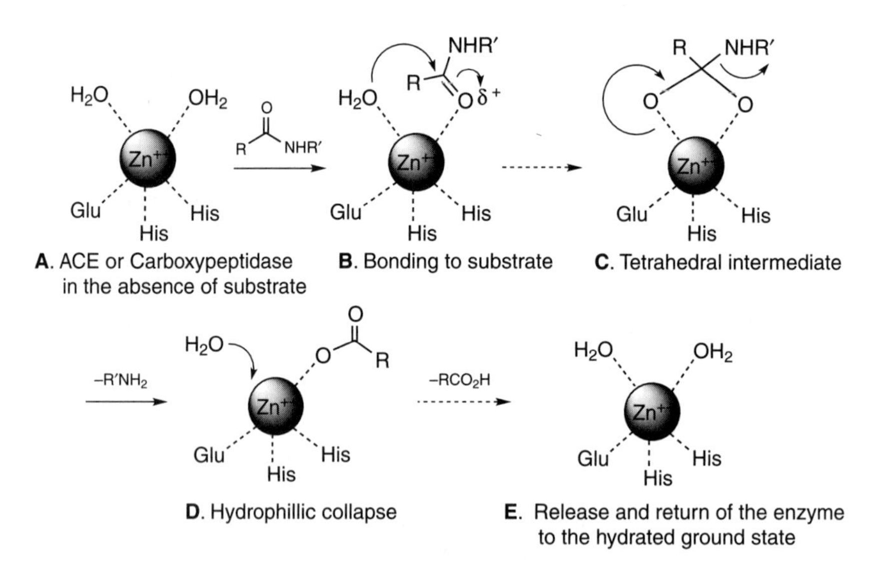

Figure 3.8 Peptide bond hydrolysis by zinc-metalloproteases.

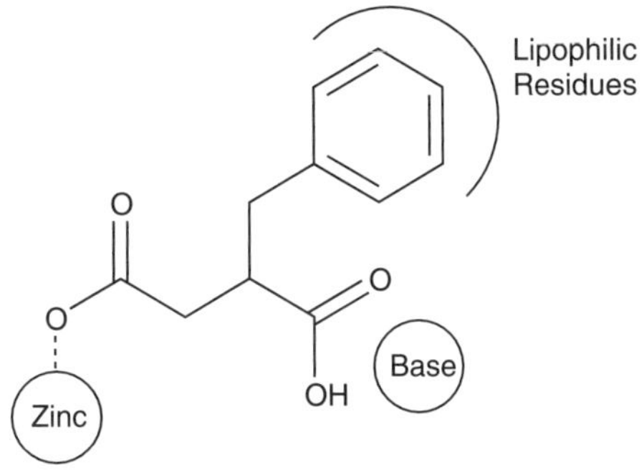

Lipophilic Residues

O O O OH Base Zinc

Scheme 3.2 Putative binding mode of benzylsuccinic acid to carboxypeptidase A.

Work on an apparently analogous enzyme, carboxypeptidase A (CPA), also known to be a zinc-based peptidase, had previously shown that comparatively simple compounds, especially 2-benzylsuccinic acid, would inhibit the enzyme. In this simplified system (illustrated in Scheme 3.2) a docking carboxyl group was present in the first position so as to align the enzyme with its ligand, and a second carboxyl group was present down the chain to interact with the zinc atom inactivating the enzyme. Binding the zinc atom to the carboxyl group of the inhibitor produces a much weaker nucleophile and displaces the water that would normally function to cleave the peptide bond. Interrupting this process by providing an alternative ligand inhibits the enzyme. The benzyl moiety evidently fits into a lipophilic pocket and/or helps align the rest of the carbon chain. The success of this inhibition served as a model for the ACE project.

3.4.10 Adaptation to Inhibition of ACE

In contrast to carboxypeptidase A, which cleaves a single amino acid at a time from the end of a peptide chain, ACE cleaves a dipeptide. The linker separating the docking carboxyl and the zinc-interacting carboxyl would therefore have to be longer. Thus, a series of dicarboxylic acids were prepared with a variety of aliphatic spacers between the two carboxyl groups to see if this idea would work with ACE. These were all found to be weak inhibitors of ACE and were not very stable. Only much later did it become apparent that the carboxypeptidase A model was wrong in some respects. In any case, to enhance potency, the distal carboxyl group was exchanged for a sulfhydryl group, as this was proposed to be a stronger ligand for the zinc atom than is a carboxylate. This bioisosteric exchange proved promising, even though the potency was still unsatisfactory, so a further series of analogs was prepared in which additional systematic structural variations were explored (Scheme 3.3).

This type of analoging of a lead substance to reveal the influence of the various functional groups and selecting the best embodiments is the soul of drug seeking and is called establishing a structure–activity relationship. In this particular instance, it was quickly found that the initial carboxyl group, preferably attached to a proline moiety, an alkyl group with the appropriate stereochemistry adjacent to the amide (peptide bond), the space between the docking carboxyl and the zinc ligand, and a thiol rather than a carboxyl chelating to the zinc were all important. In particular, it was noted that an analog ending in a thiol at the left end was more than 1000 times more effective than one with a carboxyl in the analogous position.

Analog	Relative k_i vs. ACE	Remark
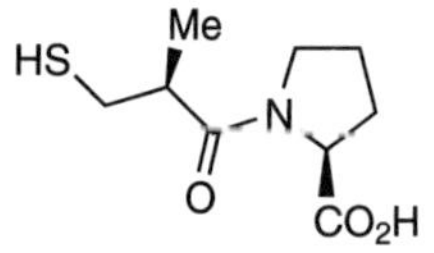 HS, Me, N, O, CO_2H	1.0	Captopril (standard) k_i ACE = 1.7×10^{-9}M CPA = 6.2×10^{-4}M CPB = 2.5×10^{-4}M (5 orders of magnitude more selective)
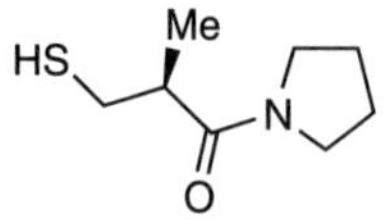 HS, Me, N, O	12,500	Docking carboxyl essential
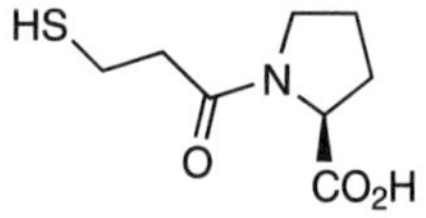 HS, N, O, CO_2H	10	"Greasy" methyl is helpful
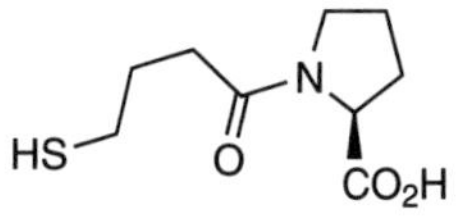 HS, N, O, CO_2H	12,000	Distance between amide carbonyl and the zinc ligand is important
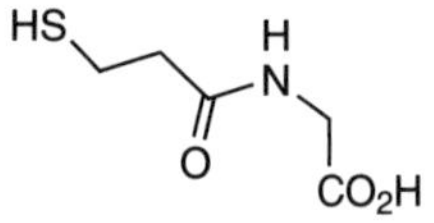 HS, H, N, O, CO_2H	120	The pyrrolidine ring is helpful
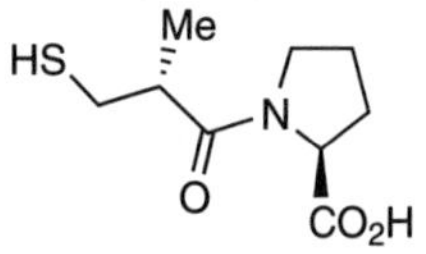 HS, Me, N, O, CO_2H	120	The stereochemistry of the "greasy" methyl is important
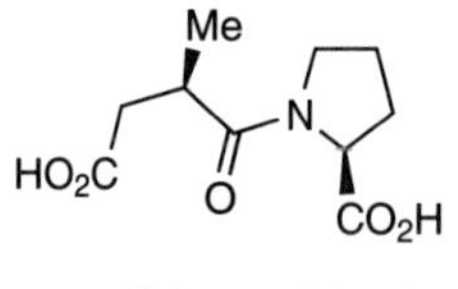 HO_2C, Me, N, O, CO_2H	1100	A sulfhydryl group is more useful than a carboxyl

Scheme 3.3 Structure – activity relationships (SAR) of captopril analogs.

The best of the analogs prepared in this work, now known as *captopril*, is a very effective drug for reducing high blood pressure. Its interaction with the enzyme was schematized at the time as shown in Scheme 3.4, in comparison with angiotensin I substrate. This sort of scheme is commonly used in advance of knowledge of the molecular interactions actually occuring. It is a convenient way to summarize the structure–activity findings but must not be taken literally. Among other problems with a literal interpretation, the reader will note that the chemical bond angles are all wrong in such a drawing. In fact, the present picture has been established recently by x-ray determination of the enzyme and its various inhibitors and differs in some significant details from the early cartoon.

The carboxyl and the sulfhydryl play their anticipated key role in successful inhibition. The carboxyl associates with an amine in the enzyme, and this is the docking interaction.

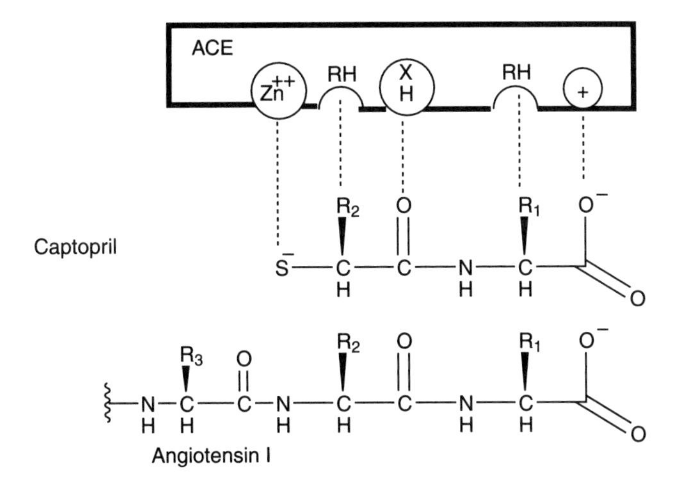

Scheme 3.4 Early beliefs about the interactions between captopril and angiotensin I with angiotensin-converting enzyme (ACE).

The other interactions take place next and establish potency and selectivity. Of great interest is the finding that captopril's very significant potency against ACE ($K_i = 1.7 \times 10^{-9} M$) is dramatically higher than its potency against carboxypeptidase A ($K_i = 6.2 \times 10^{-4} M$) or carboxypeptidase B ($K_i = 2.5 \times 10^{-4} M$). (Carboxypeptidase comes in more than one isomeric form.)

The implication of the preference for proline at the first position is that the specific conformation imposed by this amino acid is particularly well suited for the needs of ACE. One notes that the normal substrate does not begin with this amino acid at all, so the normal substrate probably must adopt an unusual conformation when bound to ACE. With proline as the lead amino acid, the inhibitors are putatively preorganized in the most suitable conformation for inhibition of the enzyme. This idea was reinforced by the finding that a number of dipeptides terminating in proline inhibited the enzyme somewhat and that the strongly inhibitory snake venom peptides frequently ended in proline.

Captopril was subsequently introduced into the clinic as a very useful and popular orally active antihypertensive agent. This is the most gratifying proof of principle demonstrating that a peptidomimetic inhibitor of the renin–angiotensin–aldosterone system would be an excellent way to control hypertension. As a further benefit, ACE inhibitors can be combined with diuretics to give even more powerful control of blood pressure.

This brief recounting leaves out a number of false starts, as the pathway to success is usually obscure before the fact and often requires the synthesis of hundreds or thousands of analogs before finishing. Many false trails are followed for a time, and the effort usually requires much successive hypothesis formulation and synthesis and evaluation cycling before a useful final result emerges.

3.4.11 Success Inspires Competition

Captopril's acceptance inspired others to enter the field with competing molecules. Proof of concept was now at hand and ACE had become a well-validated therapeutic target. To be able to compete successfully with captopril, it was essential that new contenders possess significant advantages. In this context, physicians soon noted that acceptability of captopril was diminished by the incidence of rashes and a comparative loss of taste sensations. These side effects were not severe but were disincentives to compliance, and a number of laboratories attempted to find agents retaining the desirability of captopril as a hypotensive but with side effects reduced or eliminated. By analogy to another drug that had been marketed for another indication but that also had these untoward properties, a research group at Merck hypothesized that the taste and rash were associated with the presence of the sulfhydryl group.

3.4.12 Taking a Different Approach

If this hypothesis were correct, it would be necessary to get rid of the sulfhydryl moiety. After examination of a number of potential alternatives, it was therefore decided to go back to the carboxyl group as a zinc ligand, but it was necessary to enhance potency by further analoging, as this route had proven very much less successful in the precedent work at Squibb. It is sometimes found in medicinal chemical studies that this can be accomplished by introducing another molecular feature into a lead substance that can interact with a feature in the enzyme that is not utilized by the initial lead. When this works, the added binding site gives additional energy of association and specificity. It also gives much greater inhibitory power. This must not make the molecular weight too high or introduce otherwise unsatisfactory interactions.

There were other problems to address as well. A sulfur atom is roughly twice as large as an oxygen atom (strictly speaking, it is more bioisoelectronic than bioisosteric), so the spacer group would probably have to be made somewhat longer than had been done with captopril. Adding an additional methylene unit to accomplish this was helpful, as was use of an alanine linker arm, but both together were not enough (Scheme 3.5). Partly counterbalancing the polarity of the NH unit by adding a methyl gave a rather significant enhancement of potency. After significant empirical experimentation, projecting an arm ending in an aromatic ring in the form of a homophenylalanine residue apparently associated with a greasy pocket in the enzyme and enhanced bonding dramatically. Enalaprilat was found to be very potent, even more potent than captopril itself in vitro. The putative interaction of enalaprilat with ACE is analogous to that of captopril, as illustrated in Scheme 3.4, except that a third greasy pocket would need to be added to the left in the diagram to accomodate this finding.

Enalaprilat was not very effective when given orally, however, due to poor absorption. This was judged to be a consequence of the possession of two carboxyl groups, making it too ionizable for efficient oral activity. Thus, it was decided to modify the structure so that the distal group was present as an ethyl ester. This solved the polarity problem, and this comparatively well-absorbed analog was given the name *enalapril*. The ester function, however, greatly decreased inhibition of the enzyme in vitro, but after oral absorption, esterases cleaved enalapril to produce the much more active enalaprilat in the blood where ACE is found. Thus, one had the best of both worlds. Enalaprilat inhibits ACE at 1.2 n*M*! The ethyl ester of enalaprilat (enalapril) has found great acceptance in oral treatment of human hypertension. The alteration of a functional group so as to enhance absorption with the intention that the blocking group be removed by enzymic action in the body to regenerate the active drug is known as *prodruging*.

3.4.13 Analoging to Enhance Absorption

Another way to overcome the absorption problem, but without resort to prodruging, with its requirement that the patient cooperate by activating the drug (not everyone is capable of doing this

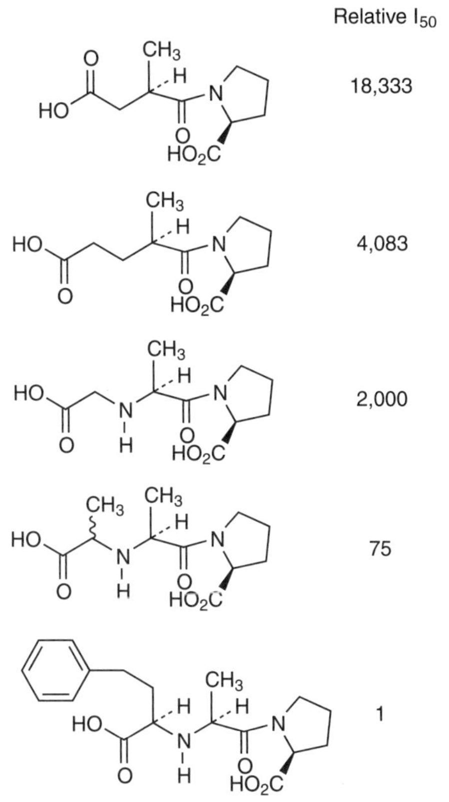

Relative I_{50}

18,333

4,083

2,000

75

1

Scheme 3.5 Synopsis of some of the structure – activity alterations leading to enalaprilat (bottom structure).

efficiently), is to alter the isoelectric point of the drug so that its degree of ionization (acidity) is reduced sufficiently that it passes through the gut wall at an acceptable level. *Lisinopril* (Scheme 3.6) embodies this concept. In this important drug, an amino group has been introduced into a side chain of the molecule to balance the acidity of one of the two carboxyl groups.

One popular way to investigate drugs such as these is intellectually to split them into component parts each of which can be synthesized readily and then assembled. In this manner, enalaprilat and lisinopril could be envisioned as being a combination of three parts, A–B–C. A thorough investigation would amount to analoging A, B, and C and then joining them together. Hopefully, the best embodiment of A would add its virtues to the best embodiments of B and C as well.

Lisinopril apparently emerged from such as exercise in which the central alanine of enalaprilat (B in the example of preceding paragraph) was replaced by a series of amino acid residues. It had been postulated that the function of the methyl group of alanine was to dictate the most useful conformation, but it was not clear from this whether further benefits could be achieved by analoging at this position. From this group of analogs, lisinopril's relative potency leaps out (Scheme 3.6, item 8). Its specific potency is equal to that of enalaprilat. Even more important, lisinopril is orally effective at nearly the same dose as its ethyl ester, so prodruging is not necessary to achieve control of high blood pressure. Lisinopril has become an extremely important drug.

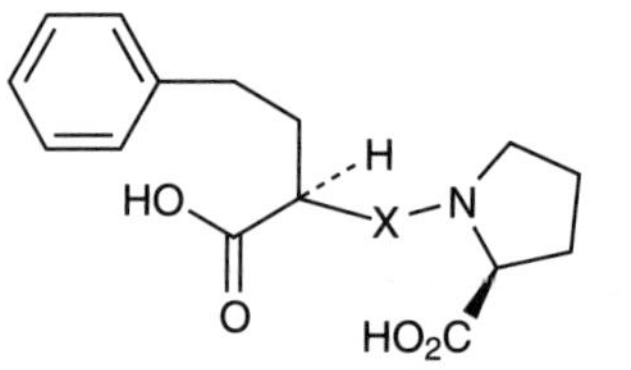

Analog	X =	Relative I_{50}
1	Gly	192
2	L-Ala	3.2
3	Alpha-MeAla	2083
4	N-Me-L-Ala	83
5	Val	65
6	His	62
7	Phe	48
8	Lys	1

Scheme 3.6 Amino acid substitutions leading to lisinopril analog 8.

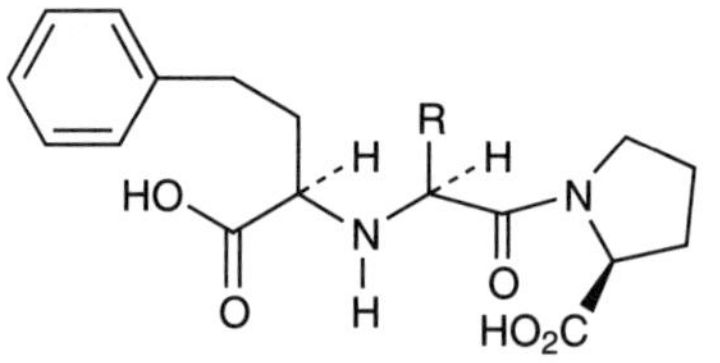

R =	Relative I_{50}
CH_2NH_2	480
$CH_2CH_2NH_2$	38
$CH_2CH_2CH_2NH_2$	4.5
$CH_2CH_2CH_2CH_2NH_2$	1
$CH_2CH_2CH_2CH_2NMe_2$	4.3
$CH_2CH_2CH_2CH_2NHAc$	7.6
$CH_2CH_2CH_2CH_2CH_2NH_2$	13

Scheme 3.7 Investigation of lisinopril analogs for comparative potency.

Lysine is particularly convenient since it is a natural amino acid. To see whether it was the best amine to use, a number of analogs were investigated systematically. It was indeed the best at that position. Thus, its contribution to activity is not only a consequence of its polarity, but it apparently finds a compatible partner in the enzyme to which it binds particularly well. The data in Scheme 3.7 support this conclusion. It is also now known that lisinopril benefits from active uptake by a peptide transporter.

This abbreviated exposition of their development of ACE inhibitors illustrates the systematic work that goes into drug design and discovery. It shows why advances achieved as a consequent of so much effort are not abandoned without great reluctance.

3.4.14 Clinical SAR

The great clinical acceptance of these three drugs led to intense competition by a variety of firms, and a substantial number of additional analogs are now available. As usual, as time elapses and the patent literature becomes increasingly congested, the various analogs introduced bear a less and less obvious molecular resemblance to captopril, where the story began. This is

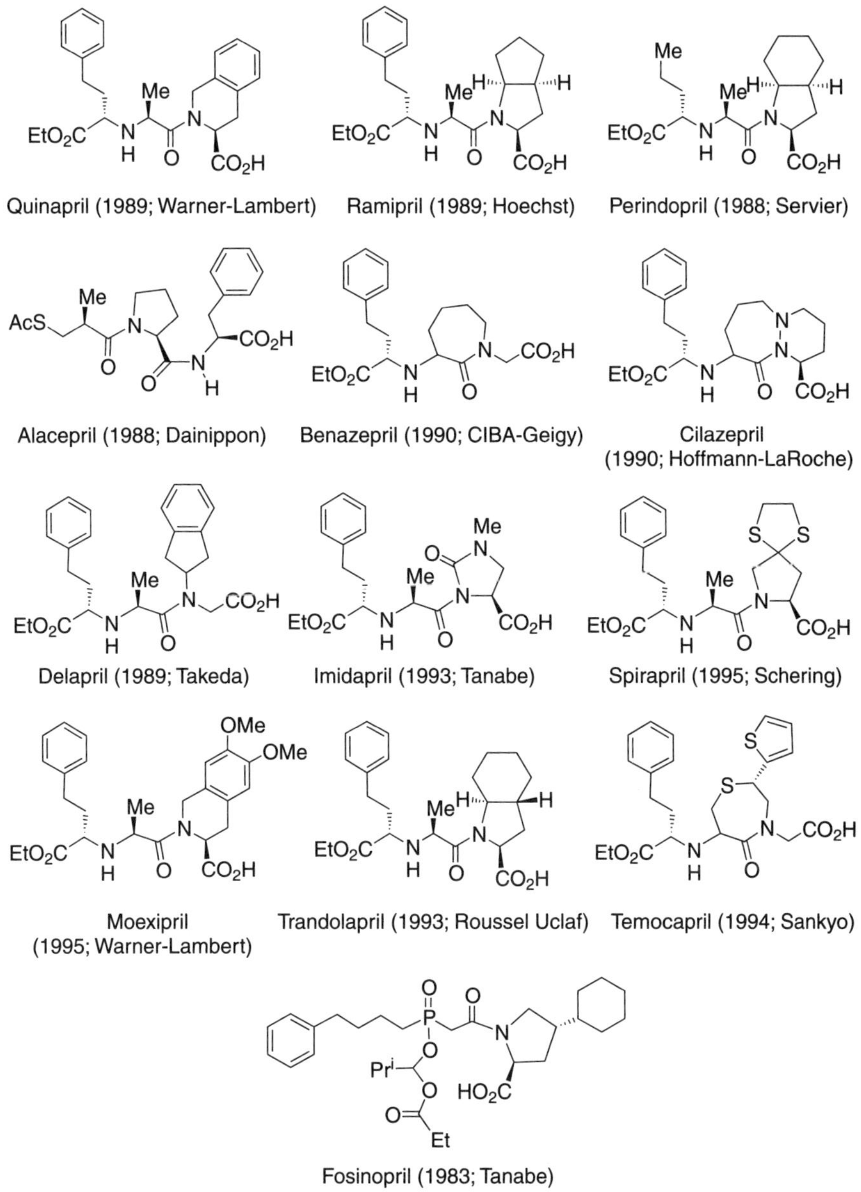

Quinapril (1989; Warner-Lambert) Ramipril (1989; Hoechst) Perindopril (1988; Servier)

Alacepril (1988; Dainippon) Benazepril (1990; CIBA-Geigy) Cilazepril (1990; Hoffmann-LaRoche)

Delapril (1989; Takeda) Imidapril (1993; Tanabe) Spirapril (1995; Schering)

Moexipril (1995; Warner-Lambert) Trandolapril (1993; Roussel Uclaf) Temocapril (1994; Sankyo)

Fosinopril (1983; Tanabe)

Scheme 3.8 Clinical competitors for captopril, enalapril, and lisinopril.

clear from inspection of the structural formulas of these agents (Scheme 3.8). In time it was found that the docking carboxyl-bearing ring could be other than proline (component C in the figure) but with the exception of temocapril, the bridging alanine unit (B) has generally been preserved, with a few exception, as has been the homophenylalanine ester moiety (A).

Most of these analogs use the prodrug concept and compete for a more and more fragmented market. As often happens, the first drugs introduced continue to generate the biggest rewards to their firms. First is best, and good enough, soon enough, are mottoes that one hears frequently in discussions between medicinal chemists.

3.4.15 More Recent Work

Later work demonstrated that successful manipulation of the renin–angiotensin–aldosterone system could also be achieved by interfering with the productive association of angiotensin II with its receptors. This stratagem allows the body to produce angiotensin II unimpeded but interferes with binding of the hormone to its receptor and decreases blood pressure in this way. Recounting this exciting story would also be profitable but would take us away from illustrating our main theme.

3.4.16 Résumé

This account benefits from some retrospective wisdom in order to make a coherent story from a complex sequence of events. More important, however, it serves not only to illustrate many of the principles set forth in the first part of this chapter but also as one of the comparatively few examples of the rational conversion of a parenteral peptide lead into an orally active peptidomimetic drug.

3.5 CONCLUSIONS

It should now be amply clear that drug design and optimization is a complex activity requiring an understanding of cellular biochemistry, pathology, and synthetic chemistry. The molecules that emerge from drug-seeking campaigns are often only a very small percentage of those prepared and tested. These molecules are not abandoned readily. They are, however, all too often not yet ready for the market because a practical synthesis must yet be found. This requires a high level of skill on the part of the developmental chemist. The remainder of this book is devoted to this important topic.

ADDITIONAL READING

The reader who is interested in digging deeper into a few of the topics covered in this survey can profit from consulting the following articles prepared by authors actively engaged in the work synopsized.

Cody, V., Galitsky, N., Rak, D., Luft, J. R., Pangborn, W., and Queener, S. F. (1999). Ligand-induced conformational changes in the crystal structures of *Pneumocystis carinii* dihydrofolate reductase complexes with folate and NADP, *Biochemistry* 38:4303–4312.

Lipinski, C. A., Lombardo, F., Dominy, B. W., Feeney, P. J. (2001). Experimental and computational approaches to estimate solubility and permeability in drug discovery and development settings, *Adv. Drug Deliv. Rev* 46:3–26.

Ondetti, M. A. (1981). Inhibitors of angiotensin-converting enzyme, in *Biochemical Regulation of Blood Pressure*, R. L. Sofer, Ed., Wiley-Interscience, New York, pp. 165–204.

Patchett, A. A. (1993). Enalapril and lisinopril, in *Chronicles of Drug Discovery*, Vol. 3, D. Lednicer, Ed., ACS Professional Reference Books, American Chemical Society, Washington, DC, pp. 125–162.

Veber, D. F., Johnson, S. R., Cheng, H.-Y., Smith, B. R., Ward, K. W., and Kopple, K. D. (2002). Molecular properties that influence the oral bioavailability of drug candidates, *J. Med. Chem.* 45:2615–2623.

4

COMBINATORIAL CHEMISTRY IN THE DRUG DISCOVERY PROCESS

IAN HUGHES
GlaxoSmithKline Pharmaceuticals
Harlow, Essex, UK

4.1 INTRODUCTION

What is combinatorial chemistry, and what is its value to medicinal chemists engaged in the drug discovery process? During the decades leading up to the 1990s, approaches to synthetic chemistry remained virtually unchanged. Obviously, the tools available to the chemist improved considerably, albeit incrementally, over the years: new, milder reaction conditions, improvements to purification and analytical methods, and the ability to synthesize ever more complex molecules. However, during this period a single medicinal chemist would be expected to make, on average, around 50 new molecules for screening per year and would typically work on only one or two molecules at a time. In the early 1990s, the medicinal chemistry community witnessed the start of a paradigm shift in the way they would work in the future. A variety of new techniques that had been developing in the field of peptide synthesis, with little impact on most synthetic chemists, were shown to be applicable to small molecule synthesis. The following years witnessed a flurry of activity as the new technologies that now constitute the field of combinatorial chemistry were developed. Driving this surge of interest was the potential to increase the speed and efficiency of drug discovery with the obvious economic advantages, as well as opportunities to commercialize the technologies themselves.

Our goal in this chapter is to review key aspects of the "combinatorial revolution" and their application to the drug discovery process. It would be foolish to attempt comprehensive coverage of such a broad field in a few pages, so the highlights selected are accompanied by a bibliography of reviews and books giving more in-depth coverage of the subjects at the end of the chapter. As with any field of science, a terminology has developed

Drug Discovery and Development, Volume 1: Drug Discovery, Edited by Mukund S. Chorghade

to communicate the new concepts, and a useful glossary has been published.[1] A similar glossary of medicinal chemistry terms might also be useful for the reader.[2] The selection of topics covered is based on the author's experience and perspective and should in no way be taken to imply inferiority of any approaches omitted. Again, to limit the material covered, the primary focus is on the application of combinatorial techniques to small molecule synthesis, as opposed to that of biopolymers (oligopeptides, oligonucleosides, and oligosaccharides).

4.1.1 The Birth of Combinatorial Chemistry

When and how did combinatorial chemistry begin? Lebl gives an interesting perspective of some "classical" papers in the area.[3] Some of the key milestones are summarized in Table 4.1 and discussed below. It is interesting to note that many of the main developments were first applied to peptide chemistry with translation to small molecule synthesis often

TABLE 4.1 Key Developments in Combinatorial Chemistry

Year	Class[a]	Contributors	Development	Ref.
1963	P	Merrifield	Solid-phase peptide synthesis	4
1970	SM	Leznoff	Early nonpeptide solid-phase synthesis	6
1984	P	Geysen	Multipins for parallel synthesis	7
1985	P	Houghten	Teabags	9
1988	P	Furka	Mix-and-split synthesis	10, 11
1991	P	Fodor	Light-directed spatially addressable parallel synthesis	12
	P	Houghten	Screening of mixtures	13
	P	Lam	On-bead screening: one bead, one peptide	22
1992	P	Houghten	Positional scanning	14, 15
	P	Brenner, Lerner	Oligonucleotide tags for encoding	23, 24
	SM	Ellman	Solid-phase synthesis of benzodiazepines	25, 26
1993	SM	De Witt	Diversomers; parallel solid-phase synthesis on resin	27
	P	Ohlmeyer	Haloaromatic tags for binary encoding	28
1994	SM	Smith	Indexed libraries	16
1995	P	Deprez	Orthogonal libraries	17
	P	Nicolaou	Radio-frequency tags	32
	P/SM	Janda	Liquid-phase combinatorial synthesis	36
1996	P	Ni	Secondary amine tags for encoding	29
	P/SM	Geysen	Mass and isotopic encoding	30
	SM	Curran	Fluorous tags for reagents and substrates	39, 42
	SM	Cheng, Boger	Mixtures by solution chemistry	44, 45
	SM	Kaldor	Supported reagents	46, 47
1997	SM	Lipinski	Design: developability	51
1999	SM	Ley	Multistep solution synthesis with supported reagents	49, 50

[a]P, peptides; SM, small molecules.

occurring a number of years subsequently. In this context, few would question the pioneering contribution of Merrifield in utilizing functionalized cross-linked polystyrene beads as a solid support for the synthesis of peptides.[4] Key attributes of solid-phase chemistry in terms of using forcing conditions (excess reagents), simplified reaction workup (filtration), and ease of automation are now well known. In a similar fashion, Letsinger applied the concept to solid-phase oligonucleotide synthesis.[5] Solid-phase approaches were also applied to nonpeptide molecules during the 1970s,[6] but the potential utility of this pioneering work was not fully appreciated and did not find widespread acceptance.

Geysen's use of an array of polyethylene rods (*multipins*) as supports for peptide synthesis[7] introduced the concept of improving the speed and efficiency of synthesis by performing many reactions in parallel. A similar process, using synthesis resin in the wells of a microtiter plate (MTP), combined with automated liquid handling was reported several years later.[8]

In 1995, Houghten introduced numbered polypropylene mesh packets of resin ("*tea bags*") as an alternative means of achieving efficiency of synthetic effort, allowing many different substrates to be subjected to the same reaction conditions in a single reaction vessel.[9] Furka, who in 1988 reported the first *split-and-mix synthesis* (also referred to as *pool/split, portion/mix*, and *divide, couple, recombine*), took the idea a stage further.[10] In this process the resin is split into batches, each of which is reacted separately with a single diversity reagent, then the batches are mixed together thoroughly. Assuming the use of n, m, p diversity reagents in successive cycles of splitting, reacting, and mixing, the process results in $n \times m \times p$ products in only $n + m + p$ synthetic steps. A subsequent publication described the split-and-mix synthesis of a mixture of 180 pentapeptides in 15 coupling cycles.[11] The potential for efficiency of synthetic operations was now enormous, and the concept fired the imagination of chemists, as witnessed by the escalation of publications over the following years.

Around the same time, Fodor's group extended and miniaturized the multipin concept by making use of photolithography techniques from the semiconductor industry.[12] In combination with photolabile protecting groups, the synthesis of an array of 1024 peptides grafted onto the surface of a glass microscope slide in an area of only 1.6 cm^2 was achieved. Densities of up to 250,000 synthesis sites per square centimeter were considered realistic goals. The approach above was somewhat specialized in its application and its enabling technologies. In contrast, the ready availability of solid-phase beads and their ease of handling in a conventional laboratory led to their continued widespread use.

4.1.2 Development of Screening Strategies for Libraries

It soon became apparent that the full latent power of combinatorial chemistry would benefit the drug discovery community only if effective methods for screening the libraries were developed. In 1991, in the same issue of *Nature*, groups led by Houghten and Lam described two distinct strategies, which were to form the basis of future approaches to screening libraries. The first strategy, described by Houghten, comprised an iterative deconvolution approach for ascertaining the active component(s) in a large library of 34 million hexapeptides.[13] In the first round of screening the library was divided into 324 pools, each having a unique combination of residues in the first two positions but all possible variations in the remaining four positions. Having ascertained the most potent residues in positions 1 and 2, a second set of libraries were prepared where these two residues were fixed and position 3 was held constant while positions 4 to 6 were varied. The active library thus defined the optimal residue at position 3. The process was repeated until the preferred residue at each position was discovered.

Houghten developed the approach further in a process known as *positional scanning*, in which six hexapetide libraries were prepared, each composed of 18 mixtures with a single position defined as one of 18 natural amino acids.[14,15] Hence, by screening the individual pools, the optimum monomer at each position could be determined. Several alternatives to this technique were to be reported over the next few years: for example, Smith's indexed libraries[16] and the orthogonal combinatorial libraries of Deprez.[17] These all rely on synthesizing subsets of a library in several different combinations in such a way that the active subsets will share a single compound. These methods all utilize mixtures of compounds released from the solid-phase supports into solution. Although the solutions thus obtained are amenable to virtually all existing assay systems, it is always possible that optimal compounds may be missed as a result of averaging of activities, or combinations of agonist and antagonist molecules within a mixture. The statistical implications of pooling and deconvolution strategies have been considered by several authors,[18–20] and tested experimentally.[21]

The second key strategy, reported by Lam's group,[22] capitalized on a consequence of the mix-and-split process: namely, that each bead in the final mixture bears a single product, since the bead has only been exposed to a single diversity reagent at each synthetic step. So, instead of cleaving the products from the beads, a library of several million pentapetides was screened by exposing the beads to a solution of acceptor molecules covalently tagged with a reporter system (e.g., the enzyme alkaline phosphatase, or fluorescene). Beads bearing products that bound strongly to the acceptor became stained and could be identified visually, removed, washed free of acceptor, and microsequenced to determine the peptide structure of interest.

This *one-bead, one-peptide approach* was subsequently made more powerful, and extended beyond the realms of peptide chemistry, by the concept of encoded libraries. The key idea here is, at each synthetic step, to introduce a "readable" chemical moiety onto the bead, such that this code or *tag* encodes the diversity reagents used at each step. Of course, a key property of a coding system is that the tag molecules can be identified more easily than the products they encode. For example, because of the small amount of material present on a single bead, Brenner and Lerner used oligonucleotide sequences to encode peptide libraries.[23,24] Amplification of the tag by the polymerase chain reaction (PCR) then allowed the sequence to be read, and this could be related back to the peptide product structure.

4.1.3 From Peptides to Small Molecule Synthesis

The focus of the discussion so far has concerned peptide chemistry, and much of this was probably considered an academic curiosity by many medicinal chemists. In 1992, however, Ellman demonstrated the multistep solid-phase synthesis of a 1,4-benzodiazepine library,[25] highlighting the fact that such methods need not be restricted to oligomeric species. Subsequently, the chemistry was used in the synthesis of a library of 192 benzodiazepine derivatives using Geysen's multipin approach.[26] These publications established that solid-phase chemistry and the associated technologies could be transferred to the small molecule medicinal chemistry arena and inspired a rapid expansion of interest.

Shortly afterward, De Witt described a simple apparatus comprising a matrix of gas dispersion tubes, each containing resin, for the simultaneous synthesis of an array of 40 products using solid-phase chemistry.[27] This *Diversomer* approach, coupled with simple automation, was applied to the parallel synthesis of hydantoin and benzodiazepine arrays.

That this innovation had come from an industrial rather than an academic laboratory may have been the signal to many industrial chemists that combinatorial chemistry tools had genuine applicability.

Two fundamental philosophies were now established (split-and-mix and parallel synthesis), and each continued to be developed independently. Further innovations were made in techniques for encoding split-and-mix bead libraries. A binary encoding system, introduced by Ohlmeyer et al.[28] in 1993 made use of combinations of chemically inert haloaromatic tags to cap off 1% of free amino groups on the growing peptide prior to adding the encoded amino acid. Following screening of the bead-bound peptides, photolytic cleavage of the tags on the active beads allowed rapid assignment by electron-capture gas chromatography. Alternative encoding strategies introduced subsequently include the use of secondary amine tags,[29] mass and isotope tags readable by mass spectrometry,[30] and a tag-free system where libraries are designed such that each component has a unique molecular weight and is thus directly identifiable by mass spectroscopy.[31]

In 1995, Houghtens' teabag idea was developed into a more robust process that could be automated. Nicolaou reported[32] the use of porous microreactors that contained not only the synthesis resin, but also a small SMART (single or multiple addressable radio-frequency tag) semiconductor unit. Because the synthetic history of each microreactor could be recorded on the chip, they could be combined in pools for common reactions, then resorted for subsequent steps by electronic scanning. Hence, the efficiencies of split-and-mix synthesis were combined with the ability to synthesize useful quantities (ca. 10 to 20 mg) of discrete compounds.

4.1.4 Beyond Solid-Phase Chemistry

Although solid-phase chemistry was the driving force behind the developments described so far, there was an underlying penalty to be paid. Because the behavior of molecules when tethered to a solid support is influenced by the type of support and the linker used to tether the molecule, it has been necessary to optimize reaction conditions on solid phase for reactions that were already well understood in solution. The plethora of publications in this area over recent years (see Table 4.7 for reviews) indicate that many reactions can indeed be transferred to solid phase, sometimes advantageously when excess reagents can be used to force reactions, but sometimes with limited scope compared to the solution counterpart. In most cases a route development phase is essential prior to embarking on the library synthesis if high-quality products are desired, because of the inability to perform purification during a solid-phase synthesis. This has necessitated the development of new procedures for the analysis of bead-bound intermediates and products.[33–35] Moreover, a necessity of solid-phase chemistry is a convenient handle on the target molecule through which it can be linked to the solid support. With peptides this was typically the terminal carboxylic acid group. Much effort has been expended in identifying suitable linker groups for the wide range of functional groups present in druglike molecules, as well as traceless linkers for situations where no convenient functional group is available. (See Table 4.7 for reviews.)

Notwithstanding these considerable efforts, it was clear that many medicinal chemistry targets did not readily lend themselves to solid-phase synthesis, or that such routes were precluded due to lengthy validation phases. To retain some advantageous features of solid-phase chemistry, yet utilize a homogeneous reaction mixture, Janda's group introduced the concept of liquid-phase combinatorial synthesis in 1995.[36] Substrates were linked to polyethylene glycol monomethyl ether, a soluble linear polymer that could be precipitated with ether to give a solid that was filtered and washed free of excess reagents. Similarly, methanol was used to precipitate substrates on non-cross-linked chloromethylated polystyrene.[37,38] A related strategy, introduced by Curran, was to capitalize on the insolubility of fluorocarbon fluids in both aqueous and organic solvents. Products tagged with a perfluoro group were fully soluble during synthesis but could be separated from products by a triphasic (aqueous, organic, fluorous) workup.[39–41] In a reverse strategy, reagents bearing fluorous groups can be separated from nonfluorous products.[42,43]

Since 1996 there has been growing attention to truly solution-phase combinatorial chemistry, and Boger[44,45] described the synthesis of mixtures by applying split-and-mix approaches to solution chemistry. Key to the success of this approach was a carefully designed strategy for application of acid–base workup procedures to purify the libraries at intermediate stages. A more recent innovation of great impact is the use of solid-supported scavengers,[46,47] which are introduced at the end of a solution-phase synthesis step to sequester unreacted building blocks or reagents. For example, aminomethyl polystyrene can be used as a scavenger for acylating agents. In tandem, a renaissance in the use of solid-supported reagents has essentially reversed the solid-phase synthesis paradigm; reaction workup is still by simple filtration, but now the product remains in solution while reagents and scavenged materials are removed on the solid support. In combination, solid-supported reagents and scavengers are not only enjoying widespread application in high-throughput parallel solution-phase library synthesis,[48] but are also proving invaluable in multistep drug[49] and natural product[50] syntheses.

Library design (discussed in detail below) has always been an important prelude to library synthesis to maximize interactions with target proteins and thus produce molecules of increasing potency and selectivity toward their target. An oft-quoted publication by Lipinski[51] in 1997 highlighted the additional need to constrain certain physical properties within ranges defined by his *rule of five* if oral activity is to be achieved.

Alongside the events noted above has been the development of commercial hardware for implementing the ideas and concepts of combinatorial chemistry. Solid-phase synthesizers, originally developed for peptide chemistry, were adapted for wider use. Heating and cooling were added as well as the ability to handle slurries (e.g., resins) and the provision of inert, moisture-free atmospheres. Simple block-based systems have also been developed for both solution and solid-phase chemistry and have become commonplace in most laboratories.

Following synthesis, the downstream processes of workup, liquid handling, evaporation, analysis, quantification, and purification have all been made more efficient by the application of automated or parallel techniques. Interestingly, the most successful hardware developments have come about through close collaborations between the users (industry) and the vendors. Additionally, most combinatorial chemistry groups have implemented software systems to manage the deluge of data generated during the various phases of library production.

The last decade has seen major advances in the tools that medicinal chemists have at their disposal to apply to the problems of drug discovery. As the field has developed, there have been a number of noticeable trends, outlined in Table 4.2, in the drivers and applications of combinatorial methods. After a period of innovation during which not all dreams became reality, we have now entered an era in which the application of appropriate methods can truly have an impact on productivity within the drug discovery process. It is aspects of these applications that are discussed in more detail in the remainder of the chapter.

TABLE 4.2 Trends in Combinatorial Chemistry

From:	To:
Quantity (numbers)	Quality
Few large libraries	Many smaller libraries
Innovation (and failure)	Application (and focus)
Chemistry driven	Biology (target) driven
Peptides	Small molecules
Mixtures	Singles
Solid-phase chemistry	Any chemistry
Synthesizability	Developability
Specialist tools	General tools
Speculative	Routine
Manual	Automated
Fragmented	Integrated

4.2 THE ROLE OF COMBINATORIAL CHEMISTRY IN DRUG DISCOVERY

Although the drug discovery process is highly complex, it may be considered to consist of four key stages, the first being the discovery and definition of a *biological target*. In the majority of cases this is a protein with some known or unknown involvement in a disease process. A consequence of the success of the Human Genome Project in providing a blueprint of the human genome, in combination with the complementary approach of proteomics, is the unprecedented rate of discovery and understanding of new targets. This is set to have a profound effect on drug discovery in the new millennium.[52,53] Once a relevant target has been identified, assays are developed to identify molecules that bind to the target and modify its behavior in the desired manner.

The process now enters the *lead generation stage*. Here, the goal is to identify a lead compound whose properties (principally, but not exclusively its potency against the target) make it a suitable candidate for more in-depth exploration. In most cases this involves a high-throughput screening exercise, where large numbers of compounds are screened against the target in the assay developed previously. Clearly, the greater the number of appropriate compounds available for screening, the greater the chance of success. The production of lead discovery libraries was an early driver for combinatorial chemistry and continues to be an important application. Libraries for this purpose are typically of significant size, from a few thousand to tens or hundreds of thousands of compounds and may exist in a number of formats that match the screening protocols in use.[54] For example, libraries of discrete compounds would typically be of general utility and enter a standard screening system along with a company's historical compound collection. Soluble mixture libraries might fulfill a similar role, but with the added necessity of deconvolution procedures to pinpoint the active component. Libraries on solid supports require alternative screening strategies,[55] which may be particularly suited to certain specific target classes but are unlikely to be as generally applicable as solution assays.

When knowledge of the three-dimensional structure of a protein is known, either through x-ray diffraction studies or through homology modeling, de novo design of a library of compounds is a possibility. Computer modeling of the interactions between candidate small molecules and the known (or assumed) active site of the protein can provide a starting point for library design. Such an approach will often take place in parallel with high-throughput screening exercises.

Once a suitable lead is identified, the compound enters the third stage of the process, *optimization*. The goal here is to modify the lead structure such that a number of criteria for progression into the final, *development*, *stage* are satisfied. Important criteria, the actual values of which will be set on a case-by-case basis, include high potency against the target as well as selectivity against related targets. The development candidate will also exhibit appropriate absorption, distribution, metabolism, and excretion (ADME) and toxicity properties and will have a positive profile against P450 enzymes. Screening for these properties at this stage, whether by in vitro, in vivo, or computational protocols is becoming an essential filter prior to entering the costly development phases.[56,57]

The key role of combinatorial chemistry at this point is to provide sets of compounds with which to derive an understanding of the contribution of structural elements of the molecule toward its binding to the protein target, so that structure–activity relationships (SAR) can be developed. Lead optimization libraries typically consist of hundreds or a few thousand compounds at most. As the iterative cycles of library design, synthesis, and screening progress and more is known about SAR, the library size might fall to a handful of compounds focused on fine-tuning some aspect of the overall profile.

Eventually, the optimized product will enter the final, although most lengthy and costly stage, that of development into a marketable drug, including optimization of synthetic routes for production-scale synthesis, stringent trials in patients to prove clinical efficacy and safety, regulatory approval, and marketing; It is only once sales have begun that the massive research and development costs can start to be recovered, and this is possible only until the patent life of the compound expires. Any time savings in the overall process therefore represent opportunities for increased revenue, and combinatorial chemistry clearly has an important role to play in this respect.

Some of the characteristic features of lead generation and lead optimization libraries are summarized in Table 4.3. These classifications are extremes, and there are frequently cross-overs. For example, generic libraries are lead generation libraries designed with the intent

TABLE 4.3 Characteristic Features of Lead Generation and Lead Optimization Libraries

Lead Generation Libraries[a]	Lead Optimization Libraries[b]
Populate screening collection	Modify leads to improve profile
Provide leads for range of targets	Optimize leads for specific target
Wide coverage of chemical space	Focus on defined area of chemical space
Larger arrays	Smaller arrays
Small scale (few milligrams)	Larger scale (tens or hundreds of milligrams)
High quality preferred	High quality essential
General druglike properties	Specific potency, selectivity, ADME properties
Compounds made in one batch	Compounds made in iterative cycles with feedback from biology
Significant route development (often, new chemistry)	Minimal route development time (often, known chemistry)
Specialized groups	Medicinal chemists

[a]Also called prospecting or generic libraries.

[b]Also called focused or directed libraries.

of interacting with some general class of target (e.g., protease, kinase, G-protein-coupled receptor). Conversely, the products from highly focused optimization libraries may ultimately join the screening collection for future lead generation operations.

4.3 DESIGNING COMBINATORIAL LIBRARIES

There have been several estimates of the number of druglike molecules that could theoretically be synthesized,[58,59] ranging from 10^{50} to 10^{200}. The impossibility of even contemplating such a goal is evident when one considers the availability of materials (the total mass of the observable universe is around 10^{52} kg) or the time required (the Earth is only 10^{17} seconds old). The key aim of library design is to select from this overwhelming number of possible compounds a subset that can be synthesized efficiently and which has a high chance of either producing new leads or improving the properties of existing leads.[60] The wide range of design methodologies described in the literature bears testament to the observation that there is no definitive solution to this problem.[61–64] Some commercial software systems for this purpose have been reviewed.[65]

Although beyond the scope of the current discussion, some of the techniques described below have been applied to selecting compounds (either individuals or library sets) for purchase to augment existing screening collections. Also, when screening capacity is less than the number of compounds available, these methods can be used to select representative or biased subsets of the entire collection for screening.

In most cases, the starting point in library design is a *virtual library*, which defines all the conceivable compounds accessible by a particular synthetic route. Frequently, the virtual library is defined as a fixed template (scaffold, core structure) with a number of variable (diversity) sites. The potential set of substituents (R-groups, fragments) at each site is derived from diversity reagents (building blocks, monomer sets) defined in scope by their availability and suitability for the particular synthetic route. Alternatively, the virtual library may be defined as a series of transforms (reactions) applied to the monomer sets. This transform approach is more intuitive to the medicinal chemist and can represent the products of certain reactions, such as the Diels–Alder reaction, more satisfactorily than the template–substituent approach.[66] However, the latter is more efficient computationally and hence encountered most commonly.

4.3.1 Describing and Measuring Diversity

A major emphasis in the literature has been methodology for selecting the diversity reagents that will be used to decorate the template. In this respect, an inescapable concept, with many definitions, is diversity.[67–69] A diverse subset of compounds is considered to be one that represent the chemical space occupied by the complete virtual library. Defining chemical space, and hence measuring diversity, requires descriptors that represent properties of compounds in a computer-friendly format (numbers, matrices, or bit strings).[70,71] A large number of descriptors have been reported (see Table 4.4 for examples)[72] and are frequently used in combination, resulting in a multidimensional chemical space. Imaginative pictorial representations such as flower plots[73] are useful for visualizing many properties simultaneously. However, when data sets are large, the dimensionality is often reduced by techniques such as principal component analysis[69,73] to simplify both computation and visualization of the results.

TABLE 4.4 Descriptors Used in Library Design

Descriptor	Comments
Molecular weight	Whole molecule property; use as simple filter
Hydrogen-bond donor, acceptor counts	Whole molecule property; simple indication of number of potential binding sites
Rotatable bonds	Indication of molecular flexibility
$c \log P$	Calculated $\log P$ (measure of lipophilicity)
CMR	Calculated molar refractivity (measure of size)
Structural keys	Presence of small structural fragments
Two-dimensional fingerprints	Bit-string representation of atom paths through molecule
Atom pairs	Atom type and number of bonds separating pairs of atoms
Topological torsion	Atom types in four atom paths
Topological indices	Single value representing molecular shape and connectivity
BCUT descriptors	Burden–CAS–University of Texas; use atomic properties related to strength of intermolecular interactions
Three-dimensional screens	Encode distances and angles between features in molecule
Pharmacophore keys	Various approaches; commonly, triplets of pharmacophoric groups defined by their types and separation

Because of the size of virtual libraries, the descriptors for each compound should be easy and quick to compute since, by definition, we cannot use measured properties for molecules that do not exist. Ideally, the descriptors used should in some way represent potential interactions between the molecules and their biological targets. The BCUT descriptors of Pearlman,[74,75] as well as various fragment keys and fingerprint descriptors,[64] fulfill this role. Descriptors representing bulk molecular properties such as molecular weight, lipophilicity, or number of rotatable bonds have greater value as filters (see below) than for diversity assessment.

Clearly, molecules and their sites of interaction with targets are not two-dimensional entities, so intuitively three-dimensional descriptors should better represent their intermolecular interactions, although this remains the subject of debate.[76–78] However, new issues, such as how to align diverse sets of molecules for meaningful comparison and how many of the multiple conformations of a flexible molecule to incorporate must be addressed carefully and will significantly challenge computational resources. It is likely that two-dimensional descriptors are adequate for lead generation libraries, while the specificity needs of an optimization library are better catered for by three-dimensional descriptors.[79]

An important class of three-dimensional descriptor is based on the types of pharmocophoric groups in a molecule and the distances between them.[80–82] Pharmocophores are features, such as acidic or basic centers, hydrogen bond (H-bond) donors and acceptors, aromatic ring centroids, and hydrophobic groups, known to be important in drug–receptor interactions. The relative positions of pharmocophores must, of course, be calculated for each likely conformation of a flexible molecule.

There has been considerable debate as to whether the less computationally intensive evaluation of reagent diversity, rather than product diversity, is an acceptable approach, and the outcome appears to be descriptor dependent.[83,84] Certainly, analysis of a few thousand reagents is considerably faster than that of many millions of enumerated products and may suffice in some cases. However, product evaluation is likely to be essential when three-dimensional descriptors are used.

A number of different algorithms have been developed and applied to the sampling of the chemical space occupied by the virtual library, of which two major types, clustering and bin-based partitioning, will be described. Clustering algorithms aim to group together those compounds with some degree of similarity to each other, and will be illustrated by two widely used methods.

Jarvis–Patrick clustering[85] groups molecules according to the number of nearest neighbors they have in common. The method is rapid and suited to large libraries, but requires careful setting of near-neighbor thresholds if clustering of diverse compounds or excessive numbers of singletons (clusters of one) are to be avoided. Being a nonhierarchical method, a fixed number of clusters are produced with no specific relationships between the clusters.

In contrast, Ward's agglomerative hierarchical clustering method[86] progressively combines the most similar molecules into related groups, and the results can be viewed as a dendrogram, similar to a family tree, showing the relationships between clusters and allowing any number of clusters to be selected. Although slower, Ward's method has been shown to outperform Jarvis–Patrick and other clustering methods[76] and in our hands has produced chemically intuitive results and proven very useful for the selection of reagent sets. A limitation of clustering methods is that they give no information on the actual coverage of compound space by the library.

Bin-based partitioning methods, on the other hand, subdivide multidimensional chemical space along the axes into hypercubes or *bins*.[87] Compounds falling into the same bin volumes are deemed to have similar chemical and, by assumption, biological properties. Empty bins indicate areas of compound space not represented by the library under investigation. Bin-based methods are useful for comparing libraries and for identifying "holes" in a collection to be filled by future libraries or compound acquisitions.

Whatever descriptors or algorithms are applied to the selection of a subset of compounds from a virtual library, one additional criterion should be applied if the compounds are to be synthesized efficiently by combinatorial methodologies, whether split-and-mix or parallel array chemistry. It is not sufficient simply to select the most diverse subset of compounds (called *cherry picking*); the compounds selected must be related through common sets of building blocks. So, as an example, the problem is not to select the 96 most diverse compounds but the most diverse 12×8 array of compounds. This problem is most easily solved by selection of diverse reagent sets. However, it has been solved successfully at the product level using a number of algorithms.[83,88–90]

4.3.2 A More Focused Approach

The discussion so far has focused on designing libraries that give a good representation of chemical space. We now consider what the bounds of this space are and two important methodologies to limit the space that we desire to fill. The first approach is to limit chemical space to those property ranges that are likely to give druglike molecules; the second is to restrict attention to structures likely to interact with specific targets or target classes. Like *diversity*, the term *druglike* has a range of definitions, but essentially it is an attempt to confer those properties on a molecule that will increase its chances of being orally absorbed, nontoxic, avoiding first-pass metabolism, and in an ideal world, passing unhindered through costly development phases into clinical use.

The publication of Lipinski's rule of five[51] in 1997 reawakened the awareness of some chemists that the drive for large numbers of compounds was not the ultimate aim in drug discovery. Lipinski analyzed a set of drugs known to be well absorbed and discovered that the vast majority satisfied most of the following criteria: molecular weight (MW) below 500, calculated log P (c log P)[91] below 5, five or fewer H-bond donors and 10 or fewer H-bond acceptors. Compounds failing more than one of these tests were predicted to be poorly absorbed unless by some active transport mechanism.

Since this paper, there have been many other publications attempting to distinguish drugs and nondrugs.[92] One very simple filter to consider is the removal of compounds with reactive functional groups (likely to be hydrolyzed or to react with biological nucleophiles) or those known to contribute to toxicity.[60] Analysis of drug databases allows the frequency of occurrence of molecular frameworks and side chains to be ranked and used as a guide in library design.[93,94] Another approach is to use neural networks to perform the drug/nondrug classification.[95,96] This technique has also been applied to the more specific problem of determining the features necessary for penetration of the blood–brain barrier, an essential requirement for central nervous system–active drugs.[97]

The results of any of these studies should be considered guidelines rather than rules since none of the methods are much more than 80% successful. Furthermore, the methods are based on historic data, and it is unlikely that all future drugs will fall within their scope. So although it makes sense that a large percentage of proposed library members should be kept within the bounds of current wisdom, it would be foolish not to explore new areas to some extent.

A further observation[98] is that the process of optimizing a lead is often associated with an increase in molecular weight and lipophilicity. Hence, instead of designing lead generation libraries with druglike properties (typically, $MW > 350$, $c \log P > 3$) which may already be close to the limits allowed for these properties, attention should be focused on leadlike property ranges ($MW < 350$, $c \log P < 3$), leaving scope for further optimization. Another factor to take into consideration is the demonstrated tendency of molecular complexity to increase during the lead optimization process.[99]

The second approach to chemical space limitation is in generic or focused library design, where some degree of prior knowledge of structural requirements reduces the chemical space of interest. A focused library is considered to be one directed toward a specific target receptor or enzyme. Its design may be de novo (based on knowledge of receptor structure only), or based on a series of compounds with known SAR against the specific target (lead optimization library), or a combination of both. The former is becoming increasingly applicable as the number of x-ray crystal structures and reliable homology models of biomolecules grows. In structure-based design,[100–103] members of a virtual library are screened in silico against their ability to bind to the active site of the receptor, using programs such as DOCK,[104] GRID,[105] and LUDI.[106] An early example was described by Kick et al.[107] in which the crystal structure of the pepstatin/cathepsin D complex was used to reduce a virtual library of more than a billion statinelike compounds to a set of 1000 for synthesis (see below). The performance of this set against a second library of 1000 optimized for diversity without consideration of the target structure showed the former to give a significantly better hit rate against cathepsin D, especially in the high-potency range.

The design of a generic library may use similar methods, but applied to a set of related targets [e.g., aspartic proteases or G-protein-coupled receptors (GPCRs)]. Probably the most common approach here is based on the concept of *privileged structures*,[108,109] defined as structural types that provide high-affinity ligands for more than one type of receptor or enzyme. Libraries based around privileged structures are likely to give faster identification of novel ligands for new members of known receptor or enzyme families. An example given by Mason is to include the privileged structure as an additional "dummy" pharmacophore in a four-point pharmacophore model.[81]

A design criterion likely to receive increased attention is that of synthetic accessibility. The RECAP technique[110] has been used to fragment biologically active molecules in a retrosynthetic manner into building blocks rich in biologically recognized substructures. However reagent sets may be chosen, their structural diversity will frequently also equate to variations in their reactivity. A quantitative structure–property relationship (QSPR) model has been applied to predict the reactivity of carboxylic acids toward the acylation of an amine.[111] Classification,[112] modeling,[113] and simulation[114] of organic reactions will also provide valuable predictive tools. At an experimental level, factorial design methods have been applied to the optimization of reaction conditions for library synthesis.[115,116]

4.4 TOOLS FOR SYNTHESIS OF COMBINATORIAL LIBRARIES

In the preceding section we focused on defining the compounds to be synthesized in a library and how the design criteria were matched to the desired function of the library. We now consider the choice of tools available to facilitate library synthesis and how these, too, must be matched to the type of library and the environment in which it is to be synthesized. The equipment may achieve its goal of increased productivity per chemist in a number of ways. At one end of the spectrum one finds relatively simple, low-cost apparatus for manually performing operations in parallel. At the other extreme, sophisticated, fully automated synthesizers offer round-the-clock unattended operation.[117] Several factors must be considered when selecting the most appropriate platform: suitability for solution or solid-phase chemistry, throughput, modular or self-contained system, synthesis scale, footprint, ease of use, cost, and chemical performance. Representative examples of currently available hardware are assembled in Table 4.5. For a more detailed discussion, the reader is referred to review articles in Table 4.7.

4.4.1 Nonautomated Tools

Perhaps the simplest devices for increasing synthetic throughput are those designed to support, heat, and stir a number (6 to 24) of round-bottomed flasks, bottles, or test tubes simultaneously. These low-cost devices occupy a small footprint and thus are suitable for use alongside conventional equipment in a standard fumehood and are applicable to a range of reaction scales from a few milligrams to several grams.

Manual systems of somewhat higher throughput are the reactor blocks based on the ubiquitous microtiter plate format so well known in biological laboratories. Throughput ranges from 96 down to 12 compounds per block, corresponding to 2 to 16 mL well volumes. Blocks may be made of polypropylene, PTFE, or glass, they can be sealed, and they often incorporate sintered frits to allow separation of soluble products from supported reagents and sequestering agents. Agitation is usually by means of vibrating or oscillating

TABLE 4.5 Selection of Commercially Available Equipment for High-Throughput Chemistry

Application	Examples	URLs
Stirring blocks (with heating or cooling)	Carousel, greenhouse	www.radleys.com
	Reacto-stations	www.stemcorp.com
	KEM-Prep, reaction blocks	www.jkem.com
Reactor blocks	MiniBlock	www.bohdan.com
	Calypso system	www.charybtech.com
	FlexChem system	www.robsci.com
Liquid handling	Microlab, Star workstations	www.hamilton.com
	MiniPrep, Genesis	www.tecan-us.com
	Quad-Z 215	www.gilson.com
Mix-and-sort	AccuTag-100	www.irori.com
	SynPhase crowns and lanterns	www.mimotopes.com
XYZ synthesizers	ACT 496, Benchmark	www.advancedchemtech.com
	Myriad Discoverer, Myriad MCS	www.mtmyriad.com
	Sophas	www.zinsser-analytic.com
	Neptune	www.bohdan.com
	Automated synthesis workstations	www.chemspeed.com
Fluidic synthesizers	Nautilus	www.argotech.com
	Domino blocks	www.torviq.com
	Quest	www.argotech.com
Other synthesizers	Zymate	www.zymark.com
	Trident automated synthesizer	www.argotech.com
Microwave synthesizers	Smith synthesizer, Creator	www.personalchemistry.com
	Discover Systems	www.cem.com
	Ethos SYNTH	www.milestonesci.com
Delivery of solids	Redi, Lipos	www.zinsser-analytic.com
	Titan resin loader	www.radleys.com
	Resin and powder dispenser system	www.anachem.co.uk
Reaction workup	ALLEX	www.mtmyriad.com
	Trident sample processing station	www.argotech.com
	Lollipop	www.radleys.com
Evaporation	SpeedVac	www.thermo.com
	HT systems, megasystems	www.genevac.com
	Syncore PolyVap	www.buchi.com
	TurboVap	www.zymark.com
	IR-Dancer	www.hettlab.ch
	ALPHA-RVC, BETA-RVC	www.matinchrist.de
Flash chromatography	Flashmaster	www.joneschrom.com
	Quad3	www.biotage.com
	CombiFlash, Optix 10	www.isco.com
Preparative HPLC	Flex/Parallex	www.biotage.com
	FractionLynx autopurification system	www.waters.com
	1100 Series purification system	www.chem.agilent.com
	Discovery VP preparative system	www.shimadzu.com
	High-throughput purifier	www.hii.hitachi.com
Weighing	AWS	www.bohdan.com
	Calli, Moss	www.zinsser-analytic.com

shakers. In some cases, temperature control may be achieved by circulating cooled or heated fluids through channels in the block assemblage. Multichannel handheld pipettes facilitate reagent delivery, but this can be enhanced by the use of simple liquid-handling robots. The Diversomer technology of Parke-Davis was interfaced to an XYZ robot for liquid delivery.[118]

4.4.2 Mix-and-Sort Systems

Interesting and widely used applications of the teabag approach to split-and-mix synthesis have been developed by Irori and Mimotopes. Although applicable to small arrays, the real power of these technologies lie in multistep syntheses of hundreds or thousands of products. The Irori AccuTag-100 system comprises small mesh-sided microreactors that the user loads with solid-phase resin and a radio-frequency (RF) tag.[119] A range of microreactors can accommodate from approximately 30 mg of resin (which yields about 10 to 20 mg of product on cleavage) up to 10 times that quantity. Prior to the first synthetic step, the microreactors are scanned and the unique code of the RF tag is assigned to one of the library components. This code is used to sort the microreactors before each synthetic step and enables a mix-and-sort strategy requiring exactly one microreactor for each planned product. Chemistry is performed in conventional labware (flasks, bottles, etc.) and culminates in an archival process where the microreactors are sorted into a 96-well plate format prior to cleavage, again directed by the RF tag codes. Scanning and sorting can be manual or by using an automated tag reader/sorter. Hence the synthetically efficient mix-and-sort process is combined with the production of useful quantities of discrete cleaved products.

The Mimotope SynPhase system circumvents the use of resin beads by using linker groups specially grafted onto easily handled support structures, with loadings from 4 to 35 µmol and again capable of delivering milligram quantities of products. In addition to RF tagging strategies similar to the Irori system, product identification can be achieved by the use of color coding or by creating spatially addressed arrays in 96-well MTP format.

4.4.3 Automated Synthesizers

Automation of synthesis comes at a higher cost than the manual tools described above, and the equipment typically fills a standard fumehood and may even require a custom-built installation. The advantages include the possibility of continuous, unattended operation, reproducibility, and reduced exposure of operators to toxic chemicals. The first successful automation was for solid-phase peptide chemistry, which is particularly well suited since only two reactions (coupling and deprotection) are involved, both having been highly optimized to give quantitative yields at room temperature. Moreover, a limited set of well-characterized amino acid building blocks are used, and standardized deprotection, coupling, and wash cycles are easily programmed.

Subsequently, modified peptide synthesizers were marketed for general solid-phase organic chemistry where fewer consecutive steps, but much more variation in reaction conditions and reagent, are the norm. The addition of slurry movement made split-and-mix synthesis possible and heating–cooling capabilities, as well as improved provision of inert atmospheres, extended the range of chemistry possible. Recently, the first fully automated oligosaccharide synthesis was reported.[120]

Several types of automated synthesizers have evolved. The most commonplace are those based around the XYZ liquid handler. These are capable of delivering reagent solutions to a grid of reaction vessels or a reactor block, and can also be programmed to move solutions from one vessel to another. They may have multiple probes and in most cases are sealed by a septum on the reaction vessel, which is pierced to deliver reagents and maintains an inert atmosphere. Alternatively, a twist cap mechanism may be used, which avoids the need for small-diameter needles for reagent delivery, allowing the use of wider-bore positive-displacement pipettes.

Using fluidic systems, where reagents and solvents flow from reservoirs, through a permanent network of tubes and valves, to the reactors can largely eliminate moving parts. Because such systems are entirely closed, control of atmosphere is straightforward, but the pipework can be complex and may be prone to blockages if suspensions or poorly soluble reagents are used. Other systems are available that use a multiarticulated robotic arm to move vessels from one workstation to another, much as a human operator might.

A multistep synthesis need not be constrained to a single synthesizer. As an alternative to truly parallel synthesis, in which every step is performed separately for every product, a split–split strategy has been proposed[121] in which each diversity step is carried out only once, and the products are split for subsequent diversity steps. Because the reaction scale decreases at each stage, the use of a range of synthesizers is necessary to match both the scale and number of vessels required for each step.

A recent development is the use of microwaves as a source of heat energy for accelerating chemical reactions.[122] Reactions that might ordinarily take many hours can be completed in a few minutes. Hence, increased throughput is by way of time compression rather than parallel operation. Early systems were based on modified domestic microwave ovens and suffered from uneven and unpredictable heating. Recent systems have purpose-designed single-vessel cavities where the power is controlled and delivered uniformly, and allow real-time monitoring of both temperature and pressure, resulting in reproducible results. Coupling such a reactor with simple robotics enables multiple reactions to be run unattended.

Delivery of soluble reagents to any of these synthesis systems requires simple liquid-handling robots, either stand alone or, more often, integrated with the synthesizer. Delivery of solids, in particular synthesis resins, polymer-supported reagents, and scavengers is less straightforward, although robotic dispensers have been developed. The use of isopycnic suspensions in appropriate solvent mixes facilitates pipetting and is the technique used in some automated synthesizers. The use of resins prepackaged in capsules[123] or in tablet form[124] is attractive but suffers from the need to dissolve away the capsule or binder material. This is inappropriate for the supported reagents and scavengers of solution chemistry, where additional impurities would be introduced, but may be suitable for distribution of resins for solid-phase chemistry. A recent publication introduced sintered resin blocks, which are proposed as an alternative to microreactors or crowns but might also find use in delivering reagents and scavengers.[125] Monolithic disks, which can be crafted in a variety of shapes and sizes, offer an alternative means of conveniently handling solid supports.[126–128] Resin dispensers based on simultaneous volumetric measurement and delivery of 96 aliquots of resin to a MTP format have been developed in our own laboratories and commercialized by others.

4.4.4 Postsynthesis Processing

Even modest increases in synthesis throughput have a knock-on effect on downstream sample processing. Routine events such as reaction workup, solvent evaporation, product purification, quantification, and analysis all require high-throughput counterparts. As stated previously, reaction workup of solid-phase syntheses involves simple filtration. In a similar way, supported reagents frequently facilitate the workup of solution-phase chemistry, and scavenger resins remove excess reagents. Conventional liquid–liquid extraction procedures have also been automated. The interface between the two liquid layers may be located by dead reckoning of solvent volumes, or proprietary interface detection technologies can be used. A simple but elegant concept is to freeze the aqueous layer and separate it from the liquid organic layer.

High-throughput solvent evaporation is facilitated by vacuum centrifuges, which are available in many sizes and can accommodate tens or hundreds of tubes or MTPs in a range of formats. An alternative means to prevent bumping under vacuum is rapid agitation. Parallel evaporation can also be achieved by a suitably angled gas flow over the solvent surface, although some means of solvent capture (condensation) is also required.

A variety of tools are available for parallel purification. Solid-phase extraction (SPE), in which product solutions are passed through plugs of sorbent (in filter plates or syringe barrels), is in widespread use.[129] In the simplest strategy, the sorbent retains the impurities and the product is collected. Alternatively, the product is retained on the first pass and subsequently eluted by a change in solvent. SPE is often performed in direct conjunction with reactor blocks but has also been automated.[130]

In many instances, true chromatographic purification methods are necessary, and when gram quantities of material are involved, flash chromatography systems are appropriate. These allow automated parallel or serial chromatography of 10 or 12 samples and may have ultraviolet (UV) detectors for recording or triggering fractions. In our hands these systems have proven very useful for purifying building blocks and other intermediates.

Smaller-scale purifications are most frequently effected by preparative high-performance liquid chromatography (HPLC). A number of automated systems are commercially available, sometimes operating multiple columns in parallel to boost throughput. Depending on the system, fractionation may be triggered by UV[131] or mass detection.[132,133] The latter is appealing in that only a single fraction needs to be collected per product (i.e., that with the correct MW). However, error recovery is impossible unless separate waste streams are collected for each sample. In contrast, the more robust method of collecting all significant peaks by UV requires downstream structure confirmation as well as effective fraction management.[134] Supercritical fluid chromatography (SFC) using pressurized carbon dioxide as the major eluent offers the advantage of small fraction volumes since the eluent evaporates instantly and avoids the costs and environmental issues of high solvent use.[135]

If the products are collected in preweighed tubes or vials, the amount of product may be quantified simply by weighing on an automated weighing station. This is not possible if products are collected directly into MTPs. Alternative techniques such as evaporative light-scattering detection (ELSD)[136,137] and chemiluminescent nitrogen detection (CLND)[138,139] are becoming popular, especially when used in conjunction with analytical HPLC or HPLC-MS. The latter is the most widely used technique for structure confirmation, and high throughput has been achieved by combining four or eight LC columns with a single multiplexed mass spectrometer.[140] Multiple MS or HPLC-MS results can be evaluated rapidly using color-coded plate maps and other visualization tools.[141,142]

Flow nuclear magnetic resonance (NMR) systems are making high-throughput NMR analysis of products in MTP format routine[143] and deliver much more structural information than MS techniques. However, interpretation of the many spectra produced is currently a bottleneck that will only be resolved as spectrum prediction software improves. The use of internal standards such as 2,5-dimethylfuran[144] allows quantification estimates to be made at the same time.

In summary, recent advances in automation and parallel processing of virtually all procedures in synthesis, purification, and analysis have rendered the high-throughput production of high-quality compound libraries a reality. Of course, as each advance is made, expectations rise, so there will doubtless be significant further developments in the future.

4.5 MANAGING THE COMBINATORIAL PROCESS

An inescapable consequence of the combinatorial chemistry revolution is the increased data-handling burden compared to traditional methods. A single chemist might be expected to design and synthesize thousands of discrete compounds in a year in contrast to 50 or so previously. Each library of compounds must be designed, the building blocks acquired, reaction quantities, molecular weights, and yields of products calculated, analytical results interpreted, products registered into databases, and so on. Additionally, the range of equipment, often robotic, in use has grown, each piece having its own operational and data input–output requirements. Furthermore, depending on circumstances, a single chemist might handle a library through all the stages of production, or increasingly commonly, the work might be shared between a group of chemists working as a team who each need access to data at different stages.

4.5.1 Specification of Combinatorial Libraries

Within the author's laboratories, the impending complexities of data and process management issues were recognized around 1996. Analysis of the repeated transfer or reentry of data between a variety of electronic and hardcopy formats revealed significant potential sources of error and tedium (time wasting). In a collaboration between medicinal chemists, computational chemists, analysts, and the newly formed Cheminformatics group, an integrated software application, RADICAL (Registration, Analysis and Design Interface for Combinatorial and Array Libraries), was developed and deployed throughout the organization.

The scope of RADICAL has been extended continually through incremental releases in order to accommodate changes to the processes and equipment in use. A further advantage of having a standard, but flexible, library management tool is that all data generated during library production is automatically captured for archival. Of particular value is the capture of synthetic routes used and their outcome, which offers great potential as a searchable source of information for subsequent library route development. In our experience, compliance with documentation and data logging is much higher if they form part of the natural workflow rather than being additional tasks. Some key features of the RADICAL environment are outlined in Table 4.6. Related systems, such as CICLOPS,[145] ADEPT,[66] and AIDD,[146] have been developed by other pharmaceutical companies; other packages, such as CHEM-X[147] and Afferent,[148] are available commercially.

4.5.2 Controlling the Automated Workflow

One disadvantage of developing a single, albeit component-based application to support high-throughput synthesis was that a new software release was required to incorporate

TABLE 4.6 Key Components of RADICAL

• Chemistry editor *Definition of synthetic route*	Define steps, reactants, products, conditions.
• Library design *Definition of building blocks*	Select reagents from databases (commercial and in-house), links to library design packages. Acquire reagents (suppliers, in-house inventory). Calculate reagent quantities.
• Library results *Enumerated products matrix*	View and enter compound-specific data. Track status of compounds. Link to registration and screening.
• Plate manager *Management of subsets of library*	Map products to plates. Submit to quality control and report results using color mapping.

frequent changes necessitated by the purchase and modification of commercial hardware such as synthesizers, analytical instruments, and balances. Also, the variety of equipment in use required that data exchange be performed at the lowest common denominator of text files, resulting in the users having to exchange files and become familiar with a number of interfaces on the various pieces of hardware. A particular limitation was the lack of support for tracking physical racks or plates of compounds around the laboratory. The decision was therefore made to separate the definition of chemistry specification and other structurally related matters (RADICAL's role) from the hardware, data, and process management-dependent issues. The requirements for the latter role were to be achieved through codevelopment of a separate but integrated software application, ACE (Automated Chemistry Environment) with the Technology Partnership, Cambridge, UK.

The role of ACE is to define and manage the workflow that the compounds defined in RADICAL will follow during their synthesis, analysis, and purification in the laboratory. The library definition from RADICAL is mapped onto an ACE workflow which is stored on the ACE manager (a server application with a database containing knowledge of all compounds in process and their associated data), which is networked to the PCs controlling the automated equipment around the laboratory. The operating software of each piece of equipment is concealed from the user within an agent, which acts as an interface between the ACE manager and the equipment. All agents have a similar appearance, which includes a listing (queue) of all jobs currently available to run on that machine, so users only have to be familiar with one interface type. Wherever possible, the agent automatically provides the equipment with the necessary data and parameters to run the current job, then guides the user through setup or loading procedures, runs the equipment, and returns any data collected to the ACE manager. As a set of compounds (e.g., a rack of vials, a microtiter plate) completes one process within its workflow (e.g., synthesis robot) the "job" will appear on the queue of the agent of the next process (e.g., workup robot). All racks, plates, and vials are bar-coded, which allows confirmation that the correct job is being processed and facilitates the identification of any sample within the laboratory.

ACE provides agents for both automated and manual steps, the latter often simply presenting a means of indicating that the job is complete. Additionally, "decision" agents have been developed to allow the user to inspect analytical data and indicate whether the compound is of sufficient quality and/or quantity to progress. When the workflow is complete, the appropriate data collected by ACE is transferred back to RADICAL for archival and registration into the corporate database. The advantage of this approach is that whenever a new piece of hardware is incorporated into the process, a relatively small change to the software environment is required.

However good the tools provided to aid in the management of combinatorial chemistry, whether for library design or production, the overall complexity of the process should not be trivialized. When errors occur, recovery is much more complicated than in traditional medicinal chemistry environments, if only because of the number of entities involved. We implement regular appraisal of processes, seeking opportunities for further streamlining and error reduction. Furthermore, it is important to provide adequate support for chemists, in terms of training, assistance in error recovery, and capturing feedback regarding future enhancements to the systems.

4.6 FROM SPECIALIST DISCIPLINE TO STANDARD TOOL

The following general remarks are based on published commentaries,[149–151] discussions at conferences, and our own experiences. In the early 1990s, the potential of the newly developing discipline of combinatorial chemistry to accelerate, and indeed revolutionize, the drug discovery process was recognized throughout the pharmaceutical industry. Most major companies established small groups of interested persons to investigate this potential and bring the necessary tools in-house. The initial remit of such groups was typically to review and develop new technologies and apply these to the synthesis of lead generation libraries to increase corporate screening collections. This was driven in part by the newly developed high-throughput screening capabilities,[152] with capacities to screen many more compounds than were available at the time.

Within this fairly generic model, each company had its own approach. Differences included the decision whether to use solid-phase or solution chemistry (or both) as their major platform and whether to rely on commercially available technologies or develop proprietary methods. A further choice was necessary between preparing discrete products or the synthetically more attractive mixtures.

Additionally, an increasingly important part of the combinatorial groups' brief was to move the technologies out into the mainstream medicinal chemistry population for application to lead optimization. In this respect a number of issues made the uptake less rapid than anticipated. The scope of documented chemistry suitable for high-throughput synthesis was restrictive, and in the case of solid-phase chemistry, the lengthy route development times were much less acceptable for the optimization phase, where time and resources were often at a premium. Many of the techniques developed, such as mix and split and encoding, offered the greatest efficiencies when applied to large libraries, whereas optimization libraries were much smaller. The equipment associated with combinatorial chemistry tended to be too bulky to accommodate readily in standard fumehoods and was operated via computer interfaces that required a greater degree of familiarity than would be acquired by occasional use outside the specialist groups.

Many of these hurdles have subsequently been overcome to a large extent, as a result of changes to technology, processes, and mindsets. The recent explosion in the range of solid-supported reagents and scavengers has had a dramatic influence. The workup of many types of solution-phase reactions can now be simplified by the application of these principles. Interestingly, these ideas are also finding imaginative applications in mainstream multistep organic synthesis.[49]

The restrictions in the scope of solid-phase chemistry are now far fewer than in the pioneering days, when "any compound as long as it's an amide" was the perceived dogma. The number of publications in the field continues to grow, and considerable databases of knowledge have been accumulated both publicly and within companies. Also, the inability to purify compounds at intermediate stages of multistep solid-phase syntheses contributed to the need for long development phases in route optimization. With the advent of high-throughput purification techniques, recovery of products from suboptimal synthetic sequences became possible, and route development times could be compressed significantly. Automated purification of final products as well as intermediates obviously also benefits solution-phase protocols.

A further change has been the development of automated systems better suited to the rigors of general organic chemistry. Careful control of reaction conditions, heating and cooling, and maintenance of inert atmospheres have been significant improvements. Appropriate system size has also been addressed; a number of systems designed for specialist high-throughput environments have undergone modification to provide related systems for the more modest needs of medicinal chemistry laboratories, and their control systems are becoming more straightforward to use. A number of tools that rely less on automation and more on parallel manual processing are very popular with chemists wishing to perform 10 to 100 simultaneous reactions. The need for tools to support the process downstream of synthesis has also been addressed. We have modified our own laboratories to include automation workstations including synthesis, workup, evaporation, and purification capabilities.

However well suited the equipment, an essential component in successful implementation of high-throughput techniques in medicinal chemistry is training and support. In our own laboratories, a short intensive training program was made available to 48 chemists per year. This involved a simple one-step solution-phase synthesis of 192 compounds. Over a period of two to three days, chemists were introduced to several pieces of robotic and ancillary equipment as well as process philosophies, analytical tools, and the previously mentioned RADICAL application. As well as providing training, the compounds produced have augmented the corporate compound collection and have produced several interesting leads from high-throughput screening. In addition to this short familiarization exercise, a number of medicinal chemists have joined the high-throughput chemistry group for a more prolonged exposure to the technologies in a series of six-month secondments.

4.7 APPLICATION OF COMBINATORIAL CHEMISTRY IN DRUG DISCOVERY

Comprehensive surveys[153–156] indicate the large number of combinatorial libraries that have been documented in the literature, many of which are accompanied by biological data. Furthermore, publications describing the efforts of medicinal chemistry groups regularly include reference to part of the work benefiting from rapid compound synthesis using parallel or other combinatorial methods, and recent successful applications have been reviewed.[157] Four case histories are presented below to illustrate some of the ways in which high-throughput methodologies are contributing to a more rapid identification and optimization of biologically active compounds.

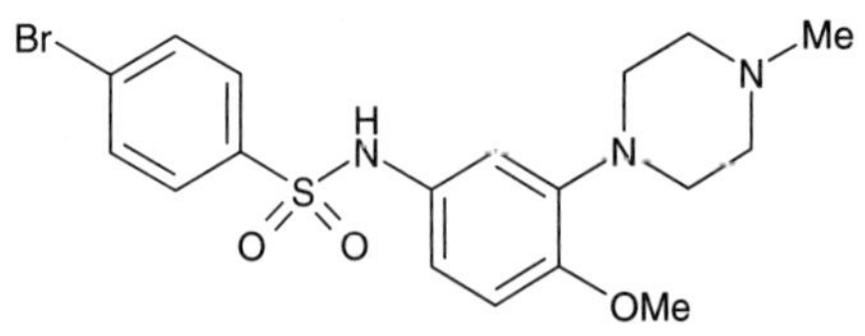

1

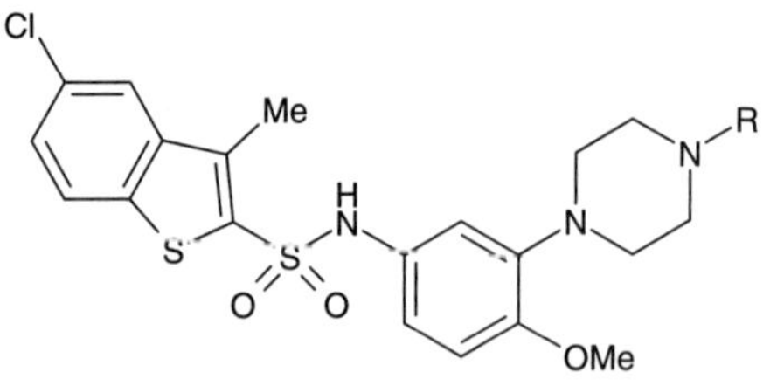

2 R = Me
3 R = H

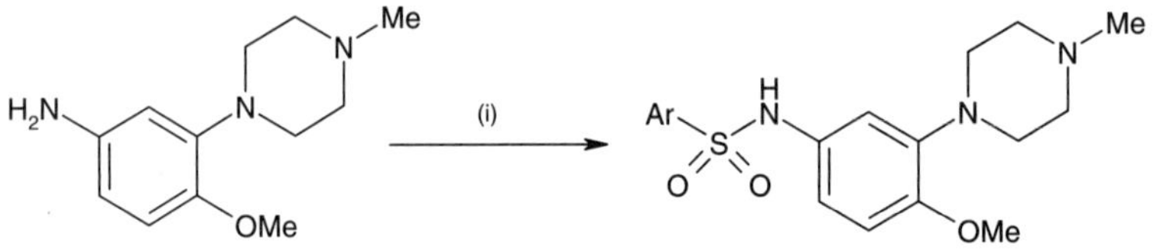

(i) $ArSO_2Cl$, acetone, rt, 18 h

Scheme 4.1

4.7.1 Case History 1

The first example illustrates the successful application of high-throughput screening of a diverse compound collection followed by lead optimization using parallel synthesis techniques.[158] The distribution of the 5-HT_6 receptor in the brain, in association with its high affinity for a range of drugs used in psychiatry, made the search for selective antagonists of this G-protein-coupled receptor an attractive proposition. The excellent affinity of bisaryl sulfonamide (**1**; Scheme 4.1) for the 5-HT_6 receptor (pK_i 8.3) was discovered following high-throughput screening of the SmithKline Beecham compound bank. Subsequent studies demonstrated the compound to have no appreciable affinity for a range of over 50 receptors, enzymes, and ion channels and to be moderately brain penetrant (25%) but with low oral availability (12%) as a result of rapid blood clearance. Using parallel solution synthesis techniques (Scheme 4.1) a series of analogs were rapidly prepared to investigate SAR around this lead compound, culminating in the identification of benzothiophene derivative (**2**) with subnanomolar 5-HT_6 receptor affinity. The compound demonstrated improved selectivity, reduced blood clearance, but was metabolically N-dealkylated to the piperazine (**3**). Subsequent synthesis and testing of **3** showed that it maintained high affinity (pK_i 8.9) and selectivity. Furthermore, pharmakokinetic studies in rats demonstrated moderate brain penetrancy (10%), low blood clearance, and, importantly, excellent oral bioavailability (>80%). The piperazine (**3**) has entered phase I clinical trials for the treatment of cognitive disorders.[159]

4.7.2 Case History 2

The second example illustrates the successful iterative deconvolution of a combinatorial (mixture) library based around a compound discovered by database searching of a compound collection.[160] Somatostatin is a widely distributed tetradecapeptide whose variety

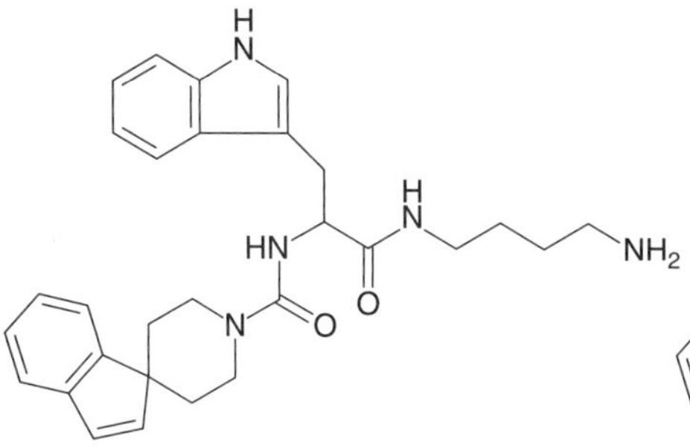

4

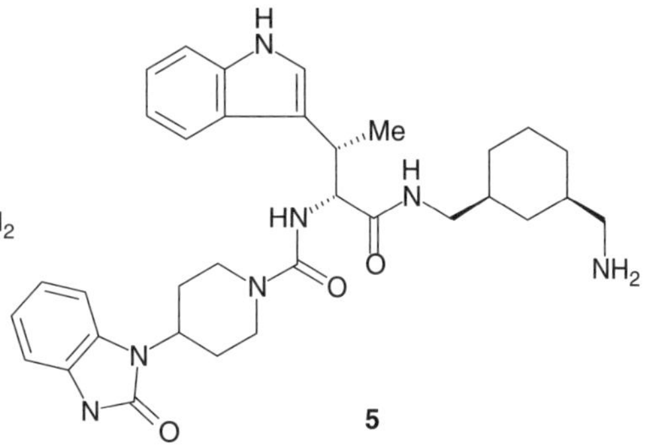

5

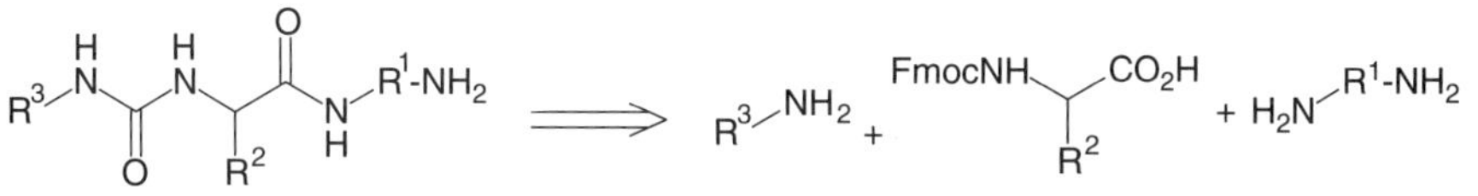

Scheme 4.2

of actions are mediated through high-affinity membrane-associated receptors (GPCRs). Five human somatostatin receptors (hSSTR1–5) have been cloned and characterized, for which a number of moderately subtype-selective petide agonists have been identified. Using a model based on the conformation of cyclic hexapeptide agonist c(Pro-Tyr-D-Trp-Lys-Thr-Phe), a three-dimensional similarity search of the Merck compound sample collection identified 75 compounds for screening, of which **4** had high affinity ($K_i = 200\,\mathrm{n}M$) in the mouse SSTR2 assay. A library was designed around this structure, synthesized by a split-and-mix strategy according to Scheme 4.2 from 20 diamines, 20 amino acids, and 79 amines. Following the introduction of diamine and amino acid building blocks, portions of resin were archived in readiness for the deconvolution process. When stereo- and regioisomers were taken into account, the library consisted of 131,670 entities in 79 pools of 1330 or 2660 members. The first deconvolution experiments focused on hSSTR2, against which, reassuringly, the most potent mixture was the one containing compound **4**. A second, equipotent mixture was also deconvoluted and yielded the very potent ($K_i = 0.04\,\mathrm{n}M$) and selective benzimidazolonylpiperidine (**5**). In a similar manner, subtype selective compounds for hSSTR1, hSSTR4, and hSSTR5 were discovered. The experiments also furnished useful structure–activity relationships that were subsequently exploited in further library designs.[161]

4.7.3 Case History 3

The next example demonstrates the application of combinatorial chemistry to the discovery and lead optimization phases using both solid-phase[162] and solution-phase[163] chemistry. P-glycoprotein (Pgp) provides an active efflux mechanism for the clearance of toxic substances. The overexpression of Pgp in cancer cells is responsible for the intrinsic or acquired immunity of tumor cells to a wide range of chemotherapeutic agents, a phenomenon known as *multidrug resistance* (MDR). Unfortunately, toxicity issues have limited the clinical use of compounds known to desensitize resistant tumor cells. Using solid-phase

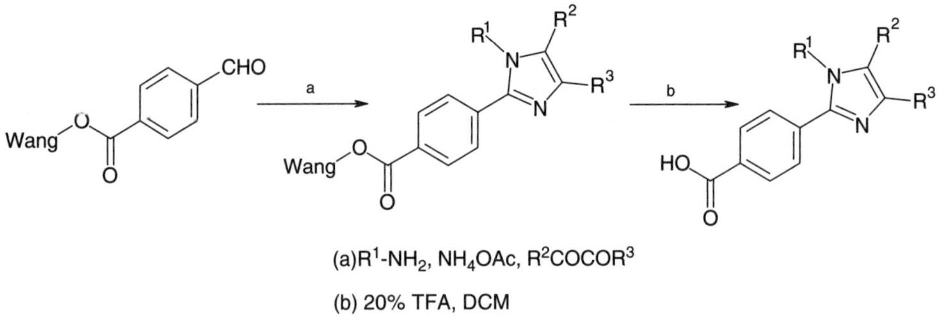

(a) R^1-NH_2, NH_4OAc, R^2COCOR^3

(b) 20% TFA, DCM

Scheme 4.3 Solid-phase synthesis.

synthesis, exemplified in Scheme 4.3, a library of 500 imidazoles was prepared, with characteristics typical of known Pgp modulators and substrates: namely, hydrophobicity and multiple amine groups. The SAR discovered from screening this library, followed by some preliminary solution-phase optimization, showed the cinnamic methyl ester (**6**), and subsequently, the less metabolically liable ether (**7**) to have good activities ($ED_{50} = 300$ and $90\,nM$, respectively). While these modulators had excellent potency against a variety of resistant cell lines, pharmacokinetic measurements indicated that the dimethylamino groups were rapidly metabolized. Using solution-phase combinatorial chemistry (Scheme 4.4) a range of alternative aminoaryl groups were investigated as 3- and 4-substituents on the imidazole ring, of which the bis-isopropylamino compound (**8**) proved most potent and had a good half-life in the dog ($t_{1/2}$ 2.79 h). This compound has been progressed as a clinical candidate and has shown virtually no adverse side effects in healthy male volunteers.

4.7.4 Case History 4

The final example was mentioned briefly in the library design section and demonstrates the power of coupling combinatorial chemistry with structure-based design.[107] A second library based on diversity considerations only was prepared as a control. The goal was to develop small-molecule nonpeptidic inhibitors of cathepsin D, an aspartyl protease implicated in tumor metastasis and Alzheimer's disease, based on a stable mimetic of the tetrahedral intermediate of peptide hydrolysis by this enzyme. The basic premise was to construct molecules of type **9** using amines (R^1) and acylating agents (R^2, R^3) as building blocks to decorate the well-known (hydroxyethyl)amine isostere, according to the disconnection in Scheme 4.5. A search for suitable commercially available building blocks revealed around 700 amines and 1900 acylating agents, which in combination would give rise to more than a billion products. Two strategies were used to reduce this number to two libraries of 1000 components each.

In the directed library design, the scaffold was modeled into the enzyme active site. Each of the building blocks was added to the scaffold independently, and a full range of conformations were assessed for intramolecular clashes with the scaffold and overlap with cathepsin D, and the 50 best scoring building blocks at each position were retained. Following removal of high-cost reagents, the remainder were clustered, and three sets of 10 building blocks were selected from unique clusters. In the diverse library design, the original building block lists were simply clustered using the Jarvis–Patrick algorithm,

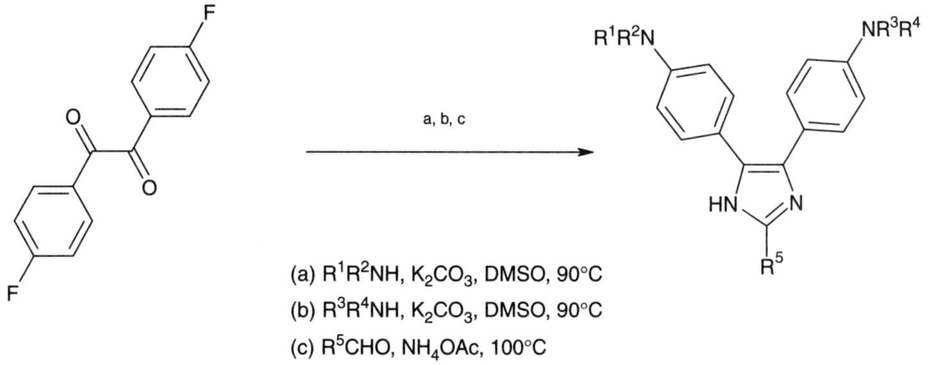

a, b, c

(a) R^1R^2NH, K_2CO_3, DMSO, 90°C
(b) R^3R^4NH, K_2CO_3, DMSO, 90°C
(c) R^5CHO, NH_4OAc, 100°C

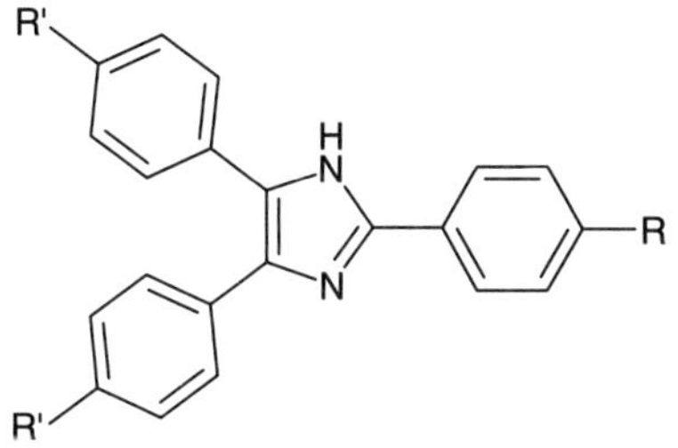

(**6**) R = $CH{=}CHCO_2Me$, R' = NMe_2
(**7**) R = $CH{=}CHCH_2OEt$, R' = Me_2N
(**8**) R = $CH{=}CHCH_2OEt$, R' = i-PrNH

Scheme 4.4 Solution-phase synthesis.

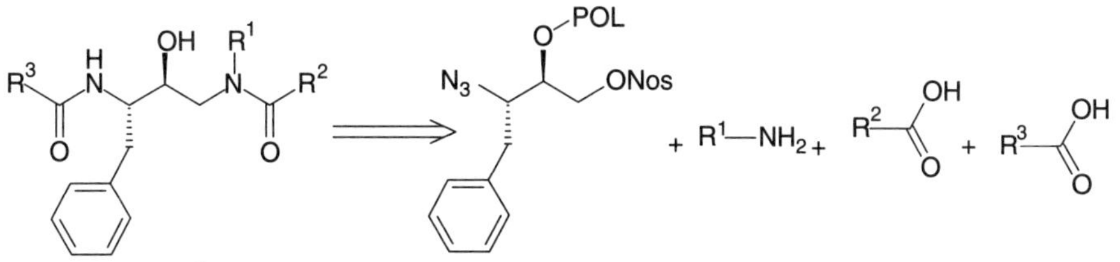

9

POL = Polystyrene resin with tetrahydropyran linker

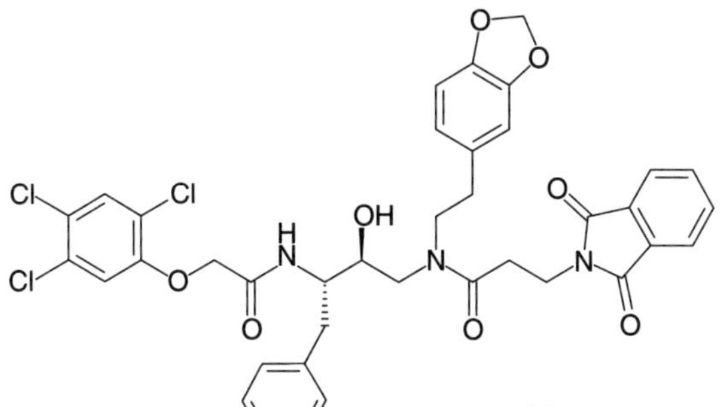

10

Scheme 4.5

and diverse sets of 10 were selected, with additional consideration of cost and balance of functional groups in each set. Both libraries were prepared using parallel solid-phase synthesis methods in a 96-well filter apparatus. When screened against cathepsin D at 1 μ*M* concentration, 67 compounds from the directed library showed inhibitions greater than 50%, compared to 26 compounds from the diverse library. At 100 n*M*, the corresponding figures were 7 (directed, including **10** with $K_i = 73\,\mathrm{n}M$) and 1 (diverse). The authors concluded that computational methods could be applied effectively to the reduction of large virtual libraries to practical numbers for synthesis and evaluation. In the absence of target information, however, diverse libraries remain an important strategy.

4.8 THE FUTURE OF COMBINATORIAL CHEMISTRY

The past decade has seen a number of developments, many of which have passed through several generations of evolution. Some, at first exciting and promising a great deal, have been replaced by alternatives, and overall strategies have been reviewed in light of the available technologies. Doubtless, much of what has been described above will continue to evolve with incremental improvements contributing to further ease of use, reliability, reproducibility, elimination of process bottlenecks, faster cycle times, improved library design processes (as our knowledge base expands), a greater range of suitable chemistry, and so on.[164] The specific roles of combinatorial chemistry may well evolve as information derived from genomics leads to new strategies.[165] It is probable that many high-throughput operations will move away from the conventional laboratory environment to a "drug discovery factory" employing industrial-scale engineering.[117,166,167] Further lessons may be learned from nature, with more libraries being designed around natural products,[168,169] or by employing biocatalysts in the execution of their synthesis.[170] It is also likely that there will be some surprises as new technologies are developed that shift the paradigm significantly. Two such ideas that are already emerging are outlined below.

4.8.1 Dynamic Combinatorial Libraries

A concept that has developed during recent years is that of *dynamic combinatorial libraries* (DCLs).[171–173] A DCL is an equilibrating pool of reagents and products (e.g., amines, aldehydes, imines) in which the products are not intended to be stable and isolable, but rather, continuously dissociate and re-form in different combinations. The idea is that in the presence of a receptor, the combinations that have the highest affinity will be formed preferentially, and their concentration will increase relative to other combinations. Hence, library generation and screening are combined in one process. Although the principle has been elegantly demonstrated in a number of simple systems, there are obstacles to overcome before this approach achieves sufficient generality to make a significant contribution to drug discovery.

4.8.2 Miniaturization

Some degree of miniaturization of chemistry is already evident—the very idea of working on a scale of 1 to 2 mL and routinely generating only a few milligrams of product would have seemed unlikely a decade or two ago. High-throughput screening is already moving to much smaller volumes; 384-, 1536- and even 3456- well plates occupy the same footprint as do traditional 96-well MTPs but use much smaller quantities of scarce biological reagents.[174] Of course, there is also a concomitant reduction in the quantity of test compound needed. Indeed, some fluorescence detection methods give measurable responses from single molecular interactions.[152]

The very basis of split-and-mix procedures relies on synthesis beads acting as tiny reactors. However, even if beads are segregated for screening purposes, they are handled in bulk quantities during synthesis using conventional labware. There are a number of physical hurdles encountered in moving to significantly smaller synthesis scales. Increased surface area relative to volume exacerbates moisture exclusion and evaporation problems, and capillary action and surface tension become dominant forces. Hence, microscale synthesis is unlikely to be performed in open wells, and a move toward enclosed fluidic systems is probable.

Fabrication techniques such as photolithography, chemical etching, and laser microforming have been applied to create networks of channels (typically, 50 to 300 μm) and reactors in materials such as silicon, quartz, glass, and plastic, allowing the manufacture of microfluidic devices.[175] Applied initially to chromatographic and analytical devices,[176] these techniques have already been used to construct microfluidic devices capable of performing solid-phase or solution chemistry.[177,178] Advantages offered by miniaturized systems include exquisite control of reaction conditions, low reagent and solvent consumption, and potential integration of synthesis, analysis, and screening into a single device.

4.9 CONCLUSIONS

Against an ever-changing scientific, economic, and political backdrop, medicinal chemists must continue to rise to the challenge of developing better, safer, more effective drugs more quickly and more economically than ever before, using whatever technologies are most appropriate. The concepts and technologies of combinatorial chemistry described in this chapter will doubtless contribute greatly toward this goal, as they are incorporated into everyday working practices. Perhaps as important as providing new tools, the development of combinatorial chemistry has opened chemists' minds to the possibilities of searching for further innovative means of attaining their goal. No single technology is likely to be a universal panacea, however convincingly it may be marketed. Medicinal chemists must continue to use judgment and scientific reasoning to assess the possible contributions new technologies might make, and how best to apply them. Similarly, manufacturers must continue to focus on the needs of the chemists. The field of combinatorial chemistry has undergone a rapid development in a relatively short period of time and revolutionized the modus operandi of the medicinal chemist. In whatever directions drug discovery may evolve in the future, we should anticipate combinatorial chemistry, in one form or another, to make significant further impact through many new and exciting developments.

In addition to the leading references provided throughout the chapter, a selection of recent review articles and books are assembled in Table 4.7 to aid the interested reader in exploring the subject further.

TABLE 4.7 Recent Reviews and Further Reading

Subject	Reviews	Books
General		179–184
Glossaries, internet resources	1, 2, 185	
Library design general, descriptors	61–63, 68, 69, 79, 82	186–188
Property-based, druglike	92, 189, 190	
Reaction prediction	191	
Solid-phase synthesis	192–197	198–201
Solid-phase supports	202, 203	
Linkers for solid-phase chemistry	204–206	
Encoding methods	207, 208	
Liquid-phase synthesis	209, 210	
Solution synthesis of libraries	211, 212	
Supported reagents and scavengers	213	
Solid-phase extraction	129	
Separation strategies	214	
Analytical methods	34, 35, 135, 215, 216	217
Automation	117, 218–221	222
Application to medicinal chemistry	153–155, 157, 223, 224	

REFERENCES

1. Maclean, D.; Baldwin, J. J.; Ivanov, V. T.; Kato, Y.; Shaw, A.; Schneider, P.; Gordon, E. M. Glossary of terms used in combinatorial chemistry. *Pure Appl. Chem.* 1999, *71*, 2349–2365.
2. Wermuth, C. G.; Ganellin, C. R.; Lindberg, P.; Mitscher, L. A. Glossary of terms used in medicinal chemistry. *Pure Appl. Chem.* 1998, *70*, 1129–1143.
3. Lebl, M. Parallel personal comments on "classical" papers in combinatorial chemistry. *J. Comb. Chem.* 1999, *1*, 3–24.
4. Merrifield, R. B. Solid phase peptide synthesis. 1. The synthesis of a tetrapeptide. *J. Am. Chem. Soc.* 1963, *85*, 2149–2154.
5. Letsinger, R. L.; Mahadevan, V. Oligonucleotide synthesis on a polymer support. *J. Am. Chem. Soc.* 1965, *87*, 3526–3527.
6. Leznoff, C. C. The use of insoluble polymer supports in general organic synthesis. *Acc. Chem. Res.* 1978, *11*, 327–333.
7. Geysen, H. M.; Meleon, R. H.; Barteling, S. J. Use of peptide synthesis to probe viral antigens for epitopes to a resolution of a single amino acid. *Proc. Natl. Acad. Sci. U.S.A.* 1984, *81*, 3998–4002.
8. Schnorrenberg, G.; Gerhardt, H. Fully automatic simultaneous multiple peptide synthesis in micromolar scale: rapid synthesis of series of peptides for screening in biological assays. *Tetrahedron* 1989, *24*, 7759–7764.
9. Houghten, R. A. General method for the rapid solid-phase synthesis of large numbers of peptides: specificity of antigen–antibody interaction at the level of individual amino acids. *Proc. Natl. Acad. Sci. U.S.A.* 1985, *82*, 5131–5135.
10. Furka, A.; Sebestyen, F.; Asgedom, M.; Dibo, G. Cornucopia of peptides by synthesis. *Abstr. 14th Int. Congr. Biochem. Prague, Czechoslovakia* 1988, 47.
11. Furka, A.; Sebestyen, F.; Asgedom, M.; Dibo, G. General method for rapid synthesis of multicomponent peptide mixtures. *Int. J. Peptide Protein Res.* 1991, *37*, 487–493.

12. Fodor, S. P. A.; Read, J. L.; Pirrung, M. C.; Stryer, L.; Lu, A. T.; Solas, D. Light-directed, spatially addressable parallel chemical synthesis. *Science* 1991, *251*, 767–773.
13. Houghten, R. A.; Pinilla, C.; Blondelle, S. E.; Appel, J. R.; Dooley, C. T.; Cuervo, J. H. Generation and use of synthetic peptide combinatorial libraries for basic research and drug discovery. *Nature* 1991, *354*, 84–86.
14. Pinilla, C.; Appel, J. R.; Blanc, P.; Houghten, R. A. Rapid identification of high affinity peptide ligands using positional scanning synthetic peptide combinatorial libraries. *BioTechniques* 1992, *13*, 901–905.
15. Dooley, C. T.; Houghten, R. A. The use of positional scanning synthetic peptide combinatorial libraries for the rapid determination of opioid receptor ligands. *Life Sci.* 1993, *52*, 1509–1517.
16. Smith, P. W.; Lai, J. Y. Q.; Whittington, A. R.; Cox, B.; Houston, J. G.; Stylli, C. H.; Banks, M. N.; Tiller, P. R. Synthesis and biological evaluation of a library containing potentially 1600 amides/esters: a strategy for rapid compound generation and screening. *Bioorg. Med. Chem. Lett.* 1994, *4*, 2821–2824.
17. Deprez, B.; Williard, X.; Bourel, L.; Coste, H.; Hyafil, F.; Tartar, A. Orthogonal combinatorial chemical libraries. *J. Am. Chem. Soc.* 1995, *117*, 5405–5406.
18. Burgess, K.; Liaw, A. I.; Wang, N. Combinatorial technologies involving reiterative division/coupling/recombination: statistical considerations. *J. Med. Chem.* 1994, *37*, 2985–2987.
19. Topiol, S.; Davies, J.; Vijayakuma, S.; Wareing, J. R. Computer aided analysis of split and mix combinatoral libraries. *J. Comb. Chem.* 2001, *3*, 20–27.
20. Konings, D. A. M.; Wyatt, J. R.; Ecker, D. J.; Freier, S. M. Deconvolution of combinatorial libraries for drug discovery: theoretical comparison of pooling strategies. *J. Med. Chem.* 1996, *39*, 2710–2719.
21. Wilson-Lingardo, L.; Davis, P. W.; Ecker, D. J.; Hebert, N.; Acevedo, O.; Sprankle, K.; Brennan, T.; Schwarcz, L.; Freier, S. M.; Wyatt, J. R. Deconvolution of combinatorial libraries for drug discovery: experimental comparison of pooling strategies. *J. Med. Chem.* 1996, *39*, 2720–2726.
22. Lam, K. S.; Salmon, S. E.; Hersh, E. M.; Hruby, V. J.; Kazmierski, W. M.; Knapp, R. J. A new type of synthetic peptide library for identifying ligand-binding activity. *Nature* 1991, *354*, 82–84.
23. Brenner, S.; Lerner, R. A. Encoded combinatorial chemistry. *Proc. Natl. Acad. Sci. U.S.A.* 1992, *89*, 5381–5383.
24. Nielsen, J.; Brenner, S.; Janda, K. D. Synthetic methods for the implementation of encoded combinatorial chemistry. *J. Am. Chem. Soc.* 1993, *115*, 9812–9813.
25. Bunin, B. A.; Ellman, J. A. A general and expedient method for the solid-phase synthesis of 1,4-benzodiazepine derivatives. *J. Am. Chem. Soc.* 1992, *114*, 10997–10998.
26. Bunin, B. A.; Plunkett, M. J.; Ellman, J. A. The combinatorial synthesis and chemical and biological evaluation of a 1,4-benzodiazepine library. *Proc. Natl. Acad. Sci. U.S.A.* 1994, *91*, 4708–4712.
27. DeWitt, S. H.; Kiely, J. S.; Stankovic, C. J.; Schroeder, M. C.; Reynolds Cody, D. M.; Pavia, M. R. "Diversomers": an approach to nonpeptide, nonoligomeric chemical diversity. *Proc. Natl. Acad. Sci. U.S.A.* 1993, *90*, 6909–6913.
28. Ohlmeyer, M. H. J.; Swanson, R. N.; Dillard, L. W.; Reader, J. C.; Asouline, G.; Kobayashi, R.; Wigler, M.; Still, W. C. Complex synthetic chemical libraries indexed with molecular tags. *Proc. Natl. Acad. Sci. U.S.A.* 1993, *90*, 10922–10926.
29. Ni, Z.-J.; Maclean, D.; Holmes, C. P.; Murphy, M. M.; Ruhland, B.; Jacobs, J. W.; Gordon, E. M.; Gallop, M. A. Versatile approach to encoding combinatorial organic syntheses using chemically robust secondary amine tags. *J. Med. Chem.* 1996, *39*, 1601–1608.

30. Geysen, H. M.; Wagner, C. D.; Bodnar, W. M.; Markworth, C. J.; Parke, G. J.; Schoenen, F. J.; Wagner, D. S.; Kinder, D. S. Isotope or mass encoding of combinatorial libraries. *Chem. Biol.* 1996, *3*, 679–688.

31. Hughes, I. Design of self-coded combinatorial libraries to facilitate direct analysis of ligands by mass spectrometry. *J. Med. Chem.* 1998, *41*, 3804–3811.

32. Nicolaou, K. C.; Xiao, X.-Y.; Parandoosh, Z.; Senyei, A.; Nova, M. P. Radiofrequency encoded combinatorial chemistry. *Angew. Chem. Int. Ed.* 1995, *34*, 2289–2291.

33. Yan, B. Monitoring the progress and the yield of solid-phase organic reactions directly on resin supports. *Acc. Chem. Res.* 1998, *31*, 621–630.

34. Swali, V.; Langley, G. J.; Bradley, M. Mass spectrometric analysis in combinatorial chemistry. *Curr. Opin. Chem. Biol.* 1999, *3*, 337–341.

35. Shapiro, M. J.; Wareing, J. R. NMR methods in combinatorial chemistry. *Curr. Opin. Chem. Biol.* 1998, *2*, 372–375.

36. Han, H.; Wolfe, M. M.; Brenner, S.; Janda, K. D. Liquid-phase combinatorial synthesis. *Proc. Natl. Acad. Sci. U.S.A.* 1995, *92*, 6419–6423.

37. Chen, S.; Janda, K. D. synthesis of Prostaglandin E2 methyl ester on a soluble-polymer support for the construction of prostanoid libraries. *J. Am. Chem. Soc.* 1997, *119*, 8724–8725.

38. Chen, S.; Janda, K. D. Total synthesis of naturally occurring prostaglandin F2alpha on a non-cross-linked polystyrene support. *Tetrahedron Lett.* 1998, *39*, 3943–3946.

39. Struder, A.; Hadida, S.; Ferritto, R.; Kim, S.-Y.; Jeger, P.; Wipf, P.; Curran, D. P. Fluorous synthesis: a fluorous-phase strategy for improving separation efficiency in organic synthesis. *Science* 1997, *275*, 823–826.

40. Struder, A.; Jeger, P.; Wipf, P.; Curran, D. P. Fluorous synthesis: fluorous protocols for the Ugi and Biginelli multicomponent condensations. *J. Org. Chem.* 1997, *62*, 2917–2924.

41. Struder, A.; Curran, D. P. A strategic alternative to solid-phase synthesis: preparation of a small isoxazoline library by "fluorous synthesis". *Tetrahedron* 1997, *53*, 6681–6696.

42. Curran, D. P.; Hadida, S. Tris(2-(perfluorohexyl)ethyl)tin hydride: a new fluorous reagent for use in traditional organic synthesis and liquid phase combinatorial synthesis. *J. Am. Chem. Soc.* 1996, *118*, 2531–2532.

43. Curran, D. P. Combinatorial organic synthesis and phase separation: back to the future. *Chem-Tracts Org. Chem.* 1996, *9*, 75–87.

44. Boger, D. L.; Tarby, C. M.; Myers, P. L.; Caporale, L. H. Generalized dipeptidomimetic template: Solution phase parallel synthesis of combinatorial libraries. *J. Am. Chem. Soc.* 1996, *118*, 2109–2110.

45. Cheng, S.; Comer, D. D.; Williams, J. P.; Myers, P. L.; Boger, D. L. Novel solution phase strategy for the synthesis of chemical libraries containing small organic molecules. *J. Am. Chem. Soc.* 1996, *118*, 2567–2573.

46. Kaldor, S. W.; Siegel, M. G.; Fritz, J. E.; Dressman, B. A.; Hahn, P. J. Use of solid supported nucleophiles and electrophiles for the purification of non-peptide small molecule libraries. *Tetrahedron Lett.* 1996, *37*, 7193–7196.

47. Kaldor, S. W.; Fritz, J. E.; Tang, J.; McKinney, E. R. Discovery of antirhinoviral leads by screening a combinatorial library of ureas prepared using covalent scavengers. *Bioorg. Med. Chem. Lett.* 1996, *6*, 3041–3044.

48. Caldarelli, M.; Habermann, J.; Ley, S. V. Synthesis of an array of potential matrix metalloproteinase inhibitors using a sequence of polymer-supported reagents. *Bioorg. Med. Chem. Lett.* 1999, *9*, 2049–2052.

49. Baxendale, I. R.; Ley, S. V. Polymer-supported reagents for multi-step organic synthesis: application to the synthesis of Sildenafil. *Bioorg. Med. Chem. Lett.* 2000, *10*, 1983–1986.

50. Ley, S. V.; Schucht, O.; Thomas, A. W.; Murray, P. J. Synthesis of the alkaloids (+/−)-oxomaritidine and (+/−)-epimaritidine using an orchestrated multi-step sequence of polymer supported reagents. *J. Chem. Soc., Perkin Trans. 1* 1999, 1251–1252.
51. Lipinski, C. A.; Lombardo, F.; Dominy, B. W.; Feeney, P. J. Experimental and computational approaches to estimate solubility and permeability in drug discovery and development settings. *Adv. Drug Del. Rev.* 1997, *23*, 3–25.
52. Ohlstein, E. H.; Ruffolo, R. R.; Elliott, J. D. Drug discovery in the next millennium. *Annu. Rev. Pharmacol. Toxicol.* 2000, *40*, 177–191.
53. Drews, J. Drug discovery: a historical perspective. *Science* 2000, *287*, 1960–1969.
54. Venton, D. L.; Woodbury, C. P. Screening combinatorial libraries. *Chemom. Intell. Lab. Syst.* 1999, *48*, 131–150.
55. Lam, K. S.; Lebl, M.; Krchnak, V. The "one-bead–one-compound" combinatorial library method. *Chem. Rev.* 1997, *97*, 411–448.
56. Eddershaw, P. J.; Beresford, A. P.; Bayliss, M. K. ADME/PK as part of a rational approach to drug discovery. *Drug Discov. Today* 2000, *5*, 409–414.
57. Brennan, M. B. Drug discovery. Filtering out failured early in the game. *Chem. Eng. News* 2000, *78*, 63–73.
58. Czarnik, A. W. Why combinatorial chemistry, why now (and why you should care). *Chem-Tracts Org. Chem.* 1995, *8*, 13–18.
59. Martin, Y. C. Challenges and prospects for comutational aids to molecular diversity. *Perspect. Drug Discov. Des.* 1997, *7/8*, 159–172.
60. Walters, W. P.; Stahl, M. T.; Murcko, M. A. Virtual screening: an overview. *Drug Discov. Today* 1998, *3*, 160–178.
61. Drewry, D. H.; Young, S. S. Approaches to the design of combinatorial libraries. *Chemom. Intell. Lab. Syst.* 1999, *48*, 1–20.
62. Bajorath, J. Selected concepts and investigations in compound classification, molecular descriptor analysis, and virtual screening. *J. Chem. Inf. Comput. Sci.* 2001, *41*, 233–245.
63. Leach, A. R.; Hann, M. M. The in silico world of virtual libraries. *Drug Discov. Today* 2000, *5*, 326–336.
64. Gorse, D.; Lahana, R. Functional diversity of compound libraries. *Curr. Opin. Chem. Biol.* 2000, *4*, 287–294.
65. Warr, W. A. Commercial software systems for diversity analysis. *Perspect. Drug Discov. Des.* 1997, *7/8*, 115–130.
66. Leach, A. R.; Bradshaw, J.; Green, D. V. S.; Hann, M. M. Implementation of a system for reagent selection and library enumeration, profiling, and design. *J. Chem. Inf. Comput. Sci.* 1999, *39*, 1161–1172.
67. Martin, Y. C. Diverse viewpoints on computational aspects of molecular diversity. *J. Comb. Chem.* 2001, *3*, 1–20.
68. Mason, J. S.; Hermsmeier, M. A. Diversity assessment. *Curr. Opin. Chem. Biol.* 1999, *3*, 342–349.
69. Agrafiotis, D. K.; Myslik, J. C.; Salemme, F. R. Advances in diversity profiling and combinatorial series design. *Mol. Diversity* 1999, *4*, 1–22.
70. Livingstone, D. J. The characterization of chemical structures using molecular properties: a survey. *J. Chem. Inf. Comput. Sci.* 2000, *40*, 195–209.
71. Katritzky, A. R.; Maran, U.; Lobanov, V. S.; Karelson, M. Structurally diverse quantitative structure–property relationship correlations of technologically relevant physical properties. *J. Chem. Inf. Comput. Sci.* 2000, *40*, 1–18.
72. Brown, R. D. Descriptors for diversity analysis. *Perspect. Drug Discov. Des.* 1997, *7/8*, 31–49.

73. Martin, E. J.; Blaney, J. M.; Siani, M. A.; Spellmeyer, D. C.; Wong, A. K.; Moos, W. H. Measuring diversity: experimental design of combinatorial libraries for drug discovery. *J. Med. Chem.* 1995, *38*, 1431–1436.
74. Pearlman, R. S.; Smith, K. M. Novel software tools for chemical diversity. *Perspect. Drug Discov. Des.* 1998, *9/10/11*, 339–353.
75. Pearlman, R. S.; Smith, K. M. Software for chemical diversity in the context of accelerated drug discovery. *Drugs Future* 1998, *23*, 885–895.
76. Brown, R. D.; Martin, Y. C. Use of structure–activity data to compare structure-based clustering methods and descriptors for use in compound selection. *J. Chem. Inf. Comput. Sci.* 1996, *36*, 572–584.
77. Patterson, D. E.; Cramer, R. D.; Ferguson, A. M.; Clark, R. D.; Weinberger, L. E. Neighborhood behavior: a useful concept for validation of "molecular diversity" descriptors. *J. Med. Chem.* 1996, *39*, 3049–3059.
78. Matter, H.; Potter, T. Comparing 3D pharmacophore triplets and 2D fingerprints for selecting diverse compound subsets. *J. Chem. Inf. Comput. Sci.* 1999, *39*, 1211–1225.
79. Van Drie, J. H.; Lajiness, M. S. Approaches to virtual library design. *Drug Discov. Today* 1998, *3*, 274–283.
80. Pickett, S. D.; Mason, J. S.; McLay, I. M. Diversity profiling and design using 3D pharmacophores: pharmacophore-derived queries (PDQ). *J. Chem. Inf. Comput. Sci.* 1996, *36*, 1214–1223.
81. Mason, J. S.; Morize, I.; Menard, P. R.; Cheney, D. L.; Hulme, C.; Labaudiniere, R. F. New 4-point pharmacophore method for molecular similarity and diversity applications: overview of the method and applications, including a novel approach to the design of combinatorial libraries containing privileged substructures. *J. Med. Chem.* 1999, *42*, 3251–3264.
82. Beno, B. R.; Mason, J. S. The design of combinatorial libraries using properties and 3D pharmacophore fingerprints. *Drug Discov. Today* 2001, *6*, 251–258.
83. Jamois, E. A.; Hassan, M.; Waldman, M. Evaluation of reagent-based and product-based strategies in the design of combinatorial library subsets. *J. Chem. Inf. Comput. Sci.* 2000, *40*, 63–70.
84. Gillet, V. J.; Nicolotti, O. Evaluation of reactant-based and product-based approaches to the design of combinatorial libraries. *Perspect. Drug Discov. Des.* 2000, *20*, 265–287.
85. Jarvis, R. A.; Patrick, E. A. Clustering using a similarity measure based on shared near neighbors. *IEEE Trans. Comput.* 1973, *C22*, 1025–1034.
86. Ward, J. H. Heirarchical grouping to optimize an objective function. *J. Am. Stat. Assoc.* 1963, *58*, 236–244.
87. Mason, J. S.; Pickett, S. D. Partition-based selection. *Perspect. Drug Discov. Des.* 1997, *7/8*, 85–114.
88. Gillet, V. J.; Willet, P.; Bradshaw, J. The effectiveness of reactant pools for generating structurally-diverse combinatorial libraries. *J. Chem. Inf. Comput. Sci.* 1997, *37*, 731–740.
89. Stanton, R. V.; Mount, J.; Miller, J. L. Combinatorial library design: maximizing model-fitting compounds within matrix synthesis constraints. *J. Chem. Inf. Comput. Sci.* 2000, *40*, 701–705.
90. Bravi, G.; Green, D. V. S.; Hann, M. M.; Leach, A. R. PLUMS: a program for the rapid optimization of focused libraries. *J. Chem. Inf. Comput. Sci.* 2000, *40*, 1441–1448.
91. Leo, A. J. Calculating log P_{oct} from structures. *Chem. Rev.* 1993, *93*, 1281–1306.
92. Walters, W. P.; Ajay; Murcko, M. A. Recognizing molecules with drug-like properties. *Curr. Opin. Chem. Biol.* 1999, *3*, 384–387.
93. Bemis, G. W.; Murcko, M. A. The properties of known drugs. 1. Molecular frameworks. *J. Med. Chem.* 1996, *39*, 2887–2893.
94. Bemis, G. W.; Murcko, M. A. Properties of known drugs. 2. Side chains. *J. Med. Chem.* 1999, *42*, 5095–5099.

95. Ajay; Walters, W. P.; Murcko, M. A. Can we learn to distinguish between "drug-like" and "non-drug-like" molecules? *J. Med. Chem.* 1998, *41*, 3314–3324.

96. Sadowski, J.; Kubinyi, H. A scoring scheme for discriminating between drugs and nondrugs. *J. Med. Chem.* 1998, *41*, 3325–3329.

97. Ajay; Bemis, G. W.; Murcko, M. A. Designing libraries with CNS activity. *J. Med. Chem.* 1999, *42*, 4942–4951.

98. Teague, S. J.; Davis, A. M.; Leeson, P. D.; Oprea, T. The design of leadlike combinatorial libraries. *Angew. Chem. Int. Ed.* 1999, *38*, 3743–3748.

99. Hann, M. M.; Leach, A. R.; Harper, G. Molecular complexity and its impact on the probability of finding leads for drug discovery. *J. Chem. Inf. Comput. Sci.* 2001, *41*, 856–864.

100. Kuntz, I. D. Structure-based strategies for drug-design and discovery. *Science* 1992, *257*, 1078–1082.

101. Li, J.; Murray, C. W.; Waszkowycz, B.; Young, S. C. Targeted molecular diversity in drug discovery: integration of structure-based design and combinatorial chemistry. *Drug Discov. Today* 1998, *3*, 105–112.

102. Kirkpatrick, D. L.; Watson, S.; Ulhaq, S. Structure-based drug design: combinatorial chemistry and molecular modeling. *Comb. Chem. High Throughput Screen.* 1999, *2*, 211–221.

103. Bohm, H.-J.; Stahl, M. Structure-based library design: molecular modelling merges with combinatorial chemistry. *Curr. Opin. Chem. Biol.* 2000, *4*, 283–286.

104. DesJarlais, R. L.; Sheridan, R. P.; Seibel, G. L.; Dixon, J. S.; Kuntz, I. D. Using shape complementarity as an initial screen in designing ligands for a receptor binding site of known three-dimensional structure. *J. Med. Chem.* 1988, *31*, 722–729.

105. Goodford, P. J. A computational procedure for determining energetically favorable binding sites on biologically important macromolecules. *J. Med. Chem.* 1985, *28*, 849–857.

106. Bohm, H.-J. The computer program LUDI: a new method for the de novo design of enzyme inhibitors. *J. Comput.- Aided Mol. Des.* 1992, *6*, 61–78.

107. Kick, E. K.; Roe, D. C.; Skillman, A. G.; Lui, G.; Ewing, T. J. A.; Sun, Y.; Kuntz, I. D.; Ellman, J. A. Structure-based design and combinatorial chemistry yield low nanomolar inhibitors of cathepsin D. *Chem. Biol.* 1997, *4*, 297–307.

108. Evans, B. E.; Rittle, K. E.; Bock, M. G.; DiPardo, R. M.; Freidinger, R. M.; Whitter, W. L.; Lundell, G. F.; Veber, D. F.; Anderson, P. S.; Chang, R. S. L.; Lotti, V. J.; Cerino, D. J.; Chen, T. B.; Kling, P. J.; Kunkel, K. A.; Springer, J. P.; Hirshfield, J. Methods for drug discovery: development of potent, selective, orally effective cholecystokinin antagonists. *J. Med. Chem.* 1988, *31*, 2235–2246.

109. Patchett, A. A.; Nargund, R. P. Privileged structures: an update. *Annu. Rep. Med. Chem.* 2000, *35*, 289–298.

110. Lewell, X. Q.; Judd, D. B.; Watson, S. P.; Hann, M. M. RECAP-retrosynthetic combinatorial analysis procedure: a powerful new technique for identifying privileged molecular fragments with useful applications in combinatorial chemistry. *J. Chem. Inf. Comput. Sci.* 1998, *38*, 511–522.

111. Braban, M.; Pop, I.; Willard, X.; Horvath, D. Reactivity prediction models applied to the selection of novel candidate building blocks for high-throughput organic synthesis of combinatorial libraries. *J. Chem. Inf. Comput. Sci.* 1999, *39*, 1119–1127.

112. Satoh, H.; Sacher, O.; Nakata, T.; Chen, L.; Gasteiger, J.; Funatsu, K. Classification of organic reactions: similarity of reactions based on changes in the electronic features of oxygen atoms at the reaction sites. *J. Chem. Inf. Comput. Sci.* 1998, *38*, 210–219.

113. Gasteiger, J.; Hondelmann, U.; Rose, P.; Witzenbichler, W. Computer-assisted prediction of the degradation of chemicals: hydrolysis of amides and benzoylphenylureas. *J. Chem. Soc., Perkin Trans. 1*, 1995, 193–204.

114. Hollering, R.; Gasteiger, J.; Steinhauer, L.; Schulz, K.-P.; Herwig, A. Simulation of organic reactions: from the degradation of chemicals to combinatorial synthesis. *J. Chem. Inf. Comput. Sci.* 2000, *40*, 482–494.

115. Kuo, P. Y.; Du, H.; Corkan, L. A.; Yang, K.; Lindsey, J. S. A planning module for performing grid search, factorial design, and related combinatorial studies on an automated chemistry workstation. *Chemom. Intell. Lab. Syst.* 1999, *48*, 219–234.

116. Pilipauskas, D. R. Can the time from synthesis design to validated chemistry be shortened? *Med. Res. Rev.* 1999, *19*, 463–474.

117. Hird, N. W. Automated synthesis: new tools for the organic chemist. *Drug Discov. Today* 1999, *4*, 265–274.

118. DeWitt, S. H.; Czarnik, A. W. Combinatorial organic synthesis using Parke-Davis's Diversomer method. *Acc. Chem. Res.* 1996, *29*, 114–122.

119. Xiao, X.-Y.; Li, R.; Zhuang, H.; Ewing, B.; Karunaratne, K.; Lillig, J.; Brown, R.; Nicolaou, K. C. Solid-phase combinatorial synthesis using MicroKan reactors, rf tagging, and directed sorting. *Biotech. Eng.* 2000, *71*, 44–50.

120. Plante, O. J.; Palmacci, E. R.; Seeberger, P. H. Automated solid-phase synthesis of oligosaccharides. *Science* 2001, *291*, 1523–1527.

121. Brooking, P.; Doran, A.; Grimsey, P.; Hird, N. W.; MacLachlan, W. S.; Vimal, M. Split-split: a multiple synthesiser approach to efficient automated parallel synthesis. *Tetrahedron Lett.* 1999, *40*, 1405–1408.

122. Larhed, M.; Hallberg, A. Microwave-assisted high-speed chemistry: a new technique in drug discovery. *Drug Discov. Today* 2001, *6*, 406–416.

123. Mahar, L. Resin capsules for solid-phase organic synthesis. *Spec. Chem.* 1998, 15–15.

124. Hird, N. W.; Hughes, I.; Re, V.; North, N. C. *Formulation.* 1999, WO 99/11676.

125. Atrash, B.; Bradley, M.; Kobylecki, R.; Cowell, D.; Reader, J. Revolutionizing resin handling for combinatorial synthesis. *Angew. Chem. Int. Ed.* 2001, *40*, 938–941.

126. Hird, N.; Hughes, I.; Hunter, D.; Morrison, M. G. J. T.; Sherrington, D. C.; Stevenson, L. Polymer disks: an alternative support format for solid phase synthesis. *Tetrahedron* 1999, *55*, 9575–9584.

127. Tripp, J. A.; Stein, J. A.; Svec, F.; Frechet, J. M. J. "Reactive filtration": use of functionalized porous polymer monoliths as scavengers in solution-phase synthesis. *Org. Lett.* 2000, *2*, 195–198.

128. Vaino, A. R.; Janda, K. D. Euclidean shape-encoded combinatorial chemical libraries. *Proc. Natl. Acad. Sci. U.S.A.* 2000, *97*, 7692–7696.

129. Nilsson, U. J. Solid-phase extraction for combinatorial libraries. *J. Chromatogr. A* 2000, *885*, 305–319.

130. Siegel, M. G.; Hahn, P. J.; Dressman, B. A.; Fritz, J. E.; Grunwell, J. R.; Kaldor, S. W. Rapid purification of small molecule libraries by ion exchange chromatography. *Tetrahedron Lett.* 1997, *38*, 3357–3360.

131. Schultz, L.; Garr, C. D.; Cameron, L. M.; Bukowski, J. High throughput purification of combinatorial libraries. *Bioorg. Med. Chem. Lett.* 1998, *8*, 2409–2414.

132. Kiplinger, J. P.; Cole, R. O.; Robinson, S.; Roskamp, E. J.; Ware, R. S.; O'Connell, H. J.; Brailsford, A.; Batt, J. Structure-controlled automated purification of parallel synthesis products in drug discovery. *Rapid Commun. Mass Spectrom.* 1998, *12*, 662.

133. Diggelmann, M.; Sporri, H.; Gassmann, E. Preparative LC/MS technology: a key component of the existing high speed synthesis at Syngenta. *Chimia* 2001, *55*, 23–25.

134. Hughes, I. Separating the wheat from the chaff: high throughput purification of chemical libraries. *J. Assoc. Lab. Autom.* 2000, *5*, 69–71.

135. Ripka, W. C.; Barker, G.; Krakover, J. High-throughput purification of compound libraries. *Drug Discov. Today* 2001, *6*, 471–477.

136. Hsu, B. H.; Orton, E.; Tang, S.-Y.; Carlton, R. A. Application of evaporative light scattering detection to the characterization of combinatorial and parallel synthesis libraries for pharmaceutical drug discovery. *J. Chromatogr. A* 1999, *725*, 103–112.

137. Fang, L.; Wan, M.; Pennacchio, M.; Pan, J. Evaluation of evaporative light-scattering detector for combinatorial libray quantitation by reversed phase HPLC. *J. Comb. Chem.* 2000, *2*, 254–257.

138. Fitch, W. L.; Szardenings, A. K.; Fujinari, E. M. Chemiluminescent nitrogen detection for HPLC: an important new tool in organic analytical chemistry. *Tetrahedron Lett.* 1997, *38*, 1689–1692.

139. Shah, N.; Gao, M.; Tsutsui, K.; Lu, A.; Davis, J.; Scheuerman, R.; Fitch, W. L. A novel approach to high-throughput quality control of parallel synthesis libraries. *J. Comb. Chem.* 2000, *2*, 453–460.

140. Biasi, V.; Haskins, N.; Organ, A.; Bateman, R.; Giles, K.; Jarvis, S. High throughput liquid chromatography/mass spectrometric analyses using a novel multiplexed electrospray interface. *Rapid Commun. Mass Spectron.* 1999, *13*, 1165–1168.

141. Gorlach, E.; Richmond, R.; Lewis, I. High-throughput flow injection analysis mass spectroscopy with networked delivery of color-rendered results. 2. Three-dimensional spectral mapping of 96-well combinatorial chemistry racks. *Anal. Chem.* 1998, *70*, 3227–3234.

142. Tong, H.; Bell, D.; Tabei, K.; Siegel, M. M. Automated data massaging, interpretation, and e-mailing modules for high throughput open access mass spectrometry. *J. Am. Soc. Mass Spectron.* 1999, *10*, 1174–1187.

143. Keifer, P. A.; Smallcombe, S. H.; Williams, E. H.; Salomon, K. E.; Mendez, G.; Belletire, J. L.; Moore, C. D. Direct-injection NMR (DI-NMR): a flow NMR technique for the analysis of combinatorial chemistry libraries. *J. Comb. Chem.* 2000, *2*, 151–171.

144. Gerritz, S. W.; Sefler, A. M. 2,5-Dimethylfuran (DMFu): an internal standard for the "traceless" quantitation of unknown samples via 1H NMR. *J. Comb. Chem.* 2000, *2*, 39–41.

145. Gobbi, A.; Poppinger, D.; Rohde, B. Developing an in-house system to support combinatorial chemistry. *Perspect. Drug Discov. Des.* 1997, *7/8*, 131–158.

146. Manly, C. J. Managing laboratory automation: integration and informatics in drug discovery. *J. Autom. Methods Manage. Chem.* 2000, *22*, 169–170.

147. Davies, K.; White, C. Managing combinatorial chemistry information. In *Molecular Diversity in Drug Design*, Dean, P. M.; Lewis, A. L. (Eds), Kluwer Academic Publishers, Dordrecht, The Netherlands, 1999, pp. 175–196.

148. Goodman, B. A. Managing the workflow of a high-throughput organic synthesis laboratory: a marriage of automation and information management technologies. *J. Assoc. Lab. Autom.* 1999, *4*, 48–52.

149. Merritt, A. T. Uptake of new technology in lead optimization for drug discovery. *Drug Discov. Today* 1998, *3*, 505–510.

150. Labaudiniere, R. F. RPR's approach to high-speed parallel synthesis for lead generation. *Drug Discov. Today* 1998, *3*, 511–515.

151. Coates, W. J.; Hunter, D. J.; MacLachlan, W. S. Successful implementation of automation in medicinal chemistry. *Drug Discov. Today* 2000, *5*, 521–527.

152. Herzberg, R. P.; Pope, A. J. High-throughput screening: new technology for the 21st century. *Curr. Opin. Chem. Biol.* 2000, *4*, 445–451.

153. Dolle, R. E. Comprehensive survey of combinatorial library synthesis: 1999. *J. Comb. Chem.* 2000, *2*, 383–433.

154. Dolle, R. E.; Nelson, K. H. Comprehensive survey of combinatorial library synthesis: 1998. *J. Comb. Chem.* 1999, *1*, 235–282.

155. Dolle, R. E. Comprehensive survey of chemical libraries yielding enzyme inhibitors, receptor agonists and antagonists, and other biologically active agents: 1992 through 1997. *Mol. Diversity* 1998, *3*, 199–233.

156. Dolle, R. E. Comprehensive survey of combinatorial libraries with undisclosed biological activity: 1992-1997. *Mol. Diversity* 1998, *4*, 233–256.

157. Golebiowski, A.; Klopfenstein, S. R.; Portlock, D. E. Lead compounds discovered from libraries. *Curr. Opin. Chem. Biol.* 2001, *5*, 273–284.

158. Bromidge, S. M.; Brown, A. M.; Clarke, S. E.; Dodgson, K.; Gager, T.; Grassam, H. L.; Jeffrey, P. M.; Joiner, G. F.; King, F. D.; Middlemiss, D. N.; Moss, S. F.; Newman, H.; Riley, G.; Routledge, C.; Wyman, P. 5-Chloro-N-(4-methoxy-3piperazin-1-ylphenyl)-3-methyl-2-benzothiophenesulfonamide (SB-271046): a potent, selective, and orally bioavailable 5-HT6 receptor antagonist. *J. Med. Chem.* 1999, *42*, 202–205.

159. Bromidge, S. M.; Clarke, S. E.; Gager, T.; Griffith, K.; Jeffrey, P.; Jennings, A. J.; Joiner, G. F.; King, F. D.; Lovell, P. J.; Moss, S. F.; Newman, H.; Riley, G.; Rogers, D.; Routledge, C.; Serafinowska, H.; Smith, D. R. Phenyl benzenesulfonamides are novel and selective 5-HT6 antagonists: identification of *N*-(2,5-dibromo-3-fluorophenyl)-4-methoxy-3-piperazin-1-ylbenzenesulfonamide (SB-357134). *Bioorg. Med. Chem. Lett.* 2001, *11*, 55–58.

160. Berk, S. C.; Rohrer, S. P.; Degrado, S. J.; Birzin, E. T.; Mosley, R. T.; Hutchins, S. M.; Pasternak, A.; Schaeffer, J. M.; Underwood, D. J.; Chapman, K. T. A combinatorial approach toward the discovery of non-peptide, subtype-selective somatostatin receptor ligands. *J. Comb. Chem.* 1999, *1*, 388–396.

161. Rohrer, S. P.; Birzin, E. T.; Mosley, R. T.; Berk, S. C.; Hutchins, S. M.; Shen, D.-M.; Xiong, Y.; Hayes, E. C.; Parmar, R. M.; Foor, F.; Mitra, S. W.; Degrado, S. J.; Shu, M.; Klopp, J. M.; Cai, S.-J.; Blake, A.; Chan, W. W. S.; Pasternak, A.; Yang, L.; Patchett, A. A.; Smith, R. G.; Chapman, K. T.; Schaeffer, J. M. Rapid identification of subtype-selective agonists of the somatostatin receptor through combinatorial chemistry. *Science* 1998, *282*, 737–740.

162. Sarshar, S.; Zhang, C.; Moran, E. J.; Krane, S.; Rodarte, J. C.; Benbatoul, K. D.; Dixon, R.; Mjalli, A. M. M. 2,4,5-Trisubstituted imidazoles: novel nontoxic modulators of P-glycoprotein mediated multidrug resistance. Part 1. *Bioorg. Med. Chem. Lett.* 2000, *10*, 2599–2601.

163. Zhang, C.; Sarshar, S.; Moran, E. J.; Krane, S.; Rodarte, J. C.; Benbatoul, K. D.; Dixon, R.; Mjalli, A. M. M. 2,4,5-Trisubstituted imidazoles: novel nontoxic modulators of P-glycoprotein mediated multidrug resistance. Part 2. *Bioorg. Med. Chem. Lett.* 2000, *10*, 2603–2605.

164. Borman, S. Combinatorial chemistry: redefining the scientific method. *Chem. Eng. News* 2000, *78*, 53–65.

165. Thorpe, D. S. Forecasting roles of combinatorial chemistry in the age of genomically derived drug discovery targets. *Comb. Chem. High Throughput Screen.* 2000, *3*, 421–436.

166. Harrison, W. Changes in scale in automated pharmaceutical research. *Drug Discov. Today* 1998, *3*, 343–349.

167. Archer, R. The drug discovery factory: an inevitable evolutionary consequence of high-throughput parallel processing. *Nat. Biotechnol.* 1999, *17*, 834–834.

168. Wessjohann, L. A. Synthesis of natural-product-based compound libraries. *Curr. Opin. Chem. Biol.* 2000, *4*, 303–309.

169. Nicolaou, K. C.; Pfefferkorn, J. A.; Roecker, A. J.; Cao, G.-Q.; Barluenga, S.; Mitchell, H. J. Natural product-like combinatorial libraries based on privileged structures. 1. General principles and solid-phase synthesis of benzopyrans. *J. Am. Chem. Soc.* 2000, *122*, 9939–9953.

170. Krstenansky, J. L.; Khmelnitsky, Y. Biocatalytic combinatorial synthesis. *Bioorg. Med. Chem.* 1999, *7*, 2157–2162.

171. Lehn, J.-M. Dynamic combinatorial chemistry and virtual combinatorial libraries. *Chem. Eur. J.* 1999, *5*, 2455–2463.

172. Cousins, G. R. L.; Poulsen, S.-A.; Sanders, J. K. M. Molecular evolution: dynamic combinatorial libraries, autocatalytic networks and the quest for molecular function. *Curr. Opin. Chem. Biol.* 2000, *4*, 270–279.

173. Huc, I.; Nguyen, R. Dynamic combinatorial chemistry. *Comb. Chem. High Throughput Screen.* 2001, *4*, 53–74.

174. Wolke, J.; Ullmann, D. Miniaturized HTS technologies: uHTS. *Drug Disc. Today* 2001, *6*, 637–646.

175. Barrow, D.; Cefai, J.; Taylor, S. Shrinking to fit. *Chem. Ind.* 1999, 591–594.

176. Cowen, S. Chip service. *Chem. Ind.* 1999, 584–586.

177. DeWitt, S. H. Microreactors for chemical synthesis. *Curr. Opin. Chem. Biol.* 1999, *3*, 350–356.

178. Haswell, S. J.; Middleton, R. J.; O'Sullivan, B.; Skelton, V.; Watts, P.; Styring, P. The application of micro reactors to synthetic chemistry. *Chem. Commun.* 2001, 391–398.

179. Terrett, N. K. *Combinatorial Chemistry*, Oxford University Press, Oxford, 1998.

180. Miertus, S.; Fassina, G. (Eds), *Combinatorial Chemistry and Technology: Principles, Methods and Applications*, Marcel Dekker, New York, 1999.

181. Jung, G. (Ed.), *Combinatorial Chemistry: Synthesis, Analysis, Screening*, Wiley-VCH, Weinheim, Germany, 1999.

182. Fenniri, H. (Ed.), *Combinatorial Chemistry*, Oxford University Press, Oxford, 2000.

183. Seneci, P. *Solid Phase Synthesis and Combinatorial Technologies*, Wiley-Interscience, New York, 2000.

184. Bannwarth, W.;Felder, E. (Eds.), *Combinatorial Chemistry: A Practical Approach*, Wiley-VCH, Weinheim, 2000.

185. Nakayama, G. R. Combinatorial chemistry web alert. *Curr. Opin. Chem. Biol.* 2001, *5*, 239–240.

186. Gordon, E. M.; Kerwin, J. F. (Eds.), *Combinatorial Chemistry and Molecular Diversity in Drug Discovery*, Wiley, New York, 1998.

187. Dean, P. M.; Lewis, R. A. (Eds.), *Molecular Diversity in Drug Design*, Kluwer Academic Publishers, Dordrecht, The Netherlands, 1999.

188. Ghose, A. K.; Viswanadhan, V. N. (Eds.), *Combinatorial Library Design and Evaluation. Principles, Software Tools, and Applications in Drug Discovery*, Marcel Dekker, New York, 2001.

189. Clark, D. E.; Pickett, S. D. Computational methods for the prediction of "drug-likeness." *Drug Discov. Today* 2000, *5*, 49–58.

190. Van de Waterbeemd, H.; Smith, D. A.; Beaumont, K.; Walker, D. K. Property-based design: optimization of drug absorption and pharmacokinetics. *J. Med. Chem.* 2001, *44*, 1313–1333.

191. Gasteiger, J.; Pfortner, M.; Sitzmann, M.; Hollering, R.; Sacher, O.; Kostka, T.; Karg, N. Computer-assisted synthesis and reaction planning in combinatorial chemistry. *Perspect. Drug Discov. Des.* 2000, *20*, 245–264.

192. Hermkens, P. H. H.; Ottenheijm, H. C. J.; Rees, D. Solid-phase organic reactions: a review of the recent literature. *Tetrahedron* 1996, *52*, 4527–4554.

193. Hermkens, P. H. H.; Ottenheijm, H. C. J.; Rees, D. Solid-phase organic reactions II: a review of the literature Nov. 95–Nov. 96. *Tetrahedron* 1997, *53*, 5643–5678.

194. Booth, S.; Hermkens, P. H. H.; Ottenheijm, H. C. J.; Rees, D. C. Solid-phase organic reactions III: a review of the literature Nov. 96–Dec. 97. *Tetrahedron* 1998, *54*, 15385–15443.

195. Lorsbach, B. A.; Kurth, M. J. Carbon-carbon bond forming solid-phase reactions. *Chem. Rev.* 1999, *99*, 1549–1581.

196. Franzen, R. G. Recent advances in the preparation of heterocycles on solid support: a review of the literature. *J. Comb. Chem.* 2000, *2*, 195–214.

197. Sammelson, R. E.; Kurth, M. J. Carbon-carbon bond-forming solid-phase reactions. Part II. *Chem. Rev.* 2001, *101*, 137–202.

198. Obrecht, D.;Villalgordo, J. M. (Eds.), *Solid-Supported Combinatorial and Parallel Synthesis of Small-Molecular-Weight Compound Libraries*, Pergamon Press, Oxford, 1998.

199. Bunin, B. A. *The Combinatorial Index*, Academic Press, San Diego, CA, 1998.

200. Dorwald, F. Z. *Organic Synthesis on Solid Phase: Supports, Linkers, Reactions*, Wiley-VCH, Weinheim, Germany, 2000.

201. Burgess, K. (Ed.), *Solid-Phase Organic Synthesis*, Wiley-Interscience, New York, 2000.

202. Hudson, D. Matrix assisted synthetic transformations: a mosaic of diverse contributions. 1. The pattern emerges. *J. Comb. Chem.* 1999, *1*, 333–360.

203. Hudson, D. Matrix assisted synthetic transformations: a mosaic of diverse contributions. 2. The pattern is completed. *J. Comb. Chem.* 1999, *1*, 403–457.

204. James, I. W. Linkers for solid phase synthesis. *Tetrahedron* 1999, *55*, 4855–4946.

205. Guillier, F.; Orain, D.; Bradley, M. Linkers and cleavage strategies in solid-phase organic synthesis and combinatorial chemistry. *Chem. Rev.* 2000, *100*, 2091–2157.

206. Comely, A. C.; Gibson, S. E. Tracelessness unmasked: a general linker nomenclature. *Angew. Chem. Int. Ed.* 2001, *40*, 1012–1032.

207. Barnes, C.; Balasubramanian, S. Recent developments in the encoding and deconvolution of combinatorial libraries. *Curr. Opin. Chem. Biol.* 2000, *4*, 346–350.

208. Affleck, R. L. Solutions for library encoding to create collections of discrete compounds. *Curr. Opin. Chem. Biol.* 2001, *5*, 257–263.

209. Gravert, D. J.; Janda, K. D. Organic synthesis on soluble polymer supports: liquid-phase methodologies. *Chem. Rev.* 1997, *97*, 489–509.

210. Toy, P. H.; Janda, K. D. Soluble polymer-supported organic synthesis. *Acc. Chem. Res.* 2000, *33*, 546–554.

211. An, H.; Cook, P. D. Methodologies for generating solution-phase combinatorial libraries. *Chem. Rev.* 2000, *100*, 3311–3340.

212. Baldino, C. M. Perspective articles on the utility and application of solution-phase combinatorial chemistry. *J. Comb. Chem.* 2000, *2*, 89–103.

213. Ley, S. V.; Baxendale, I. R.; Bream, R. N.; Jackson, P. S.; Leach, A. G.; Longbottom, D. A.; Nesi, M.; Scott, J. S.; Storer, R. I.; Taylor, S. J. Multi-step organic synthesis using solid-supported reagents and scavengers: a new paradigm in chemical library generation. *J. Chem. Soc., Perkin Trans. 1* 2000, 3815–4196.

214. Curran, D. P. Strategy-level separations in organic synthesis: from planning to practice. *Angew. Chem. Int. Ed.* 1998, *37*, 1175–1196.

215. Hughes, I.; Hunter, D. Techniques for analysis and purification in high-throughput chemistry. *Curr. Opin. Chem. Biol.* 2001, *5*, 243–247.

216. Kassel, D. B. Combinatorial chemistry and mass spectrometry in the 21st century drug discovery laboratory. *Chem. Rev.* 2001, *101*, 255–267.

217. Swartz, M. (Ed.), *Analytical Techniques in Combinatorial Chemistry*, Marcel Dekker, New York, 2000.

218. Hardin, J. H.; Smietana, F. R. Automating combinatorial chemistry: a primer on benchtop robotic systems. *Mol. Diversity* 1995, *1*, 270–274.

219. Cargill, J. F.; Lebl, M. New methods in combinatorial chemistry: robotics and parallel synthesis. *Curr. Opin. Chem. Biol.* 1997, *1*, 67–71.

220. Antonenko, V. V. Automation in combinatorial chemistry. In *Combinatorial Chemistry and Technology*, Miertus, S.; Fassina, G. (Eds.), Marcel Dekker, New York, 1999, pp. 205–232.

221. Bondy, S. S. The role of automation in drug discovery. *Curr. Opin. Drug Discov. Dev.* 1998, *1*, 116–119.

222. Hoyle, W. (Ed.), *Automated Synthetic Methods for Speciality Chemicals*, Royal Society of Chemistry, Cambridge, 2000.

223. Floyd, C. D.; Leblanc, C.; Whittaker, M. Combinatorial chemistry as a tool for drug discovery. *Prog. Med. Chem.* 1999, *36*, 91–168.

224. Houghten, R. A.; Pinilla, C.; Appel, J. R.; Blondelle, S. E.; Dooley, C. T.; Eichler, J.; Nefzi, A.; Ostresh, J. M. Mixture-based synthetic combinatorial libraries. *J. Med. Chem.* 1999, *42*, 3743–3778.

5

PARALLEL SOLUTION-PHASE SYNTHESIS

NORTON P. PEET AND HWA-OK KIM
CreaGen Biosciences, Inc.
Woburn, Massachusetts

5.1 INTRODUCTION

The concept of performing tasks in parallel rather than in series is simple. Parallel processing was the central concept of industrialization as it dawned in this country, and assembly lines made it possible to produce commercial goods in multiples rather than one at a time. The concept as applied to organic and medicinal chemistry debuted in the 1990s and is termed *parallel solution-phase synthesis* (PSPS). In this chapter we trace some personal history with this very useful technique and highlight important contributions that have been made by others to the methodology of PSPS.

It is very clear that PSPS is an extremely versatile tool for the simultaneous construction of related molecules. A variety of equipment has been employed for performing this commonsense chemistry, ranging from nonautomated procedures with multiple reaction vessels to fully automated programmable medicinal chemistry synthesis instruments. In between these extremes is semiautomated equipment such as reaction blocks (96-well and smaller) and partially programmable devices. The latter are perhaps the most useful to the medicinal chemist who is intent on building relatively small, focused libraries. It is not in the scope of this review to describe the equipment that has been developed and used for PSPS, but rather, to present some of the chemistry that has been enabled with these tools.

5.2 AHEAD OF OUR TIME

Several reports from our laboratories from the 1980s demonstrate that we were performing parallel chemistries in solution phase. The enactment of these chemistries was clearly more

Drug Discovery and Development, Volume 1: Drug Discovery, Edited by Mukund S. Chorghade

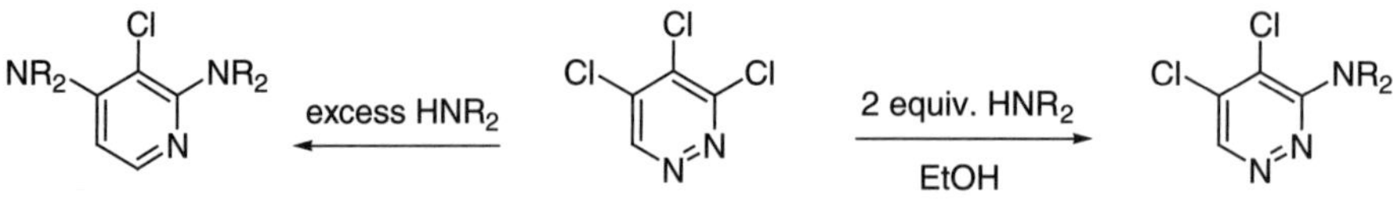

Cl NR2 NR2 N 3 excess HNR2 Cl Cl Cl N N 1 2 equiv. HNR2 EtOH Cl Cl NR2 N N 2

Scheme 5.1

behavioral than technological. Perhaps it was this early mindset, however, in our laboratories and those of others, that led to the widespread use of automated and semiautomated techniques for doing parallel chemistries in solution in the next decade.

Our early PSPS experiments were earmarked by simple shortcuts that we employed. These did not include special equipment or reagent-choosing software, but they did involve the parallel use of standard glassware and notebook entries in which all experimental and analytical data for a series of compounds were reported in tabular form in one entry rather than as single entries.

To investigate specially substituted 1,2,4-triazolo[4,3-*b*]pyridazines with potential bronchodilator activity, we treated 3,4,5-trichloropyridazine (**1**) with secondary amines to prepare requisite starting materials.[1] The reactive intermediate (**1**) underwent a regiospecific monodisplacement reaction at the 3-position to give compounds (**2**) with two equivalents of amine and double displacement with excess amine to provide the 3,5-disubstituted products (**3**), as shown in Scheme 5.1.

Purity ranged from 78 to 100% for these conversions and was determined by ^{1}H nuclear magnetic resonance (NMR) spectroscopy prior to purification. Since the positions for disubstitution reactions of **1** with other nucleophiles produced products with substituents on adjacent positions, the structure of **3** (NR_2 = 1-pyrrolidinyl) was unequivocally determined by hydrogenolysis of the chloro group and demonstrating by ^{1}H NMR spectroscopy that the pyridazine protons were meta-coupled.

Another early example of PSPS from our laboratories was the preparation of a series of 4-(1,3,4-oxadiazol-2-yl)-*N,N*-dialkylbenzenesulfonamides (**7**). This was an example of a $2 \times 2 \times 4$ cross PSPS. Carboxybenzenesulfonyl chlorides (**4**) were treated with two different secondary amines to produce benzenesulfonamides (**5**), which were then esterified to give the corresponding ethyl esters (**6**). Treatment of the esters with hydrazine hydrate gave the hydrazides (**7**), which were converted to oxadizoles (**8**) upon cyclization with four different orthesters,[2] as shown in Scheme 5.2.

Another early example of PSPS from our laboratories involved the synthesis of several key intermediates, which were new reactive intermediates for the preparation of tricyclic systems, using a novel double displacement reaction.[3] The reactive intermediates (**10**) were prepared by treating benzoyl chloride (**8**) with *S*-methyl cyclic thioureas (**9**). Subsequent double displacement reactions with hydrazines gave the imidazo[2,1-*b*]quinozolin-5-(3*H*)-ones (**11**) as shown in Scheme 5.3.

This novel cyclization was an example of a $3 \times 2 \times 5$ cross PSPS. In addition, it was shown that other aroyl chlorides would substitute for the benzoyl chlorides in this scheme to produce a variety of tricyclic systems. Selected compounds that were prepared in this PSPS protocol were evaluated as musculatropic bronchodilator agents in an artificially insufflated guinea pig animal model.

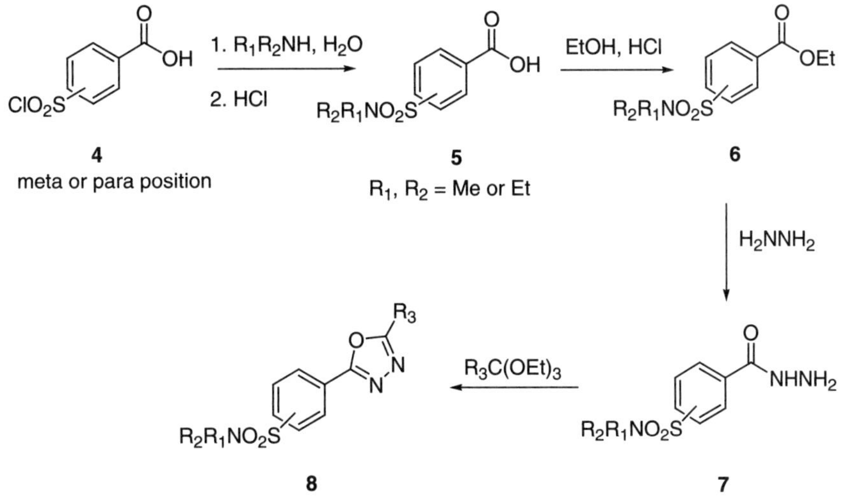

1. R_1R_2NH, H_2O
2. HCl

EtOH, HCl

ClO_2S — OH — **4** — meta or para position

$R_2R_1NO_2S$ — OH — **5** — R_1, R_2 = Me or Et

$R_2R_1NO_2S$ — OEt — **6**

H_2NNH_2

$R_2R_1NO_2S$ — $NHNH_2$ — **7**

$R_3C(OEt)_3$

$R_2R_1NO_2S$ — O — N — N — R_3 — **8**

Scheme 5.2

R_2 — Cl — R_1 — **8** — R_2 = F, NO_2; R_1 = H, Cl, CH_3

+

SMe — HN — N — n — **9** — n = 1, 2

pyridine

CH_2Cl_2

R_2 — O — SMe — N — N — R_1 — n — **10**

$H_2NNR_3R_4$
diglyme or neat

R_2 — O — $HNNR_3R_4$ — N — N — R_1 — n

NR_2 = 4-morpholinyl, 1-piperidinyl, 1-homopiperidinyl, dimethylamino, benzylamino

R_2 — O — N — N — N — N — R_3 — R_4 — n — **11**

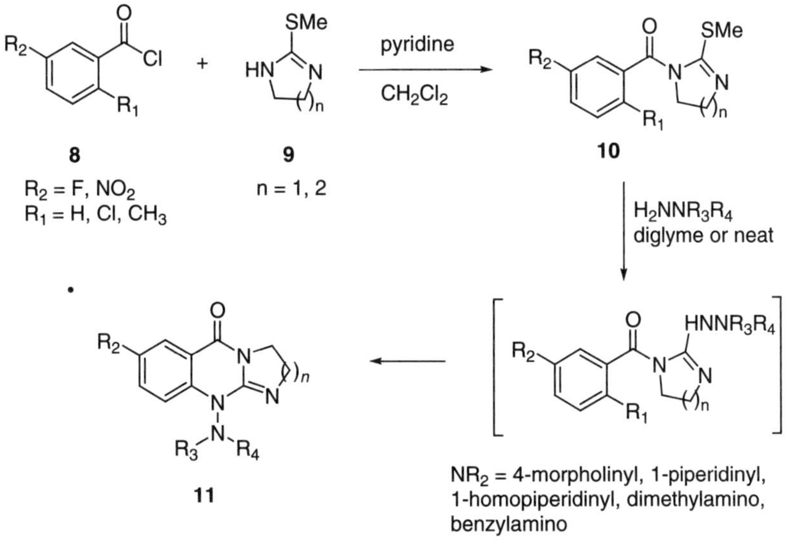

Scheme 5.3

We also prepared compounds for respiratory research that were very effective in a rat passive cutaneous anaphylaxis (PCA) model. These compounds were active as mediator release inhibitors in this model,[4] and one of these agents (MDL 427)[5] was studied extensively in phase I human clinical trials.

Again, 2-nitrobenzoyl chlorides (**12**) were starting materials of choice for the preparation of a 6×4 cross of quinazolinones bearing a tetrazole substituted at the 3-position, all of

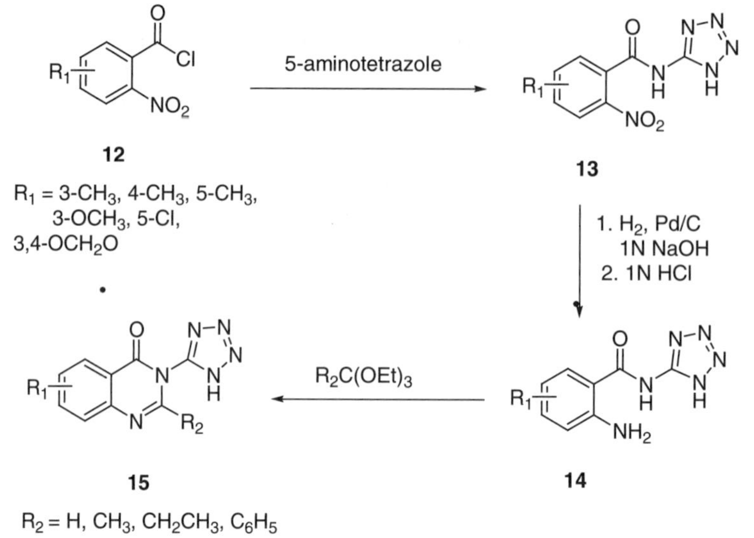

5-aminotetrazole

12

13

R_1 = 3-CH_3, 4-CH_3, 5-CH_3, 3-OCH_3, 5-Cl, 3,4-OCH_2O

1. H_2, Pd/C 1N NaOH
2. 1N HCl

$R_2C(OEt)_3$

15

14

R_2 = H, CH_3, CH_2CH_3, C_6H_5

Scheme 5.4

which were active as antiallergic agents. As shown in Scheme 5.4, acid chlorides (**12**) were treated with 5-aminotetrazole to produce the tetrazolyl amide (**13**). Catalytic reduction of the sodium salts of **13** in aqueous media cleanly gave the corresponding anthranilamides (**14**), which were cyclized with orthoesters to produce the 3-(1*H*-tetrazol-5yl)-4(3*H*)-quinazolinones (**15**). Compounds **15** were converted to sodium salts to provide water-soluble agents, which greatly facilitated their administration to animals for biological evaluation.

We were also interested in adenosine receptor ligands for the treatment of respiratory diseases and other disorders. The preparation of 8-substituted xanthines allowed us to use PSPS for the preparation of a group of designed adenosine antagonists.[6] Key intermediate **16**[7] was coupled with 16 different carboxylic acids, using the mixed anhydride method, to provide the pyrimidinones (**17**) in which the amino group at the 5-position was regioselectively acylated. To protect the chiral integrity of the acyl substituent during construction of the xanthine nucleus, we prepared imino ethers (**18**) using Meerwein's reagent under relatively mild conditions. Imino ethers (**18**) could then be cyclized, again under the same relatively mild conditions, to the (very) selective adenosine A_1 receptor antagonists (**19**). This route, which is shown in Scheme 5.5, provided a novel and very convenient route to xanthines bearing chiral substituents at the 8-position. Interestingly, the facile closure of **18** to **19** constitutes a 5-*endo-trig* anti-Baldwin ring closure.[8,9] Alternatively, when racemization of the acyl group was not an issue, we were able to effect the closure of intermediates **17** directly to **19** with ethanolic potassium hydroxide.

5.3 RECENT REPORTS OF PARALLEL SOLUTION-PHASE SYNTHESIS

One of the companies that spearheaded PSPS and built a successful business around an automated platform for performing PSPS is ArQule. Several reports from their laboratories cite

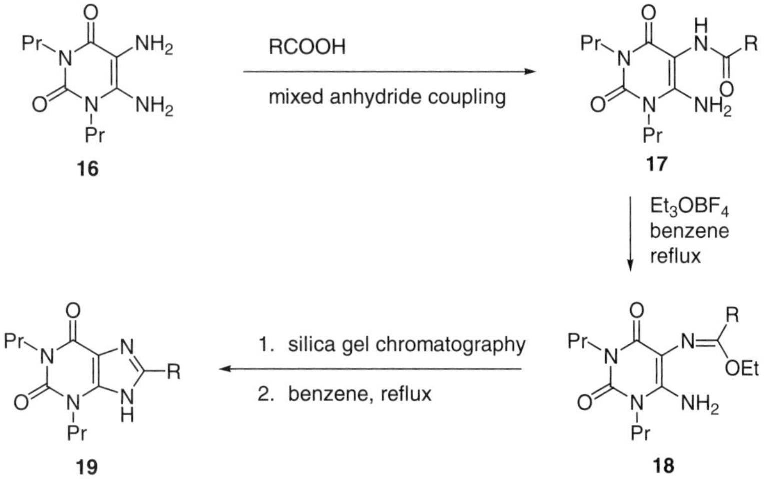

R = (R)-1-(phenylmethyl)ethyl, (S)-1-(phenylmethyl)ethyl, 1-(phenylmethyl)ethyl, (R)-1-phenylpropyl, (S)-1-phenylpropyl, 1-phenylpropyl, (R)-1-phenylethyl, (S)-1-phenylethyl, 1-phenylethyl, 1-(phenylmethyl)propyl, 1-(phenylmethyl)butyl, 2-indanyl, 1,2,3,4-tetrahydro-2-naphthyl, 1-(hydroxymethyl)-2-phenylethyl, *trans*-2-phenylcyclopentyl

Scheme 5.5

successful examples of PSPS in building diverse chemical libraries. These libraries include triazine-based compounds containing both carbohydrates and peptides,[10] chalcone-based screening libraries,[11] spiro[pyrrolidine-2,3′-oxindole] libraries,[12] and several others.[13,14] A representative and typical instructive example of convergent automated PSPS is shown in Scheme 5.6.[11] Chalcones were used as versatile, key intermediates for the construction

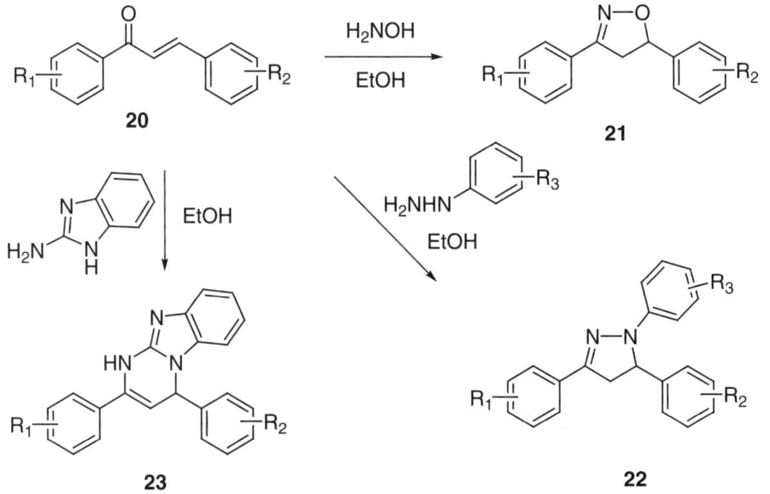

Scheme 5.6

of several heterocycles. Treatment of 1280 chalcones with hydroxylamines gave the 1280 isoxazolines (**21**). Additional diversity was added by cyclization of compounds (**20**) with substituted phenylhydrazines to produce 7680 pyrazolines **22**. Similarly, addition of substituted aminobenzamides to enones (**20**) gave a library of 7680 fused pyrimidines (**23**).

Many laboratories are presently using PSPS as an integral part of their discovery programs. In some research organizations, PSPS is part of discovery chemistry groups. In other organizations, PSPS is a part of collaborative units, such as high-throughput or high-speed chemistry groups, SWAT teams, or combinatorial chemistry groups. A few universities are now engaging and teaching students to use PSPS. There are numerous examples of successful PSPS campaigns, both published and unpublished. In this review we do not attempt to report these successes comprehensively, but instead, highlight a few representative recent examples from both our laboratories and other laboratories.

A series of antimalarial rhodacyanine dyes has recently been reported[15] employing a $3 \times 3 \times 3$ cross PSPS as displayed in Scheme 5.7. Methylthioiminium salts (**24**) were condensed with rhodamines (**25**) to give vinylogous amides (**26**). Tosylated salts (**27**) were prepared and coupled with iminium salts (**28**) to afford the final library of compounds (**29**). A representative compound with high antimalarial activity is MKH-57.

A hit compound arising from an anti-infective screen of 1500 higher plants[16] provided the rationale for building the library shown in Scheme 5.8. Dimethoxybenzoyl chlorides

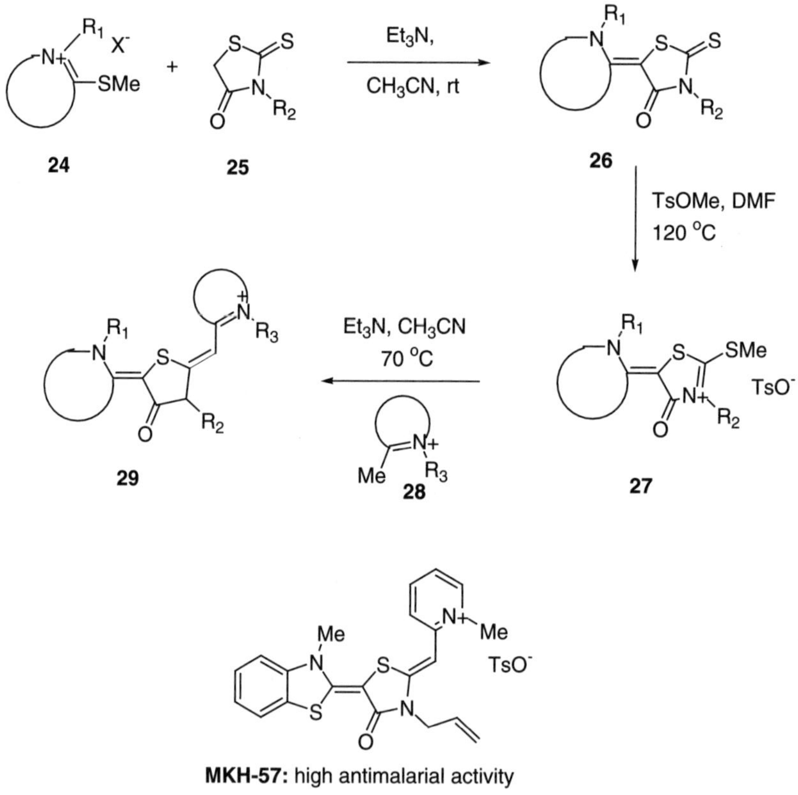

24 + 25 → (Et$_3$N, CH$_3$CN, rt) → 26 → (TsOMe, DMF, 120 °C) → 27 → (28, Et$_3$N, CH$_3$CN, 70 °C) → 29

MKH-57: high antimalarial activity

Scheme 5.7

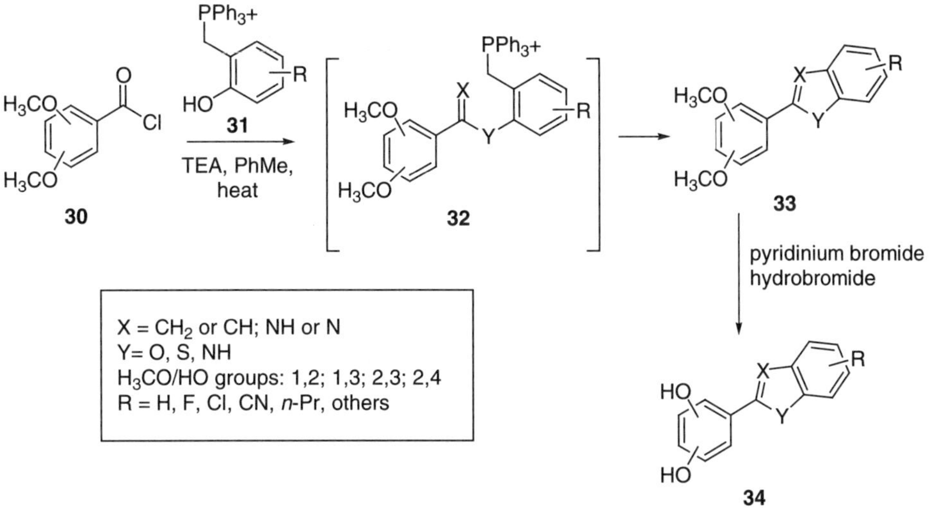
PPh3+
H3CO
H3CO
O
Cl
30
HO
31
TEA, PhMe, heat
32
33
pyridinium bromide hydrobromide
HO
HO
34
X = CH2 or CH; NH or N
Y= O, S, NH
H3CO/HO groups: 1,2; 1,3; 2,3; 2,4
R = H, F, Cl, CN, n-Pr, others

Scheme 5.8

(**30**) were treated with triphenylphosphonium ylides (**31**) to provide, after acylation and intramolecular Wittig cyclization, benzo-fused heterocycles (**33**). Subsequent treatment of **33** with pyridinium bromide hydrobromide produced the corresponding didemethylated compounds (**34**). The focused array of compounds represented by general structure **34** was evaluated in a panel of antifungal assays.

A small library of flavones was prepared[17] using PSPS in a 4×9 cross sequence as shown in Scheme 5.9 and evaluated as high-affinity benzodiazepine receptor ligands. This

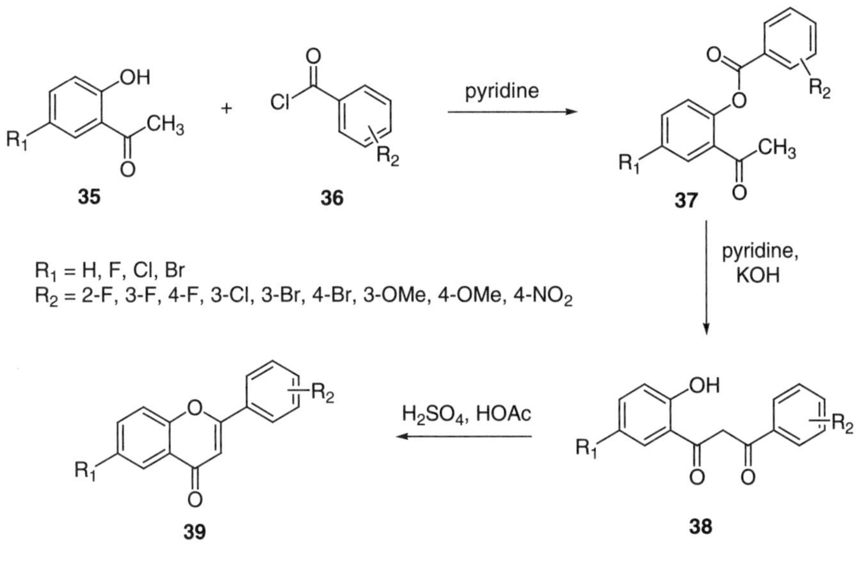
OH
CH3
R1
35
Cl
R2
36
pyridine
37
pyridine, KOH
38
H2SO4, HOAc
39
R1 = H, F, Cl, Br
R2 = 2-F, 3-F, 4-F, 3-Cl, 3-Br, 4-Br, 3-OMe, 4-OMe, 4-NO2

Scheme 5.9

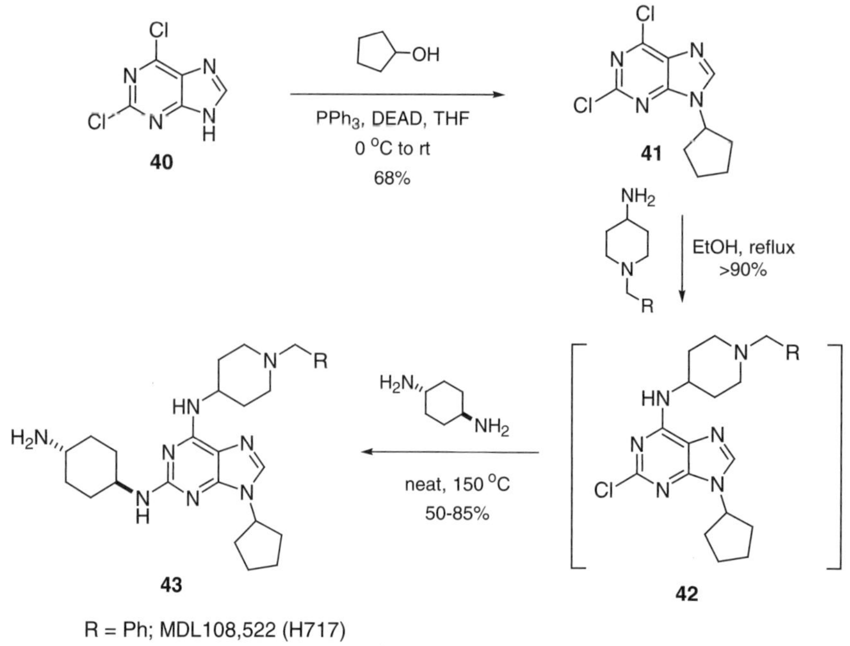

R = Ph; MDL108,522 (H717)

Scheme 5.10

library was prepared from acetophenones (**35**) and benzoyl chlorides (**36**), which produced a series of phenyl esters (**37**). Esters **37** were converted to β-diketones (**38**) with pyridine and potassium hydroxide. Subsequent cyclization of **38** under acidic conditions gave 36 flavones of general structure **39**. Interestingly, to facilitate a screening protocol, nine mixtures of four compounds were also produced, using the same synthetic protocol but employing a mixture of all four acetophenones with each acid chloride.

We have employed PSPS to prepare a substantial set of what we refer to as *adenine scaffold-derived* cyclin-dependent kinase (cdk2 and cdk4) inhibitors, as shown in Scheme 5.10. These compounds were evaluated, and many shown to be active in in vitro tumor cell proliferation models for breast, colon, and prostate cancer.[18] After defining optimal substituents for the 2- and 9-positions of the adenine scaffolds, we used PSPS to optimize the substituent at the 6N-position. We knew from modeling and x-ray crystallographic studies[19] that substituents at this position protruded out of the enzyme active site and into a solvent-accessible site, and thus referred to it as the *ADME handle*. We could, therefore, adjust this substituent to affect ADME parameters such as solubility and metabolism without greatly affecting enzyme inhibitory activity.

Our two-step one-pot PSPS products initiated from 2,6-dichloro-9-cyclopentylpurine (**41**), which was easily accessible from dichloropurine (**40**) by treatment with cyclopentanol under Mitsunobu conditions. Treatment of **41**, as shown in Scheme 5.10, with a variety of 4-amino-1-benzylpiperidines provided the 6,9-disubstituted purines (**42**) in situ. Subsequent treatment of compounds **42** with *trans*-1,4-diaminocyclohexane, after removal of solvent and heating to 150 °C, gave the target compounds **43** in quite good overall yields.

In other recent studies we have developed process chemistry and performed proof-of-principle experiments on key intermediates for the preparation of PSPS libraries. Thus,

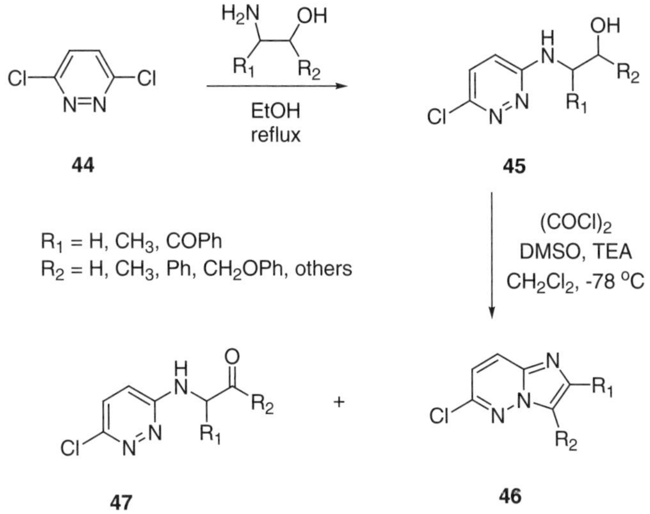

R_1 = H, CH_3, COPh
R_2 = H, CH_3, Ph, CH_2OPh, others

Scheme 5.11

under Swern conditions, we have optimized the conversion of amino alcohols (**45**), prepared from 3,6-dichloropyridazine (**44**) by displacement with simple amino alcohols, to imidazo[1,2-*b*]pyridazines (**46**) and minimized the production of ketones (**47**).[20] Imidazopyridazines (**46**) are useful starting points for PSPS, for the introduction of a wide variety of substituents at the 6-position, after nucleophilic displacement using a palladium cross-coupling procedure (Scheme 5.11).

We have also extended the Iqbal multicomponent procedure[21–23] for preparing 2-[(acetylamino)methyl]-1,3-dicarbonyl compounds to the preparation of heterocycles. Thus, compounds **49** were prepared from ethyl acetoacetate (**48**) by treatment with various substituted benzaldehydes, acetyl chloride, and acetonitrile. As proof-of- principle for the conversion of **49** to heterocycles, which can be accomplished using PSPS, we treated **49** (R = H) with phenylhydrazine to produce pyrazolone (**50**), as shown in Scheme 5.12.[24]

There are only a limited number of multicomponent condensations, which are general reactions that reliably produce the desired target compounds and are amenable to parallel synthesis. Two other multicomponent condensations that fall into this category are the Ugi[25] and the Passerini[26] reactions. The Passerini reaction involves the treatment of an isocyanide with a carboxylic acid and an aldehyde or a ketone. The Ugi reaction uses

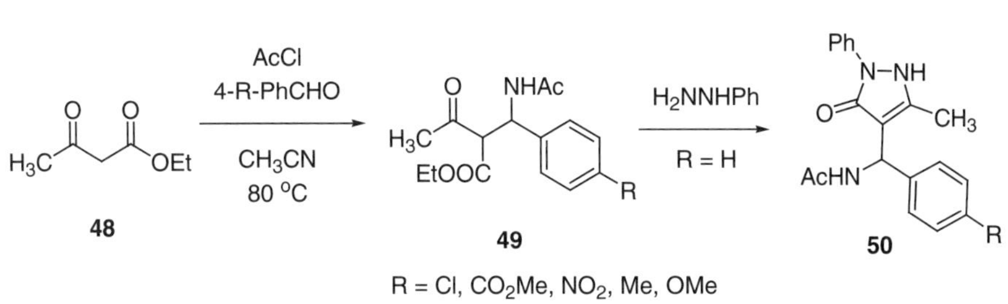

R = Cl, CO_2Me, NO_2, Me, OMe

Scheme 5.12

all of these components plus ammonia or an amine. Another multicomponent reaction is the Biginelli condensation, whose reactants are aldehydes, ketones with activated methylene groups, and ureas or thioureas.[27–29] A recent example of microwave-assisted Biginelli condensations, using a variety of catalysts to optimize yields, has been reported.[30] Microwave-assisted parallel synthesis (MAPS) is a combination of new technologies, which will undoubtedly be used for many future applications.

Several excellent review articles on PSPS have appeared. One of these articles focuses on PSPS and solid-phase strategies that use natural product templates, and covers PSPS strategies for libraries based on distamycin, flavonoids, kramerixin (benzofurans), mappicine, curacin, stipiamide, sarcodictine, and taxoids.[31] Several additional articles are cited in Section 5.4.

Many additional examples of PSPS, which have been applied to the synthesis of medium-sized target libraries with biological activity, are shown in Table 5.1. It is very clear that PSPS facilitated the time-efficient preparation of these focused libraries and has decreased the cycle time for hit-to-lead and lead-to-high value lead iteration.

5.4 SOLID SUPPORTED REAGENTS, SCAVENGERS, AND CATALYSTS

A critically important supporting technology for parallel solution-phase synthesis is that of functionalized polymers. There are three classes of functionalized polymers: solid-supported reagents (SSRs), solid-supported scavengers (SSSs), and solid-supported catalysts (SSCs). It is these reagents that allow solution-phase synthesis to produce reaction solutions and products that are ultimately quite clean, simply by allowing excess reagent or catalyst to be removed by filtration. Similarly, excess reagents that are not resin-bound can be trapped by designed SSSs and removed by filtration, as can by-products of the reaction or unreacted starting materials. Combinations of SSRs, SSSs, and SSCs can be used for PSPS steps. Indeed, the use of these reagents is what allows the execution of multistep solution-phase synthetic procedures.

In our laboratory we designed a resin-based reagent for diazo transfer,[58] related to an earlier reagent that was reported by the Rebek group.[59] Our reagent is polystyrene-supported benzenesulfonyl azide (**52**), which is thermally stable and not friction sensitive. Resin-bound reagent **52** was prepared by treatment of commercially available polymer-supported benzenesulfonyl chloride with sodium azide in aqueous dimethylformamide at room temperature. Transfer of the diazo group from **52** to a variety of active methylene compounds (**53**) was demonstrated to produce diazo compounds (**54**), as shown in Scheme 5.13.

A recent example of PSPS which employs both resin-bound reagents and scavengers is shown in Scheme 5.14 and describes the syntheses of trisubstituted benzimidazolones.[60] Thus, 2-fluoronitrobenzenes (**55**) were treated with primary amines to produce anilines (**56**) via Meisenheimer displacement. Excess 2-fluoronitrobenzenes (**55**) were removed using the polyamine scavenger **A**. Treatment of 2-nitroanilines (**56**) with Raney nickel gave phenylenediamines (**56**), which were cyclized to benzimidazolones (**58**) with 1,1′-carbonyldiimidazole (CDI). Alkylation of benzimidazolones (**58**) was achieved by treatment with resin-bound base A and an alkyl bromide, followed by removal of the excess alkyl bromide with resin-bound thiourea scavenger B. This overall procedure allowed the production of an array of compounds using purification strategies that were differentiated completely from conventional techniques.

TABLE 5.1 Parallel Solution-Phase Synthesis Libraries

Entry	Library	Structure	Biological Activity	Comments	Ref.
1	Cyclic (depsi)peptides	HUN-7293 analogs	Inhibition of cell adhesion molecule expression	VCAM-1, ICAM-1, and E-selectin are adhesion molecules that can induce inflammation.	32
2	Spiperone analogs	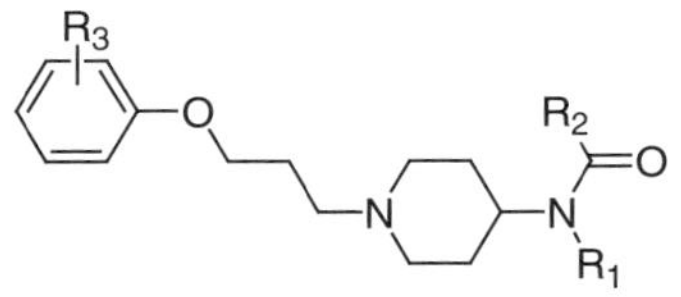	5-HT_{2A}/D2 ligands	Improved selectivity for 5-HT_{2A} with respect to spiperone was achieved.	33
3	Acylguanidines	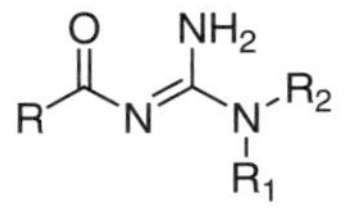	Sodium channel blockers	Selected compounds were evaluated in the audiogenic mouse antiseizure model.	34
4	*N*-(1-Phenylethyl)-5-phenylimidazole-2-amines	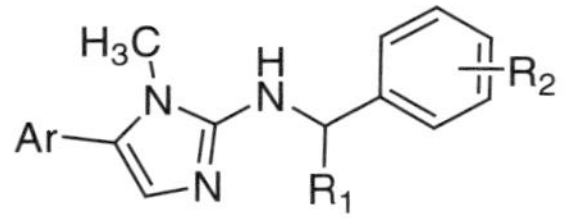	Na^+/K^+ ATPase inhibitors	Inhibitors have potential utility for congestive heart failure (CHF).	35
5	5-Carboxamido-1-benzyl-(3-dimethylamino-propyloxy-1*H*-pyrazoles	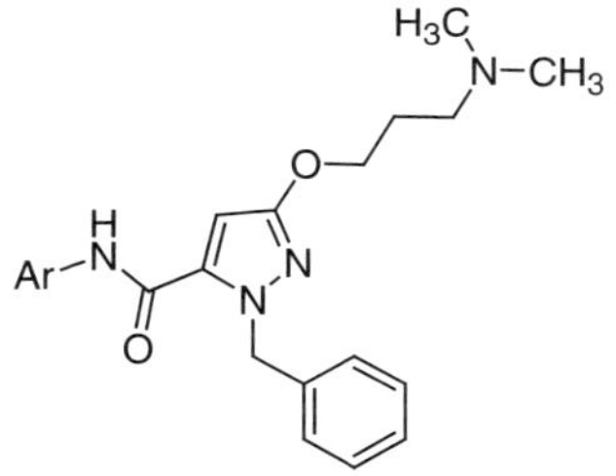	Activators of soluble guanylate cyclase	Compounds also inhibit platelet aggregation and display oral bioavailability in this lipophilicity-restrained library.	36
6	2,4,6-Trisubstituted quinazolines	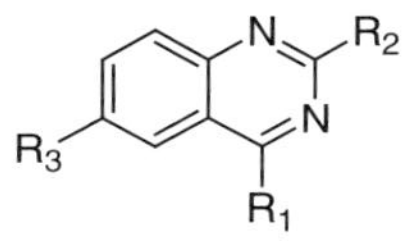	Cdk4/cyclin D1 and cdk2/cyclin E inhibitors	X-ray structure of inhibitor bound to cdk2 was obtained.	37

(*Continued*)

TABLE 5.1 (*Continued*)

Entry	Library	Structure	Biological Activity	Comments	Ref.
7	3-Substituted indoles	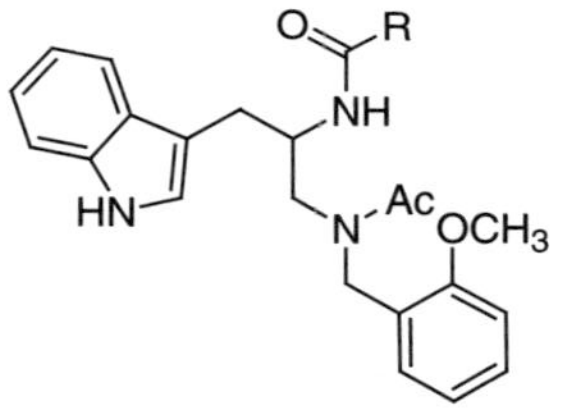	Neurokinin-1 (NK-1) antagonists	Subnanomolar affinity, orally bioavailable compound was obtained.	38
8	Proline derivatives	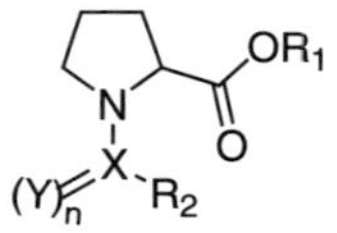	FKBP12 inhibitors	Compounds are nanomolar inhibitors of peptidyl-prolyl isomerase (PPIase or rotamase).	39
9	*N*-Aryl-*N*-oxalylanthranilic acids	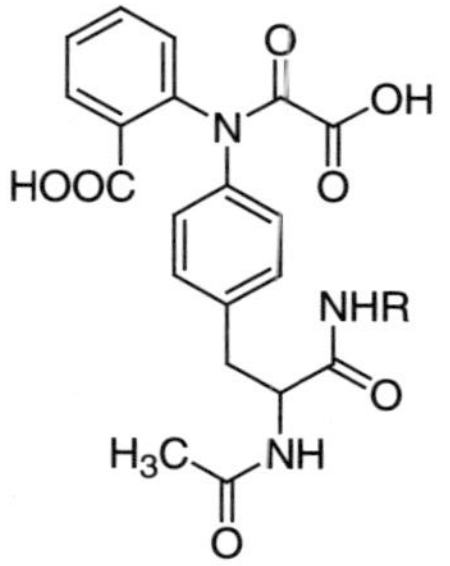	Protein tyrosine phosphatase 1B (PPT1B) inhibitors	Selectivity is observed over T-cell PTPase (TCPTP).	40
10	CC-1065 analogs	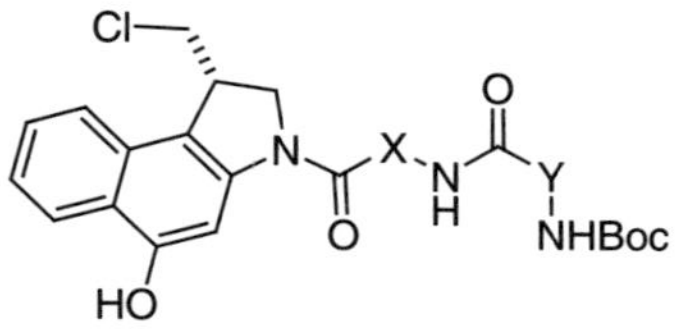	Sequence selective alkylation of duplex DNA	CC-1065 and the related duocarmycins are potent antitumor antibiotics.	41

11	Tethered dimers	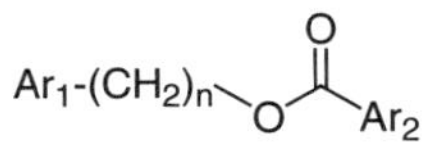	NAD synthetase inhibitors	Compounds also inhibited growth of gram (+) bacteria.	42
12	*p*-Acylthiocinnamides	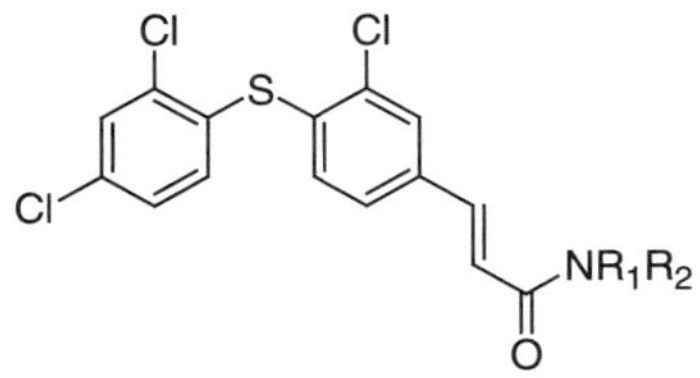	Antagonists of leukocyte function-associated antigen-1/intracellular adhesion molecules-1 (LFA/ICAM-1) interaction	Compounds have potential utility for inflammatory diseases, autoimmune disorders, tumor metastasis, allograft rejection, and reperfusion injury.	43
13	3-Imidazolylcarbolines	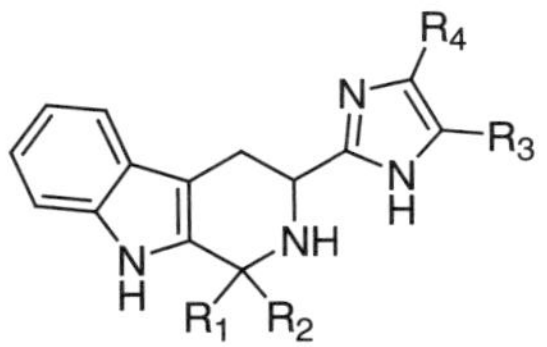	Somatostatin antagonists	Nonpeptidic antagonists are sst_3 selective with potential to treat acromegaly, neuroendocrine tumors and gastrointestinal disorders.	44
14	β-*C*-Mannosides	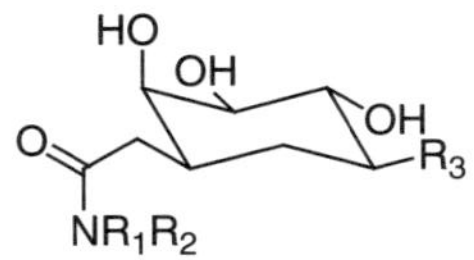	Selectin inhibitors	The tetrasaccharide sialyl Lewis^x is the carbohydrate epitope recognized by selectins E, P, and L.	45
15	Pyrrolidines	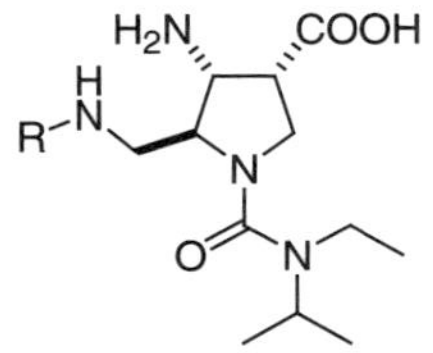	Influenza neuraminidase (NA) inhibitors	NA is necessary for virus replication and infectivity; x-ray structure determined for inhibitor A-192558.	46

(*Continued*)

TABLE 5.1 *(Continued)*

Entry	Library	Structure	Biological Activity	Comments	Ref.
16	*N*-(Pyrroldinylmethyl) hydroxamic acids	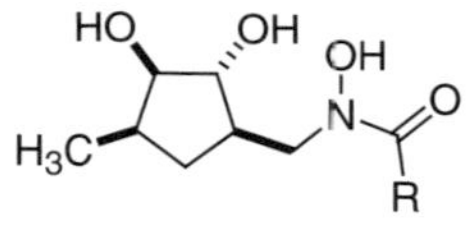 HO, OH, OH, N, O, H_3C, R	Transition-state inhibitors of fucosyltransferases	Potential therapeutic areas include inflammatory diseases and cancer metastasis.	47
17	Mercaptomethyl ketones	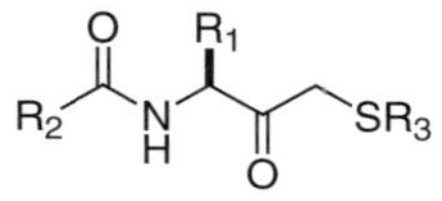 O, R_1, R_2, N, H, SR_3, O	Mechanism-based inhibitors of cysteine proteinases	A seven-step synthesis of substrate-based ketone inhibitors was developed; calpains are implicated in neurodegenerative diseases and osteoporosis; caspases are involved with programmed cell death.	48
18	Kerolides	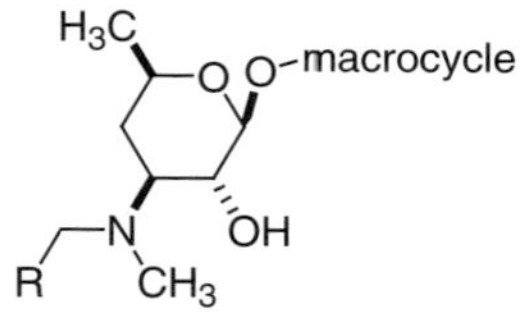 H_3C, O, O-macrocycle, N, OH, R, CH_3	These semisynthetic erythromycins have potential for treating macrolide MLS_B resistance	Purification of library was accomplished using ion exchange.	49
19	Tetrahydrofurans	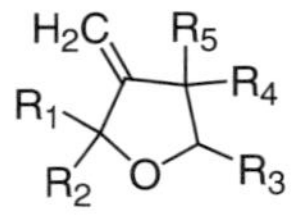 H_2C, R_5, R_4, R_1, R_2, O, R_3	Lead discovery library	One-step construction of diverse heterocycles via [3 + 2] cycloaddition reaction.	50
20	Oxazoles	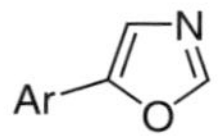 N, Ar, O	Lead discovery library	One-step construction of heterocycles using *p*-toluenesulfonylmethylisocyanide (TosMIC).	51
21	1,2-Phenethyldiamines	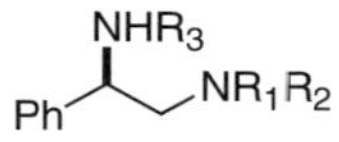 NHR_3, Ph, NR_1R_2	Lead generation library	A wide range of biological activities has been reported for this class of compound.	52

22	Imidazolinones	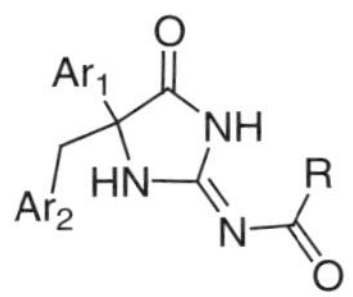	None reported	Structures produced resemble the natural product creatine.	53
23	Spiropyrrolopyrroles	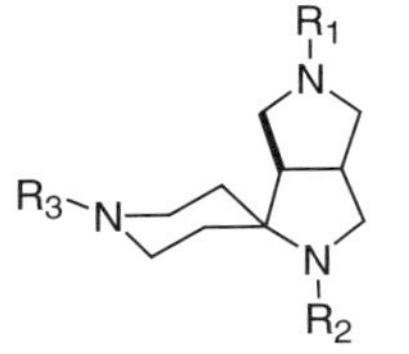	Neurokinin (NK) receptor antagonists	Low nanomolar affinities for the NK-1 receptor were achieved.	54
24	3,4-Dihydroquinoxalin-2-ones	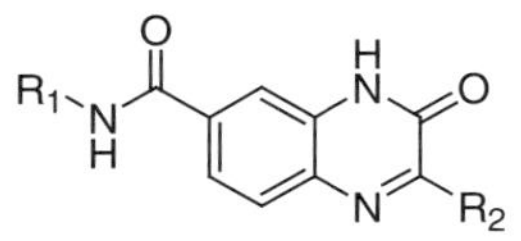	Aldose reductase inhibitor, antagonism of AMPA and angiotensin II receptors	Combination of solution- and solid-phase synthesis was used.	55
25	Purines	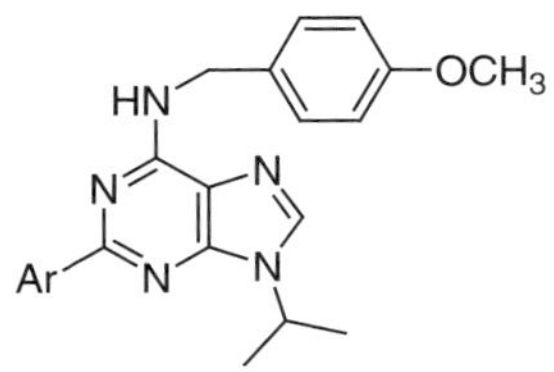	Potential inhibitors of ATP-dependent proteins	In addition, Suzuki coupling of aryl boronic acids produced N^9-aryl derivatives.	56
26	3-Arylindoles	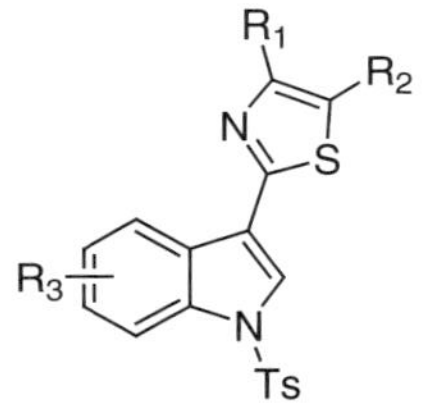	Potential inhibitors of topoisomerase, related to BE10988	Suzuki coupling strategy was used.	57

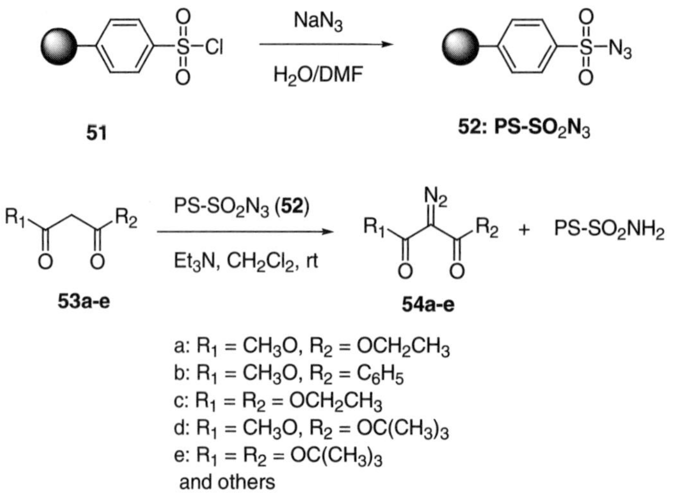

51 → 52: PS-SO_2N_3 (NaN_3, H_2O/DMF)

53a-e → 54a-e + PS-SO_2NH_2 (PS-SO_2N_3 (52), Et_3N, CH_2Cl_2, rt)

a: $R_1 = CH_3O$, $R_2 = OCH_2CH_3$
b: $R_1 = CH_3O$, $R_2 = C_6H_5$
c: $R_1 = R_2 = OCH_2CH_3$
d: $R_1 = CH_3O$, $R_2 = OC(CH_3)_3$
e: $R_1 = R_2 = OC(CH_3)_3$
and others

Scheme 5.13

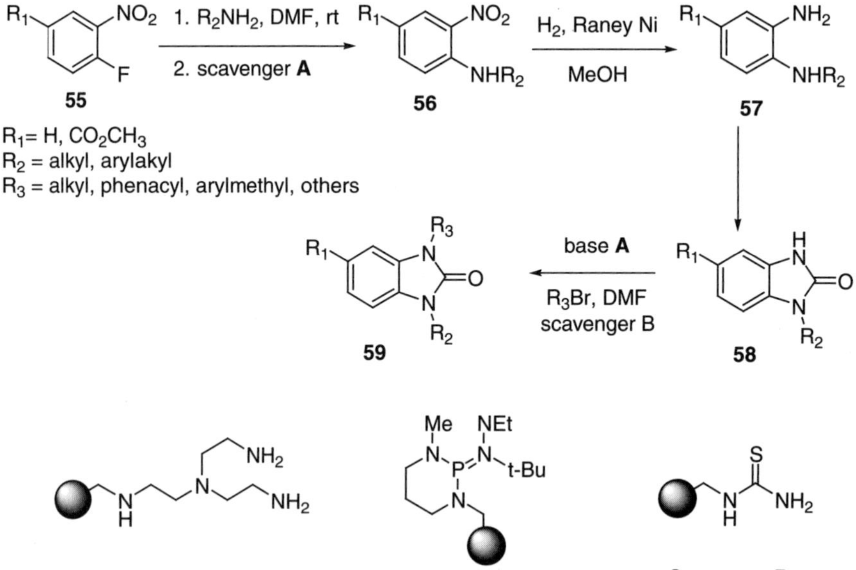

55 → 56 (1. R_2NH_2, DMF, rt; 2. scavenger **A**) → 57 (H_2, Raney Ni, MeOH) → 58 → 59 (base **A**, R_3Br, DMF, scavenger B)

R_1= H, CO_2CH_3
R_2 = alkyl, arylakyl
R_3 = alkyl, phenacyl, arylmethyl, others

Scavenger **A** base **A** Scavenger **B**

Scheme 5.14

Another recent report, which used polymer-bound reagents, bases, and scavengers in concert with PSPS, described the preparation of peptidomimetic pyrazinone antithrombotics (Scheme 5.15).[61] These pyrazinones were shown to be inhibitors of tissue factor VIIa complex. Treatment of pyrazinone (**60**) with HOBt (**61**) in the presence of resin-bound carbodiimide (reagent A) gave the activated pyrazinone (**62**), which then was coupled with primary amines to yield amides (**63**). The reaction mixture was treated with scavenger A

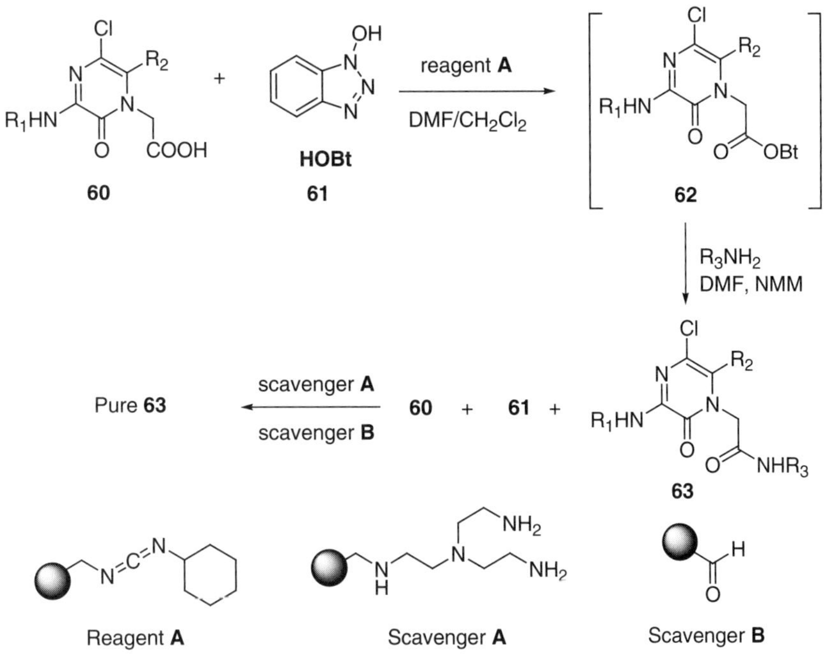

Scheme 5.15

to sequester unreacted **60** and scavenger B to capture excess HOBt (**61**). The sequestering agents in this and related protocols allowed the preparation of hundreds of pure compounds without chromatographic purification.

In Table 5.2 are compiled representative resin-bound agents that have been employed in PSPS formats. From inspection of the classification column it is evident that a wide variety of processes can be mediated with these resin-bound agents. Included are resin-bound reagents such as a wide variety of oxidizing and reducing agents; nucleophiles of various kinds for displacement reactions; coupling reagents such as Wittig, and peptide coupling reagents; hydrazine delivery reagents, using a catch and release method; and many others. Deprotection, dehydration, isomerization, and catalysis are additional processes that are mediated by these resin-bound agents.

Resin-bound acids and bases are employed for initiating reactions, absorbing bases or acids that are generated by PSPS processes, or for scavenging excess or unreacted starting materials or by-products that are produced. A variety of other scavengers has been developed, such as immobilized isocyanates for removal of reactive amines in case of acid sensitivity and immobilized thiols for the removal of ketones.

Interestingly, PSPS reagents or scavengers can be tagged for removal by resin-bound agents. Entries 20 and 21 in Table 5.2 exemplify these processes. Thus, a soluble carbodiimide reagent can be tagged with a tertiary amine, which can be removed by an ion-exchange resin (e.g., a resin-bound sulfonic acid). Similarly, tetrafluorophthalic anhydride, which is used to scavenge excess secondary amines, produces a tagged species after scavenge occurs. The benzoic acid that is produced by opening the anhydride with the secondary amine is a tagged scavenger, which can be removed with a resin-bound primary amine.

TABLE 5.2 Resin-Bound Agents Used for PSPS

Entry	Compound Class Prepared	Resin-Bound Agent Used	Classification of Agent	Comments	Refs.
1	Carbamates	S, O, N–OH, O	Reagent (activation)	Catch and release reagent.	62
2	Secondary amines	CrO_3	Reagent (oxidation)	Tandem three-phase reaction.	63
3	Secondary amines	BH_4^-	Reagent (reduction)	Tandem three-phase reaction.	63
4	Viscinol diols	O, N, O	[4 + 2] Cycloaddition removal	9-Anthrylmethyl ester tag employed for capture.	64
5	2-Pyrazolines	NCO	Scavenger (amine)	Excess intermediate pyrazoline with free NH was scavenged.	65
6	α-Ketothiazole	$(N^+Me_3)_2\ S_2O_3^{-2}$	Reagent (reduction)	Excess Dess–Martin reagent is reduced by thiosulfate resin.	66
7	(+)-Plicamine	OAc, I, OAc	Reagent (oxidation)	Natural product plus its enantiomer were prepared using thirteen resin-bound agents, including the solid supported iodonium diacetate.	67

8	Carponone	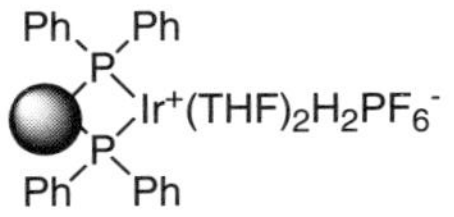 Ph, Ph, P, $Ir^+(THF)_2H_2PF_6^-$, P, Ph, Ph	Catalyst (isomerization)	Reagent is an immobilized version of Felkin's iridium catalyst.	68
9	(R)-Salmeterol	 $-PPh_2$	Reagent (activation)	Used with CBr_4 to convert alcohol to alkyl bromide.	69
10	Bicyclo[2.2.2]octanes	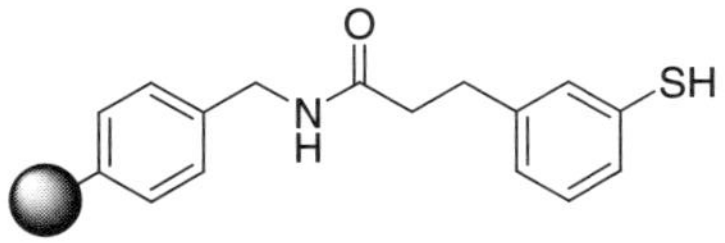 O, N, H, SH	Scavenger (ketone)	Used in conjunction with resin-bound diisopropylethylamine.	70
11	1,2,3-Thiadiazoles	 $-SO_2NHNH_2$	Reagent (hydrazine delivery)	Catch and release reagent.	71
12	Olefins	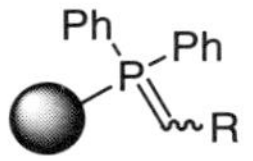 Ph, Ph, P, R	Reagent (coupling)	Wittig reaction.	72–74
13	4-Phenyl-2*H*-phthalazin-1-one	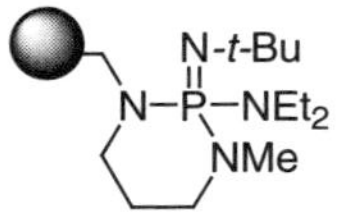 N-*t*-Bu, N–P–NEt_2, NMe	Reagent (deprotonation)	PBEMP is one of several bases that can be used for *N*-alkylation of weakly acidic *N*-heterocycles.	75
14	α-Bromoketones; vicinal dibromides	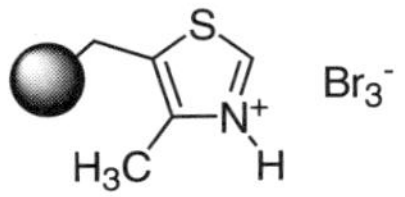 S, N^+, H, H_3C, Br_3^-	Reagent (halogenation)	PM5VTHT.	76
15	α-Azidoalcohols; alkyl azides	 $N^+Me_3\ N_3^-$	Reagent (nucleophile)	Basic ion-exchange resins can be loaded with both inorganic and organic anions.	77, 78

(*Continued*)

TABLE 5.2 (*Continued*)

Entry	Compound Class Prepared	Resin-Bound Agent Used	Classification of Agent	Comments	Refs.
16	Alkyl and aryl nitriles	PPh_2/CCl_4	Reagent (dehydration)	Both carboxamides and aryl oximes can be converted to nitriles with this reagent.	79
17	α,β-Unsaturated esters and nitriles	NMe_3 X $P(OR)_2$ O	Reagent (Horner-Emmons)	This ion-pair reagent was the first reported polymer-supported reagent.	80
18	α,β-Unsaturated esters	Ph n OEt O–P COOEt O	Reagent (Horner–Emmons)	Barton's base used to generate anion.	81
19	Bromohydrins and iodohydrins	PPh_2X_2 X= Br, I	Reagent (Mitsunobu)	Polymer-bound triarylphosphine used with DEAD.	82, 83
20	Aryl ethers	O O N=N OCH_3 O	Reagent (Mitsunobu)	Polymer-bound DEAD used with PPh_3.	84–87
21	Derivatives of secondary amines	F F F F O O O	Reagent (tagging)	This tagged carboxy amide is removed with NH_2	88, 89

22	Aldehydes and ketones	Cl^- Me_2HN^+–CH_2CH_2–N=C=N–Et "tagged" by (resin)–SO_3H	Reagent (tagging)	Tagged carbodiimide reagent is removed by ion exchange resin.	90
23	Perhydro-3-oxo-1,4-diazepinium derivatives	Wang aldehyde HL resin	Scavenger (amine)	Microwave irradiation was employed to increase scavenger reactivity.	91
24	Flavanoid cycloadducts	9-anthracenyl–CH_2CH_2–C(=O)–NH–(resin)	[4 + 2] Cycloaddition removal	Compounds prepared are related to prenylflavanoid Diels–Alder natural products.	92
25	Amines (α-substituted)	(resin)–SO_3H	Capture (amine)	Capture of the product amines, derived from alcoholysis of corresponding sulfonamides, was microwave-assisted.	93
26	Terminal acetylenes	(resin)–CH_2–O–P(=O)(OEt)–C(=N_2)–C(=O)–CH_3; (resin): ROMP	Reagent (Horner–Wadsworth–Emmons olefination)	This ROMP gel-supported ethyl 1-diazo-2-oxopropylphosphonate is one example of many ROMP agents.	94

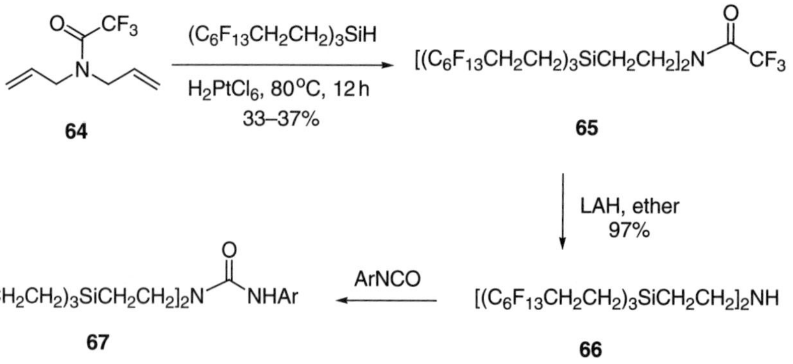

$[(C_6F_{13}CH_2CH_2)_3SiH$; H_2PtCl_6, 80°C, 12 h; 33–37%; LAH, ether 97%; ArNCO; 64; 65; 66; 67]

Scheme 5.16

Another approach to PSPS, which is closely related to the tagging methodology, is the fluorous synthesis technique that was introduced in 1997.[95] This technique adds another tool to the armamentarium of scavenging and capturing techniques represented by attachment to polymer support and linkage to ionizable functionality. The fluorous synthesis technique, simply described, is a third liquid phase (e.g., perfluorohexanes), which is immiscible with both water and common organic solvents.[96–98]

Typical organic compounds have virtual no solubility in the fluorous phase. They can, however, be rendered soluble in the fluorous phase by attachment of a fluorous tag. In a number of fluorous synthetic techniques, the tris(perfluorohexylethyl)silyl group has been employed as a convenient fluorous phase tag.[99,100] The synthesis of fluorous amine scavenger (**66**) has been described, as shown in Scheme 5.16. The tris(perfluorohexylethyl)silyl group was doubly appended to *N,N′*-diallyltrifluoroacetamide (**64**), and the resulting new trifluoroacetamide (**65**) was converted to an amine scavenger (**66**) by treatment with lithium aluminum hydride. Compound **66** was shown to be a good scavenger for arylisocyanates, for example, to produce ureas (**67**), which had been used in excess as reagents in the PSPS synthesis of a library of ureas.[101]

Fluorous tethered reagents have also been developed. A Staudinger protocol using a fluorous tethered triarylphosphine (**69**) has been developed. In a one-pot, two-step procedure, 4-azidobenzoic acid (**68**) was converted quantitatively to 4-aminobenzoic acid (**70**) in a 4-hour time frame, as shown in Scheme 5.17. Purification was accomplished by Fluoro*Flash*™ chromatography.[102] This Staudinger protocol was then applied successfully to a structurally diverse group of azides. The use of fluorous tethered reagents with Fluoro*Flash*™ chromatography has been reviewed.[103,104]

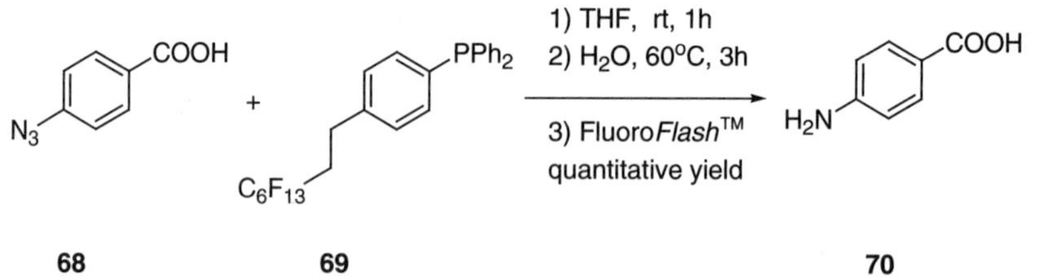

Scheme 5.17

Entry 26 describes one ROMP (ring-opening metathesis polymerization) reagent that has been used for the preparation of terminal acetylenes from aldehydes.[94] ROMP reagents, which are prepared from monomeric scavengers or reagents by a metathesis polymerization process, include an *N*-hydroxysuccinimide coupling agent,[105] a primary alcohol for scavenging a variety of electrophiles,[106] sulfonyl chlorides for scavenging a variety of nucleophiles,[107] and bis-acid chlorides as nucleophile scavengers.[108] In addition, ROMP oligomers have been used as solid supports for multistep reaction sequences[109] and also as capture–release agents for the synthesis of amines and alkyl hydrazines.[110] An older topic, which is not covered in this review, is the use of organic soluble supports and reagents. This topic has been reviewed previously.[111,112]

A number of extensive reviews have appeared that cover the use of resin-bound agents in PSPS. A comprehensive review by Kirschning et al.[113] covers all classifications of immobilized agents, regardless of their function. This reference serves as the premier leading reference for the reader who wants a cursory view of any subgroup of agent and who requires additional references to consult for more detail. In addition, other more specialized review articles deal with the topic of immobilized reagents,[114–116] catalysts,[117–119] and scavenging agents.[120]

5.5 THE FUTURE

It has become increasingly clear that the production of large random chemical libraries using combinatorial chemistry techniques has not been productive for the efficient generation of hit compounds. Problems inherent in this approach include (1) lack of chemical diversity of the compound sets; (2) the relatively high cost of screening protocols for meaningful biological assays; and (3) the inattention to important compound properties such as solubility, cLogP, polar surface area, and number of rotatable bonds.

Thus, the continuing and increasing need is established for the production of focused, carefully designed libraries with a high likelihood of generating high-quality hit or lead compounds. These relatively small libraries, numbering in the hundreds rather than the thousands, will contain leadlike[121–123] or druglike[124,125] molecules.

Semiautomated PSPS methods will be increasingly useful tools to generate these required focused libraries. To minimize the need for purification, which always is an issue with solution-phase chemistry protocols, resin-bound reagents and scavengers will also become increasingly useful (and less expensive).

Thus, in the future we will see a definite shift to the production of small, thought-driven target libraries for hit generation. Many of these will initiate from natural product scaffolds, which will introduce new chemical diversity and inherent biological activities. Iterations of small, focused libraries will produce higher-value compounds in a faster time frame than was achieved from large random libraries.

REFERENCES

1. N. P. Peet, Synthesis of 7-(1-pyrrolidinyl)-1,2,4-triazolyl[4,3-*b*]pyridazines, *J. Heterocycl. Chem.*, *21*, 1389–1392 (1984).
2. N. P. Peet and S. Sunder, Factors which influence the formation of oxadiazoles from anthranilhydrazides and other benzoylhydrazines, *J. Heterocycl. Chem.*, *21*, 1807–1816 (1984).

3. N. P. Peet, J. Malecha, M. E. LeTourneau, and S. Sunder, Preparation of imidazo[2,1-*b*]quinazolin-5(3*H*)-ones and related tricyclic systems using a novel, double displacement reaction, *J. Heterocycl. Chem., 26*, 257–264 (1989).
4. N. P. Peet, L. E. Baugh, S. Sunder, and J. E. Lewis, Synthesis and antiallergic activity of some quinolinones and imidazoquinolinones, *J. Med. Chem., 28*, 298–302 (1985).
5. N. P. Peet, L. E. Baugh, S. Sunder, J. E. Lewis, E. H. Matthews, E. L. Olberding, and D. N. Shah, 3-(1*H*-Tetrazol-5-yl-4(3*H*)-quinazolinone sodium salt (MDL 427): a new antiallergic agent, *J. Med. Chem., 29*, 2403–2409 (1986).
6. N. P. Peet, N. L. Lentz, M. W. Dudley, A. M. L. Ogden, D. R. McCarty, and M. M. Racke, Xanthines with C^8 chiral substituents as potent and selective adenosine A_1 antagonists, *J. Med. Chem., 36*, 4015–4020 (1993).
7. J. W. Daly, W. L. Padgett, M. T. Shamim, P. Butts-Lamb, and J. Waters, 1,3-Dialkyl-8(*p*-sulfophenyl)xanthines: potent, water-soluble antagonists for A_1- and A_2-adenosine receptors, *J. Med. Chem., 28*, 487–492 (1985).
8. J. E. Baldwin, Rules for ring closure, *J. Chem. Soc., Chem. Commun.*, 734–736 (1976).
9. J. E. Baldwin, J. Cutting, W. Dupont, L. Kruse, L. Silberman, and R. C. Thomas, 5-Endo-trigonal reactions: a disfavoured ring closure, *J. Chem. Soc., Chem. Commun.*, 736–738 (1976).
10. G. R. Gustafson, C. M. Baldino, M. E. O'Donnell, A. Sheldon, R. J. Tarsa, C. J. Verni, and D. L. Coffen, Incorporation of carbohydrates and peptides into large triazine-based screening libraries using automated parallel synthesis, *Tetrahedron, 54*, 4051–4065 (1998).
11. D. G. Powers, D. S. Casebier, D. Fokas, W. J. Ryan, J. K. Troth, and D. L. Coffen, Automated parallel synthesis of chalcone-based screening libraries, *Tetrahedron, 54*, 4085–4096 (1998).
12. D. Fokas, W. J. Ryan, D. S. Casebier, and D. L. Coffen, Solution phase synthesis of a spiro[pyrrolidine-2,3′-oxindole] library via a three component 1,3-dipolar cycloaddition reaction, *Tetrahedron, 39*, 2235–2238 (1998).
13. C. M. Baldino, Perspective articles on the utility and application of solution-phase combinatorial chemistry, *J. Comb. Chem., 2*, 89–103 (2000).
14. C. M. Baldino, D. S. Casebier, J. Caserta, G. Slobodkin, C. Tu, and D. L. Coffen, Convergent parallel synthesis, *Synlett*, 488 (1997).
15. K. Takasu, H. Terauchi, H. Inoue, H.-S. Kim, Y. Wataya, and M. Ihara, Parallel synthesis of antimalarial rhodacyanine dyes by the combination of three components in one pot, *J. Comb. Chem., 5*, 211–214 (2003).
16. R. A. Fecik, K. E. Frank, E. J. Gentry, L. A. Mitscher, and M. Shibata, Use of combinatorial and multiple parallel synthesis methodologies for the development of anti-infective natural products, *Pure Appl. Chem., 71*, 559–564 (1999).
17. M. Marder, H. Viola, J. A. Bacigaluppo, M. I. Colombo, C. Wasowski, C. Wolfman, J. H. Medina, E. A. Ruveda, and A. C. Paladini, Detection of benzodiazepine receptor ligands in small libraries of flavone derivatives synthesized by solution phase combinatorial chemistry, *Biochem. Biophys. Res. Commun., 249*, 481–485 (1998).
18. P. W. Shum, N. P. Peet, P. M. Weintraub, T. B. Le, Z. Zhao, F. Barbone, B. Cashman, J. Tsay, S. Dwyer, P. C. Loos, E. A. Powers, K. Kropp, P. S. Wright, A. Bitonti, J. Dumont, and D. R. Borcherding, The design and synthesis of purine inhibitors of CDK2: part 3, *Nucleosides Nucleotides, 20*, 1067–1078 (2001).
19. M. K. Dreyer, D. R. Borcherding, J. A. Dumont, N. P. Peet, J. Tsay, P. S. Wright, A. J. Bitonti, J. Shen, and S.-H. Kim, Crystal structure of human cyclin-dependent kinase 2 in complex with the adenine-derived inhibitor H717, *J. Med. Chem., 44*, 524–530 (2001).
20. P. Raboisson, B. Mekonnen, and N. P. Peet, *Tetrahedron Lett., 44*, 2919–2921 (2003).
21. M. M. Reddy, B. Bhatia and J. Iqbal, A one pot synthesis of β-acetamido ketones and furans by cobalt(II) catalyzed coupling of 1,3- or 1,4-dicarbonyl compounds and aldehydes: a remarkable role of dioxygen, *Tetrahedron Lett., 36*, 4877–4880 (1995).

22. M. Mukhopadhyay, B. Bhatia, and J. Iqbal, Cobalt catalyzed multiple component condensation route to β-acetamido carbonyl compound libraries, *Tetrahedron Lett.*, *38*, 1083–1086 (1997).
23. B. Bhatia, M. M. Reddy, and J. Iqbal, Cobalt-catalyzed three-component coupling involving ketones or ketoesters, aldehydes and acetonitrile: a novel one-pot synthesis of β-acetamido ketones, *J. Chem. Soc., Chem. Commun.*, *6*, 713–714 (1994).
24. S. M. Antonelli, M. Tandon, and N. P. Peet, Mechanistic and preparative studies of a multicomponent condensation reaction, *Org. Lett.*, manuscript submitted for publication.
25. I. Ugi, Neuere Methoden der präparativen organischen Chemie IV. Mit sekundär-reaktionen Gekoppelte α-Additionen von Immonium-Ionen und Anionen an Isonitrile, *Angew. Chem. Int. Ed. Engl.*, *1*, 8–21 (1962).
26. M. Passerini, Passerini reaction, *Gazz. Chim. Ital.*, *51*, 126–181 (1921).
27. P. Biginelli, Aldehyde–urea derivatives of aceto- and oxaloacetic acids, *Gazz. Chim. Ital.*, *23*, 360–413 (1893).
28. C. O. Kappe, 100 Years of the Biginelli dihydropyrimidine synthesis, *Tetrahedron*, *49*, 6937–6963 (1993).
29. C. O. Kappe, Recent advances in the Biginelli dihydropyrimidine synthesis: new tricks from an old dog, *Acc. Chem. Res.*, *33*, 879–888 (2000).
30. A Stadler and C. O. Kappe, Automated library generation using sequential microwave-assisted chemistry: application toward the biginelli multicomponent condensation, *J. Comb. Chem.*, *3*, 624–630 (2001).
31. D. G. Hall, S. Manku, and F. Wang, Solution- and solid-phase strategies for the design, synthesis and screening of libraries based on natural product templates: a comprehensive survey, *J. Comb. Chem.*, *3*, 125–150 (2001).
32. Y. Chen, M. Bilban, C. A. Foster, and D. L. Boger, Solution-phase parallel synthesis of a pharmacophore library of HUN-7293 analogues: a general chemical mutagenesis approach to defining structure–function properties of naturally occurring cyclic (depsi)peptides, *J. Am. Chem. Soc.*, *124*, 5431–5440 (2002).
33. H. C. Hansen, R. Olsson, G. Croston, and C.-M. Andersson, Multistep solution-phase parallel synthesis of spiperone analogues, *Bioorg. Med. Chem. Lett.*, *10*, 2435–2439 (2000).
34. S. Padmanabhan, R. C. Lavin, P. M. Thakker, J. Guo, L. Zhang, D. Moore, M. E. Perlman, C. Kirk, D. Daly, K. J. Burke-Howie, T. Wolcott, S. Chari, D. Berlove, J. B. Fischer, W. F. Holt, G. J. Durant, and R. N. McBurney, Solution-phase, parallel synthesis and pharmacological evaluation of acylguanidine derivatives as potential sodium channel blockers, *Bioorg. Med. Chem. Lett.*, *11*, 3151–3155 (2001).
35. B. E. Blass, C. T. Huang, R. M. Kawamoto, M. Li, S. Liu, D. E. Portlock, W. M. Rennells, and M. Simmons, Parallel synthesis and evaluation of *N*-(1-phenylethyl)-5-phenylimidazole-2-amines as Na^+/K^+ ATPase Inhibitors, *Bioorg. Med. Chem. Lett.*, *10*, 1543–1545 (2000).
36. D. L. Selwood, D. G. Brummell, R. C. Glen, M. C. Goggin, K. Reynolds, M. A. Tatlock, and G. Wishart, Solution-phase parallel synthesis of 5-carboxamido 1-benzyl-3-(3-dimethylaminopropyloxy)-1*H*-pyrazoles as activators of soluble sodium guanylate cyclase with improved oral bioavailability, *Bioorg. Med. Chem. Lett.*, *11*, 1089–1092 (2001).
37. T. M. Sielecki, T. L. Johnson, J. Liu, J. K. Muckelbauer, R. H. Grafstrom, S. Cox, J. Boylan, C. R. Burton, H. Chen, A. Smallwood, C.-H. Chang, M. Boisclair, P. A. Benfield, G. L. Trainor, and S. P. Seitz, Quinazolines as cyclin dependent kinase inhibitors, *Bioorg. Med. Chem. Lett.*, *11*, 1157–1160 (2001).
38. J. E. Fritz, P. A. Hipskind, Karen L. Lobb, J. A. Nixon, P. G. Threlkeld, B. D. Gitter, C. L. McMillian, and S. W. Kaldor, Expedited discovery of second generation NK-1 antagonists: identification of a nonbasic aryloxy substituent, *Bioorg. Med. Chem. Lett.*, *11*, 1643–1646 (2001).
39. C. Choi, J.-H. Li, M. Vaal, C. Thomas, D. Limburg, Y.-Q. Wu, Y. Chen, R. Soni, C. Scott, D. T. Ross, H. Guo, P. Howorth, H. Valentine, S. Liang, D. Spicer, M. Fuller, J. Steiner, and

G. S. Hamilton, Use of parallel-synthesis combinatorial libraries for rapid identification of potent FKBP12 inhibitors, *Bioorg. Med. Chem. Lett.*, *12*, 1421–1428 (2002).

40. Z. Xin, T. K. Oost, C. Abad-Zapatero, P. J. Hajduk, Z. Pei, B. G. Szczepankiewicz, C. W. Hutchins, S. J. Ballaron, M. A. Stashko, T. Lubben, J. M. Trevilly, M. R. Jirousek, and G. Liu, Potent, selective inhibitors of protein tyrosine phosphatase 1B, *Bioorg. Med. Chem. Lett.*, *13*, 1887–1890 (2003).
41. D. L. Boger, H. W. Schmitt, B. G. Fink, and M. P. Hedrick, Parallel synthesis and evaluation of 132 (+)-1,2,9,9*a*-tetrahydrocyclopropa[*c*]benz[*e*]indol-4-one (CBI) analogues of CC-1065 and the duocarmycins defining the contribution of the DNA-binding domain, *J. Org. Chem.*, *66*, 6654–6661 (2001).
42. S. E. Velu, W. A. Cristofoli, G. J. Garcia, C. G. Brouillette, M. C. Pierson, C.-H. Luan, L. J. DeLucas, and W. J. Brouillette, Tethered dimers as NAD synthetase inhibitors with antibacterial activity, *J. Med. Chem.*, *46*, 3371–3381 (2003).
43. G. Liu, J. T. Link, Z. Pei, E. B. Reilly, S. Leitza, B. Nguyen, K. C. Marsh, G. F. Okasinski, T. W. von Geldern, M. Ormes, K. Fowler, and M. Gallatin, Discovery of novel *p*-arylthio cinnamides as antagonists of leukocyte function-associated antigen-1/intracellular adhesion molecule-1 interaction. 1: identification of an additional binding pocket based on an anilino diaryl sulfide lead, *J. Med. Chem.*, *43*, 4025–4040 (2000).
44. L. Poitout, P. Roubert, M.-O. Contour-Galcéra, C. Moinet, J. Lannoy, J. Pommier, P. Plas, D. Bigg, and C. Thurieau, Identification of potent non-peptide somatostatin antagonists with sst_3 selectivity, *J. Med. Chem.*, *44*, 2990–3000 (2001).
45. N. Kaila, L. Chen, B. E. Thomas, D. Tsao, S. Tam, P. W. Bedard, R. T. Camphausen, J. C. Alvarez, and G. Ullas, β-*C*-Mannosides as selectin inhibitors, *J. Med. Chem.*, *45*, 1563–1566 (2002).
46. G. T. Wang, Y. Chen, S. Wang, R. Gentles, T. Sowin, W. Kati, S. Muchmore, V. Giranda, K. Stewart, H. Sham, D. Kempf, and W. G. Laver, Design, synthesis and structural analysis of influenza neuraminidase inhibitors containing pyrrolidine cores, *J. Med. Chem.*, *44*, 1192–1201 (2001).
47. M. Takayanagi, T. Flessner, and C.-H. Wong, A strategy for the solution-phase synthesis of *N*-(pyrrolidinylmethyl)hydroxamic acids, *J. Org. Chem.*, *65*, 3811–3815 (2000).
48. A. Lee and J. A. Ellman, Parallel solution-phase synthesis of mechanism-based cysteine protease inhibitors, *Org. Lett.*, *3*, 3707–3709 (2001).
49. A. Denis and C. Renou, Novel *N*-demethylation of ketolide: application to the solution phase parallel synthesis of *N*-desosaminyl-substituted ketolides using ion exchange resins, *Tetrahedron Lett.*, *43*, 4171–4174 (2002).
50. M. Cavicchioli, X. Marat, N. Monteiro, B. Hartmann, and G. Balme, Solution-phase parallel tetrahydrofuran synthesis with propargyl alcohols and benzylidene-(or alkylidene-)malonates, *Tetrahedron Lett.*, *43*, 2609–2611 (2002).
51. B. A. Kulkarni and A. Ganesan, Solution-phase parallel oxazole synthesis with TosMIC, *Tetrahedron Lett.*, *40*, 5637–5638 (1999).
52. C. T. Lowden and J. S. Mendoza, Solution phase parallel synthesis of 1,2-phenethyldiamines, *Tetrahedron Lett.*, *43*, 979–982 (2002).
53. L. Varga, T. Nagy, I. Kovesdi, J. Benet-Buchholz, G. Dorman, L. Urge, and F. Darvas, Solution-phase parallel synthesis of 4,6-diarylpyrimidine-2-ylamines and 2-amino-5,5-disubstituted-3,5-dihydroimidazol-4-ones via a rearrangement, *Tetrahedron*, *59*, 655–662 (2003).
54. K. H. Bleicher, Y. Wuthrich, G. Adam, T. Hoffmann, and A. J. Sleight, Parallel solution- and solid-phase synthesis of spiropyrrolopyrazoles as novel neurokinin receptor ligands, *Bioorg. Med. Chem. Lett.*, *12*, 3073–3076 (2002).
55. E. Laborde, B. T. Peterson, and L. Robinson, Traceless, self-cleaving solid- and solution-phase parallel synthesis of 3,4,7-trisubstituted 3,4-dihydroquinoxalin-2-ones, *J. Comb. Chem.*, *3*, 572–577 (2001).

56. G. M. Green, N. P. Peet, and W. A. Metz, Polystyrene-supported benzenesulfonyl azide: a diazo transfer reagent that is both efficient and safe, *J. Org. Chem., 66*, 2509–2511 (2001).
57. S. Ding, N. S. Gray, Q. Ding, and P. G. Shultz, Expanding the diversity of purine libraries, *Tetrahedron Lett., 42*, 8751–8755 (2001).
58. N. Zou, J.-F. Liu, and B. Jiang, Solution-phase synthesis of a thiazoyl-substituted indolyl library via Suzuki cross-coupling, *J. Comb. Chem., 5*, 754–755 (2003).
59. W. R. Roush, D. Feitler, and J. Rebek, Polymer-bound tosyl azide, *Tetrahedron Lett.*, 1391–1392 (1974).
60. B. Raju, N. Nguyen, and G. W. Holland, Solution-phase parallel synthesis of substituted benzimidazoles, *J. Comb. Chem., 4*, 320–328 (2002).
61. J. J. Parlow, B. L. Case, T. A. Dice, R. L. Fenton, M. J. Hayes, D. E. Jones, W. L. Neumann, R. S. Wood, R. M. Lachance, T. J. Girard, N. S. Nicholson, M. Clare, R. A. Stegeman, A. M. Stevens, W. C. Stallings, R. G. Kurumbail, and M. S. South, Design, parallel synthesis, and crystal structures of pyrazinone antithrombotics as selective inhibitors of the tissue factor VIIa complex, *J. Med. Chem., 46*, 4050–4062 (2003).
62. H. Sumiyoshi, T. Shimizu, M. Katoh, Y. Baba, and M. Sodeoka, Solution-phase parallel synthesis of carbamates using polymer-bound *N*-hydroxysuccinimide, *Org. Lett., 4*, 3923–3926 (2002).
63. J. C. Pelletier, A. Khan and Z. Tang, A tandem three-phase reaction for preparing secondary amines with minimal side products, *Org. Lett., 4*, 4611–4613 (2002).
64. X. Wang, J. J. Parlow and J. A. Porco, Parallel Synthesis and Purification Using Anthracene-Tagged Substrates, *Org. Lett., 2*, 3509–3512 (2000).
65. U. Bauer, B. J. Egner, I. Nilsson, and M. Berghult, Parallel solution phase synthesis of *N*-substituted 2-pyrazoline libraries, *Tetrahedron Lett., 41*, 2713–2717 (2000).
66. M. S. South, T. A. Dice, T. J. Girard, R. M. Lachance, A. M. Stevens, R. A. Stegeman, W. C. Stallings, R. G. Kurumbail, and J. J. Parlow, Polymer-assisted solution-phase (PASP) parallel synthesis of an α-ketothiazole library as tissue factor VIIa inhibitors, *Bioorg. Med. Chem. Lett., 13*, 2363–2367 (2003).
67. I. R. Baxendale, S. V. Ley, and C. Piutti, Total synthesis of the amaryllidaceae alkaloid (+)-plicamine and its unnatural enantiomer by using solid-supported reagents and scavengers in a multistep sequence of reactions, *Angew. Chem. Int. Ed. Engl., 41*, 2194–2197 (2002).
68. I. R. Baxendale, A. Lee, and S. V. Ley, A concise synthesis of the natural product carpanone using solid-supported reagents and scavengers, *Synlett*, 1482–1484 (2001).
69. R. N. Bream, S. V. Ley, and P. A. Procopiou, Synthesis of the β_2-agonist (*R*)-salmeterol using a sequence of supported reagents and scavenging agents, *Org. Lett., 4*, 3793–3796 (2002).
70. S. V. Ley and A. Massi, Polymer supported reagents in synthesis: preparation of bicyclo[2.2.2]octane derivatives via tandem michael addition reactions and subsequent combinatorial decoration, *J. Comb. Chem., 2*, 104–107 (2000).
71. Y. Hu, S. Baudart, and J. A. Porco, Parallel synthesis of 1,2,3-thiadiazoles employing a "catch and release" strategy, *J. Org. Chem., 64*, 1049–1051 (1999).
72. M. Bernard, W. T. Ford, and E. C. Nelson, Synthesis of ethyl retinoate with polymer-supported wittig reagents, *J. Org. Chem., 48*, 3164–3168 (1983).
73. S. D. Clarke, C. R. Harrison, and P. Hodge, Phase transfer catalyzed polymer-supported Wittig reactions, *Tetrahedron Lett., 21*, 1375–1378 (1980).
74. J. Castells, J. Font and A. Virgili, Reaction of dialdehydes with conventional and polymer-supported Wittig reagents. *J. Chem. Soc., Perkin Trans. 1*, 1–6 (1979).
75. S. Shinkai, H. Tsuji, H. Yoichiro, and O. Manabe, Polymer-bound dimethylaminopyridine as a catalyst for facile ester synthesis, *Bull. Chem. Soc. Jpn., 54*, 631–632 (1981).
76. A. Badadjamian and A. Kessat, Bromination of various unsaturated ketones and olefins with poly(4-methyl-5-vinylthiazolium) hydrotribromide, *Synth. Commun., 25*, 2203–2209 (1995).

77. A. Hassner and M. Stern, Synthesis of alkyl azides with a polymeric reagent, *Angew. Chem. Int. Ed. Engl.*, *25*, 478–479 (1986).

78. M. Lakshman, D. V. Narkarni, and R. E. Lehr, Regioselective ring opening of polycyclic aromatic hydrocarbon epoxides by polymer-supported azide anion, *J. Org. Chem.*, *55*, 4892–4897 (1990).

79. C. R. Harrison, P. Hodge, and W. J. Rodgers, Conversion of carboxamides and oximes to nitriles or imidoyl chloride using a polymer-supported phosphine and carbon tetrachloride, *Synthesis*, 41–43 (1977).

80. G. Cainelli, M. Contento, F. Manescalchi, and R. Regnoli, Polymer-supported phosphonates: olefins from aldehydes, ketones, and dioxolanes by means of polymer-supported phosphonates, *J. Chem. Soc., Perkin Trans.*, *1*, 2516–2519 (1980).

81. A. G. M. Barrett, S. M. Cramp, R. S. Roberts, and F. J. Zeeri, Horner–Emmons synthesis with minimal purification using ROMPGEL: a novel high-loading matrix for supported reagents, *Org. Lett.*, *1*, 579–582 (1999).

82. A. R. Tunoori, D. Dutta, and G. I. Georg, Polymer-bound triphenylphosphine as traceless reagent for Mitsunobu reactions in combinatorial chemistry: synthesis of aryl ethers from phenols and alcohols, *Tetrahedron Lett.*, *39*, 8751–8754 (1998).

83. J. C. Pelletier and S. Kincaid, Mitsunobu reaction modifications allowing product isolation without chromatography: application to a small parallel library, *Tetrahedron Lett.*, *41*, 797–800 (2000).

84. L. D. Arnold, H. I. Assil, I. Hanaa, and J. C. Vederas, Polymer-supported alkyl azodicarboxylates for Mitsunobu reactions, *J. Am. Chem. Soc.*, *111*, 3973–3976 (1989).

85. M. C. Desai and L. M. S. Stramiello, Polymer-bound EDC (P-EDC): a convenient reagent for formation of an amide bond, *Tetrahedron Lett.*, *34*, 7685–7688 (1993).

86. M. Adamczyk, J. R. Fishpaugh, and P. G. Mattingly, An easy preparation of hapten active esters via solid supported EDAC, *Tetrahedron Lett.*, *36*, 8345–8346 (1995).

87. M. Adamczyk and J. R. Fishpaugh, A solid supported synthesis of thiol esters, *Tetrahedron Lett.*, *37*, 4305–4308 (1996).

88. J. J. Parlow and D. L. Flynn, Solution-phase parallel synthesis of a benzoxazinone library using complementary molecular reactivity and molecular recognition (CMR/R) purification technology, *Tetrahedron*, *54*, 4013–4031 (1998).

89. J. J. Parlow, W. Naing, M. S. South, and D. L. Flynn, In situ chemical tagging: tetrafluorophthalic anhydride as a "sequestration enabling reagent" (SER) in the purification of solution-phase combinatorial libraries, *Tetrahedron Lett.*, *38*, 7959–7962 (1997).

90. D. L. Flynn, J. Z. Crich, R. V. Devraj, S. L. Hockerman, J. J. Parlow, M. S. South, and S. Woodard, Chemical library purification strategies based on principles of complementary molecular reactivity and molecular recognition, *J. Am. Chem. Soc.*, *119*, 4874–4881 (1997).

91. I. Masip, C. Ferrandiz-Huertas, C. Garcia-Martinez, J. A. Ferragut, A. Ferrer-Montiel, and A. Messeguer, Synthesis of a library of 3-oxopiperazinium and perhydro-3-oxo-1,4-diazepinium derivatives and identification of bioactive compounds, *J. Comb. Chem.*, *6*, 135–141 (2004).

92. X. Lei and J. A. Porco, Synthesis of a polymer-supported anthracene and its application as a dienophile scavenger, *Org. Lett.*, *6*, 795–798 (2004).

93. T. Mukade, D. R. Dragoli, and J. A. Ellman, Parallel solution-phase asymmetric synthesis of α-branched amines, *J. Comb. Chem.*, *5*, 590–596 (2003).

94. A. G. M. Barrett, B. T. Hopkins, A. C. Love, and L. Tedeschi, Parallel synthesis of terminal alkynes using a ROMPgel-supported ethyl 1-diazo-2-oxopropylphosphonate, *Org. Lett.*, *6*, 835–837 (2004).

95. A. Studer, S. Hadida, R. Ferritto, S.-Y. Kim., P. Jeger, P. Wipf, and D. P. Curran, Fluorous synthesis: a fluorous-phase strategy for improving separation efficiency in organic synthesis, *Science*, *275*, 823–826 (1997).

96. I. T. Horvath and J. Rabai, Facile catalyst separation without water: fluorous biphase hydroformylation of olefins, *Science, 266*, 72–75 (1994).

97. B. Cornils, Fluorous biphase systems: the new phase-separation and immobilization technique, *Angew. Chem. Int. Ed. Engl., 36*, 2057–2059 (1997).

98. I. T. Horvath, Fluorous biphase chemistry, *Acc. Chem. Res., 31*, 641–650 (1998).

99. A. Studer, P. Jeger, P. Wipf, and D. P. Curran, Fluorous synthesis: fluorous protocols for the Ugi and Biginelli multicomponent condensations, *J. Org. Chem., 62*, 2917–2924 (1997).

100. A. Studer and D. P. Curran, A strategic alternative to solid phase synthesis: preparation of a small isoxazoline library by "Fluorous Synthesis," *Tetrahedron, 53*, 6681–6696 (1997).

101. B. Linclau, A. K. Sing, and D. P. Curran, Organic-fluorous phase switches: a fluorous amine scavenger for purification in solution phase parallel synthesis, *J. Org. Chem., 64*, 2835–2842 (1997).

102. C. W. Lindsley, Z. Zhao, R. C. Newton, W. H. Leister, and K. A. Strauss, A general Staudinger protocol for solution-phase parallel synthesis, *Tetrahedron Lett., 43*, 4467–4470 (2002).

103. D. P. Curran, Fluorous reverse phase silica gel: a new tool for preparative separations in synthetic organic and organofluorine chemistry, *Synlett, 9*, 1488–1496 (2001).

104. Z. Lou, J. Williams, R. W. Read, and D. P. Curran, Fluorous Boc (FBoc) carbamates: new amine protecting groups for use in fluorous synthesis, *J. Org. Chem., 66*, 4261–4266 (2001).

105. A. G. M. Barrett, S. M. Cramp, R. S. Roberts, and F. J. Zecri, A ROMPGel supported *N*-hydroxysuccinimide: a host of acylations with minimal purification, *Org. Lett., 2*, 261–264 (2000).

106. J. D. Moore, A. M. Harned, J. Henle, D. L. Flynn, and P. R. Hanson, Scavenger-ROMP-filter: a facile strategy for soluble scavenging via norbornenyl tagging of electrophilic reagents, *Org. Lett., 4*, 1847–1849 (2002).

107. J. D. Moore, R. H. Herpel, J. R. Lichtsinn, D. L. Flynn, and P. R. Hanson, ROMP-generated oligomeric sulfonyl chlorides as versatile soluble scavenging agents, *Org. Lett., 5*, 105–107 (2003).

108. J. D. Moore, R. J. Byrne, P. Vedantham, D. L. Flynn, and P. R. Hanson, High-load, ROMP-generated oligomeric bis-acid chlorides: design of soluble and insoluble nucleophile scavengers, *Org. Lett., 5*, 4241–4244 (2003).

109. A. M. Harned, S. Mukherjee, D. L. Flynn, and P. R. Hanson, Ring-opening metathesis phase-trafficking (ROMPpt) synthesis: multistep synthesis on soluble ROM supports, *Org. Lett., 5*, 15–18 (2003).

110. S. Murkherjee, K. W. C. Poon, D. L. Flynn, and P. R. Hanson, Capture-ROMP-release: application to the synthesis of amines and alkyl hydrazines, *Tetrahedron Lett., 44*, 7187–7190 (2003).

111. D. J. Gravert and K. D. Janda, Organic synthesis on soluble polymer supports: liquid-phase methodologies, *Chem. Rev., 97*, 489–509 (1997).

112. P. H. Toy and K. D. Janda, Soluble polymer-supported organic synthesis, *Acc. Chem. Res., 33*, 546–554 (2000).

113. A. Kirschning, H. Monenschein, and R. Wittenberg, Functionalized polymers: Emerging versatile tools for solution-phase chemistry and automated parallel synthesis, *Angew. Chem. Int. Ed. Engl., 40*, 650–679 (2001).

114. D. H. Drewry, D. M. Coe, and S. Poon, Solid-supported reagents in organic synthesis, *Med. Res. Rev., 19*, 97–148 (1999).

115. S. J. Shuttleworth, P. M. Allin, and K. Sharma, Functionalized polymers: recent developments and new applications in synthetic organic chemistry, *Synthesis*, 1217–1239 (1997).

116. S. W. Kaldor and M. G. Siegel, Combinatorial chemistry using polymer-supported reagents, *Curr. Opin. Chem. Biol., 1*, 101–106 (1997).

117. B. Jandeleit, D. J. Schaefer, T. S. Powers, H. W. Turner, and W. H. Weinberg, Combinatorial materials science and catalysis, *Angew. Chem. Int. Ed. Engl., 38*, 2494–2532 (1999).

118. E. Lindner, T. Schneller, F. Auer, and H. A. Mayer, Chemistry in interphases: a new approach to organometallic syntheses and catalysis, *Angew. Chem. Int. Ed. Engl.*, *38*, 2154–2174 (1999).
119. J. H. Clark and D. J. Macquarries, Environmentally friendly catalytic methods, *Chem. Soc. Rev.*, *25*, 303–310 (1996).
120. J. J. Parlow, R. V. Devraj, and M. S. South, Solution-phase chemical library synthesis using polymer-assisted purification techniques, *Curr. Opin. Chem. Biol.*, *3*, 320–336 (1999).
121. S. J. Teague, A. M. Davis, P. D. Leeson, and T. I. Oprea, The design of leadlike combinatorial libraries, *Angew. Chem. Int. Ed. Engl.*, *24*, 3743–3748 (1985).
122. T. I. Oprea, A. M. Davis, S. J. Teague, and P. D. Leeson, Is there a difference between leads and drugs? A historical perspective, *J. Chem. Inf. Comput. Sci.*, *41*, 1308–1315 (2001).
123. T. I. Oprea, I. Zamora, and A.-L. Ungell, Pharmacokinetically based mapping device for chemical space navigation, *J. Comb. Chem.*, *4*, 258–266 (2002).
124. J. Sadowski and H. Kubinyi, A scoring scheme for discriminating between drugs and nondrugs, *J. Med. Chem.*, *41*, 3325–3329 (1998).
125. Ajay, W. P. Walters, and M. A. Murcko, Can we learn to distinguish between "drug-like" and "nondrug-like" molecules?, *J. Med. Chem.*, *41*, 3314–3324 (1998).

6

TIMING OF ANALOG RESEARCH IN MEDICINAL CHEMISTRY

JÁNOS FISCHER AND ANIKÓ GERE
Gedeon Richter Ltd.
Budapest, Hungary

6.1 INTRODUCTION

The term *analog* is used according to the IUPAC recommendations,[1] where an incremental innovation[2] differentiates the drug from the original one. This overview focuses on analogs, but bioisosteres[3] are also mentioned if appropriate. Since not all analogs fit into a general formula, we depict all the formulas individually. We would like to demonstrate that analog research can be successful both in the early phase, when no product has yet been introduced to the market, and also after a drug has been launched successfully. The analogs afford in most cases an incremental and in some cases an essential innovation. We propose a classification of drug discoveries according to their timing: *early phase analogs* and *drug analogs*. With the help of some significant examples from the past few decades, we show how analogs from both approaches can contribute to drug discovery.

6.2 EARLY PHASE ANALOGS

Early phase analogs are structurally similar drugs discovered before the original drug is launched. As a result of early phase parallel research, the discovery dates of such derivatives are often very close to each other.

6.2.1 ACE Inhibitors

The first successful angiotensin-converting enzyme (ACE) inhibitor was captopril.[4] The pioneer discovery by Ondetti and Cushman was achieved by replacing the carboxyl group

Drug Discovery and Development, Volume 1: Drug Discovery, Edited by Mukund S. Chorghade

TABLE 6.1 ACE Inhibitors

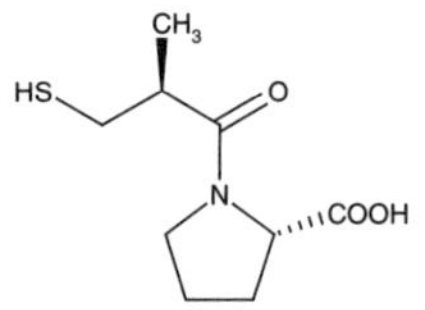

captopril 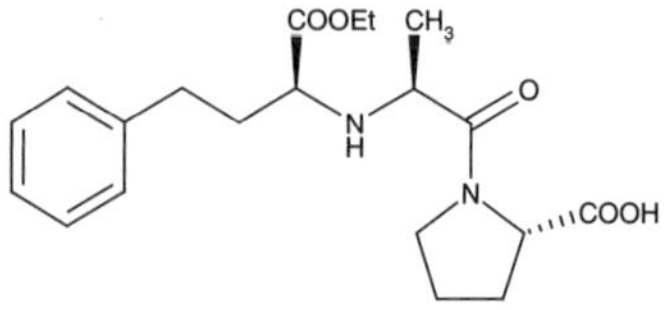 enalapril 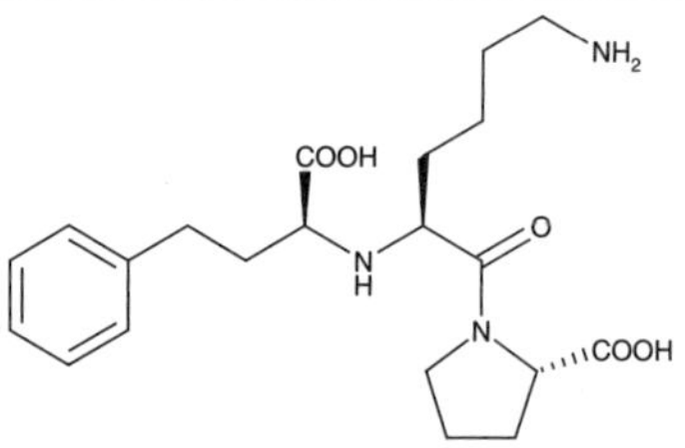 lisinopril

Name	Originator	Basic Patent	Launch
Captopril	Squibb (Bristol-Myers Squibb)	1976	1980
Enalapril	Merck & Co.	1978	1984
Lisinopril	Merck & Co.	1978	1987

of the analogous carboxyalkanoyl-L-proline by an SH group. To get more active and mercapto-free analogs, Patchett started from an homologous carboxyalkanoyl-L-proline, and among *N*-carboxyalkyl dipeptides, enalapril[5] and lisinopril proved to be long-acting ACE inhibitors. In a parallel research activity many analogous ACE inhibitors were discovered before enalapril appeared on the market to create a class of drugs (Table 6.1), which can be used in the treatment of hypertension and congestive heart disease.

6.2.2 AT_1 Antagonists

In 1982, hypotensive imidazole-5-acetic acid derivatives were published,[6] which antagonized angiotensin II evoked vasoconstriction. The clinical breakthrough came with losartan[7] and its analogs, which except for eprosartan and telmisartan, have a biphenyltetrazole moiety (Table 6.2). The angiotensin II antagonists are competitive with the ACE inhibitors, and further clinical trials will decide which class is more effective.

6.2.3 Proton Pump Inhibitors

Omeprazole,[8] which is the first successfully introduced H^+/K^+-ATPase inhibitor, was followed by novel analogs (Table 6.3). They are irreversible blockers of the proton pump that is responsible for acid secretion by the gastric parietal cells. According to the mechanism of action,[9] omeprazole itself is inactive, but it is transformed into a sulfenamide, which is the active inhibitor in vivo. A comparison of omeprazole, lansoprazole, and pantoprazole demonstrated differences in pharmacokinetics and drug interaction profile.[10]

6.2.4 Insulin Sensitizers: Glitazones

The first member of the thiazolidine-2,4-diones, ciglitazone,[11] reduced plasma glucose after oral administration in several insulin-resistant animal models, but a more potent compound was needed. The clinical breakthrough was troglitazone,[12] which was introduced in 1997. In contrast to the three classes of drugs noted above, where the first member of the class remained successful even after the introduction of their analogs, the case of the "glitazones" shows an opposite situation. Troglitazone was withdrawn[13] from the market

TABLE 6.2 Angiotensin II Antagonists

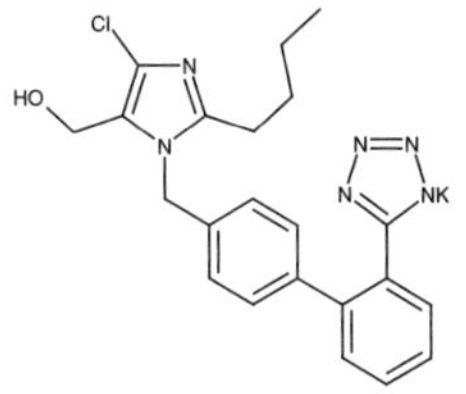

losartan 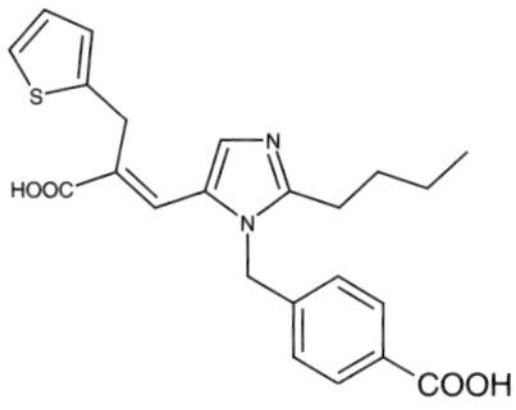 eprosartan 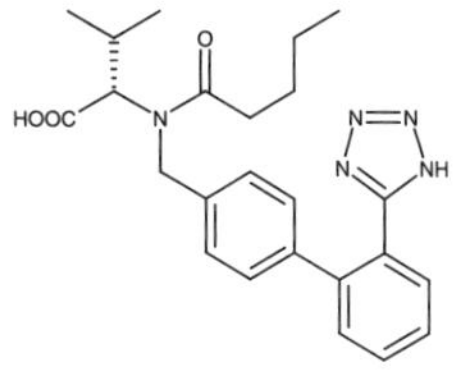 valsartan

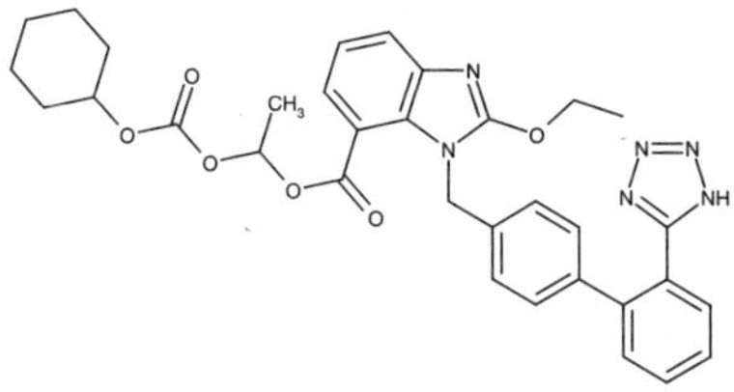

candesartan 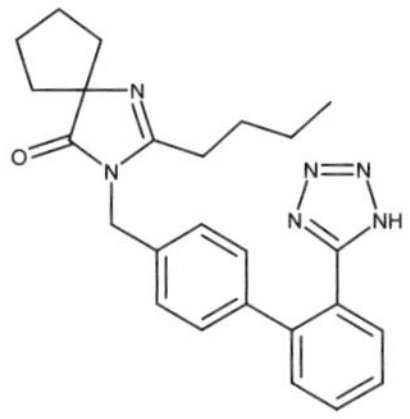 irbesartan 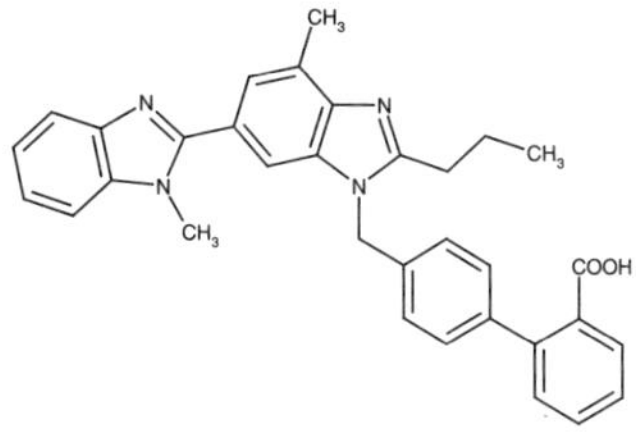 telmisartan

Name	Originator	Basic Patent	Launch
Losartan	DuPont	1986	1994
Eprosatran	SmithKline Beecham (GlaxoSmithKline)	1989	1997
Valsartan	Ciba-Geigy (Novartis)	1990	1996
Candesartan	Takeda	1990	1999
Irbesartan	Sanofi	1990	1997
Telmisartan	Boehringer-Ingelheim	1991	1999

TABLE 6.3 Omeprazole and Its Analogs

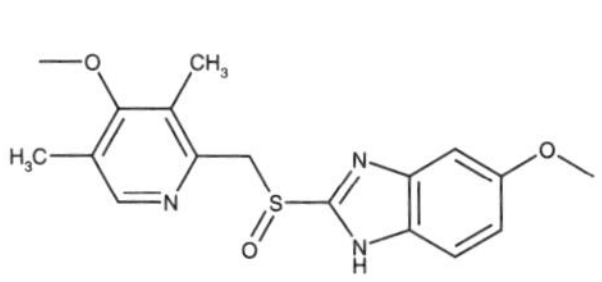

omeprazole 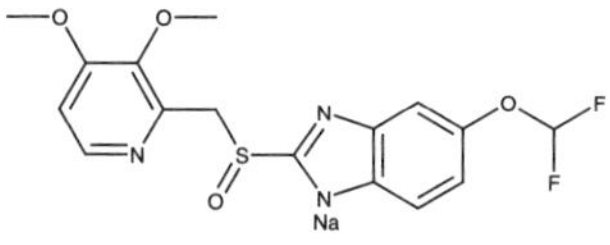 pantoprazole 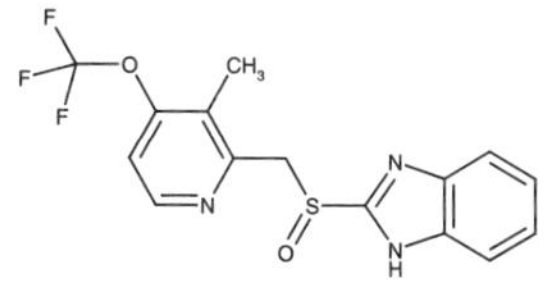 lansoprazole

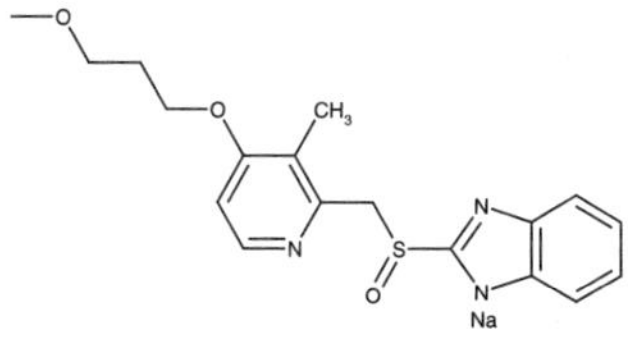

rabeprazole

Name	Originator	Basic Patent	Launch
Omeprazole	Hässle (AstraZeneca)	1978	1988
Pantoprazole	Byk-Gulden	1983	1994
Lansoprazole	Takeda	1984	1991
Rabeprazole	Eisai	1986	1997

TABLE 6.4 Glitazones

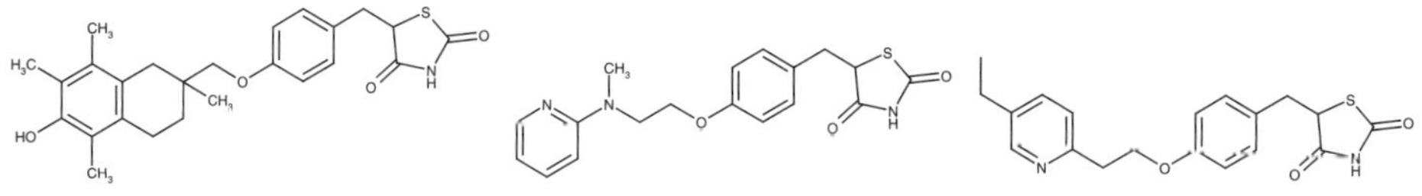

troglitazone rosiglitazone pioglitazone

Name	Originator	Basic Patent	Launch
Troglitazone	Sankyo	1983	1997
Pioglitazone	Takeda	1985	1999
Rosiglitazone	SmithKline Beecham (GlaxoSmithKline)	1987	1999

in 2000 because of liver toxicity in humans (Table 6.4). The clinical application of rosiglitazone and pioglitazone is carried out according to the U.S. Food and Drug Administration (FDA) labeling, including the need for liver enzyme monitoring before the start of therapy and periodically during the treatment. The glitazones exert their insulin sensitizer activity via stimulation of peroxisome proliferator activated receptor gamma subtype (PPAR-γ).[14]

6.2.5 HMG-CoA Reductase Inhibitors

Mevastatin (compactin), a fungal metabolite[15] and a potent inhibitor of hydroxymethylglutaryl (HMG) CoA reductase, initiated a series of *statins* for treatment of lipoprotein disorders (Table 6.5). The clinical breakthrough was lovastatin,[16] followed by simvastatin and pravastatin. Their therapeutic field is the treatment of hypercholesterolemia. As a result of intense activity in the design of synthetic analogs of the foregoing statins, new analogs were obtained where the decalin moiety was replaced by different heterocyclic rings bearing almost the same substituents, such as 3,5-dihydroxyheptanoic acid derivative, 4-fluorophenyl, and isopropylsubstituents. The first member of these heterocyclic statins was fluvastatine sodium, followed by atorvastatin, which lower both cholesterol and triglyceride levels. The lactone forms are prodrugs, which are metabolized to the corresponding active hydroxy acid form.[17]

6.2.6 Antimigraine Drugs

The first representative of drug with 5-$HT_{1B/1D}$ agonist mechanism was sumatriptan,[18] which proved to be a useful drug for the treatment of acute attacks of migraine. Among the analogs 5-HT_{1D} selectivity and pharmacokinetic parameters play an important role (Table 6.6). Rizatriptan showed both an increased oral bioavailability and more rapid absorption to oral sumatriptan.[19]

6.3 DRUG ANALOGS

A drug analog is a structurally similar drug which was discovered later (much later) than the launch of the original drug. There are some lonely drugs, such as aspirin, levodopa,

TABLE 6.5 Statins

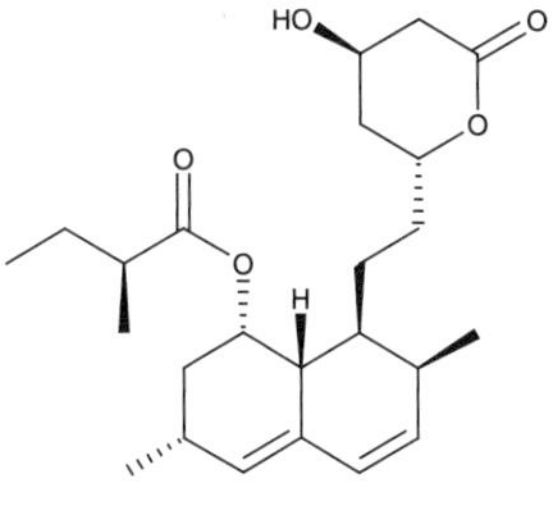

lovastatin 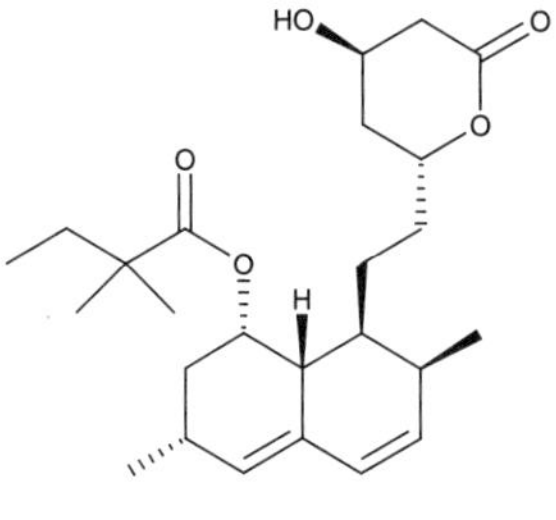 | simvastatin 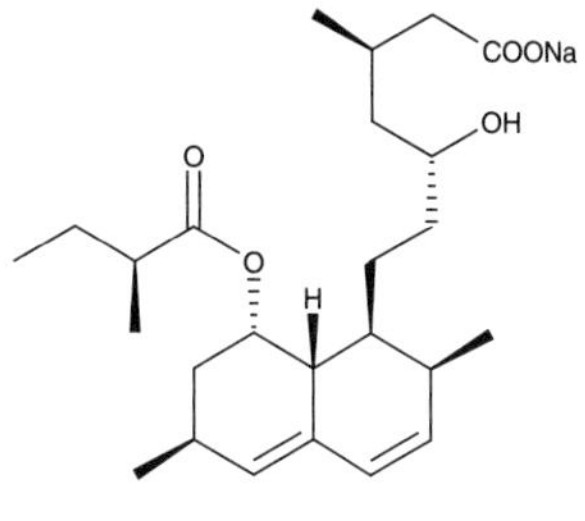 | pravastatin

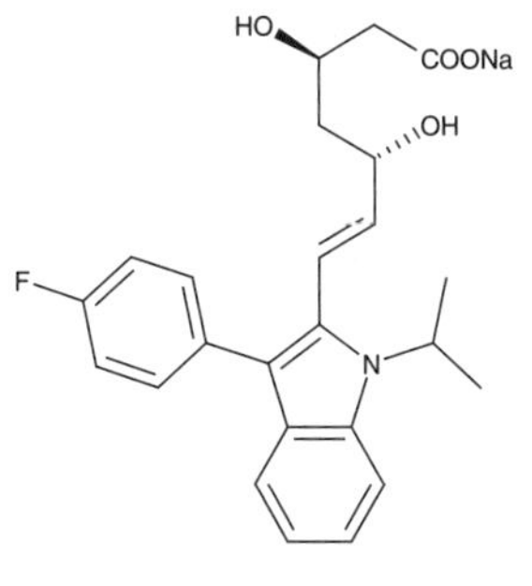

fluvastatin 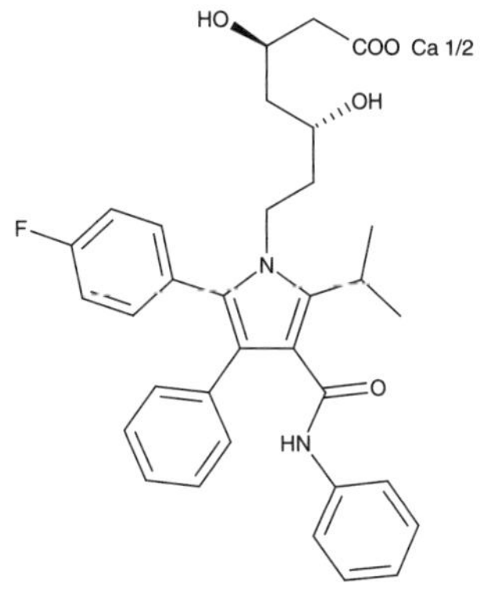 | atorvastatin

Name	Originator	Basic Patent	Launch
Lovastatin	Merck & Co.	1979	1987
Simvastatin	Merck & Co.	1980	1989
Pravastatin	Sankyo	1980	1989
Fluvastatin	Sandoz (Novartis)	1982	1994
Atorvastatin	Pfizer	1986	1997

methyldopa, metformin, PAS, and colchicine, without analogs, but they represent only a minority group of the drugs. In most cases several examples of successful analog-research based on pioneer drugs can be observed.

6.3.1 Metoclopramide Analogs

The history of these drugs amounts to about four decades (Table 6.7). Metoclopramide was discovered in 1961. It is a centrally acting antiemetic agent,[20] but its mechanism of action has not been fully elucidated. Antidopaminergic properties at both D_1 and D_2 receptor subtypes play an important role in its activity,[21] but its extrapyramidal side effects are also correlated with this mechanism. Twenty years later analogous cisapride was dicovered, which does not exhibit potent dopamine receptor antagonist activity. Its main therapeutic

TABLE 6.6 Antimigraine Drugs

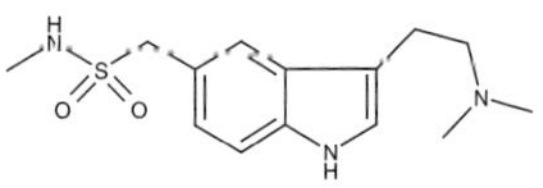

sumatriptan 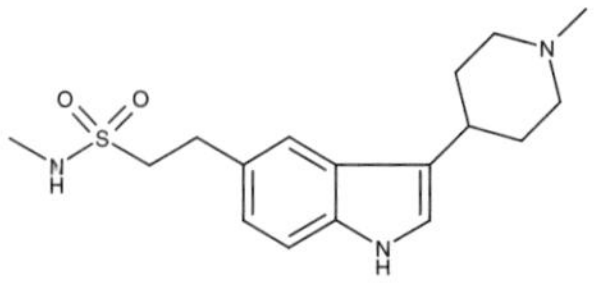 naratriptan 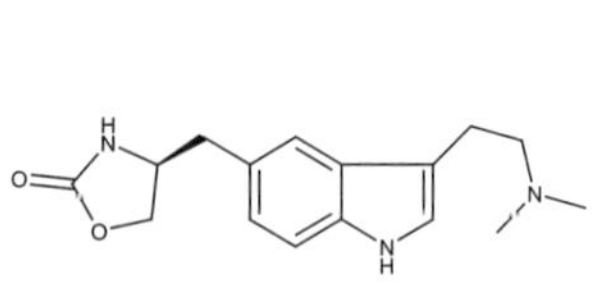 zolmitriptan

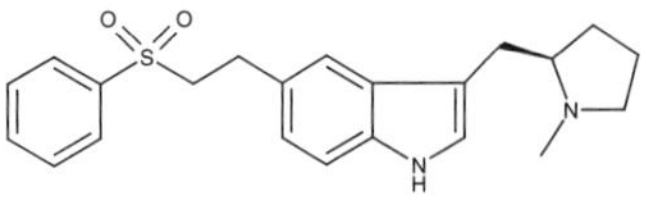

eletriptan 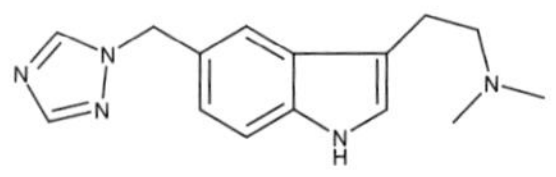 rizatriptan 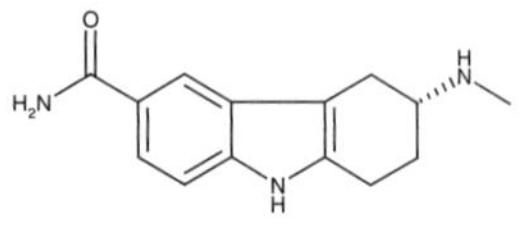 frovatriptan

Name	Originator	Basic Patent	Launch
Sumatriptan	Glaxo (GlaxoSmithKline)	1982	1991
Naratriptan	Glaxo (GlaxoSmithKline)	1987	1997
Zolmitriptan	Wellcome (GlaxoSmithKline)	1990	1997
Eletriptan	Pfizer	1990	1999[a]
Rizatriptan	Merck & Co.	1991	1998
Frovatriptan	SmithKline Beecham (GlaxoSmithKline)	1991	2000

[a]

use is the treatment of gastroesophageal reflux disease and it was one of the most successful drugs of the last decade. Agonistic action at 5-HT_4 receptors, and thereby facilitation of cholinergic excitatory neurotransmission, has been suggested as the mechanism by which these agents enhance gastric motility.[22] Current data suggest that concomitant administration of cisapride and certain azole-derivatives (e.g., ketoconazole) can result in

TABLE 6.7 Gastroprokinetic Drugs

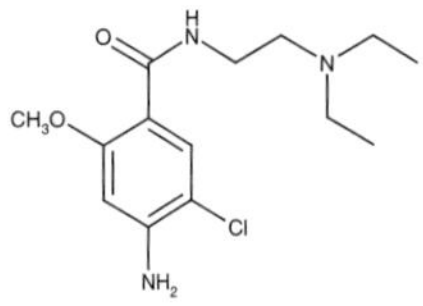

metoclopramide 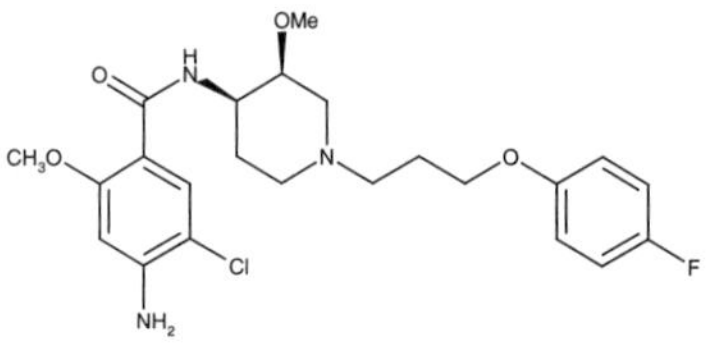 cisapride 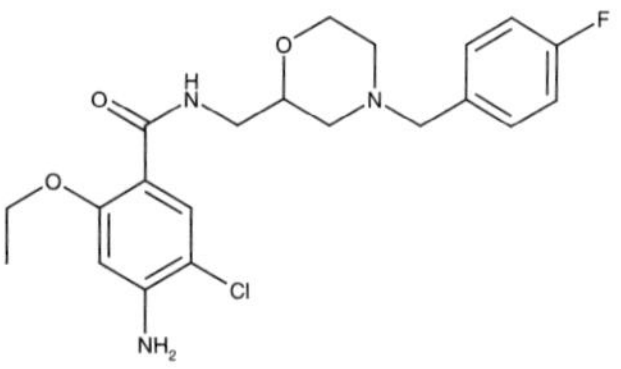 mosapride

Name	Originator	Basic Patent	Launch
Metoclopramide	Société d'Études Scientifiques et Industrielles de l'Île-de-France	1961	1964
Cisapride	Janssen	1981	1988
Mosapride	Dainippon	1986	1998

TABLE 6.8 Azatadine Analogs

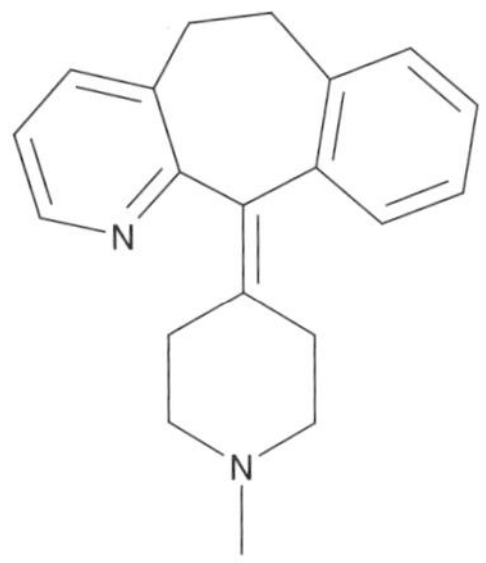

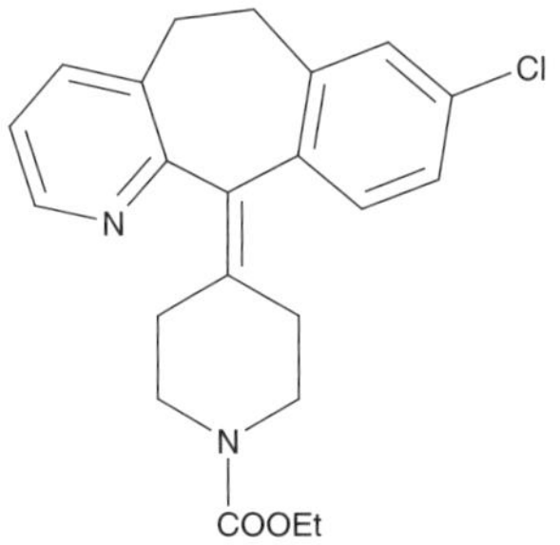

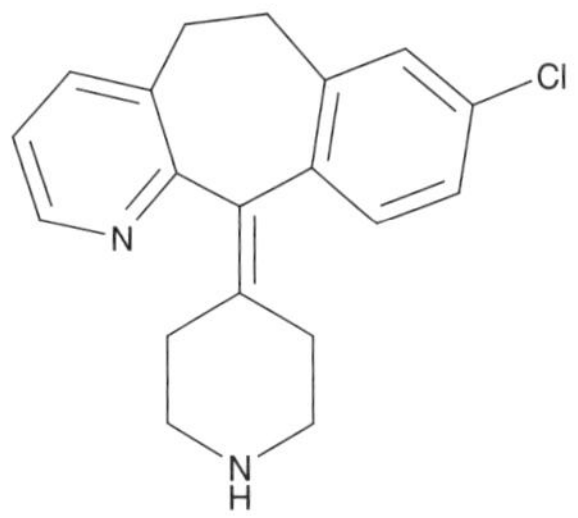

azatadine loratadin desloratadine

Name	Originator	Basic Patent	Launch
Azatadine	Schering-Plough	1963	1977
Loratadine	Schering-Plough	1980	1988
Desloratadine	Sepracor/Schering-Plough	1984	2001

prolongation of the QT interval. The marketing of cisapride was terminated in the United States in 2000,[23] but further research is going on in this field. Mosapride was discovered in 1986. The mode of action of mosapride on gastrointestinal motor activity was clearly different from that of cisapride, which stimulates motor activity in all sites of the gastrointestinal tract.[24] Further clinical trials are needed to evaluate the drug interaction profile of mosapride.

6.3.2 Azatadine Analogs

It was a popular view a generation ago that a nonsedating H_1-receptor antagonist is unobtainable.[25] There was a lack of validated methods to predict the sedative liability of a new antihistamine. Terfenadine served as a clinical breakthrough, which proved to be such an agent in 1978.[26] This initiated researchers at Schering-Plough to carry out drug analog research, whose lead molecules were terfenadine and azatadine.[27] A selected battery of central nervous system (CNS) tests in guinea pigs and mice using terfenadine and azatadine as reference drugs helped them to screen analogs. The carbamate analog of azatidine removed its central nervous system (CNS) activity while retaining much of its antihistamine potency. Further optimizing the carbamate azatadine afforded the 8-chloro derivative loratadine,[28] with a longer duration of action (Table 6.8). Its active metabolite, desloratadine was launched in 2001.

6.3.3 Miconazole Analogs

Miconazole is used primarily as a topical antifungal agent. Ketoconazole and fluconazole can also be given orally. The terminal half-life of fluconazole is approximately three times higher than that of ketoconazole. Oral fluconazole with a single dose produces clinical cure of uncomplicated vulvovaginal candidiasis. All of these agents are fungistatic by inhibiting

TABLE 6.9 Miconazole Analogs

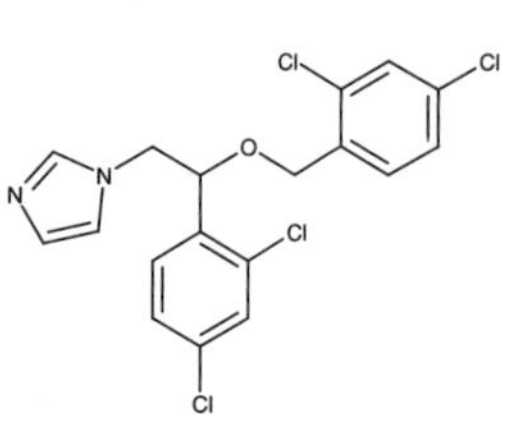

miconazole

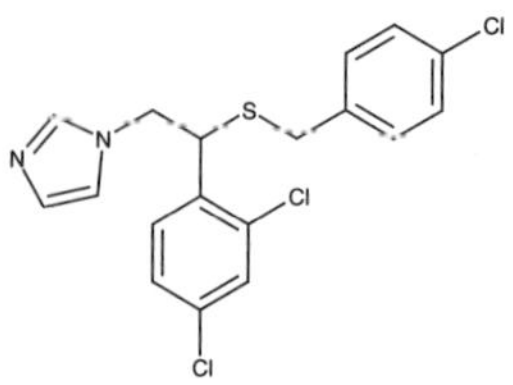

sulconazole

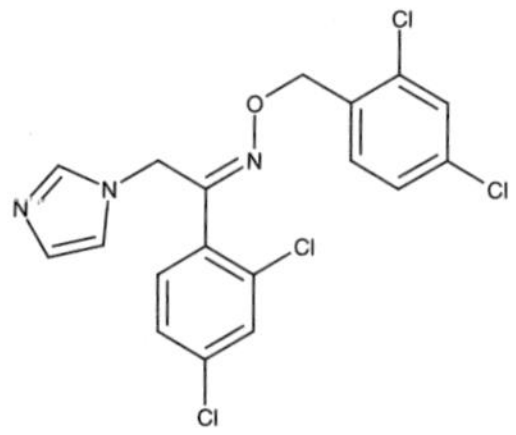

oxiconazole

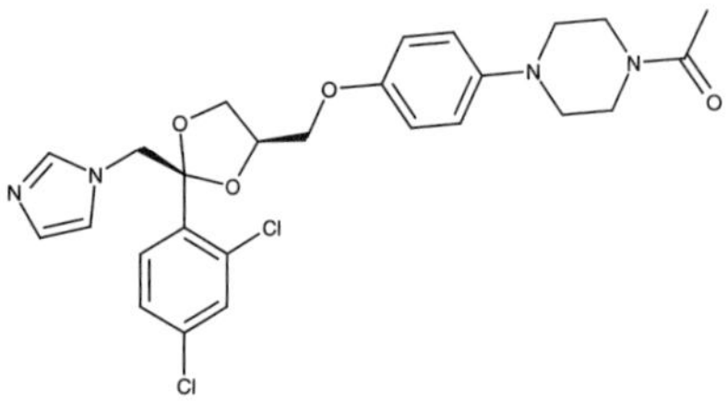

ketoconazole

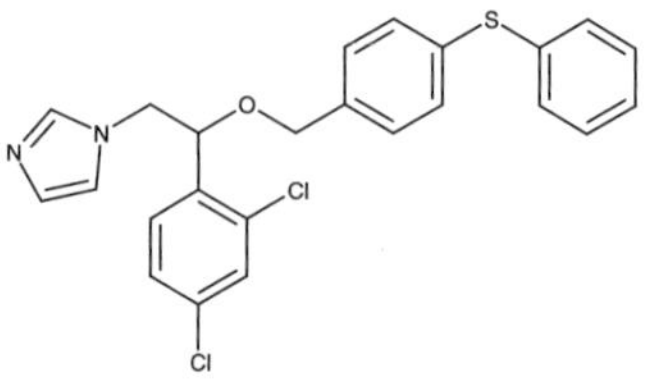

fenticonazole

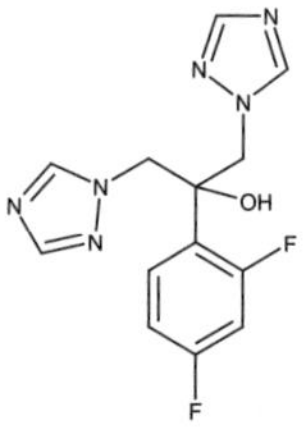

fluconazole

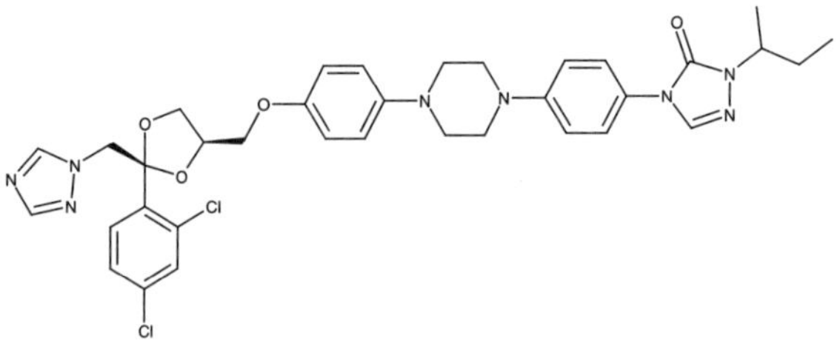

itraconazole

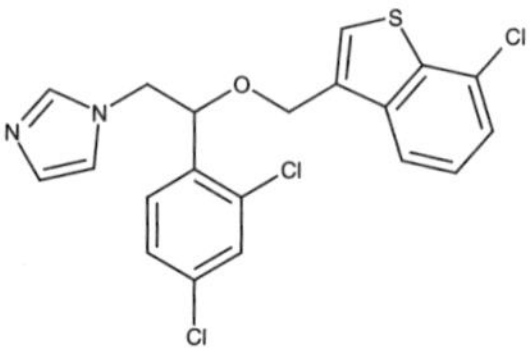

sertraconazole

Name	Originator	Basic Patent	Launch
Miconazole	Janssen	1968	1971
Sulconazole	Syntex	1974	1985
Oxiconazole	Siegfried	1975	1983
Ketoconazole	Janssen	1977	1981
Fenticonazole	Recordati	1978	1987
Fluconazole	Pfizer	1981	1988
Itraconazole	Janssen	1983	1988
Sertraconazole	Ferrer	1984	1992

the biosynthesis of ergosterol. Clinical failure of antifungal therapy due to resistance to existing agents is spreading rapidly and is often multifactorial[29] (Table 6.9).

6.3.4 Nifedipine Analogs

The calcium channel blocking mechanism was discovered by Fleckenstein[30] in 1967. Nifedipine,[31] the first member of this class, has a short duration of action. Many structural

TABLE 6.10 Calcium Channel Blockers

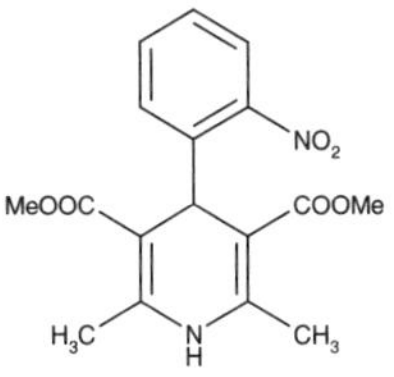

nifedipine 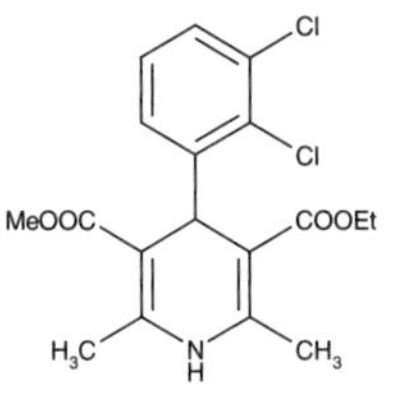 felodipine 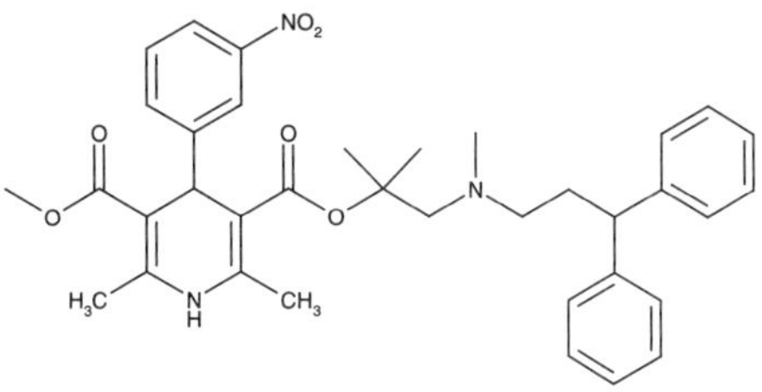 lercanidipine

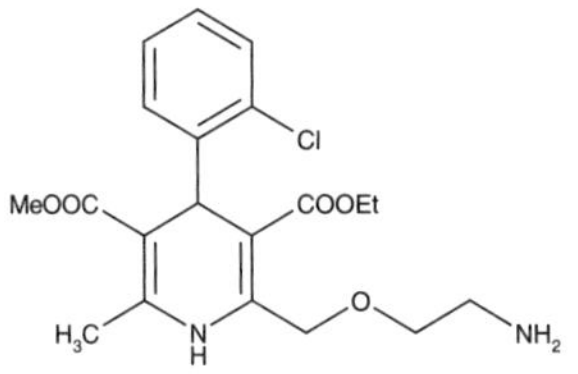

amlodipine 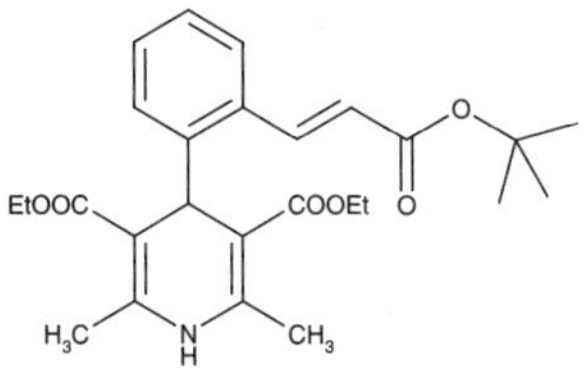 lacidipine

Name	Originator	Basic Patent	Launch
Nifedipine	Bayer	1967	1975
Felodipine	Hässle (AstraZeneca)	1978	1988
Lercanidipine	Recordati	1984	1997
Lacidipine	GlaxoSmithKline	1985	1991
Amlodipine	Pfizer	1986	1990

analogs were developed with a better pharmacokinetic profile (gradual onset, long duration of action). These drugs are used for the treatment of mild and moderate hypertension (Table 6.10).

6.3.5 Propranolol Analogs

Propranolol is the first nonselective β-adrenergic blocking agent with no intrinsic sympathomimetic activity. It is used for the treatment of arrhythmias, angina pectoris, and hypertension. Because of its ability to block β-receptors in bronchial smooth muscle, the drug is generally not used in people with bronchial asthma. As a consequence, there has been a search for β-adrenergic blocking agents that are cardioselective. Practolol, discovered in 1966, was the first such agent, but it was withdrawn because of its toxic side effects. Table 6.11 summarizes the β_1-selective (cardioselective) antagonists, which contributed an essential improvement to the therapy.[32]

6.3.6 Clodronate Analogs

Bisphosphonates are powerful bone resorption inhibitors that have been found to be clinically useful in the treatment of osteoporosis. Bisphosphonates are generally very poorly absorbed when given orally, but once absorbed they are taken up preferentially in bones.

TABLE 6.11 β_1-Adrenergic Blocking Agents

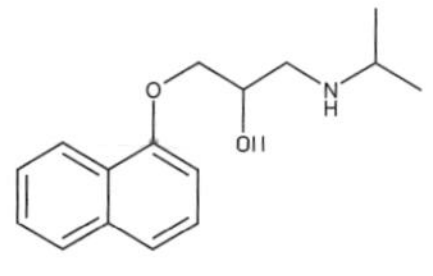

propranolol 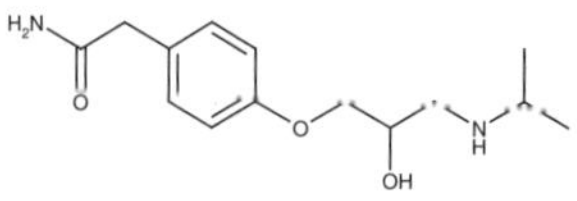 atenolol 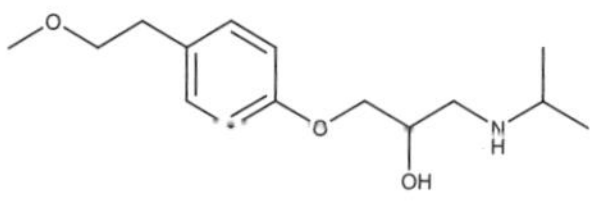 metoprolol

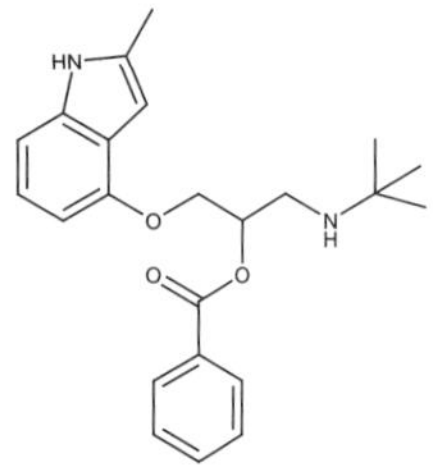

bopindolol 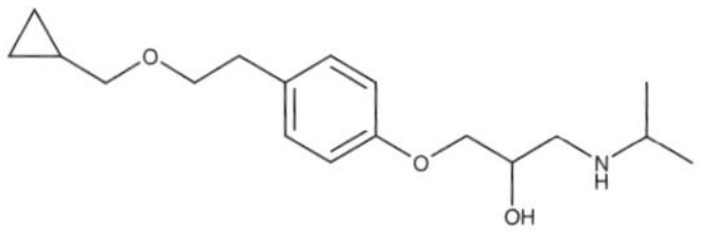 betaxolol 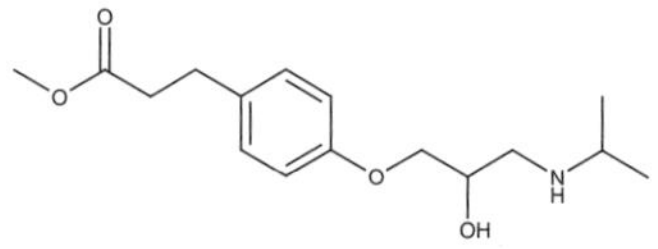 esmolol

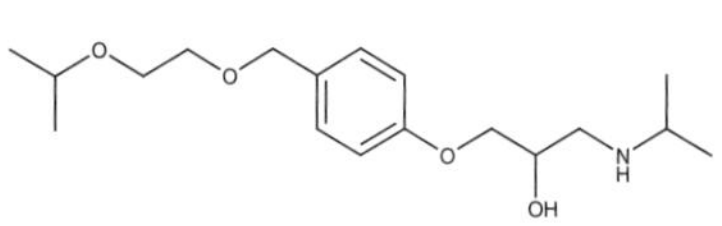

bisoprolol 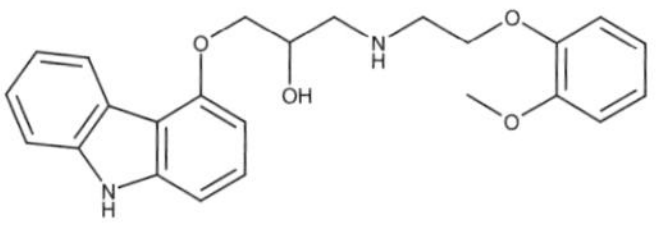 carvedilol

Name	Originator	Basic Patent	Launch
Propranolol	ICI (AstraZeneca)	1962	1964
Atenolol	ICI (AstraZeneca)	1969	1975
Metoprolol	Hässle (AstraZeneca)	1970	1975
Bopindolol	Sandoz (Novartis)	1975	1985
Betaxolol	Synthélabo	1975	1983
Esmolol	American Hospital Supply (DuPont)	1980	1987
Bisoprolol	Merck (Damstadt)	1976	1986
Carvedilol	Boehringer-Mannheim (Roche)	1978	1991

Clodronate disodium and etidronate sodium were used for the treatment of Paget's disease.[33] The second-generation products (alendronate and pamidronate) are much more effective. A third-generation agent, the risedronate, seems to have fewer esophageal side effects[34] (Table 6.12).

6.4 SUMMARY

This short overview of analog research in medicinal chemistry has defined two main directions according to the timing: early phase analogs and drug analogs. A breakthrough discovery in medicinal chemistry initiates parallel early phase research activities in several research centers of the world. The marketing of new drugs, however, takes on average 10

TABLE 6.12 Clodronate Analogs

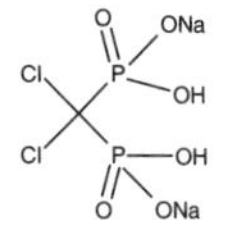

clodronate disodium 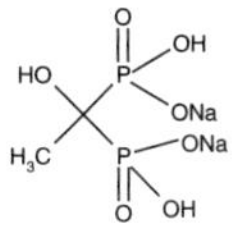　etidronate disodium 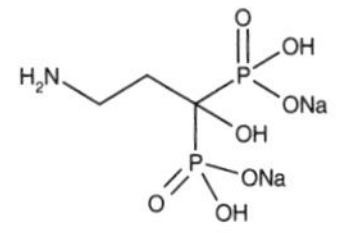　pamidronate disodium 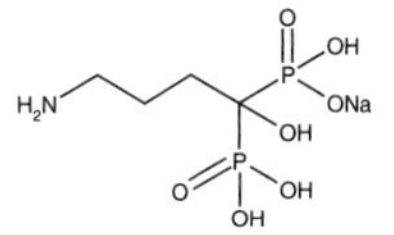　alendronate sodium

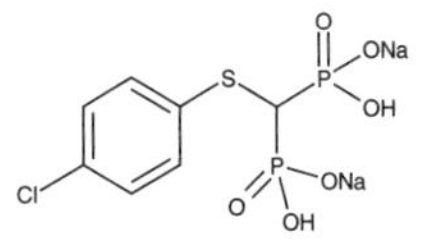

tiludronate disodium 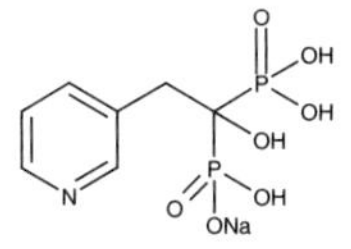　risedronate sodium 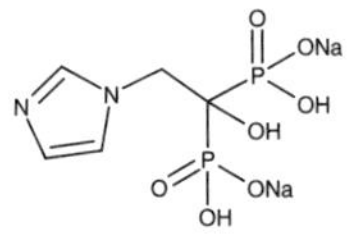　zoledronate disodium 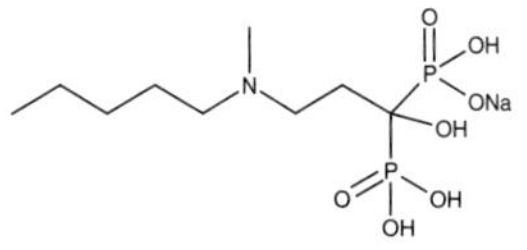　ibandronate sodium

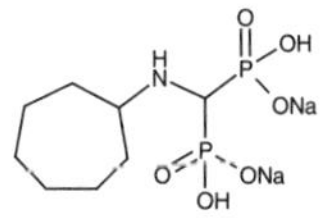

incadronate sodium

Name	Originator	Basic Patent	Launch
Clodronate	Procter & Gamble	1963	1986
Etidronate	Procter & Gamble	1966	1977
Pamidronate	Gador and Henkel	1971	1987
Alendronate	Gentili/Merck	1982	1993
Tiludronate	Sanofi-Synthélabo	1982	1993
Risedronate	Procter & Gamble	1984	1998
Zoledronate	Novartis	1986	2000
Ibandronate	Roche	1986	1996
Incadronate	Yamanouchi	1988	1997

to 15 years, and during this long period several similar lead compounds are identified to give at the end several early phase analogs on the market. In the case of drug analogs the situation is different. A drug is the end product of a long optimizing process in research and development, nevertheless, during their clinical trials, side effects, drug interactions, and other weak points can be observed, which stimulates researchers to make drug analogs which also provide remarkable achievements, as shown with the examples above.

These two approaches overlap in some cases. If a parallel research activity is starting in a period when the clinical results of a drug candidate are published (phase III), the products of the analog research will be regarded as a drug analog because of the long development process of today. There are no general rules as to whether it is preferable to undertake early phase or a drug analog research. It depends on the marketing conditions, the company strategy, the medicinal chemistry possibilities, and last but not least, on the inventive capacities of the people involved.

ACKNOWLEDGMENTS

The authors thank Professor C. R. Ganellin for his helpful comments and S.Lévai for technical assistance.

REFERENCES AND NOTES

1. IUPAC Recommendations: glossary of terms used in medicinal chemistry, *Pure Appl. Chem.* 70 (1998) 1129–1143.
2. Riefberg V., Pinkus G., *In Vivo* (1996) 18–23.
3. Wermuth C. G. (Ed.), *The Practice of Medicinal Chemistry*, Academic Press, San Diego, CA (1996) p. 207.
4. Ondetti M. A., Rubin B., Cushman D. W., *Science* 196 (1977) 441; U.S. patent 4,046,889 (1976).
5. Patchett P. A., Harris E., Tristram E. W., Wyvratt M. J., Wu M. T., Taut D., Peterson E. R., Ikeler T. J., ten Broeke J., Payne L. G., Ondeyka D. L., Thorsett E. D., Greenlee W. J., Lohr N. S., Hoffsommer R. D., Joshua H., Ruyle W. V., Rothrock J. W., Aster S. D., Maycock A. L., Robinson F. M., Hirschmann R., Sweet C. S., Ulm E. H., Gross D. M., Vassil T. C., Stone C. A., *Nature* 288 (1980) 280; European patent 12,401 (1978).
6. Furakawa Y., Kishimoto S., Nishikawa K. (Takeda Chemical Industries Ltd.), U.S. patents 4,340,598 and 4,355,040 (1982).
7. Carini D. J., Duncia J. V., Aldrich P. E., Chiu A. T., Johnson A. L., Pierce M. E., Price W. A., Santella J. B. III, Wells G. J., Wexler R. R., Wong P. C., Yoo S.-E., Timmermans P. B. M., *J. Med. Chem.* 34 (1991) 2525–2547; European patent 253,310 (1986).
8. Sjöstrand S. E., Junggren U. K. (Hässle Läkemedel AB), European patent 5129 (1978).
9. Lindberg P., Nordberg P., Alminger T., Bränström A., Wallmark B., *J. Med. Chem.* 29 (1986) 1327–1329.
10. Zech K., Steinijans V. W., Huber R., Radtke H. W., *Int. J. Clin. Pharmacol. Ther.* 34 (Suppl. 1) (1996) 3–6.
11. Sohda T., Mizuno K., Imamiya E., Suguyama Y., Fujita T., Kawamatsu Y., *Chem. Pharm. Bull.* 30 (1982) 3580–3600.
12. Yoshioka T., Fujita T., Kanai T., Aizawa Y., Hasegawa K., Horikoshi H., *J. Med. Chem.* 32 (1989) 421–428.
13. Warner-Lambert decided to discontinue marketing troglitazone (Rezulin) for the treatment of type II diabetes for safety and efficacy reasons (Warner-Lambert news release in *Prous Science Daily Essentials*, March 22, 2000).
14. Lehmann J. M., Moore L. B., Smith-Oliver T. A., Wilkison W. O., Willson T. M., Kliewer S. A., *J. Biol. Chem.* 270 (1995) 12953–12956.
15. Endo A., Kuroda M., Tsujita Y., *J. Antibiot.* 29 (1976) 1346.
16. Alberts A. W., Chen J., Kuron G., Hunt V., Huff J., Hoffman C., Rothrock J., Lopez M., Joshua H., Harris E., Patchett A., Monaghan R., Currie S., Stapley E., Albers-Schonberg G., Hensens O., Hirschfield J., Hoogsteen K., Liesch J., Springer J., *Proc. Natl. Acad. Sci. USA* 77 (1980) 3957.
17. Todd P. A., Goa K. L., *Drugs* 40 (1990) 583–607.
18. Heuring R. E., Peroutka S. J., *J. Neurosci.* 7 (1987) 894–903.
19. Sciberras D. G., Polvino W. J., Gertz B. J., Cheng H., Stepanavage M., Wittreich J., Olah T., Edwards M., Mant T., *Br. J. Clin. Pharmacol.* 43 (1997) 49–54.
20. French patent 1 313 758 (Societé d'Études Scientifiques et Industrielles de l'Île-de-France, 1961).

21. MacDonald T. M., *Eur. J. Clin. Pharmacol.* 40 (1991) 225–230.
22. Craig D. A., Clarke D. E., *J. Pharmacol. Exp. Ther.* 252 (1990) 1378–1386.
23. Janssen decided to stop marketing cisapride (Propulsid) in the United States as of July 14, 2000 (Janssen news release in *Prous Science Daily Essentials*, March 28, 2000).
24. Yoshida N., Ito T., Karasawa T., Itoh Z., *J. Pharmacol. Exp. Ther.* 257 (1991) 2572–2578.
25. Barnett A., Green M. J., in Lednice D. (Ed.), *Chronicles of Drug Discovery*, Vol. 3, ACS Professional Reference Book, Washington, DC 1993, p. 83.
26. Clarke C. H., Nicholson A. N., *Br. J. Clin. Pharmacol.* 6 (1978) 31–35.
27. U.S. patent 3,301,863 (Schering Corp., 1963).
28. Villani F. J., Wefer E. A., Mann T. A., Peer L., Levy A. S., *J. Heterocycl. Chem.* 9 (1972) 1203–1207.
29. Watkins W. J., Renau T. E., *Annu. Rep. Med. Chem.* 35 (2000) 157–166.
30. Fleckenstein A., *Arzneim.-Forsch.* (Drug Res.) 22 (1967) 22–33.
31. U.S. patent 3,485,847 (Bayer, 1967).
32. Harting J., Becker K. H., Bergmann R., Bourgois R., Enenkel H. J., Fuchs A., Jonas R., Lettenbau K., Minck K. O., Schelling P., Schulze E., *Arzneim.-Forsch.* (Drug Res.) 36 (1986) 200–208.
33. Meunier C., Chapuy M. C., Courpron P., Vignon E., Edouard C., Bernard J., *Rev. Rheum.* 42 (1975) 699–705.
34. Lanza F. L., Hunt R. H., Thomson A. B. R., Provenza J. M., Blank M. A., *Gastroenterology*, 119 (2000) 631–638.

7

POSSIBLE ALTERNATIVES TO HIGH-THROUGHPUT SCREENING

CAMILLE G. WERMUTH
Prestwick Chemical, Inc.
Illkirch, France

7.1 INTRODUCTION

A program for therapeutic discovery typically starts with the search for *hits*: molecules that exhibit a certain affinity for a target. Identifying hits for a new target usually involves the screening of a wide range of structurally diverse small molecules in an in vitro bioassay. Alternatively, small molecules can be screened for their potential for modulation of a biological process that is considered to be critical in a disease or in which the target is thought to play a major role. Thanks to miniaturization and robotics, the number of compounds that can be screened has greatly increased, and several thousand compounds can be screened in a single day. Once a hit has been identified, its activity must be confirmed and validated. The following are typical hit validation criteria: (1) the activity of the hit must be reproducible; (2) the hit must not display activity against many diverse targets; (3) the analogs of the hit must display SAR; and (4) the hit must not contain chemically reactive functions.[1] Only then does it become a lead substance, commonly referred to as the *lead*.

The pharmaceutical industry has come to rely heavily on high-throughput screening (HTS), and virtually every pharmaceutical company in the world has established HTS as an integral part of its discovery process. High-throughput screening technology can currently be regarded as being the preferred method for identification of new hits. There are many accounts in the literature of successes based on this approach, including the discovery of insulin mimetics[2] or of the new opioid ORL1 receptor agonists.[3] This strategy for drug discovery has several limitations, however, that are in the main due to the nature of the chemical libraries that are input to the robots. These libraries are usually huge and typically contain from 100,000 to 1,000,000 compounds. They are most often assembled by parallel

Drug Discovery and Development, Volume 1: Drug Discovery, Edited by Mukund S. Chorghade

or combinatorial chemistry. Some of the limitations exhibited by these libraries include inadequate diversity (which leads to a decrease in the chances of success), rather moderate or even low hit rates, and finally, the low biopharmaceutical quality of the hits ("A hit is not a drug!"). Present trends in HTS are focused on reducing the size of the libraries, on making them more effective, and on excluding ADME (absorption, distribution, metabolism, and elimination)-, and toxicological-inadequate candidates at as early a stage as possible. Computational chemists can help to reduce the number of compounds that are selected for HTS.

HTS technology has overshadowed all other drug discovery paradigms, and its almost exclusive use as a method for discovering drugs can lead to rather disappointing results. The aim of the present chapter is to draw the attention of scientists involved in drug discovery to the existence of some valuable alternative approaches. The first of these is analog design, a second discusses the usefulness of physiopathological hypotheses, and a third deals with contributions from clinical observations. Finally, in the last part, opportunities to derive new drugs from old drugs that are provided by the selective optimization of side activities (SOSA) approach are considered.

7.2 ANALOG DESIGN

7.2.1 Definitions

The term *analogy*, which is derived from the Latin and Greek *analogia*, has been used in natural sciences since 1791 to describe structural *and* functional similarity.[4] When extended to apply to drugs, this definition implies that the analog of an existing drug molecule shares chemical and therapeutic similarities with the original compound. In formal terms, this definition means that three categories of drug analogs may be anticipated: (1) analogs that exhibit chemical *and* pharmacological similarity, (2) analogs that exhibit only chemical similarity, and (3) analogs that display similar pharmacological properties but exhibit totally different chemical structures.

Since they exhibit both chemical and pharmacological similarities, analogs in the first category, may be regarded as *direct* analogs. These correspond to the class of drugs that are often referred to as *me-too compounds*. They are usually improved versions that exhibit pharmacological, pharmacodynamic, or biopharmaceutical advantages over a pioneer drug. Examples are the ACE inhibitors derived from captopril, the histamine H_2 antagonists derived from cimetidine, and the HMG-CoA reductase inhibitors derived from mevinolin. There are production and marketing reasons behind such analogs which can be justified in much the same way as any other industrial products, such as laptop computers or automobiles.

The second class of analogs is made of molecules that have chemical resemblances and for which the term *structural analogs* is suggested. This class contains compounds that were originally prepared as close and patentable analogs of a novel lead, but in which biological assays have revealed totally unexpected pharmacological properties. A historical example of the emergence of a new activity is provided by the discovery of the antidepressant properties of imipramine, which was originally designed as an analog of the potent neuroleptic drug chlorpromazine. Another example which illustrates that chemical similarity does not necessarily mean biological similarity is to be found in steroid hormones: although testosterone and progesterone are chemically very similar, they have totally different biological functions. The observation of an "emergent" activity may be purely fortuitous or may be the result of a deliberate and systematic investigation.

In the third class of analogous compounds, no chemical similarity is observed, but they, compounds share common biological properties. The term *functional analogs* is suggested for such compounds. Examples include the neuroleptics chlorpromazine and haloperidol and the tranquillizers diazepam and zopiclone. Despite the fact that they have totally different chemical structures, they show similar affinities for dopamine and benzodiazepine receptors, respectively. Currently, the design of such drugs has been facilitated, thanks to virtual screening of large libraries of diverse structures.

7.2.2 Pharmacophore-Based Analog Design: Scaffold Hopping or Scaffold Morphing

One option for designing functional analogs that are of interest involves searching large virtual compound libraries for structures that are isofunctional but which are based on a different scaffold. This approach is referred to as *scaffold hopping*. The objective is to escape from a patented chemical class by identifying molecules in which the central scaffold is different but in which the essential function-determining points are retained to form the basis of a relevant pharmacophore.

In an exploratory test[5] a program called CATS (chemically advanced template search) was applied to the prediction of novel cardiac T-type Ca^{2+} channel blocking agents by exploring the Roche in-house compound depository. Mibefradil (**1**), a known T-channel blocking agent ($IC_{50} = 1.7\,\mu M$) (Scheme 7.1), was used as the seed structure for CATS. The 12 highest-ranking molecules were tested for their ability to inhibit cellular Ca^{2+} influx using a cell culture assay. Nine compounds (75%) showed significant activity ($IC_{50} < 10\,\mu M$), of which one compound (**2**, clopimozid) had an $IC_{50} < 1\,\mu M$. The IC_{50} values for the next best structures (not given in the article) were 1.7, 2.2, 3.2, and 3.5 μM.

These hits have structural scaffolds that show significance differences from the query structure **1**. Essential function-determining points are, however, retained, and these form the basis for a relevant pharmacophore pattern. A similar approach, described as *scaffold morphing*, involves a series of directed chemical transformations of the initial structure, the aim of which is to generate new chemotypes with enhanced properties such as potency, selectivity, safety, and novelty.[6] An early claim of scaffold morphing was made in 1994 by Buehlmayer et al.[7] The objective was the design of the Novartis drug valsartan (Diovan)[8] using the DuPont angiotensin II receptor antagonist (**3**, losartan, Scheme 7.2) as the starting model.

This last example can be regarded as being a particularly refined case of bioisostery in which the *n*-butyl chains overlap, as does the carbonyl with the imino dipole, and the lipophilic chlorine with the isopropyl group. In valsartan (**4**) the metabolic oxidation of the hydroxymethylene function has already been anticipated. Rational morphing strategy

Scheme 7.1 Query structure **1** (mibefradil) and a high-ranking isofunctional structure **2** (clopimozid) derived from **1** by CATS.

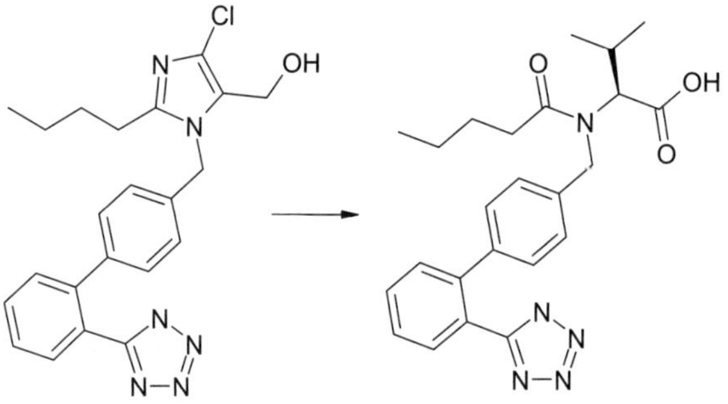

Scheme 7.2 The transition from the DuPont angiotensin II receptor antagonist losartan (**3**) to the Novartis analog valsartan (**4**) represents an example of computer-guided scaffold morphing. (From ref. 7.)

is closely associated with virtual screening.[9] Other detailed descriptions of modern computer-based hit location and optimization technologies are described in the excellent book by Kubinyi and Müller.[10]

7.2.3 Natural Compounds as Models

There are many historical examples recorded of drugs that are the result of modification, usually a simplification, of a natural model: morphinanes, benzomorphanes, and phenylpiperidines derived from morphine; hundreds of steroid analogs derived from the endogenous hormones, and procaine and various other local anesthetics derived from cocaine, for example. More recently, a second generation of statins characteristically contains only two chiral centers instead of the eight in the natural lead lovastatin (**5**) or in its close analog pravastatin (**6**). A typical example is found in rosuvastatin (**7**) (Scheme 7.3).

7.2.4 Emergence of New Activities

Imipramine It can also happen that a totally new property which is absent in the original molecule appears unexpectedly during pharmacological or clinical studies on a me-too

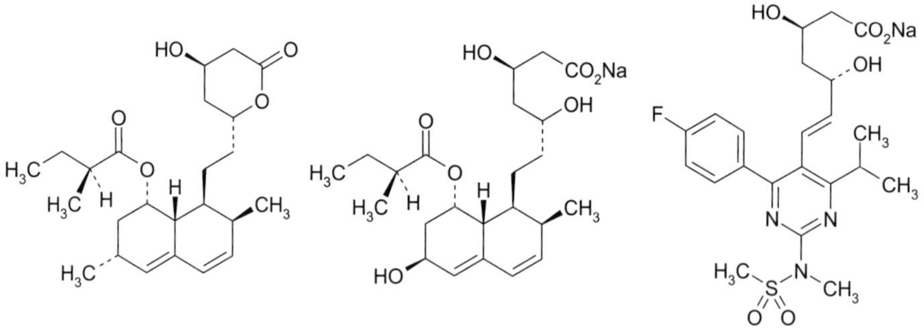

5 lovastatin (mevinolin) **6** pravastatin **7** rosuvastatin

Scheme 7.3 The more recent statins contain only two chiral centers instead of the eight in natural lovastatin.

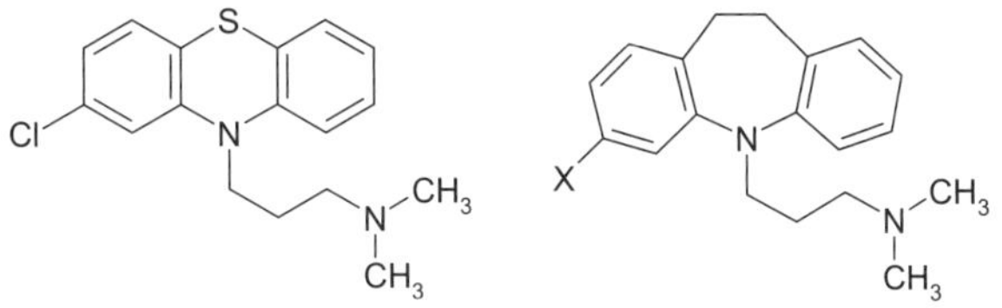

8 chlorpromazine

9 imipramine (X = H)
10 chlorimipramine (X = Cl)

Scheme 7.4 Structures of imipramine (**9**) and chlorimipramine (**10**) compared to that of chlorpromazine (**8**).

compound. The emergence of such a new activity means that the therapeutic copy in turn becomes a new lead structure. This was the case for imipramine (**9**) (Scheme 7.4), which was initially synthesized as an analog of chlorpromazine (**8**) and presented to clinical investigators to study its antipsychotic profile.[11,12] During its clinical evaluation this substance exhibited much greater activity against depressive states than against psychoses. Since 1954, imipramine has opened up genuine therapeutic avenues for the pharmacological treatment of depression.

Sildenafil A more recent example is provided by the discovery that sildenafil (**11**, Scheme 7.5; Viagra), a phosphodiesterase type 5 (PDE5) inhibitor, can be used as an efficacious, orally active agent for the treatment of male erectile dysfunction.[13,14] On its way to becoming Viagra, the compound UK-92,480, which was prepared in 1989 by Pfizer scientists in Sandwich, England, went from being at first a drug for hypertension to a drug for angina. Then it changed again when a 10-day tolerance study in Wales discovered its unusual side effect: penile erection.[15] Sildenafil was originally provided for clinical use as a hypotensive and cardiotonic substance; its usefulness in male erectile dysfunction clearly resulted from the clinical observations.

7.3 PHYSIOPATHOLOGICAL HYPOTHESES

7.3.1 Discovery of Levodopa

The amino acid L-dihydroxyphenylalanine (levodopa) was not used in medicine until the role of dopamine as a neurohormone was discovered by Carlsson et al.[16] Work carried out from the 1930s through to the 1950s identified the biosynthetic pathway (in chromaffin tissue and in adrenergic neurons) that linked dietary amino acids with catecholamines.[17,18]

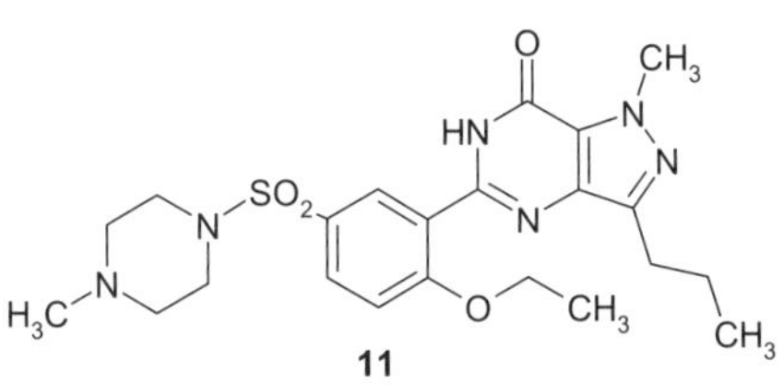

11

Scheme 7.5 Structure of the phosphodiesterase type 5 (PDE5) inhibitor sildenafil (**11**).

A multistage biosynthetic pathway was proposed for the synthesis of adrenaline from L-tyrosine (Scheme 7.6). In this pathway, L-tyrosine is converted to levodopa, which undergoes decarboxylation to form dopamine. Dopamine is the immediate precursor of noradrenaline, which is converted to adrenaline.

An enzyme known as levodopa decarboxylase was discovered in 1939 which degrades any levodopa present in mammalian tissues and which thus hindered its detection. Levodopa can, however, be administered to correct dopamine deficiency in Parkinson's disease. It behaves like a pro-drug, undergoing metabolic conversion to dopamine once it has entered the brain.

The introduction of racemic dihydroxyphenylalanine into therapeutic use occurred at the University of Vienna after Hornykiewicz[19] had gathered evidence that pointed to there being a depletion of dopamine reserves in the brains of patients with Parkinson's disease. Since dopamine was too polar to cross the blood–brain barrier, Hornykiewicz attempted to alleviate the disease by administering 50 to 150 mg of dihydroxyphenylalanine intravenously to 20 patients and used this metabolic precursor of dopamine, since the neurohormone itself could not cross over into the brain from the general circulation.[20] His results seemed favorable, as were those reported around the same time by Barbeau[21] at the University of Montreal. Barbeau administered dihydroxyphenylalanine by mouth to six patients. These findings were, however, disputed by others, and it was not until 1967 that treatment

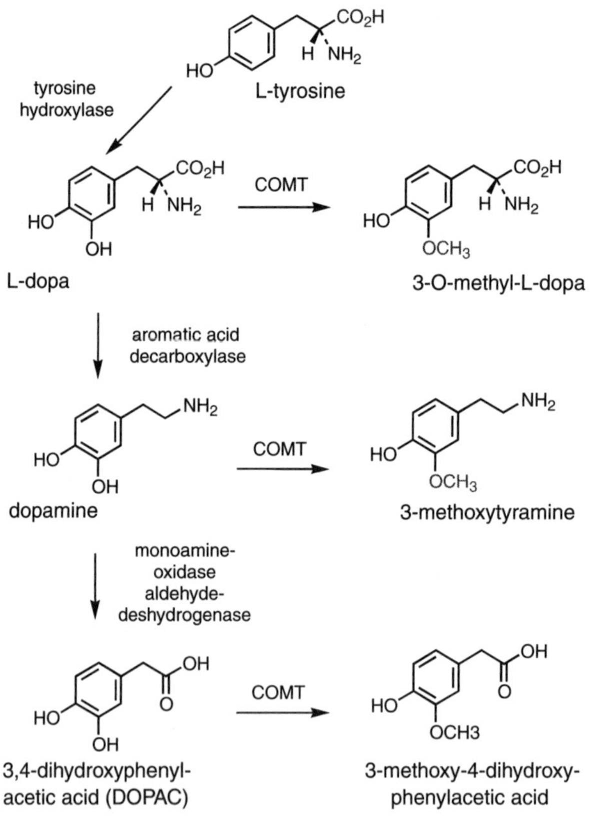

Scheme 7.6 Dopamine biosynthesis and metabolism.

protocols were perfected by Cotzias[22] and his colleagues at the Medical Research Centre of Brookhaven National Laboratory. They demonstrated that oral doses of up to 16 g each day consistently improved the general clinical condition of more than 50% of patients. This improvement lasted only while treatment continued. Because of the expense involved, racemic dihydroxyphenylalanine had been used in the early trials. Since that time, levodopa (the optically active isomer that is the metabolic precursor of dopamine) has become the universal treatment for Parkinson's disease. Several hundred thousand patients have benefited from this treatment. Of the DOPA administered by the oral route, however, 95% undergoes decarboxylation in the peripheral circulation before crossing the blood–brain barrier. To prevent the peripheral DOPA from undergoing this unwanted premature degradation, a peripheral inhibitor of DOPA–decarboxylase is usually added to the treatment. An additional improvement to the treatment involves the simultaneous addition of a catechol *O*-methyltransferase (COMT) inhibitor such as tolcapone or entacapone.

7.3.2 H_2-Receptor Antagonists

Research into the development of specific antagonists for the H_2 histamine receptor in the treatment of gastric ulcers has also been carried out using a rational physiopathological process.[23,24] Starting from the observation that the antihistaminic compounds known at the time (H_1-receptor antagonists) were not capable of acting as antagonists for the gastric secretion provoked by histamine, Black and his collaborators envisaged that an unknown subclass of the histamine receptor (the future H_2 receptor) existed. From 1964 on, they initiated a program of systematic research for specific antagonists for this receptor.

The starting point was guanylhistamine (**12**, Scheme 7.7), which exhibits weak antagonistic properties against the gastric secretion that histamine induced. Lengthening the side chain of this compound clearly increased H_2 antagonistic activity, but a residual agonist

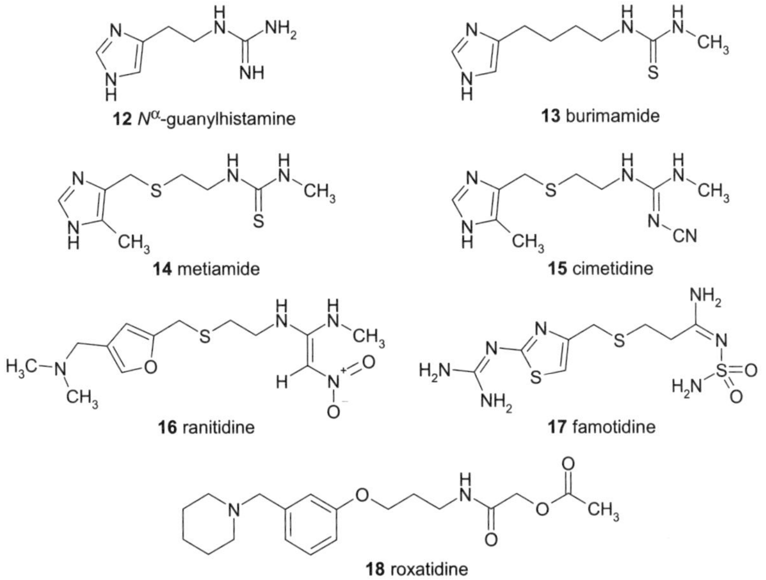

12 N^{α}-guanylhistamine

13 burimamide

14 metiamide

15 cimetidine

16 ranitidine

17 famotidine

18 roxatidine

Scheme 7.7 Structures of some key compounds in the development of H_2-receptor antagonists.

effect remained. By replacing the strongly basic guanidino function by a neutral thiourea, burimamide (**13**), was obtained. Although very active, this compound was rejected due to its low oral bioavailability. The addition of a methyl group in position 4 of the imidazolic ring, followed by the introduction of an electron-withdrawing sulfur atom in the side chain, finally led to a compound that was both very active and less highly ionized, properties that improved its absorption by the oral route. The derivative thus obtained, metiamide (**14**), was excellent and, moreover, 10 times more potent than burimamide. Because of its thiourea grouping, however, metiamide exhibited undesirable side effects (agranulocytosis, nephrotoxicity) that would limit its clinical use. Replacement of the thiourea by an isosteric group that had the same pK_a (*N*-cyanoguanidine) finally led to *cimetidine* (**15**), which became a medicine of choice for the treatment of gastric ulcers. It later appeared that the imidazolic ring that was present in histamine and in all H_2 antagonists discussed hitherto was not essential for the H_2 antagonistic activity. Thus, ranitidine (**16**), which possesses a furan ring, has proved to be even more active than cimetidine. The same proved to be true for famotidine (**17**) and roxatidine (**18**).

7.3.3 Rimonabant and Obesity

Among drug-taking communities it was well known that smoking of marijuana was followed by an increase in appetite. One consequence of this was the widespread use of smoked marijuana as a treatment for HIV-related anorexia and weight loss. This activity also suggested that orally administrated synthetic tetrahydrocannabinol (THC, dronabinol, Marinol), the main psychoactive ingredient in marijuana, could be used as an appetite stimulant. The treatment has proved, in fact, to be efficacious.[25] Numerous pharmacological studies confirmed that the CNS receptor involved was the cannabinoid CB1 receptor. Exogenously administered cannabinoid receptor agonists such as Δ^9-tetrahydrocannabinol stimulate food consumption in animals as well as in humans. Endogenous cannabinoid receptor agonists are present in the brain, and the level of these agonists in the brain increases with greater demand for food by rodents. Conversely, administration of CB1 receptor antagonists was hypothesized as being a means of reducing food intake and represented a treatment for obesity. This hypothesis prompted the scientists of SanofiAventis to start a screening program for CB1 receptor antagonists. This search culminated in the discovery of compound SR 141716A (rimonabant; **19**, Scheme 7.8). This selective CB1 receptor antagonist is in phase III clinical trials for the treatment of obesity and has been found to decrease appetite and body weight in humans.

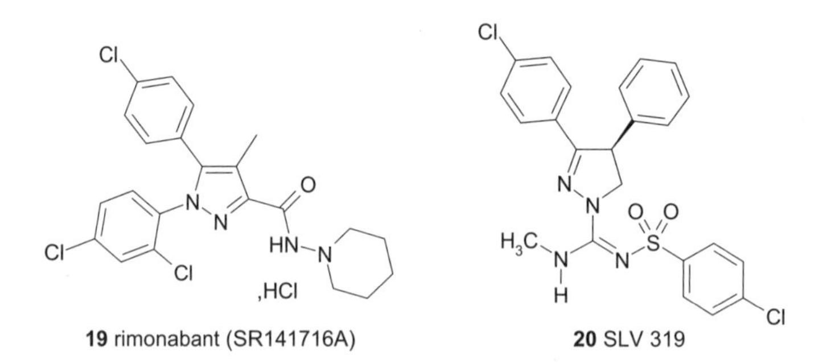

Scheme 7.8 Structures of the cannabinoid receptor CB1 antagonists rimonabant (**19**) and SLV 319 (**20**).

Cannabinoid CB1 receptor antagonists are currently the subject of intensive research, due to their highly promising therapeutic prospects. Novel chemical entities that have CB1 antagonistic properties have recently been disclosed by several pharmaceutical companies and some academic research groups. Some of these entities are close structural analogs of the lead compound rimonabant. A considerable number of these CB1 antagonists are bioisosteres that derived from rimonabant by the replacement of the pyrazole moiety with an alternative heterocycle. As well as these achiral compounds, Solvay Pharmaceuticals have disclosed a novel class of chiral pyrazolines [such as compound SLV 319 (**20**), (Scheme 7.8)] which are potent and CB1/CB2 subtype-selective cannabinoid receptor antagonists.[26]

7.4 CONTRIBUTIONS FROM CLINICAL INVESTIGATIONS

The clinical observation of entirely unexpected side effects constitutes a nearly inexhaustible source of avenues to follow in the search for lead compounds. Indeed, in addition to the desired therapeutic action, most drugs possess side effects that are either accepted from the beginning as a necessary evil or are recognized only after some years of utilization. When side effects themselves are of medical interest, the disassociation of the primary effect from the side-effect activities may become an objective: Enhance the activity that was originally considered as secondary and minimize or eliminate the activity that was initially dominant. Promethazine, for example, an antihistaminic derivative of phenothiazine, has important undesirable sedative effects. To their credit, clinicians such as Laborit[27] have promoted the utilization of this side effect and have directed research toward better profiled analogs. The emergence of chlorpromazine, the prototype for a new therapeutic series, the neuroleptics (the existence of which was previously unsuspected and which have revolutionized psychiatric practice) was the result of this impulse.[28,29] Countless other examples can be found in the literature, such as the hypoglycemic effect of some antibacterial sulfamides, the uricosuric effect of the coronaro-dilating drug benziodarone, the antidepressant effect of isoniazide, an antituberculosis drug, and the hypotensive effect of β-blocking agents.

In some cases a new clinical activity observed in an existing drug is sufficiently potent and of sufficient interest to justify the immediate use of the drug in the new indication, as will be illustrated hereafter. Amiodarone (**21**, Scheme 7.9), for example, was introduced as a coronary dilator for angina. Concern about corneal deposits, discoloration of skin that was exposed to sunlight and thyroid disorders led to the drug being withdrawn in 1967. In 1974, however, it was discovered that amiodarone was highly effective in the treatment of a rare type of arrhythmia known as the Wolff–Parkinson–White syndrome. Accordingly, amiodarone was reintroduced specifically for this purpose.[30]

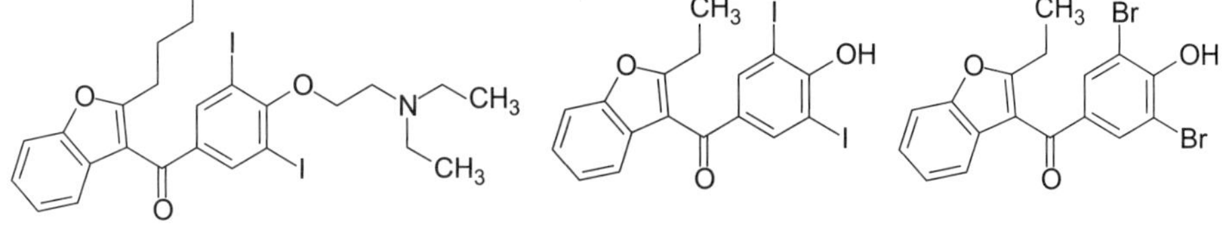

Scheme 7.9 Structures of the benzofuranic arones.

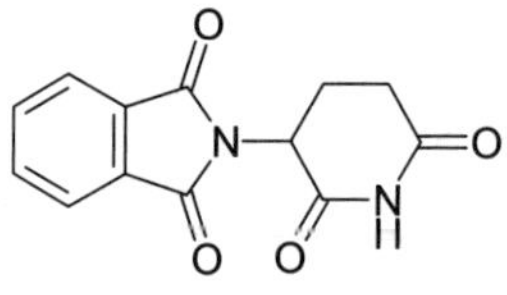

24 thalidomide

Scheme 7.10 Structure of thalidomide. The marketed compound is the racemate.

Benziodarone (**22**), used initially in Europe as a coronary dilator, later proved to be a useful uricosuric agent. It is at present withdrawn from the market as a result of there having been several cases of jaundice associated with its use.[30] The corresponding brominated analog, benzbromarone (**23**), was marketed specifically for its uricosuric properties.

Thalidomide (**24**), was initially launched as a sedative/hypnotic drug (Scheme 7.10), but withdrawn because of its extreme teratogenicity. Under restricted conditions (no administration during pregnancy or to any woman of childbearing age), it has found a new use as an immunomodulator. In particular, it appears to be efficacious in the treatment of erythema nodosum leprosum, a possible complication in the chemotherapy of leprosy.[31]

In 2001, the antimalarial drug quinacrine (**25**) and the antipsychotic drug chlorpromazine (**8**, Scheme 7.11) were shown to inhibit prion infections in cells. Prusiner and coworkers[32] identified the drugs independently and found that they inhibit conversion of normal prion protein into infectious prions and clear prions from infected cells. Both drugs can cross over from the bloodstream into the brain, where prion diseases are localized. In many therapeutic families a new generation of compounds is born from the previous generation. In the past this occurred for the sulfonamides, penicillins, steroids, prostaglandins, and tricyclic psychotropics families, and one can draw true genealogical trees that represent the progeny of these discoveries. More recent examples are to be found in fields that include statins, ACE inhibitors, and in the family of histaminergic H_2 antagonists.

Research programs based on the exploitation of side effects are of great interest in the discovery of new avenues insofar as they depend on information about activities that have been *observed directly in humans* and not in animals. On the other hand, they enable new therapeutic activities to be detected *even when no pharmacological models exist in animals*.

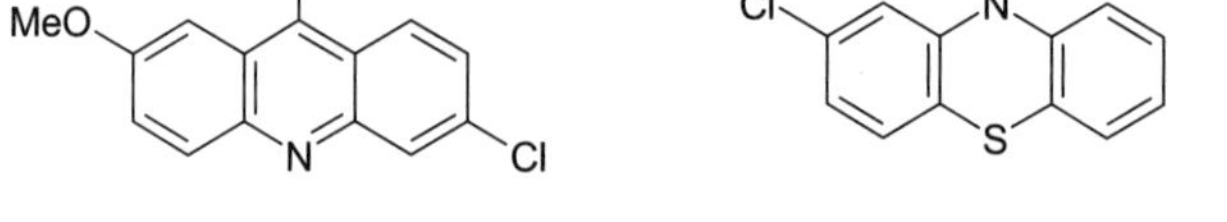

25 quinacrine

8 chlorpromazine

Scheme 7.11 Old drugs, new use: the antimalarial drug quinacrine (**25**) and the antipsychotic drug chlorpromazine (**8**) can inhibit prion infections.

7.5 NEW LEADS FROM OLD DRUGS: THE SOSA APPROACH

The SOSA (selective optimization of side activities) approach represents an original alternative to HTS.[33–35] It involves two steps:

1. Screening of newly identified pharmacological targets using a limited set (approximately 1000 compounds) of well-known drug molecules for which bioavailability and toxicity studies have already been performed and which have proven to be useful in human therapy.
2. Once a hit has been obtained with a given drug molecule, the task is then to prepare analogs of this molecule to transform the observed side activity into the main effect and to reduce significantly or eliminate the initial pharmacological activity.

7.5.1 Rationale

The rationale behind the SOSA approach is based on the fact that in addition to their main activities, almost all drugs used in human therapy exhibit one or more side effects. In other words, as well as being capable of exerting a strong interaction with the main target, they also exert weaker interactions with other biological targets. Most of these targets bear no relation to the primary therapeutic activity of the compound. The objective is then to move toward a reversal of these affinities so that the side effect identified becomes the main effect, and vice versa.

A chemical library that is available for the SOSA approach is the Prestwick Chemical Library (Prestwick Chemical, Inc., 1825 K Street NW, Suite 1475, Washington, DC 20006-1202; www.prestwickchemical.com). It contains 1120 biologically active compounds that exhibit a high degree of chemical and pharmacological diversity as well as known bioavailability and safety in humans. Over 85% of the compounds are well-established drugs, with 15% of them being bioactive alkaloids. For scientists who have an interest in drug similarities, this library most certainly fulfills their requirements in the quest for druglike leads!

7.5.2 Examples

Selective Ligands for the Endotheline ET_A Receptors The development of these ligands by scientists from Bristol-Myers Squibb provide us with an illustration of the SOSA approach.[36,37] Starting with an in-house library, the antibacterial compound sulfathiazole (**26**, Scheme 7.12) was an initial, but weak, hit ($IC_{50} = 69\,\mu M$). Testing of related sulfonamides identified the more potent sulfisoxazole (**27**) ($IC_{50} = 0.78\,\mu M$). Systematic variation finally led to the potent and selective ligand BMS-182874. In vivo, this compound was orally active and produces a long-lasting hypotensive effect.

Further optimizations that were guided by pharmacokinetic considerations led the BMS scientists to replace the naphtalene ring with a diphenyl system.[39] Among the compounds prepared **29** (BMS-193884, ET_A $K_i = 1.4\,nM$; ET_B $K_i = 18{,}700\,nM$) showed promising hemodynamic effects in a phase II clinical trial on congestive heart failure. More recent studies have led to an extremely potent antagonist (**30**; BMS-207940 ET_A $K_i = 10\,pM$), which exhibits an 80,000-fold selectivity ratio for ET_A versus ET_B. The bioavailability of **30** is 100% in rats and it exhibits oral activity even at a $3\,\mu M/kg$ dosage.[39]

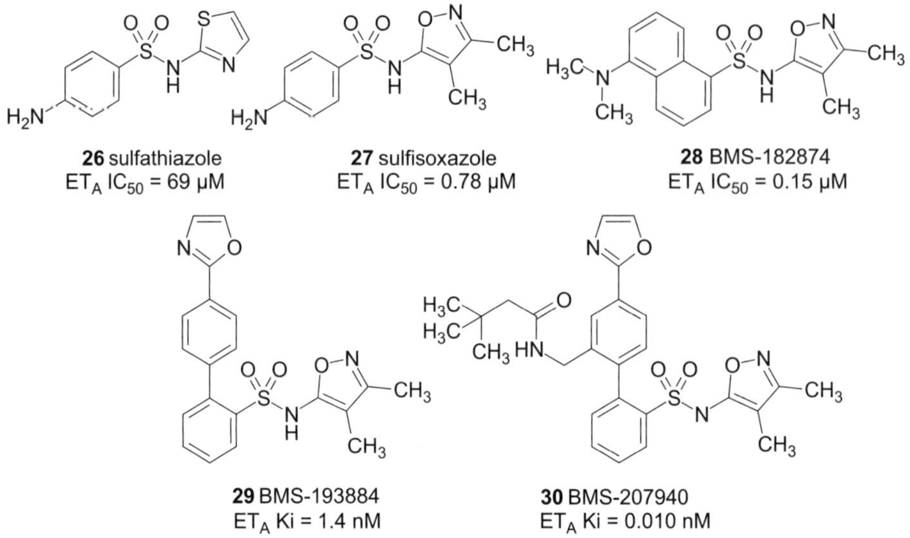

26 sulfathiazole
ET_A IC_{50} = 69 μM

27 sulfisoxazole
ET_A IC_{50} = 0.78 μM

28 BMS-182874
ET_A IC_{50} = 0.15 μM

29 BMS-193884
ET_A Ki = 1.4 nM

30 BMS-207940
ET_A Ki = 0.010 nM

Scheme 7.12 A successful SOSA approach was used in the identification of the antibacterial sulfonamide sulfathiazole (**26**) as a ligand for the endothelin ETA receptor and in its optimization to provide the selective and potent compounds BMS-182874 (**28**), BMS-193884 (**29**), and BMS-207940 (**30**). (From ref. 38 and 39.)

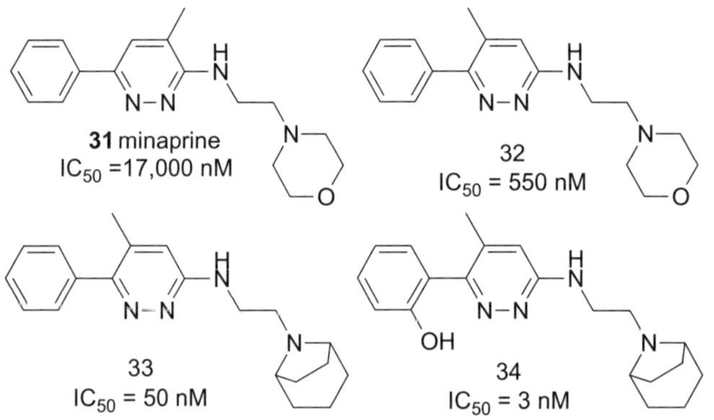

31 minaprine
IC_{50} =17,000 nM

32
IC_{50} = 550 nM

33
IC_{50} = 50 nM

34
IC_{50} = 3 nM

Scheme 7.13 Progressive transformation from minaprine to a potent and selective partial muscarinic M1 agonist. (From refs. 40 to 42.)

Cholinergic Agonists In our second example, the starting lead was the antidepressant minaprine (**31**, Scheme 7.13). In addition to properties of reinforcing serotonergic and dopaminergic transmission, this aminopyridazine possesses a weak affinity for muscarinic M_1 receptors ($K_i = 17\,\mu M$). Simple chemical variation was used to eliminate dopaminergic and serotonergic activities and to boost the cholinergic activity to nanomolar concentrations.[40–42]

Acetylcholinesterase Inhibitors Starting from the same minaprine lead, since this molecule is recognized by the acetylcholine receptors, it was conceivable that it might

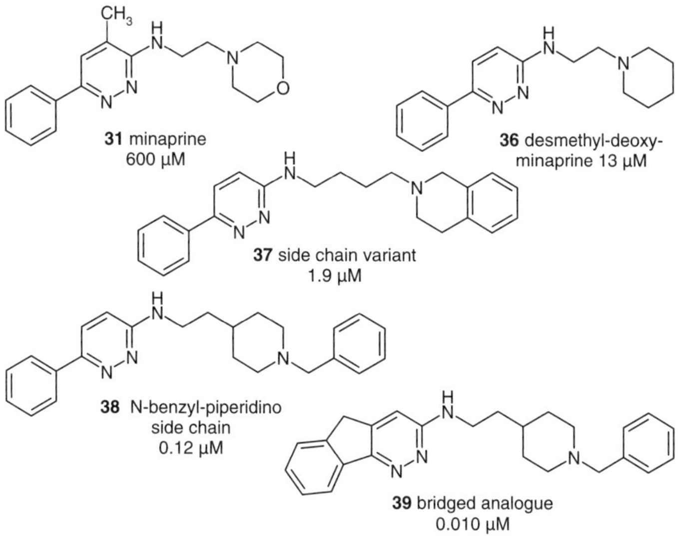

31 minaprine 600 μM

36 desmethyl-deoxy-minaprine 13 μM

37 side chain variant 1.9 μM

38 N-benzyl-piperidino side chain 0.12 μM

39 bridged analogue 0.010 μM

Scheme 7.14 IC_{50} values for acetylcholinesterase inhibition (electric eel enzyme). (From refs. 43 and 44.)

also be recognized by the acetylcholine enzyme. It turned out that minaprine had only a very weak affinity for acetylcholinesterase (600 μ*M* on electric eel enzyme). Relatively simple modifications, however (creation of a lipophilic cationic head, increasing the length of the side chain, and bridging the phenyl and the pyridazinyl rings) allowed us to obtain nanomolar affinities (Scheme 7.14).[43,44]

CRF Antagonists Another interesting switch involved the progressive transformation from desmethylminaprine (**40**) to the bioisosteric thiadiazole (**41**, Scheme 7.15) and then to bioisosteric thiazoles. Trisubstitution of the phenyl ring and replacement of the aliphatic morpholine by a pyridine led to compound **42**, which exhibited some affinity for the receptor of the 41-amino acid neuropeptide, corticotrophin releasing factor (CRF). Further optimization led to nanomolar CRF antagonists such as **43**.[45,46]

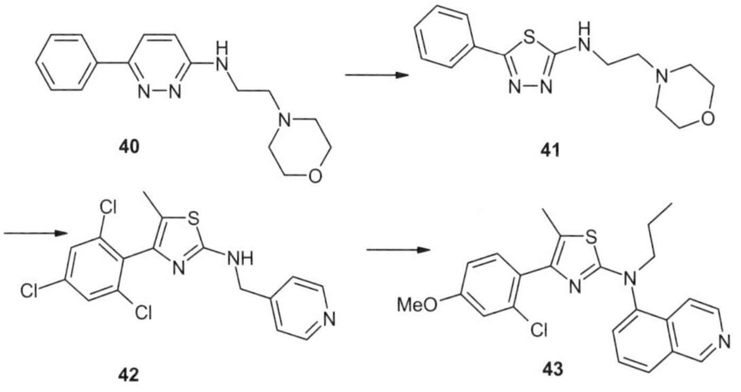

40 **41** **42** **43**

Scheme 7.15 Switch from the antidepressant molecule minaprine to the potent CRF receptor antagonist.[9] (From refs. 45 and 46.)

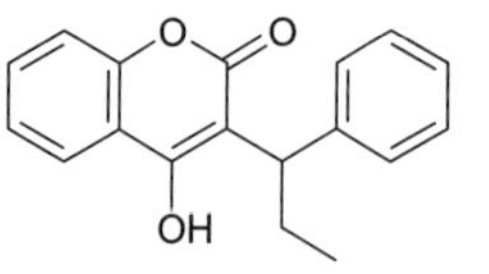

44 phenprocoumon Ki = 1 μM 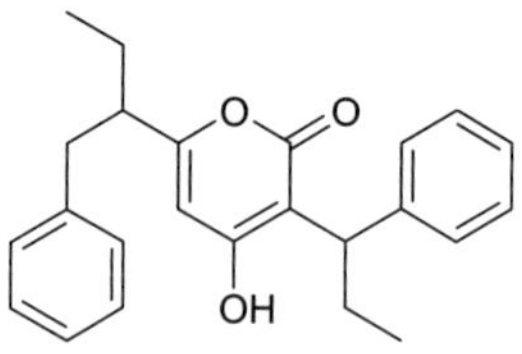 **45** PNU-96-988 Ki = 38 nM

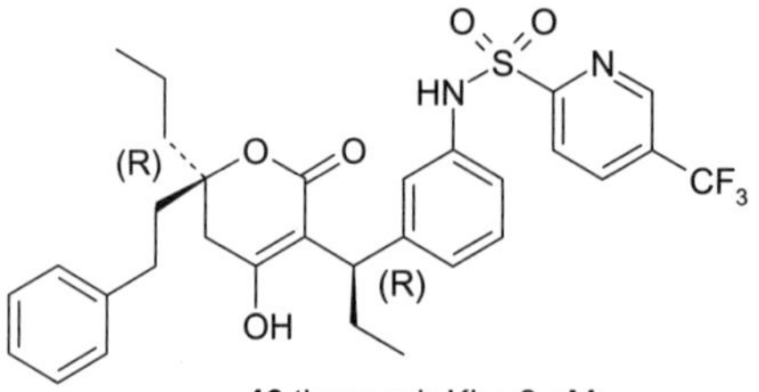

46 tipranavir Ki = 8 pM

Scheme 7.16 The anticoagulant phenprocoumon yields the HIV-protease inhibitor tipranavir.

HIV Protease Inhibitors Yielded by The Anticoagulant Phenprocoumon At Upjohn, phenprocoumon (**44**, Scheme 7.16), used therapeutically as an anticoagulant, was independently discovered to be a moderately active HIV protease inhibitor ($K_i = 1\,\mu M$).[47] Optimization produced the bis-aralkyl-substituted 4-hydroxypyrone (**45**; PNU-96 988, $K_i = 38\,nM$)[47], and finally, the picomolar inhibitor tipranavir (**46**; *R,R*-diastereomer: $K_i = 8\,pM$). Surprisingly, other diastereomers of tipranavir show a very low stereospecificity of drug action; they are also very potent HIV protease inhibitors (*R,S*-diastereomer: $K_i = 18\,pM$; *S,R*-diastereomer: $K_i = 32\,pM$; *S,S*-diastereomer: $K_i = 220\,pM$).[48]

From the Antibiotic Erythromycin to LHRH Antagonists[49] Screening of the Abbott chemical repository identified erythromycin A derivatives that bound to rat LHRH receptor with submicromolar affinity and which exhibited LHRH antagonistic properties. One of the most potent antagonists was an anilinoethyl cyclic carbamate (**47**, Scheme 7.17). Initial SAR studies, guided by overlaying compound **47** with the potent cyclic decapeptide (**48**), led to the *para*-chlorophenyl analog **49**, which exhibited a 20-fold improvement in potency relative to compound **47**. Optimization efforts based mainly on replacement of the cladinose moiety in position 3 led to compound **50**, which has 1 to 2 n*M* affinity for both rat and human LHRH receptors and is a potent in vitro inhibitor of LH release ($pA_2 = 8.76$). In vivo, compound **50** was found to produce a dose-dependent suppression of LH in castrated male rats via both intravenous end oral dosing.

7.5.3 Discussion

The SOSA approach appears to be an efficient strategy for the discovery of drugs, particularly since it is based on the screening of drug molecules and thus yields druglike hits automatically. It can represent an attractive alternative before a costly HTS campaign is started. Once the initial screening has provided a hit, this will be used as the starting point for a drug discovery program. Using both traditional medicinal chemistry and parallel synthesis, the initial side activity is transformed into the main activity, and conversely, the initial main activity is greatly reduced or eliminated. There is a high probability that this strategy will lead to *safe, bioavailable, original, and patentable* analogs.

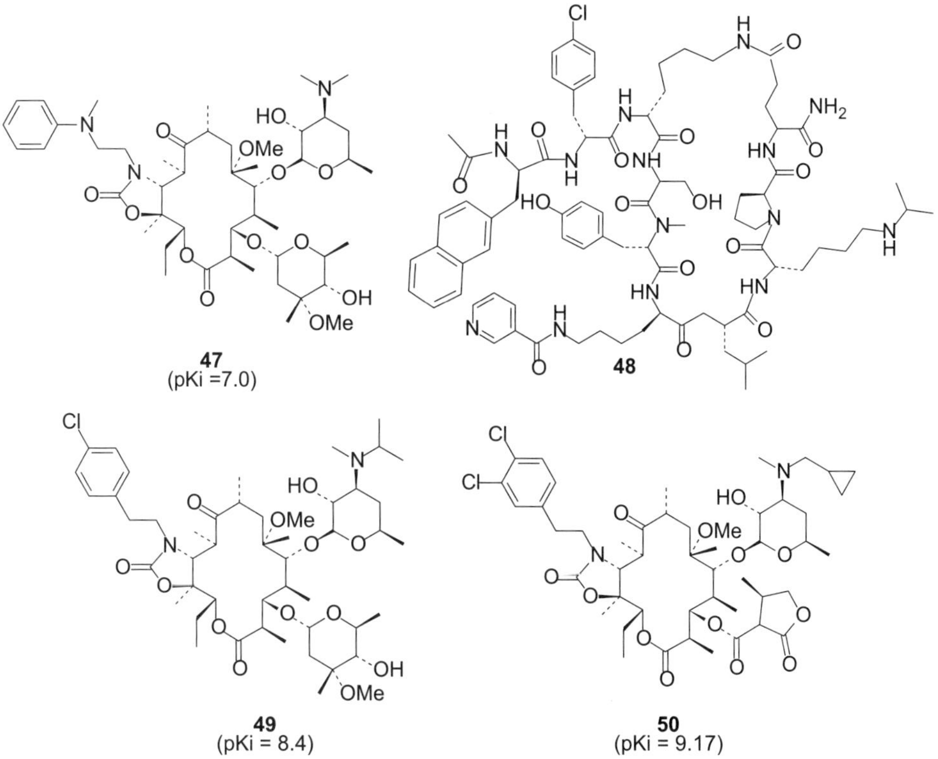

47
(pKi =7.0)

48

49
(pKi = 8.4)

50
(pKi = 9.17)

Scheme 7.17 From derivatives of the antibiotic erythromycin A to LHRH antagonists.

Safety and Bioavailability Over a period of years of practicing SOSA approaches, it has been observed that when analog synthesis is performed with a drug molecule as the lead substance, there is a notably increased probability of obtaining safe new chemical entities. In addition, most of these satisfy Lipinski's,[50] Veber's,[51] Bergström's,[52] and Wenlock's[53] observations in terms of solubility, oral bioavailability, and drug-likeness.

Patentability When a well-known drug hits with a new target, there is a possibility that several hundred or even thousands of analogs of the original drug molecule have already been synthesized by the original inventors and their competitors. These molecules are usually protected by patents or already belong in the public domain. At first glance it would appear that there is a high probability of interference. In fact, during optimization of a therapeutic profile different than that of the original inventors, medicinal chemists rapidly prepare analogs whose chemical structures are notably different from that of the original hit. As an example, a medicinal chemist interested in phosphodiesterases (PDEs) and using diazepam as a lead will inevitably synthesize compounds that are in terms of structure outside the scope of the original patents, precisely because they exhibit predominantly PDE-inhibiting properties and show practically no further affinity for the benzodiazepine receptor (Scheme 7.18).[54]

Originality The screening of a library of several hundred therapeutically diverse drug molecules sometimes ends up with very surprising results. A nice example of unexpected findings resulting from systematic screening is to be found in the tetracyclic compound

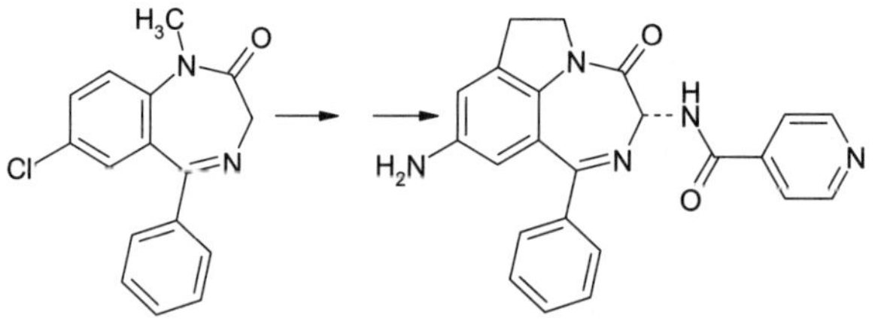

51 diazepam **52** CI-1044

Scheme 7.18 Starting from the tranquillizer diazepam (**51**), the improvement in the PDE (**4**) inhibitory activity leads to structures such as **52** that are sufficiently original to render them patentable.

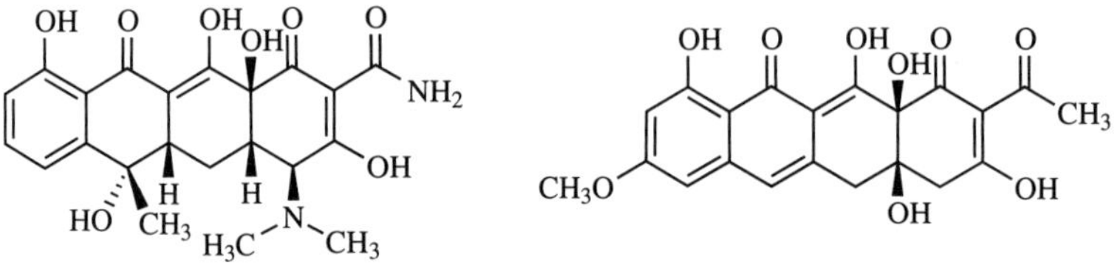

53: tetracycline **54:** BMS-192548

Scheme 7.19 Unexpected CNS activity of the tetracycline analog **54** (BMS-192548). (From ref. 55.)

54 (BMS-192548) extracted from *Aspergillus niger* WB2346 (Scheme 7.19). For any medicinal chemist or pharmacologist, the similarity of this compound to the antibiotic tetracycline is striking. No one, however, would forecast a priori that BMS-192548 would exhibit CNS activities. The compound turns out in fact to be a ligand for neuropeptide Y receptor preparations.[55]

Orphan Diseases As mentioned above, a peculiarity of this type of library that differentiates it from others is that it is made up of compounds that have already been given safely to humans. Thus, if a compound were to hit with sufficient potency on an orphan target, there is a significant chance that it could be tested rapidly on patients for proof of principle. This possibility represents another advantage of the SOSA approach.

7.6 CONCLUSIONS

The relative lack of pharmaceutical creativity over recent years is at least in part attributable to a misleading faith that identification of a new hit or lead molecule is the most important step in a drug discovery program. As mentioned earlier, "hits are not drugs." Problems in later development, such as pharmacokinetic and toxicological profiling (ADME, bioavailability, genotoxicity, etc.) or pharmaceutical formulation (salts, polymorphism, water solubility, etc.) are much too often regarded as being straightforward ancillary activities. It can currently be assumed that most pharmaceutical companies possess dozens of hits and that their main concern is to transform them into valuable drug candidates.

The four alternatives to high-throughput screening that have been proposed here (i.e., analog design, research based on physiopathological hypotheses, exploitation of clinical observation, and the identification of new leads from old drugs) share a precious common advantage: They are all derived from a knowledge of properties that are inherent in existing drug molecules of proven activity and/or usefulness in humans. Many of the drawbacks that are encountered in the development of HTS-derived molecules are therefore bypassed and fewer safety problems should arise. For all of these reasons, it is recommended that pharmaceutical company management adopt these alternatives as fruitful and complementary approaches to HTS.

REFERENCES

1. Ramesha, C. S. How many leads from HTS?—Comment. *Drug Discov. Today* 2000, *5*, 43–44.
2. Liu, K.; Xu, L.; Szalkowski, D.; Li, Z.; Ding, V. Discovery of a potent, highly selective, and orally efficacious small-molecule activator of the insulin receptor. *J. Med. Chem.* 2000, *43*, 3487–3494.
3. Roever, S.; Adam, G.; Cesura, A. M.; Galley, G.; Jenck, F. E. A. High-affinity, non-peptide agonists for the ORL1 (orphanin FQ/nociceptin) receptor. *J. Med. Chem.* 2000, *43*, 1329–1338.
4. Rey, A. *Dictionnaire historique de la langue française*, Dictionnaires Le Robert, Paris, 1992.
5. Schneider, G.; Giller, T.; Neidhart, W.; Schmid, G. "Scaffold-hopping" by topological pharmacophore search: a contribution to virtual screening. *Angew. Chem. Int. Ed.* 1999, *38*, 2894–2896.
6. Balakin, K. V.; Tkachenko, S. E.; Okun, I.; Skorenko, A. V.; Ivanenkov, Y. A.; Savchuk, N. P.; Ivashchenko, A. A.; Nikolsky, Y. Bioisosteric morphing in primary hit optimization. *Chimi. Oggi* (Chem. Today) 2004, 15–18.
7. Buehlmayer, P.; Furet, P.; Cricione, L.; de Gasparo, M.; Whitebread, S.; Schmidlin, T.; Lattman, R.; Wood, J. Valsartan, a potent, orally active angiotensin II antagonist developed from the structurally new amino acid series. *Bioorg. Med. Chem. Lett.* 1994, *4*, 29–34.
8. Croom, K. F.; Keating, G. M. Valsartan: a review of its use in patients with heart failure and/or left ventricular systolic dysfunction after myocardial infarction. *Am. J. Cardiovasc. Drugs* 2004, *4*, 395–404.
9. Bajorath, J. Integration of virtual and high-throughput screening. *Nature Rev. Drug Discov.* 2002, *1*, 882–894.
10. Kubinyi, H.; Müller, G. Chemogenomics and drug discovery. In *Methods and Principles in Medicinal Chemistry*, Mannhold, R., Kubinyi, H., Folkers, G. (Eds), Wiley-VCH, Weinheim, Germany, 2004, p. 463.
11. Kuhn, R. Über die Behandlung depressiver Zuständen mit einem Iminodibenzylderivat (G22355). *Schweiz. Med. Wochenschr.* 1957, *87*, 1135–1140.
12. Kuhn, R. The treatment of depressive states with G22355 (imipramine hydrochloride). *Am. J. Psychiatry* 1958, *115*.
13. Terret, N. K.; Bell, A. S.; Brown, D.; Ellis, P. Sildenafil (Viagra), a potent and selective inhibitor of type 5 cGMP phosphodiesterase with utility for the treatment of male erectile dysfunction. *Bioorg. Med. Chem. Let.* 1996, *6*, 1819–1824.
14. Boolell, M.; Allen, M. J.; Ballard, S. A.; Gepi-Attee, S.; Muirhead, G. J.; Naylor, A. M.; Osterloh, I.; Gingell, C. Sildenafil: an orally activetype 5 cyclic GMP-specific phosphodiesterase inhibitor for the treatment of penile erectile dysfunction. *Int. J. Impot. Res.* 1996, *8*, 47–52.
15. Kling, J. From hypertension to angina to Viagra. *Mod. Drug Discov.* 1998, 31–38.

16. Carlsson, A.; Lindquist, M.; Magnusson, T.; Waldeck, P. On the presence of 3-hydroxytyramine in the brain. *Science* 1958, *127*, 471.

17. Blaschko, H. Metabolism and storage of biogenic amines. *Experientia* 1957, *13*, 9–12.

18. Blaschko, H. The development of current concepts of catecholamine formation. *Pharmacol. Rev.* 1959, *11*, 307–316.

19. Ehringer, H.; Hornykiewicz, O. Verteilung von Noradrenalin und Dopamin (3-hydroxytyramin) im Gehirn des Menschen und ihr Verhalten bei Erkrankungen des extrapyramidalen Systems. *Wien. Klin. Wochenschr.* 1960, *38*, 1236–1239.

20. Birkmayer, W.; Hornykiewicz, O. The L-dihydroxyphenylalanine (L-dopa) effect in Parkinson's syndrome in man: on the pathogenesis and treatment of Parkinson akinesis. *Arch. Psychiatr. Nervenkr. Z. Gesamte Neurol. Psychiatr.* 1962, *203*, 560–574.

21. Barbeau, A. The use of L-dopa in Parkinson's disease: a 20 year follow-up. *Trends Pharmacol. Sci.* 1981, *2*, 297–299.

22. Cotzias, G. C.; Van Woert, M. H.; Schiffer, L. M. Aromatic amino acids and modification of parkinsonism. *New Engl. J. Med.* 1967, *276*, 374–379.

23. Black, J. W.; Duncan, W. A. M.; Durant, J. C.; Ganellin, C. R.; Parsons, M. E. Definition and antagonism of histamine H2-receptors. *Nature* 1972, *236*, 385–390.

24. Ganellin, C. R. Cimetidine. In *Chronicles of Drug Discovery*, Wiley, New York, 1982, pp. 1–38.

25. Abrams, D. I. Medical marijuana: tribulations and trials. *J. Psychoactive Drugs* 1998, *30*, 163–169.

26. Lange, J. H.; Kruse, C. G. Recent advances in CB1 cannabinoid receptor antagonists. *Curr. Opin. Drug Discov. Dev.* 2004, *7*, 498–506.

27. Laborit, H.; Huguenard, P.; Alluaume, R. Un nouveau stabilisateur végétatif, le 4560 R.P. *Presse Méd.* 1952, *60*, 206–208.

28. Thuilier, J. *Les dix ans qui ont changé la folie,* Robert Laffont, Paris, 1981 pp. 253–257.

29. Maxwell, R. A.; Eckhardt, S. B. *Drug Discovery: A Casebook and Analysis*, Humana Press, Clifton, NJ, 1990.

30. Sneader, W. *Drug Prototypes and Their Exploitation,* Wiley, Chichester, West Sussey, England, 1996, p. 242.

31. Iyer, C. G. S.; Languillon, J.; Ramanujam, K. WHO coordinated short-term double-blind trial with thalidomide in the treatment of acute lepra reactions in male lepromatus patients. *Bull. W.H.O.* 1971, *45*, 719–732.

32. Korth, C.; May, B. C. H.; Cohen, F. E.; Prusiner, S. B. Acridine and phenothiazine derivatives as pharmacotherapeutics for prion disease. *Proc. Natl. Acad. Sci. USA* 2001, *98*, 9836–9841.

33. Wermuth, C. G. Search for new lead compounds: the example of the chemical and pharmacological dissection of aminopyridazines. *J. Heterocycl. Chem.* 1998, *35*, 1091–1100.

34. Wermuth, C. G.; Clarence-Smith, K. "Drug-like" leads: bigger is not always better. *Pharm. News* 2000, *7*, 53–57.

35. Wermuth, C. G. The "SOSA" approach: an alternative to high-throughput screening. *Med. Chem. Res.* 2001, *10*, 431–439.

36. Riechers, H.; Albrecht, H.-P.; Amberg, W.; Baumann, E.; Böhm, H.-J.; Klinge, D.; Kling, A.; Muller, S.; Raschack, M.; Unger, L.; Walker, N.; Wernet, W. Discovery and optimization of a novel class of orally active non-peptidic endothelin-A receptor antagonists. *J. Med. Chem.* 1996, *39*, 2123–2128.

37. Poroikov, V.; Akimov, D.; Shabelnikova, E.; Filimonov, D. Top 200 medicines: can new actions be discovered through computer-aided prediction? *SAR QSAR Environ. Res.* 2001, *12*, 327–344.

38. Stein, P. D.; Hunt, J. T.; Floyd, D. M.; Moreland, S.; Dickinson, K. E. J.; Mitchell, C.; Liu, E. C.-K.; Webb, M. L.; Murugesan, N.; Dickey, J.; McMullen, D.; Zhang, R.; Lee, V. G.; Serdino,

R.; Delaney, C.; Schaeffer, T. R.; Kozlowski, M. The discovery of sulfonamide endothelin antagonists and the development of the orally active ET_A antagonist 5-(dimethylamino)-*N*- (3,4-dimethyl-5-isoxazoly1)-l-naphthalenesulfonamide. *J. Med. Chem.* 1994, *37*, 329–331.

39. Murugesan, N.; Gu, Z.; Spergel, S.; Young, M.; Chen, P.; Mathur, A.; Leith, L.; Hermsmeier, M.; Liu, E. C.-K.; Zhang, R.; Bird, E.; Waldron, T.; Marino, A.; Koplowitz, B.; Humphreys, W. G.; Chong, S.; Morrison, R. A.; Webb, M. L.; Moreland, S.; Trippodo, N.; Barrish, J. C. Biphenylsulfonamide endothelin receptor antagonists. 4. discovery of *N*-[[(4,5-dimethyl-3-iso xazolyl)amino]sulfonyl]-4-(2-oxazolyl)[1,1′-biphenyl]-2-yl]methyl]-*N*,3,3-trimethylbutanamide (BMS-207940, a highly potent and orally active ET_A selective antagonist. *J. Med. Chem.* 2003, *46*, 125–137.

40. Wermuth, C. G.; Schlewer, G.; Bourguignon, J.-J.; Maghioros, G.; Bouchet, M.-J.; Moire, C.; Kan, J.-P.; Worms, P.; Bizière, K. 3-Aminopyridazine derivatives with atypical antidepressant, serotonergic and dopaminergic activities. *J. Med. Chem.* 1989, *32*, 528–537.

41. Wermuth, C. G. Aminopyridazines: an alternative route to potent muscarinic agonists with no cholinergic syndrome. *Il Farmaco* 1993, *48*, 253–274.

42. Wermuth, C. G.; Bourguignon, J.-J.; Hoffmann, R.; Boigegrain, R.; Brodin, R.; Kan, J.-P.; Soubrié, P. SR 46559A and related aminopyridazines are potent muscarinic agonist with no cholinergic syndrome. *Bior. Med. Chem. Lett.* 1992, *2*, 833–836.

43. Contreras, J.-M.; Rival, Y. M.; Chayer, S.; Bourguignon, J. J.; Wermuth, C. G. Aminopyridazines as acetylcholinesterase inhibitors. *J. Med. Chem.* 1999, *42*, 730–741.

44. Contreras, J.-M.; Parrot, I.; Sippl, W.; Rival, R. M.; Wermuth, C. G. Design, Synthesis and structure-activity relationships of a series of 3-[2-(1-benzylpiperidin-4-yl)ethylamino]pyridazine derivatives as acetylcholinesterase inhibitors. *J. Med. Chem.* 2001, *44*, 2707–2718.

45. Gully, D.; Roger, P.; Valette, G.; Wermuth, C. G.; Courtemanche, G.; Gauthier, C. Dérivés alkylamino ramifiés du thiazole, leurs procédés de préparation et les compositions pharmaceutiques qui les contiennent; Elf-Sanofi: French Demande 9207736; June 24, 1992.

46. Gully, D.; Roger, P.; Wermuth, C. G. 4-Phenyl-aminothiazole derivatives: method for preparing same and pharmaceutical compositions containing said derivatives; Sanofi: World patent, Jan. 9, 1997.

47. Thaisrivongs, S.; Tomich, P. K.; Watenpaugh, K. D.; Chong, K. T.; Howe, W. J.; Yang, C. P.; Strohbach, J. W.; Turner, S. R.; McGrath, J. P.; Bohanon, M. J.; Lynn, J. C.; Mulichak, A. M.; Spinelli, P. A.; Hinshaw, R. R.; Pagano, P. J.; Moon, J. B.; Ruwart, M. J.; Wikinson, K. F.; Rush, B. D.; Zipp, G. L.; Dalga, R. J.; Schwende, F. J.; Howard, G. M.; Padbury, G. E.; Toth, L. N.; Zhao, Z.; Koeplinger, K. A.; Kakuk, T. J.; Cole, S. L.; Zaya, R. M.; Piper, R. C.; Jeffrey, P. Structure-based design of HIV protease inhibitors: 4-hydroxycoumarins and 4-hydroxy-2-pyrones as non-peptidic inhibitors. *J. Med. Chem.* 1994, *37*, 3200–3204.

48. Turner, S. R.; Strohbach, J. W.; Tommasi, R. A.; Aristoff, P. A.; Johnson, P. D.; Skulnick, H. I.; Dolak, L. A.; Sees, T. E. P.; Tomich, P. K.; Bohanon, M. J.; Horng, M. M.; Lynn, J. C.; Chong, K. T.; Hinshaw, R. R.; Watenpaugh, K. D.; Janakiraman, M. N.; Thaisrivongs, S. Tipranavir (PNU-140690): a potent, orally bioavailable nonpeptidic HIV protease inhibitor of the 5,6-dihydro-4-hydroxy-2-pyrone sulfonamide class. *J. Med. Chem.* 1998, *41*, 3467–3476.

49. Randolph, J. T.; Waid, P.; Nichols, C.; Sauer, D.; Haviv, F.; Diaz, G.; Bammert, G.; Besecke, L. M.; Segreti, J. A.; Mohning, K. M.; Bush, E. N.; Wegner, C. D.; Greer, J. Nonpeptide luteinizing hormone-releasing hormone antagonists derived from erythromycin A: design, synthesis, and biological activity of cladinose replacement analogues. *J. Med. Chem.* 2004, *47*, 1085–1097.

50. Lipinski, C. A.; Lombardo, F.; Dominy, B. W.; Feeney, P. J. Experimental and computational approches to estimate solubility and permeability in drug discovery and development settings. *Adv. Drug. Deliv. Rev.* 2001, *46*, 3–26.

51. Veber, D. F.; Johnson, S. R.; Cheng, H. Y.; Smith, B. R.; Ward, K. W.; Kopple, K. D. Molecular properties that influence the oral bioavailability of drug candidates. *J. Med. Chem.* 2002, *45*, 2615–2623.

52. Bergström, C. A. S.; Strafford, M.; Lazorova, L.; Avdeef, A.; Luthman, K.; Artursson, P. Absorption classification of oral drugs based on molecular surface properties. *J. Med. Chem.* 2003, *46*, 558–570.

53. Wenlock, M. C.; Austin, R. P.; Barton, P.; Davis, A. M.; Leeson, P. D. A comparison of physicochemical property profiles of development and marketed oral drugs. *J. Med. Chem.* 2003, *46*, 1250–1256.

54. Burnouf, C.; Auclair, E.; Avenel, N.; Bertin, B.; Bigot, C.; Calvet, A.; Chan, K.; Durand, C.; Fasquelle, V.; Féru, F.; Gilbertsen, R.; Jacobelli, H.; Kebsi, A.; Lallier, E.; Maignel, J.; Martin, B.; Milano, S.; Ouagued, M.; Pascal, Y.; Pruniaux, M.-P.; Puaud, J.; Rocher, M.-N.; Terasse, C.; Wrigglesworth, R.; Doherty, A. M. Synthesis, structure-activity relationships, and pharmacological profile of 9-amino-4-oxo-1-phenyl-3,4,6,7-tetrahydro[1,4]diazepino[6,7,1-*hi*]indoles: discovery of potent, selective phosphodiesterase type 4 inhibitors. *J. Med. Chem.* 2000, *43*, 4850–4867.

55. Shu, Y. Z.; Cutrone, J. Q.; Klohr, S. E.; Huang, S. BMS-192548, a tetracyclic binding inhibitor of neuropeptide Y receptors, from *Aspergillus niger* WB2346. II. Physico-chemical properties and structural characterization. *J. Antibiot.* 1995, *48*, 1060–1065.

8

PROTEOMICS AND DRUG DISCOVERY

Susan Dana Jones
BioProcess Technology Consultants, Inc.
Acton, Massachusetts

Peter G. Warren
Independent Biotechnology Consultant
Lexington, Massachusetts

8.1 INTRODUCTION

Drug discovery is a process that is composed of many essential steps. A key component of drug discovery is the identification of a therapeutically relevant molecular target whose activation or inhibition by a pharmaceutical agent will have an impact on the state or progression of a disease. For example, inhibition of the proton pump in the gastric mucosa is correlated directly with reduced stomach acidity and therefore reduced pain in the stomach and esophagus caused by extended low-pH conditions. Direct or indirect inhibition of this proton pump is the target of a number of successful drugs developed by companies such as Astra Zeneca, Abbott, Altana Pharma, and others. The challenge for new drug discovery efforts is to identify a novel molecular target for therapeutic intervention. In recent years, genomics and proteomics have been hailed as technologies whose application to drug discovery will revolutionize and enable novel therapeutic concepts through identification of previously unknown disease-associated targets. Herein, the tenets of these technologies are reviewed along with the ways in which data generated through genomics, proteomics, and other related approaches are currently used to improve the success and efficiency of drug discovery. More advanced applications of these technologies are on the horizon for drug discovery, and these are reviewed here as well. Throughout, the primary focus is on proteomics.

Drug Discovery and Development, Volume 1: Drug Discovery, Edited by Mukund S. Chorghade

8.2 DRUG DISCOVERY PROCESS

8.2.1 Process Overview

Target Identification and Validation The process of drug discovery starts long before the screening of compound libraries for molecules that bind to and affect the action of a protein associated with disease. Drug discovery starts with the identification of the disease-associated protein or molecular target. In the past, identification of molecular targets was a painstakingly slow process, generally carried out by investigators who could spend entire careers studying one pathway or one protein that is involved in some aspect of human development, metabolism, or disease. In the course of defining the role of a given protein in the cellular life cycle, the direct relevance of this protein to a disease state would often become clear. By removing or altering the target's function from the cell, the target would then be validated as a suitable target for drug screening. In recent times, methods to identify panels of targets that are associated with particular pathways or disease states have emerged, and the acceleration of target identification has led to a bottleneck in target validation. Methods described in this chapter are being deployed to reduce the bottleneck and increase the availability of new, useful, disease-associated molecular targets for development of new classes of effective therapeutic compounds.

Screening for Hits Once a disease-associated molecular target has been identified and validated in disease models, screening for a selective and potent inhibitor (or activator) of the target is the next step. Libraries of compounds that are either synthetic chemicals, peptides, natural or engineered proteins, or antibodies are exposed to the target in a manner that will detect and isolate those members of the library that interact with and, preferably, have an effect on the target. The compounds selected are called *hits*. Initially, screening can be performed by searching for compounds that bind to the target, but binding is not sufficient for therapeutic activity. More recent screening procedures include an activity-based readout as part of the initial screening assay. For example, if the goal is to inhibit a protein that is involved in activating the expression of a particular gene or set of genes, the assay can include a readout to determine if the expression of the gene is reduced by the compound. Such assays can be cell-based, but more often they are enzymatic assays that can be performed in a high-throughput manner.

Lead Optimization Once the initial screening is performed, a large collection of hits is obtained. These hits are then evaluated and the best ones identified in a process known as *lead optimization*. During this stage, scientists determine which, if any, of the hits selected has the appropriate properties to justify continued expenditure of resources on the development of the compound as a clinical candidate. These properties include ease of synthesis; adherence to the Lipinsky rules, which describe chemical characteristics that are predictive of biodistribution and in vivo activity; specificity for the target; and efficacy in the disease. The hits discovered during the screening process are therefore characterized by a variety of biochemical, biophysical, and biological methods to narrow the set down to a handful of compounds. Assays that measure each compound's activity directly are a useful first step at determining potency and can enable the most effective compounds to be identified. The resulting smaller set of hits, or leads, is then tested in more rigorous models of disease, including either cell-based or animal models of the disease.

Pharmacology and Toxicology Once one or more potential lead compounds emerge from the lead optimization efforts, the compound must be evaluated in multiple disease models, and if possible compared to existing therapies for the same disease. Further, properties of the compound are studied when delivered into a living organism. The classic set of properties that must be appropriate include Absorption (through the intended route of administration), distribution (what organs does it end up in?), metabolism (what are the by-products of cellular metabolism of the compound, and what potential effects do these compounds have on the organism and disease?), and excretion (how is it processed and eliminated from the body?). Collectively known as ADME, this set of properties is essential to evaluate in multiple species, in addition to measuring efficacy in sophisticated disease models. Design of ADME and efficacy experiments is a crucial activity in drug discovery and development.

All drugs that are intended for human clinical applications must be tested for toxicity using the same material that will be used in humans, manufactured by the same process. Toxicology testing is a highly regulated process that is governed by regulatory authorities such as the U.S. Food and Drug Administration (FDA). Developing a manufacturing process for a drug (which is outside the scope of this chapter) and executing the necessary toxicology program are very expensive and are therefore usually performed on only one lead compound identified during lead optimization and subsequent efficacy testing. Only if the compound fails at this stage will another compound from the same screening (if possible) be advanced into preclinical development. Different companies, and in fact countries, have different standards for which products move into human clinical testing, but in all cases the product's safety must be adequately demonstrated before regulatory approval can be obtained for advancing the product into humans.

Clinical Trials Human clinical testing follows an established process in most regulated countries worldwide. If possible, the initial administration of a compound to a human subject is performed on people who do not have the disease indication for which the product is designed. The compound is administered at a dose that is significantly lower than the intended therapeutic dose, and the trial subjects are monitored for any signs of toxicity. Most often, these adverse events or side effects consist of headache, fever, nausea, or other discomforts. If the trial subjects receiving the initial low dose have no or minimal side effects, the next cohort of subjects is treated with a higher dose. Using this dose escalation method, the aim is to determine the maximum tolerated dose (MTD) or to determine the safety of a dose level that is known to generate therapeutic benefit. Depending on the disease indication, the tolerance for adverse reactions is different. For example, almost all chemotherapeutic agents for cancer are significantly toxic, but many cancer patients are willing to suffer these effects in order to have a chance to overcome their disease and continue to live without cancer. Due to their known toxicity, oncology drugs are therefore rarely tested in normal volunteers, but in fact are often tested initially in patients who have failed all other chemotherapy or other treatment regimens and have no other option for survival. Chronic, nonfatal diseases, however, are not usually treated with agents that have severe side effects because the benefit of the drug does not outweigh the risks and discomfort of the side effects.

Once the MTD has been determined, drugs usually move into phase II testing, in which the drug's efficacy is determined in a small select group of patients who have the disease for which the drug was developed. These trials can also include a range of doses and dosing regimens (modes of delivery, frequency, etc.), in order to measure the clinical efficacy of different doses. Designing phase II trials and choosing endpoints or objectives for the therapy that will accurately reflect the compound's efficacy is a very demanding process, and many compounds can fail at this stage. Only 60% of drugs that enter phase II successfully complete this stage and move on to the pivotal phase III trial.

Phase III is considered pivotal because the drug is administered to a much larger group of patients and is evaluated for efficacy with greater rigor than in phase II. These trials can cost millions of dollars and therefore are attempted only when the phase II results are highly convincing. There are regulatory hurdles that must be crossed to enter phase III. The drug product that is used in phase III must be manufactured exactly as the first commercial product will be made. The facility or reactor that is used, the process, the analytical methods, and the formulation and vialing are all identical to the intended final product that will be made and launched upon product approval. Following the successful completion of phase III clinical testing, the owner of the compound then files an application with regulatory authorities in various countries, such as the U.S. FDA, for permission to sell the compound for the intended therapeutic indication. The FDA or other agency must then grant a license to market the compound, and the new drug is launched onto the market.

8.2.2 Motivation for Improvement

The drug discovery and development process described above follows a logical and linear path from target identification through completion of clinical trial and submission of an application to the regulatory authorities seeking approval to market a drug. However, this process is lengthy, cumbersome, and most important, generates compounds that are more likely than not to fail during clinical testing. The major hurdles in drug discovery and development today are:

- *Time.* From the initial discovery and validation of a molecular target to final marketing of an effective drug can take 10 to 15 years.
- *Efficiency.* a significant portion of the high cost of drug development is due to the very high attrition rate of compounds that enter the clinic; only 12% succeed.
- *Expense.* Drug development can cost up to $800 million per successful compound.

Therefore, improved clinical outcomes of therapeutic candidates at all stages of development would contribute significantly to reducing development costs and improving the number of candidates that succeed in obtaining approval for commercialization. New methods that enable better target identification and validation, a deeper understanding of total systems biology and the implications for any specific drug or target in development, and methods to understand the nature of molecular targets to enable more effective screening and lead identification would be extremely valuable in meeting the needs for cost-effective development of new and useful therapeutics for human disease. The emerging technologies described in this chapter aim to address the many aspects of drug development and to enable more effective understanding of targets and compounds through high-throughput high-content analysis.

8.3 HIGH-THROUGHPUT SCREENING APPROACHES TO DRUG DISCOVERY

High-throughput screening (HTS) is a term that became common in drug discovery within the past decade. The term refers to the ability to search through large compound libraries to identify those compounds that have a desired binding or activity property. Together with combinatorial chemistry, in which large chemical or biological libraries could be designed and built, HTS was envisioned as the solution to a shrinking pool of new therapeutic compounds in development. Unfortunately, the actual mechanics of HTS did not enable screening of large libraries in a reasonable amount of time, and therefore although more compounds were being screened, fewer hits were identified. For chemical libraries, the trend in the recent past has been to design smaller libraries focused on particular families of molecular targets.

During the screening process, the library is exposed to a protein target or, in some cases, a cell, and those hits that bind and/or act on the target are identified. Most screens are searching for an inhibitor of the target, and using this approach many inhibitors can readily be discovered. The problem that is faced in drug discovery is not the output of the screening procedures, but the failure of many of the compounds in subsequent efficacy studies in cells or animals. Failure can occur for many reasons, but often the cause of compound failure in a disease model is that the compound is only screened against a single target and in fact has cross-reactivity to other proteins that may be necessary for a completely different function in the organism. Understanding the exact structure and function of the target and of those molecules that are most closely related to it is the first step in increasing the identification of hits whose preclinical and clinical performance is more likely to be successful. The emerging technologies described in the rest of this chapter are those technologies that are designed to meet this need.

8.4 EMERGING TECHNOLOGIES AND APPROACHES: SCALE AND SPEED

We now turn our attention to newer, high-throughput technologies that show considerable promise for improving and accelerating drug discovery and development. The biggest shifts we consider here are in scale and focus: from primarily low-scale biological, biochemical, and pharmacological methods to high-throughput high-content system-wide data-driven approaches. These approaches can generate data on the scale of an entire biological system (cell, tissue, or organism). We begin to see the emergence of genomics, proteomics, metabolomics, and so on, leading in the direction of an integrated *systems biology* perspective. With this major shift to larger scale comes a resulting major increase in speed. Much more significant biological information can be gathered and processed in a much shorter time.

We look next briefly at genomics, then focus primarily on proteomics as a key set of technologies that can help achieve this broadening of scope. We then look at the implications of this wider focus for improving the drug discovery process.

8.5 GENOMICS

The genome of an organism is the complete genetic makeup, the entire DNA complement, of that organism. Genomics, then, is the study of entire genomes. The intention of executing the sequencing and analysis of the entire human genome was to enable more rapid and effective identification of disease-associated genes and thereby provide drug companies with prevalidated targets. The key areas of genomics study are the development and application of tools for the prediction and detection of (1) genes, (2) sequence similarity, (3) motif/domain similarity, and (4) gene expression variations (measurement of mRNA levels through microarray analysis: coexpression, regulation, etc.).

Genomics has indeed aided the drug discovery process significantly. Gene expression analysis in particular has been used extensively with some success in the search for new drug targets and for biomarkers of disease and therapeutic activity. However, in many respects, genomics has raised more questions than it has answered. This is due to a fundamental limitation: mRNA expression levels do not reliably predict protein expression levels.

It has been shown experimentally that gene expression results can be quite different from protein expression results, even when assayed on the same sample under identical conditions [Ideker et al., 2001]. There are a number of possible reasons for these discrepancies. The main differentiators are as follows:

1. *Half-life.* RNA and proteins may have very different half-lives in the cell. RNA may get degraded before translation [e.g., by RNA interference (RNAi)].
2. *Post translational modifications* (PTMs). *No* information on PTMs is available from transcripts. These are modifications that, by definition, occur after the transcript has been translated into a protein.
3. *Localization of protein.* The site of activity of a protein cannot be predicted reliably from transcripts.
4. *Protein interactions.* There is *no* information from transcripts on protein interactions with nucleotides, phospholipids, ligands, or other proteins.

Since all four of these are central to a great many biological functions, these limitations illuminate the need to study the function, structure, and interactions of the proteins themselves. It is the proteins in a living organism that are the molecular targets of most drugs on the market and in development. The study of individual or small numbers of proteins has been performed in academic laboratories worldwide for some time, but there is a growing need, both for basic science and in drug discovery, to study the function, structure, and interactions of proteins at a system-wide scale: This is the emerging field of *proteomics.* In addition, it is expected that proteomics will become more tightly linked with genomics, metabolomics, and so on, leading toward an integrated systems biology approach.

8.6 PROTEOMICS

In its widest sense, *proteomics* is the systematic study of the proteome: all the proteins that are encoded in the genome of an organism. Normally, this means studying the set of proteins in a cell or tissue. The term refers both to the study itself and to the set of techniques and methods that enable this study. Due in part to the success of high-scale gene expression studies, it is being recognized that studying proteins on the proteomic scale yields insights that are just not possible when studying proteins in isolation. A good example of such an insight is when a protein–protein interaction study results in the mapping and functional elucidation of an entire metabolic or signaling pathway, with all its constituent proteins and their interconnections.

The types of information that proteomics studies aim to produce include:

1. Protein structure and function
2. Protein expression levels

3. Posttranslational modifications
4. Subcellular localization
5. Protein–protein interactions
6. Protein–nucleotide interactions (e.g., to identify and study transcription regulating proteins, known as *transcription factors*)
7. Protein–lipid interactions (e.g., for membrane-associated proteins, very important in signal transduction pathways)

In this section we look at some of the key technologies and methods used in proteomics studies to achieve these goals.

8.6.1 Functional Areas of Proteomics

Proteomics activities normally fall in one of several functional categories. We outline these first, then look at each in detail. Note that for each area there are low-scale methods that are typically used in individual protein studies, and high-throughput methods that are more applicable for proteome-scale investigations. We outline these functional areas first, then look in detail at the methods used in each area.

1. *Fractionation and purification* (separation, isolation). Normally, protein studies begin with a cell lysate, the result of tissue fractionation, or a biological fluid (e.g., blood, urine). The proteins of interest must be isolated and separated from the mixture, which may contain many other proteins as well as metabolites, lipids, and so on.
2. *Identification* (primary sequence). The proteins that have been separated out must now be identified. This involves determining the primary amino acid sequence of the proteins or matching peptide information to known proteins (fingerprinting).
3. *Quantitation*. Frequently, it is important to know how much of a protein has been detected. In particular, the relative quantities of various proteins in a cell or tissue yield important information.
4. *Characterization*. Now the proteins can be analyzed and studied. Key areas of study include:
   a. Sequence homologies
   b. Posttranslational modifications (PTMs)
   c. Functional analysis, including interactions, pathways, and networks
   d. Structural analysis and structure–function relationships

8.6.2 Fractionation and Purification

There are many ingenious ways that have been developed to separate and purify proteins. Low-scale techniques include gel- and column-based fractionation methods, as well as differential centrifugation or centrifugation through density gradients. In all cases, the proteins are separated into bulk fractions or bands based on a single global property such as size or charge. More advanced methodologies allow the fractionation and purification of an individual protein from a protein mixture, through methods such as immunoprecipitation in which an antibody to a particular protein is used to separate that protein from all others, or affinity chromatography, in which one or a set of related proteins are captured on an immobilized ligand. These methods, although practical from an analytical point of view, do little to assist in understanding complex biological pathways and essential protein interactions with other proteins or other components of the cell.

The two most common separation methods for high-throughput applications described next.

***Two-Dimensional Polyacrylamide Gel Electrophoresis.* (*2D-PAGE*)** Standard (one-dimensional) gel electrophoresis separates proteins based on mass alone. The sample is loaded onto a gel slab and an electric current is passed through the gel. The current pulls the proteins down the gel at a rate proportional to their mass, resulting in a series of bands that represent proteins of different mass. However, separation by mass alone is often not sufficient. 2D-PAGE separates along the first dimension of the gel by isoelectric point, or pI (the pH at which the protein's net charge is zero), then along the second dimension by mass, as above. The resulting two-dimensional arrangement gives better separation and results in a pattern of spots, each representing a protein (or proteins) of specific mass and pI. It is an effective and widely used technique. However, there are some inherent problems with the technology: 2D-PAGE results do not replicate well; it does not detect proteins at low concentrations; it does not work well for hydrophobic (e.g., membrane-associated) proteins; and it is possible for several proteins to inhabit one spot on the gel, as they may have similar mass and pI.

***High-Performance Liquid Chromatography* (HPLC)** HPLC uses a tube (known as a *column*) filled with some form of treated packing material (matrix or beads) that separates proteins by selectively capturing and eluting them. A detector at the end of the column identifies the proteins as they elute. There are several types of HPLC, using different techniques to separate proteins based on the characteristic desired. These include gel filtration, ion exchange, affinity, and hydrophobic interaction. Each of these techniques captures the protein of interest and holds it in the column while the rest of the sample solution washes through. Then a different solution, which dissociates the protein of interest from the column matrix, is applied and washed through the column. This elutes the protein, carrying it past the detector, which typically registers it as a characteristic peak on a time plot, allowing it to be identified. Many types of detectors are used with HPLC, including refractive index, ultraviolet, fluorescent, radiochemical, electrochemical, near-infrared, mass spectrometry, nuclear magnetic resonance, and light scattering. HPLC has some advantages over 2D-PAGE in high-throughput environments: It is much faster, the results are more repeatable, and it is more sensitive to low-concentration proteins. Also, it can be used to separate and identify proteins by many more attributes than just mass and pI. For example, proteins can be separated by ionization, by specific affinity to an antibody or ligand, or by hydrophobicity.

8.6.3 Identification

Once the proteins are separated, they must be identified. A standard method is to use the output of 2D-PAGE directly: using visualization tools, the gel can be "read" in such a way as to determine the mass and pI of the dark protein spots. If the protein of interest is not novel, its mass and pI can then be looked up in a database and compared to known proteins to find a match. This is often sufficient, but there are problems associated with this method.

For example, posttranslational modifications can change a protein's mass, thus shifting the gel spot such that it may be mistaken for a different protein. Also, 2D-PAGE by itself cannot identify an unknown protein. For this, additional steps are needed: The protein's amino acid *sequence* must be determined.

Edman Sequencing There are two main methods for identifying amino acid sequence. *Edman sequencing* is a well-established method that degrades the protein one residue at a time and determines each residue. It is effective, especially for de novo sequencing of an unknown protein. However, it is relatively slow. *Mass spectrometry* (MS) is becoming the preferred method because it can be applied in high-throughput environments.

Mass Spectrometry Analysis Mass spectrometry has really revolutionized the study of proteins, due largely to its ability to identify proteins quickly and accurately and provide sequence information. It has become an important and ubiquitous tool in proteomic studies. There are many variants of the technology, but the basic elements are as follows, from input to output:

1. *Sample inlet and ionization chamber.* MS works by separating peptide *ions* according to mass and charge. (Most whole proteins are too large to be analyzed intact by MS.) Therefore, sample molecules must first be broken up and charged by ionization. The two most common forms of ionization are *electrospray ionization* (ESI) (used with a liquid sample such as HPLC output), and *matrix-assisted laser desorbtion ionization* (MALDI), which is used to ionize proteins by laser bombardment of solid-state or viscous sample preparations on a plate. In either case, the ionized fragments are then injected into the mass analyzer section.
2. *Mass analyzer.* This section separates the ions by *mass-to-charge ratio* (*m*/*z*). Two of the main methods to accomplish this are as follows:
   a. *Electromagnetic field separation.* Apply an electromagnetic field in such a way as to let through only those of a certain charge. A *quadropole analyzer* is an example of this type. Other types let all ions through, but use a magnetic field to deflect them according to their *m*/*z* ratio.
   b. *Time-of-flight separation.* In this case, the analyzer is essentially a long chamber. Lighter ions traverse the chamber faster than do heavier ones. A time-of-flight (TOF) analyzer exemplifies this type.
3. *Detector.* The detector registers the arrival of an ion and sends the information to a recorder to be processed and graphed for analysis.
4. *Recorder.* The recorder processes the detector's output, generating intensity versus *m*/*z* ratio results in text and graphical form.

Tandem mass spectrometry (MS/MS) is an important type of MS. It uses *two* mass analyzer sections, separated by a collision cell that takes peptide ions separated by the first MS, breaks them down into smaller ions of all possible lengths from one amino acid up to the full peptide length (these are called *b and y ions)*, then uses the second MS section to separate these so that the detector identifies the b and y ions of each length. The action of this middle chamber is called *collision-induced dissociation* (CID).

To provide peptides for MS identification, separated proteins can be obtained by excising gel spots and digesting with trypsin to give peptides. However, in high-throughput situations, HPLC is often used to fractionate the proteins, with the output of the column directly feeding the MS ionizer input. The separated peptides from the column are then ionized by ESI and run through MS/MS to be identified in a continuous flow. The output of the MS is a spectrum of mass/charge (m/z) ratios, showing intensity peaks corresponding to all the detected b and y peptide fragments. Duc to the CID that occurs between the two MS stages, the fragments correspond to cleavages at every bond between the amino acids. By analyzing the differences between one fragment's m/z and the next, the sequence can be deduced: each difference is the m/z of one amino acid in the sequence. However, there can be ambiguity, since more than one amino acid may share the same m/z. So the resulting sequence data for several peptides in the protein are used to find matches against a database of existing sequences, and the best match identifies the protein.

Peptide mass fingerprinting is a simpler and often sufficiently effective method of protein identification. MALDI/TOF is typically used for this. The protein is digested into peptides; these are ionized and fed into a single MS stage (without further fragmentation); and the resulting m/z spectrum identifies the m/z value of each peptide. This is the fingerprint: this set of m/z values are then used to search a database of known peptide m/z values, and the best match across all peptides identifies the protein.

8.6.4 Quantitation

To fully understand the relevance of a given protein to a disease state, quantitation of the protein under different conditions is essential. For an isolated protein, many methods are used. Ultraviolet (UV) absorption at a wavelength of 280 is a very rapid, simple, and effective method of measuring a protein concentration in solution, but this is accurate only if the extinction coefficient for this actual protein is known. More often, one does not have a single protein or does not know the extinction coefficient, as determining this is an expensive and time-consuming process that is usually completed only when the protein has already been shown to be important or is being used as a biopharmaceutical.

Gel-based methods described above for separation and identification of proteins can also be used to quantitate numerous proteins that are within a given sample, provided that they separate well from each other on the gel. To use gels for quantitation requires staining the gel with agents that recognize and bind to the proteins in some predictable manner and requires that several lanes be devoted to standards of known quantity for comparison. The accuracy and sensitivity of gel-based methods depends on the ability of the agent to bind indiscriminately to proteins of all amino acid compositions, and to do so efficiently within the gel matrix. Coomassie blue staining is often used and can be quantitative. Silver staining is more sensitive and will detect minor contaminant or other bands if they are present, but it is less quantitative. Recent advances in developing new reagents have made various fluorescent or light-emitting dyes available. Companies such as Amersham, Invitrogen, and others all market products that are described as accurate and sensitive at monitoring exact protein levels in a gel-or solution-based assay.

Mass spectrometry is generally not considered quantitative but is used to measure the components that are present in a mixture or in a protein. The relative peak heights of an MS measurement are not at all correlated with levels of the protein or peptides in solution. However, MS can be used for *relative* quantitation of a protein under two different conditions. Here, the protein from one sample is labeled with a stable isotope, changing its mass so that MS can detect and quantitate its presence relative to the unlabeled protein.

8.6.5 Characterization

Sometimes, identification and quantitation are end goals in themselves. Perhaps we run an experiment to determine how much of protein X exists in tissue sample Y. Often, however, these are preliminary steps, and the real goal is to *characterize* the protein(s) we find. Before proceeding, however, a few words about *protein sequence, structure, and function* are necessary to set the stage. A cardinal rule of proteins is that their *structure* determines their *function*. The final three-dimensional configuration of a protein, especially any *active site* or *binding sites* it may have, determines the protein's biological or chemical function. There are four categories of protein structure:

1. *Primary structure* refers to the sequence of amino acids that form the protein chain. The sequence is critical; it determines the three-dimensional configuration, or fold, of the protein.
2. *Secondary structure* indicates the three-dimensional structure of subsections of the protein chain, which fall into a handful of characteristic configurations, such as alpha helices and beta sheets. These are the first structures to form as the protein is being created.
3. *Tertiary structure* occurs next, when the alpha helices, beta sheets, and other secondary structures assemble into the globular protein structure.
4. *Quaternary structure* results if the individual protein chains further assemble into dimers, trimers, and so on.

The protein's structure may not be complete at the tertiary or quaternary stage, however. Often, one or more *posttranslational modifications* (PTMs) change the structure further. PTMs are important players that influence protein stability, function, localization, and interactions with other proteins, lipids, ligands, and so on. Some of the most important PTMs are described in Table 8.1.

TABLE 8.1 Significant PTMs

PTM	Description
Phosphorylation	Addition or removal of a phosphate group. Quickly modifies the three-dimensional structure of the protein or of its active site. Acts like a switch: for example, activating/deactivating an enzyme.
Proteolysis	Cleavage of proteins or peptides into smaller pieces. This cleaves off a precursor segment to produce the mature protein. Also used to destroy proteins marked for destruction (*see* Ubiquitination).
Glycosylation	Addition of carbohydrate chains to proteins. This occurs while the protein is still unfolded and emerging from the ribosome. This affects many functions, including targeting a protein to different tissues.
Cystein oxidation	Formation of disulfide bonds. Stabilizes the protein; ensures correct folding; enables certain protein–protein binding.
Ubiquitination	Attachment of a small ubiquitin protein to a target protein, thus marking the target for destruction.

There are two more terms related to structure and function that should be defined, since they are a core part of protein characterization.

1. *Domain.* A protein domain is a region of the protein that is responsible for a particular function. An example is the SH2 domain, which specifically binds to another protein's phosphorylated tyrosine area.
2. *Motif.* A motif is the amino acid sequence for a functional domain. It is conserved across a set of related proteins that share the function.

Earlier in this section we summarized four major areas of protein characterization. With this structure–function background, we can now elaborate on each.

Sequence Homologies Let us suppose that we have separated and identified a new human protein whose function is unknown. Once the protein's amino acid sequence is determined, it is possible to search among other known protein sequences in the sequence databases for *homologs* (sequences that share an evolutionary relationship). (In Section 8.8 and the Appendix, sequence analysis and domain/motif databases and software, such as BLAST, are discussed in more detail.) In sequence analysis, we look for significant similarities between regions of our new protein and those of other proteins. A region that appears with a high degree of similarity across multiple proteins identifies a motif. Given this motif, we can then check databases to determine if the motif has been identified as a functional domain. By using this kind of knowledge of the homologous proteins, we can do the following with our new protein:

- Identify its probable function
- Assign it to a known protein family
- Determine many of its structural features (e.g., α-helixes, β-sheets).

Posttranslational Modifications Characterizing proteins with PTMs essentially means determining the presence or absence of the particular PTM. Several ingenious ways have been devised to accomplish this. We discuss three ways of using affinity purification followed by MS analysis to identify and quantify proteins in the presence of PTMs. In each case, the first ligand of the pair is the *capture ligand*: It is immobilized by covalent bonding to a matrix in order to bind with specificity to the substance of interest, which is the second of the pair.

1. *Lectin/carbohydrate*, for detecting *glycosylation.* Lectin binds the sugar part of a glycoprotein directly, making lectin an effective means of directly separating out glycosylated proteins. In this case, the separation is fully complete before the MS analysis.
2. *Antibody/epitope*, for detecting any PTM. Here, an immobilized antibody is used to bind with high specificity to an epitope-tagged protein of interest. Then the protein is analyzed with MS to detect whether PTMs have occurred, since all PTMs alter either the mass or the charge of a protein, or both. For instance, a sample might have a mix of proteins, including some unknown ratio of phosphorylated to unphosphorylated protein X. If all of protein X is epitope tagged, the antibody can separate it all out and send it to MS. MS can then quantify how much of X is phosphorylated and how much is not based on the m/z difference.

3. *Receptor/ligand tag*, for detecting any PTM. An example of this is the use of a biotin-tagged protein of interest, which is captured by avidin (which has a high affinity for biotin) immobilized on the matrix. When through, only the bound biotin-tagged proteins remain. To detect PTMs on the protein, MS analysis is required to differentially quantitate the tagged protein with and without the PTM, as in the antibody/epitope discussion above.

Functional Analysis, Including Interactions, Pathways, and Networks The basis of functional analysis is the study of *protein interactions:* interactions with ligands, nucleotides, small molecules, lipids, and especially, with other proteins. All are important, both for basic science and in drug discovery. (A case study of proteome-wide protein–ligand study leading to a successful drug discovery appears in Section 8.9.7.). In the present section we concentrate on *protein–protein interactions* (PPIs).

Protein–protein interactions provide substantial insight into the functions of a protein. Proteins frequently bind together into complexes, which carry out certain functions only when combined in this way. Examples of such complexes are commonly found in signal transduction pathways and gene transcription complexes. Therefore, knowing the other proteins with which a new protein associates can help assign a putative function to the new protein and to identify a possible role in a network or pathway, such as:

- Signal transduction
- Cell cycle regulation
- Transcription control
- Immunology
- Metabolism

Much study has gone into associating certain diseases with abnormal behavior of particular pathways. Therefore, if a new protein can be identified with a pathway, it could be implicated in a disease ("guilt by association") and could then be considered a possible new drug target.

High-throughput methods have recently been applied to the study of protein–protein interactions at the proteome-wide scale. Some problems remain to be addressed, but this area holds considerable promise. There are three main techniques: yeast two-hybrid, phage display, and protein chips. A fourth, TAP, is normally a low-scale method, but it has recently been used successfully in a proteome-wide study (Gavin et al., 2002). The details:

1. *Yeast two-hybrid* (Y2H) uses two hybrids expressed in yeast to detect binding. Each is encoded by its own plasmid inserted into the yeast cell. The two hybrids are:
   a. *Bait*: a DNA binding domain fused to a protein of interest, which is to be tested for binding against a large number of....
   b. *Prey*: transcription activation domains fused to transcribed and translated *open reading frames* (ORFs). ORFs are putative genes, each one read off the genome from start codon to stop codon.

   This tests one prey ORF at a time per cell culture, and is repeated for each ORF per culture against the one bait. Once expressed, the two hybrids will bond if the bait and prey proteins interact. This brings the activation domain close to the DNA binding domain, triggering the activation of RNA polymerase and thus transcribing a reporter gene such as His3. If the reporter protein is detected (say, by growth/no growth), a positive interaction is reported (Ito et al., 2001; Uetz et al., 2000).

2. In *phage display*, a fusion gene sequence is inserted into a bacteriophage and allowed to express. This contains a sequence that codes a peptide for display, and this is usually fused to the gene for the phage surface tip protein, pIII. When expressed and translated, the phage displays the peptide on its surface. This peptide is then used as a recognition site to attract the binding sites of a number of molecules that being tested for interactions with the protein that would normally contain this displayed peptide (Danner et al., 2001).
3. *Protein chips* are microarrays, but instead of using DNA on the spots, proteins (or protein-binding molecules such as antibodies) are applied to the spots. This technology is dealt with separately in the next section.
4. *Tandem affinity purification* (TAP) uses a single-compound epitope tag of calmodulin binding peptide with protein A. This tag is used through two affinity separations in sequence. The technique separates complexes of associated proteins of interest from background proteins. At the end, the interacting proteins in the resulting complex can then be detected and analyzed (Gavin et al., 2002).

Structural Analysis and Structure–Function Relationships A thorough understanding of a protein's structure is one of the key goals of protein study. It is of fundamental importance in rational, or structure-based, drug discovery. As noted, amino acid sequence dictates structure, and structure dictates function. This three-way relationship cannot be separated. Having looked at sequence and function, let us now turn to structural analysis. There are two broad areas for determining protein structure. The first involves laboratory techniques, and the second uses computational techniques.

Laboratory Structure Determination Laboratory determination of protein structure is still considered the gold standard. For drug discovery in particular, current computational approaches are not capable of generating a precise enough model of the protein to enable optimized structure-based drug design. There are several laboratory approaches for accurately determining protein structure. The two key methods are *x-ray crystallography*, which uses x-ray diffraction to resolve the structure of crystallized proteins, and *nuclear magnetic resonance imaging* (NMR), which uses nuclear spin transitions to resolve the structure of proteins in solution. Both are capable of providing images down to a very fine resolution, and each has advantages and disadvantages. With x-ray crystallography, proteins are often crystallized with a ligand bound in the active site. This helps determine the physical configuration and energetics of the binding pocket, which are critical factors in structure-based drug design.

Laboratory methods are not without their drawbacks. For instance, a protein is not a static structure; there can be significant motion in some of its parts. (This is particularly true for an *allosteric protein*, one that changes shape based on whether or not a ligand is bound to its active site) So x-ray crystallography can only give a snapshot of a protein and may miss other important configurations. But this is the current state of the art, and despite the drawbacks it is still possible to obtain a tremendous amount of highly useful structural information with these methods.

Computational Structure Determination Although not as accurate as the laboratory techniques, computational approaches provide significant value. There are three main classes of computational approaches that attempt to predict protein structure based on amino acid sequence:

1. *Homology modeling*. This approach starts with a template protein of known structure whose sequence is homologous with the new protein of unknown structure. A model is built by "threading" the novel protein's sequence along the scaffold of the known protein. The threading is fairly complex: The growing model must be checked repeatedly for energy minimization as the sequences are aligned, residues are substituted, and loops are constructed. The configuration with globally minimized energy is assumed to be the most likely model. This method has too many uncertainties to generate high confidence, but it can help provide valuable probable structural information (Kopp et al., 2004).
2. *Ab initio fold prediction*. The goal of solving a protein's three-dimensional structure computationally from sequence data alone has been a holy grail for some time. This is partly due to the difficulties of solving structures using laboratory techniques. Considerable research has gone into this area. The annual competitions, comparing fold predictions with known (but held secret) crystallographic results, show that *ab initio* prediction algorithms are continuing to improve. The algorithms are too complicated to go into here. However, there is one, known as Rosetta (see the Appendix), which has scored much better than competitors in the competitions and has come close with several new proteins (Bonneau et al., 2001; Rohl et al., 2002). However, the current view is that even the best *ab initio* folding algorithms do not yet generate a good enough model to be used with high confidence in structure-based drug discovery.
3. *Secondary structure prediction*. An example will illustrate the use of computational secondary structure determination and the practical importance of the results. Suppose that you have the sequence of a novel protein of unknown structure and function. You note that several regions of the protein are a mixture of hydrophobic and hydrophilic residues. Looking at the sequence alone may not tell you whether these regions are likely to result in overall hydrophobic or hydrophilic areas, or neither. However, you next determine that the sequences of these regions are characteristic of alpha helices (using available structure prediction software tools such as PredictProtein or ProtScale; see the Appendix). Once these secondary structures are folded correctly, it turns out that they present all their alternating hydrophobic residues outward and therefore create an overall hydrophobic region. These structures now tell you that these regions are likely to belong to a membrane-associated protein. If you find further that you have seven such alpha helices, each about 20 amino acids long and separated by beta turns, you may deduce important *functional* information from this: You have almost certainly found a new G-protein-coupled receptor (GPCR). GPCRs are by far the most common target for drugs (Drews et al.; 2000), so with further functional analysis, you may have discovered a new drug target.

It must be stressed that computational approaches are not foolproof. They are somewhat probabilistic, indicating for instance that a particular string of amino acids *is likely* to fold into an alpha helix or a beta sheet. They are usually based on a good deal of experience, but there are always exceptions to the derived rules. So laboratory techniques should always be used to verify the computational findings. As noted, an experimentally derived three-dimensional structure (usually via x-ray crystallography) is considered the most authoritative given the current state of the art.

8.7 PROTEIN CHIP TECHNOLOGY

Protein microarrays are a new technology for highly parallel high-throughput sensitive analysis of proteins across different conditions and time frames. It is the first technology that is showing the potential for *simultaneous* very high scale (up to full proteome) analysis of protein functions, interactions, and comparative expression. Applications include protein identification, quantitation, and functional analysis, all of which can be applied to drug discovery, biomarker discovery, and pathway building. The protein microarray concept grew out of the DNA microarray, which has proven such a valuable tool in genomics studies, such as comparative gene expression studies. Protein array technology is considerably more challenging than its DNA precursor, for reasons we elaborate on below. Therefore, it not only shows great promise but also has significant hurdles to surmount. However, it is such a potentially important tool for proteomics and for drug discovery that it warrants its own section.

We review the issues that the technology aims to address: the current state of the technology, the underlying principles, different implementations, and the advantages and drawbacks of the technology as it stands today.

There are two types of protein arrays:

1. *Functional protein microarrays* (also known as *interaction arrays*). These are arrays of proteins or peptides immobilized to spots on the array substrate. The arrays are then exposed to solutions of substances of interest (usually, label tagged), such as other proteins, ligands, small molecules, nucleotides, or phospholipids to see if they bind to the arrayed proteins. Protein spots where binding has occurred are generally detected by the tag (usually fluorescent). The arrayed proteins may also be enzymes, for studying highly parallel enzyme–substrate interactions.
2. *Protein capture microarrays*. These are usually arrays of fixed capture agents such as antibodies or other compounds: for example, aptamers, which, like antibodies, can bind with specificity to a protein target. The arrayed capture agents then capture and bind proteins of interest that are applied to the finished array. As with the functional protein arrays, proteins of interest to be applied to the array are normally tagged, and detected if bound. Alternatively, a sandwich method can be used [basically, a high-scale enzyme-linked immunosorbent assay (ELISA)], where a second antibody is label tagged and applied *after* the proteins of interest have been applied to the array of immobilized first antibodies. Where the second antibodies bind to the proteins captured by the first antibodies, the labels are detected. Capture arrays are especially useful for diagnostics and comparative expression studies.

8.7.1 Issues Addressed

The chief aim of protein microarray technology is to enable highly parallel, simultaneous, high-scale, high-throughput screening and analysis of proteins: identification, quantitation, interactions, and function. The technology also enables comparative protein studies under varying conditions and as a function of time. Further, the technology provides high selectivity among the arrayed proteins, and high sensitivity to proteins at very small expression levels. These capabilities address limitations in other technologies (e.g., 2D-PAGE, immunoassays, mass spectrometry, ICAT, Y2H): in particular, lack of wide dynamic range and sensitivity to low expression levels and lack of simultaneous high-scale parallelism (Hanash, 2003; Lopez and Pluskal, 2003; Phizicky et al., 2003; Taussig et al., 2002). In addition, protein microarrays can be used with hydrophobic (i.e., membrane-associated) proteins (Lopez and Pluskal, 2003). Most other existing technologies have difficulty analyzing hydrophobic proteins. Capture arrays can in addition give direct protein expression-level information and can be used to detect different posttranslational modifications of a protein.

8.7.2 Current State of the Technology

Protein microarray technology is not yet considered a mature technology. According to Gavin MacBeath at Harvard (who is himself a core researcher in protein microarrays), most protein microarray work that has been reported to date has been *proof-of-concept*: "We're at the point now where people need to start applying this technology to real questions and use it to learn something about biology" (Gershon, 2003). A seminal 2001 study by Zhu et al. (2001) falls into that category, as does their previous study on protein kinases (Zhu et al., 2000) and MacBeath and Schreiber's (2000) own important study.

However, signs of maturation are beginning to appear. Some papers are showing interesting practical applications with potential clinical significance, and commercial developments are proliferating. One recent paper by W. Robinson et al. detailed a study using arrayed antigens to help guide tolerance-inducing therapy against autoimmune diseases, in this case a mouse model for multiple sclerosis (Robinson et al., 2003). The therapy was a "cocktail" vaccine of DNA, each component of which was aimed at one of numerous antibody responses that were detected by the array assay. The authors were able to show a significant (45%) reduction in autoimmune response, thus showing promise for application in humans with MS.

Commercially, there are a number of suppliers with chips or assay services on the market (or announced). Of particular interest in the functional array area is Michael Snyder's new company, ProtoMetrix (recently purchased by Invitrogen), which aims to commercialize and sell a nearly *full-proteome* yeast proteome chip reported on in a 2001 study (Zhu et al., 2001). Another interesting venture is Sweden's Biacore, which markets a variety of protein arrays based on a different implementation, microfluidics, and uses laser-based surface plasmon resonance to detect interactions without labels. Most of their stock sensor chips are protein capture chips, but they also make a custom chip that has powerful functional protein array capabilities.

There is a general lack of industry acceptance of the technology so far. It is not yet widely perceived as a trustworthy mainstream tool for drug discovery. For now, many protein chip companies are going with a service model, where they will do the assays for the drug companies and biotechs and return the data (Uehling, 2003). However, there is a general sense that this is probably a true breakthrough technology that when somewhat more mature, could prove to be of major importance in drug discovery, biomarker discovery, and pathway building. We will now look at the details of the two major types: functional arrays and capture arrays.

Functional Protein Arrays

Underlying Principles Functional protein microarrays are based on essentially the same principle as DNA microarrays: Bind a large number of whole or partial proteins (rather than DNA strands) to a surface in an array (typically, high-density) of spots, wells, and so on, and then probe the entire array with solutions containing one or more substances of interest. The results are measured and analyzed for identification and strength of binding. Heng Zhu of Michael Snyder's lab has provided a good summary: "In principle, the biochemical activities of proteins can be systematically probed by producing proteins in a high-throughput fashion and analyzing the functions of hundreds or thousands of protein samples in parallel using protein microarrays" (Zhu et al., 2001). Although there are a number of different ways of implementing these (see below), the basic principle remains the same: Probe an array of immobilized proteins or peptides with solutions of other proteins, small molecules, ligands, lipids, and so on, and screen for the resulting interactions.

Implementations A number of ways of implementing functional protein microarrays have been devised. First, a large number of proteins must be created, amplified, and purified. All protein chip implementations share this problem: how to successfully create the large number of proteins to be arrayed. In some high-scale cases, a recombinant technique similar to that described in Zhu et al. (2001) is used (Lopez and Pluskal, 2003). This process involves the following: Create constructs of open reading frames (ORFs), usually along with a fusion tag; amplify these with PCR and clone into expression vectors (e.g., plasmid); insert the vector into cells for expression (e.g., yeast, bacteria, or other); then extract and purify the resulting proteins. Another interesting method used in generating high-scale peptide microarrays involves direct synthesis of the peptides in situ directly on the chip using digital photolithography to achieve programmable light activation of small local chemical reactions on each spot (Pellois et al., 2002).

Next comes the implementation options of the arrays themselves. The proteins are immobilized onto a surface, often using the same type of robotic application systems as that used to spot DNA microarrays. The array surface can be glass (usually coated or derivatized), silicon, nitrocellulose, or PVDF membranes, MS plates, microtiter or other well technology, or even microbeads or other particles (Taussig et al., 2002). Zhu et al. (2001) used nickel-coated glass slides, while MacBeath and Schreiber (2001) derivatized their glass slides chemically with a cross-linking agent that reacted with primary amines on the proteins they applied. An earlier Zhu et al. study (2000) of yeast protein kinases using protein chips used arrays of microwells in silicone elastomer sheets which sat atop glass slides. Solutions containing the substances of interest (proteins, ligands, lipids, small molecules, etc.) are then applied to the chips and incubated to achieve binding.

The chips are then scanned to detect binding patterns and strengths. This detection is done in several ways. Typically, the probe molecules are labeled with fluorescent or radioactive labels so that when they bind, their signal can be read by a scanner. (An interesting non labeled technique is the laser-based surface plasmon resonance used by Biacore, discussed previously.) The scanner reads and records the intensity of the signal at each spot on the array, and these data are then analyzed in software.

Advantages Functional protein microarray technology holds considerable promise for high-scale highly parallel high-throughput assays for investigating a number of things: protein interactions with proteins, phospholipids, ligands, small molecules, nucleotides, and so on. This could become a powerful tool for drug discovery. Among its advantages over other technologies:

- Sensitivity over a wider dynamic range: in particular, for proteins at very low expression levels.
- Ideal for proteome-wide protein expression studies.
- Ability to screen large numbers of interactions at one time: high parallelism, high throughput.
- Able to accommodate hydrophobic (e.g. membrane-associated) proteins.
- Potential for enabling simultaneous small molecule screening over a wide range of human proteins. This could identify both on-target efficacy and off-target cross-reactivity (indicating negative side effects or toxicity).
- Can make use of some existing DNA array scanning technology and software.

Disadvantages

- Generating proteins from ORFs has disadvantages. Although it is an efficient way to generate a large library of expressed products, there is no guarantee that the products are in fact valid proteins, since only one of six reading frames is valid for any one gene. Also, the ORF-based approach usually generates only one isomer, and may miss other splice variants.
- In some high-scale cases, especially when using bacteria to express the proteins, the protein extraction process may denature the proteins, so they may not be properly folded and thus be nonfunctional. [However, even screening with denatured proteins has been shown to be useful for certain applications (Lopez and Pluskal, 2003; Taussig et al., 2002).]
- Problems can arise in using fusion tags for protein purification step and/or for covalently bonding the tagged proteins to the slide or other substrate. The tags may inhibit proper folding of the protein and may interfere with PTMs.
- Difficulty in generating thousands of *verifiably* valid, functional, full-length proteins. The high-throughput approach may miss some slow-developing proteins. Active sites may be blocked or facing down.

Protein Capture Arrays

Underlying Principles Capture arrays use the same basic principles as do established low-throughput immunoassay technologies. It is essentially affinity capture, like that used in fractionation and purification. A capture agent such as an antibody or aptamer with high specificity for a particular protein is immobilized on the array and then used to attract and bind that protein.

Implementations For protein capture arrays, the capture agents must first be developed, and in suitable quantity. Several types are used for arrays. Two of the most commonly used are antibodies and aptamers. Antibodies are glycoproteins which are normally used by the immunological system to selectively detect recognition sites on foreign invader proteins. Their natural selectivity, or affinity for a specific protein makes them a good choice as capture agents. Aptamers are synthetic ligands composed of DNA or RNA that are designed to bind with specificity to a particular protein. They are short oligonucleotides that fold into predictable three-dimensional configurations which complement the binding sites of the target proteins.

Antibodies can be made by conventional immunization techniques (monoclonal or polyclonal). Another common technique is to create an antibody library using phage display, then selecting from the library and using recombinant expression, generally using *E. coli* (Taussig et al., 2002). Aptamers can be created as large libraries of oligonucleotides, then selected using the SELEX procedure, which is amenable to automation for high-throughput applications (Mayer et al., 2003; Taussig et al., 2003). In both cases, specific antibodies or aptamers are selected using high-throughput capture methods that apply the protein of interest, then select out only those capture agents that have bound to that protein. These are then further purified and amplified.

The implementation options of the arrays themselves are similar to those used for functional protein arrays. Being proteins themselves, antibodies are amenable to many of the same surfaces and immobilization procedures. Elaborate methods using covalent cross-linking or tag affinity (e.g., biotin/streptavidin) are sometimes used to try to orient the immobilized proteins such that their binding sites face away from the surface and are therefore more available for analyze binding. Nucleic acid aptamers are simpler to deal with. They are less susceptible to orientation issues, and they can use the same surface technology (usually coated or derivatized glass slides) that is used in the more mature DNA microarrays.

Detection of binding on capture arrays uses many of the same techniques as are used for functional protein arrays. Fluorescent labeling of the analyze protein is most common. However, in sandwich ELISA assays, the secondary antibody carries the label tag. For instance, for two-color comparative protein expression studies, the secondary antibodies are normally tagged with Cy3 and Cy5 labels. These provide red and green colors indicating the relative binding intensity of normal vs. disease analyses, for instance. Label-free methods can also be used with capture arrays, such as the laser-based surface plasmon resonance method discussed previously.

Advantages

- High sensitivity: protein capture arrays have demonstrated successful detection as low as 10 f*M* (Bock et al., 2004).
- High selectivity and affinity if used in sandwich ELISA mode.
- Potential to enable high-scale protein expression profiling (e.g., diseased vs. normal tissues) to help determine proteins that over- or underexpress in the disease state. This can be a powerful drug target identification tool as well as a biomarker discovery tool.
- Potentially effective as diagnostic chips, to detect disease by known biomarkers in a clinical setting.
- *Multiple* biomarkers can be identified and assayed in parallel. Some diseases (e.g., cancers) are characterized by protein *signatures*, that is, changes in expression of multiple proteins vs. normal tissue expression levels.
- Can make use of some existing DNA array technology, as with functional protein arrays.

Disadvantages

- Cannot be used to discover unknown proteins. By definition, you must know what proteins you are trying to capture, then design a capture agent for those proteins.
- Scale is currently limited: It is very challenging to create thousands of unique capture agent types. Most current capture array implementations are limited to tens or at most hundreds of unique spots.
- It can be difficult to immobilize antibodies to the array while maintaining their biological activity and structure, as with functional protein arrays.
- Antibodies do not always bind with high selectivity and specificity, particularly if non target proteins are in high abundance. (But sandwich ELISA methods can improve specificity significantly, as noted above.)

Summary Protein microarrays have generated considerable excitement due to the potential use of simultaneous high-scale high-throughput assays in quantifying and characterizing all manner of protein interactions and for proteome-wide protein expression profiling. Of particular interest is their potential value in the drug discovery and development process, in biomarker discovery, and as clinical diagnostic tools. However, the technology is still quite immature and has many hurdles to surmount before it is accepted by industry as a valid, reliable tool. Protein chip companies may need to adopt a services model at first, until they can repeatedly and reliably generate useful high-quality, data for the pharmaceutical and biotech industries on a contractual basis. This should help the technology gain wider acceptance.

8.8 PROTEOMICS DATA ANALYSIS: COMPUTATIONAL BIOLOGY AND BIOINFORMATICS

The key to the success and utility of genomics and proteomics is the ability to deal with the voluminous quantities of data they generate. Perhaps the most major difference between the new high-scale approaches and standard biological methods is the shift from *hypothesis-driven* to *data-driven* exploration. Hypothesis-driven methodology will always be a core part of basic science and drug discovery. But high-scale approaches are complementing this with data-driven methodology. The activities relating to managing and analyzing these data can collectively be called *informatics*. We now look at the informatics side of proteomics. Several functions are needed to deal with these data. They can be grouped as follows:

1. *Data generation*. Laboratory instruments used in proteomics have become extremely sophisticated computer-controlled machines, including mass spectrometers, 2D-PAGE machines, robots for creating microarrays, and microarray scanners. These all require specialized software to control and monitor the process, generate and sometimes preprocess the data, and generate reports on the results.
2. *Storage, retrieval, and tracking*. Data must be retrieved from laboratory machines and stored in an appropriate format and configuration to promote access and mining for relationships among the data.
   a. *Storage*. Large quantities of storage space must be allocated. Since multiple storage servers are often required, it can be helpful to install a *storage access network* (SAN)

to help with the speed and efficiency of data access. The SAN manages a cluster of storage servers and takes over the task of deciding on which machine(s) to allocate storage for a particular dataset. All of this must be well architected and managed, and regular backups are imperative.

b. *Database management.* Considerable effort goes into the design, management, and maintenance of databases for biological data. It helps for database managers in this area to have a solid grounding in biology and biochemistry. This promotes database designs that make it easier for biologists to find biologically meaningful associations in their data mining. The databases should, if at all possible, "speak the same language." That is, for any one protein or gene, the same key (*accession number*) should be acceptable across all databases. (Among public databases, for instance, this is not always the case. Significant efforts are under way to try to coordinate the databases to fix this problem.)

c. *LIMS.* Laboratory information management systems is a specialized area that combines data storage and retrieval, database management, and data tracking for data-intensive laboratory environments. These systems combine biological databases with the means to organize and track the sources of these data and coordinate the source information with the experimental data. Thus, all data resulting from a particular tissue sample on a particular date can be effectively tracked. LIMSs are used primarily to do sample tracking and to aid in project management. In a drug discovery environment, they can be especially useful in generating, maintaining, and archiving the kinds of information required in filing for FDA approval for a drug candidate.

3. *Computation infrastructure.* High-scale genomics and proteomics methods often require massive computing power. In structural proteomics, for example, computational protein folding is an example of a particular compute-intensive area: given a primary amino acid sequence of a protein, there are an enormous number of possible three-dimensional configurations. Protein folding algorithms must explore all the feasible folds, generate a free-energy function for each, and try to find the configuration with the global minimum free energy. This is the kind of process that can take a supercomputer to accomplish in any reasonable time, and supercomputers have been used in these problems. However, supercomputer use can be quite expensive and may be out of reach for some organizations. In recent years, cluster computing has come into its own for these and many other proteomics computation requirements. Clusters of simple relatively inexpensive PC-type hardware, often running Linux, can do distributed processing on complex problems after they are appropriately broken down into smaller, parallel chunks that can be assigned to the multiple individual processors. Cluster computing has proven extremely effective and is considerably less expensive than the use of supercomputers. However, the latter still have their place, especially when symmetric multiprocessing, using one large memory space, is required.

4. *Data checking and manipulation.* These address the crucial area of *data integrity and validity.* Before the data can be used in biological analysis, manipulation and checking must usually be done to minimize sources of error and to help ensure the biological significance of the data. Microarray gene expression data provide a good example. Normally, the laboratory instruments generating the data come with some software to preprocess the raw data they generate. This may involve normalization and error checking. However, additional manipulation and checking is usually needed. For example, data from several

repeat experimental runs must be cross-checked and correlated; they may need to be normalized across all runs so that the data are comparable. Data are subjected to statistical analysis to:

a. Perform normalization.
b. Check for accuracy and precision.
c. Check for sufficient signal-to-noise ratio.
d. Calculate significance values.

5. *Analysis: software and databases.* Computational biologists and bioinformaticians create and use sophisticated software data analysis and visualization tools. In addition to the software tools, and tightly linked with them, well-curated databases have become ubiquitous and essential in genomics and proteomics analysis. Refer to the Appendix for a representative sample of some important publicly available databases and software tools. In proteomics, areas addressed by software and databases include:

a. Protein identification.
   - Analysis and visualization of 2D-PAGE data
   - Analysis and visualization of mass spectrometry peptide spectra
   - Detection of primary amino acid sequence from MS/MS spectra or Edman sequencing
   - Peptide fingerprinting from MS output for protein identification
   - Protein identification from 2D-PAGE spot localization or MS/MS peptide sequencing

b. Protein quantitation
c. Characterization
   - Sequence homologies, mapping to protein families
   - Posttranslational modifications
   - Functional analysis (construction and visualization of signaling pathways, and mapping and visualizing protein–protein interactions)
   - Structural analysis(two- and three-dimensional)
   - Structure–function correlation

The effectiveness of all this elaborate data generation, storage, handling, computing, and analysis is, of course, dependent on one key prerequisite: *proper experimental design!* Just as in standard biological methodology, it is imperative to consider carefully the intended outcome of the experiment and then to design it in such a way as to maximize the potential for that outcome and minimize the chances of spurious conclusions. One of the dangers of data-driven high-scale approaches is that the data can appear so authoritative, so definitive, that new practitioners may be tempted to accept them at face value and to assume their validity. Recent history has given ample evidence of this danger. Microarray gene expression studies, especially in the early days of this technology, roughly the late 1990s, provide a good example. In that period, many exciting results were generated, and papers published, based on single experimental runs with microarrays. Several of these results could not be replicated later and therefore had to be considered suspect. It turns out that there is often highly significant variance in the data from one experimental run to another, and that the results of a single experiment therefore proved inadequate, and in some cases, erroneous. This variance can be due to different tissue sample processing, different microarray processing, variations in quality and quantity of cDNA probes on the microarray spots, varying background noise, and so on. Therefore, it is now generally accepted procedure to perform at least three separate microarray experiments with the same probe sets. The resulting data can then be cross-correlated, normalized, and checked for errors. Outliers on one array may be discarded. Therefore, the importance of proper experimental design cannot be overstressed.

8.9 PROTEOMICS AND DRUG DISCOVERY

Earlier in this chapter, the current practices in drug discovery were outlined. Recall the issues with the current drug discovery process:

- Very long time span between initial conception and final marketing of a drug: 10 to 15 years is not uncommon.
- Very high attrition rate: about 5000 compounds in for every one drug out.
- Exceedingly costly process: a major contributor to high costs when a drug candidate fails late in the process (e.g., in phase III trials).

Therefore, it is desired to improve the process. For instance:

- More targets identified and validated, and faster
- Fail problematic drug candidates earlier and more cheaply
- Enable targeted therapeutics

Now that some background has been presented, we can consider how proteomics can be applied to help address these issues. We look at proteomics applications in the drug discovery process, both those in current use as well as potential. A number of new possibilities are opening up, due to the increases of scale and speed that result from these new technologies. For instance, there are new capabilities for:

- Systematic high-scale searches for new drug targets
- Identification of thousands of previously uncharacterized proteins through high-throughput expression studies and functional proteomics
- Biomarker detection using high-scale differential protein expression studies of pathological vs. normal samples

In this section we outline some of the key ways these capabilities can assist in each stage of the drug discovery process. Reference is made to the methods and technologies discussed earlier in the chapter. Some illustrative examples are provided along the way.

8.9.1 Target Identification

The initial stage of the drug discovery process is the identification of a molecular target, almost always a protein, which can be shown to be associated with disease. To accelerate this, more proteins need to be identified and characterized. Although the set of human genes is by now quite well understood, this is not as true for all the proteins in the human proteome. Because of splice variants and posttranslational modifications, the number of proteins far exceeds the number of genes. As of today, a great many human proteins have yet to be fully identified and characterized. The new high-throughput technologies of proteomics are helping to close this knowledge gap in a number of ways:

1. *Identify large numbers of novel proteins in the search for new drug targets.* High-throughput separation, purification, and identification methods are being used for this. 2D-PAGE is often used for separation, even though it has some drawbacks. HPLC is a common separation and purification technique, and is particularly adapted for high-throughput use when coupled directly to tandem MS for identification. These procedures produce peptide sequences that must then be compared to possible homologs in other organisms (or paralogs in humans) using informatics. Edman sequencing may be needed for sequence verification.

2. *Perform initial functional characterization of novel proteins. Initially, assign them to functional areas that might implicate some of them in disease pathology*:
   a. Metabolic pathways
   b. Transcriptional regulatory networks
   c. Cell cycle regulatory networks
   d. Signal transduction pathways

Computational sequence homology analysis can help putatively assign a novel protein to a known pathway or network that its homolog participates in. Sequence analysis also helps identify known motifs and functional domains that the new protein shares with known proteins. This information can provisionally assign membership in a protein family to the new protein.

3. *Screen a compound with known therapeutic effect against a large number of human proteins to identify the exact target.* It is not uncommon to have a compound with known but perhaps limited or suboptimal therapeutic effect against a disease, yet not know the molecular target for this compound. A proteome-scale search can pinpoint a possible target, once binding has been identified. The case study of the development of LAF389, presented at the end of this section, describes a successful example of this in detail.

4. *Perform high-throughput differential protein expression profiling, comparing diseased with normal tissue samples, to identify the biomarker proteins that are possible contributors to the disease by their over- or underexpression.* This is one of the most promising areas of application of proteomics in target identification. Several proteomics technologies can be used for expression profiling. Two-dimensional PAGE has been used, although comparing two different gels (one for normal, one for disease state) can be problematic, due to the replicability issue discussed earlier. A variant of this called *two-dimensional differential in-gel electrophoresis* (2D-DIGE) overcomes this drawback. 2D-DIGE has been used successfully to identify biomarkers that could be potential drug targets (Lee et al., 2003). As noted previously, protein chips have perhaps the greatest potential for high-throughput simultaneous differential protein expression profiling. Several successful attempts to use protein arrays for target identification have been reported (Simon et al., 2003). However, the technology is still relatively immature and is therefore further out on the horizon as far as general industry acceptance for target identification.

Example A recent study showed how an integrated approach involving proteomics, bioinformatics, and molecular imaging was used to identify and characterize disease-tissue-specific signature proteins displayed by endothelial cells in the organism (Oh et al., 2004).

Working with blood vessels in normal lungs and in lung tumors in rats, the researchers used several high-throughput affinity-based separation procedures, followed by MS and database analysis to identify and map the proteins displayed by endothelial cells that line the blood vessels. With this approach they identified proteins that are displayed only on solid tumor blood vessel walls. They then demonstrated that radioisotope-labeled antibodies can recognize these tumor-specific proteins, allowing them to be imaged. The radioisotope labeling itself also resulted in significant remission of solid lung tumors, demonstrating tissue-targeted therapeutic potential. Although individual endothelial proteins had been identified and targeted in previous studies, this is the first time a proteomics approach had been used to move toward a complete tissue-specific mapping of the proteins displayed on the blood-exposed surfaces of blood vessels. This demonstrates a new approach for identifying potential novel targets for therapy.

8.9.2 Target Validation

At this point in the drug discovery process, one or more potentially disease-related protein targets have been identified. For now, let us assume one. The next step is to validate the target. Primarily, this means that the target's relevance to disease pathology must be determined unambiguously. This involves more detailed functional characterization, more evidence for the pathway or network assignment, and modulating the protein's activity to determine its relationship to the disease phenotype. Some determination of tractability may be done in this stage as well. Proteomics can assist in several ways:

1. *Determine in what tissues and cell components it appears and in which developmental stages.* High-throughput techniques such as protein chips and 2D-DIGE can be used for proteome-scale expression studies comparing different tissue types and developmental stages. These add evidence that the putative target is found in a disease-related tissue and at the expected developmental stage. Interaction studies can help determine subcellular localization by showing binding to proteins or phospholipids in the cell that have known location. Further, sequence analysis can identify known location-specifying signal peptides on the protein. Finally, posttranslational modification analysis can identify certain PTMs that determine the destination of the protein.
2. *Understand when and for how long the target gets expressed, and when degraded.* High-throughput protein expression studies can be done in multiple runs over time, and then compared. In this case, expression patterns are assayed and then compared across multiple time points rather than normal vs. disease states. For additional evidence it would be very informative to do a time-based expression study first of normal tissue, then another one of diseased tissue, and compare the behavior *over time* of normal vs. disease tissues. This would provide a multidimensional target validation.
3. *Verify the target protein's specific role within the protein family and the pathway or network identified in the previous stage.* Initial putative functional assignments for the new protein were made in the target identification stage. In target validation, high-throughput protein–protein interaction studies can be used to strengthen the evidence for the protein family, network, or pathway involvement. Protein–phospholipid interaction assays can determine whether the new protein is membrane associated. Technologies such as protein chips are beginning to be used for these interaction studies and have great potential in this area (Walgren et al., 2004; Zhu et al., 2001). Other techniques such as

Y2H, phage display, and tandem affinity purification have also been used with success in this area.

Posttranslational modifications to the new protein can also be identified by methods discussed earlier. Knowledge of the binding partners and posttranslational modifications of a new protein goes a long way to help characterize it functionally, to solidify its assignment to a pathway, and so on. Since the association of a disease and a particular function or pathway is often already known, a solid assignment of the new protein to such a function or pathway implicates the protein in the disease. This adds evidence toward validating the target.

4. *Determine the effect of inhibiting the putative target.*
   a. Does target inhibition disrupt a disease-related pathway?
   b. Does target inhibition slow or stop uncontrolled growth? Have other effects?
   c. Does this effect validate the target as unambiguously related to the disease pathology?

Several methods can be used to answer these questions. Gene knockout studies in mice have been an effective tool for some time, but knockouts cannot be done with all putative target proteins. It is in any case a slow, laborious process to breed the knockout strain correctly and reliably and may not result in viable mice to study.

One recent alternative method has garnered a great deal of attention and is currently achieving rapid adoption in the industry due to its relative ease, rapidity, effectiveness, and lower cost. It is called *RNA interference* (RNAi) (Hannon, 2002). This technique has roots in both genomics and proteomics. *Small interfering RNAs* (siRNA) are synthetic 19- to 23-nucleotide RNA strands that fold in hairpin configuration. They elicit strong and specific suppression of gene expression; this is called *gene knockdown* or *gene silencing*. RNAi works by triggering the degradation of the mRNA transcript for the target's gene before the protein can be formed. RNAi is done in vitro in the lab to verify disease relevance of a putative target in several possible ways:

- If the target is believed to cause pathology by *overexpression*, investigators can knock down the gene for the target and observe whether the disease pathology is reduced.
- If *underexpression* is assumed to contribute to disease, the gene can be knocked down in *healthy* tissue samples to see whether this elicits the same disease phenotype.
- The pathway in which the target is believed to participate can itself be validated as disease-related by performing knockdowns of some or all of the genes in the pathway. Then the putative target gene can be knocked down, and the effect of this on the pathway's function can be observed. This both validates disease relevance and verifies the functional assignment of the target to the pathway.

Note that RNAi is also being investigated for its *therapeutic* potential. Although there are major ADME issues to be addressed, RNAi molecules, with their high specificity and efficacy in gene suppression, may themselves hold great promise as drug candidates.

8.9.3 Screening for Hits

Now we have one or more possible targets. For each target we need to screen many compounds to look for drug candidates that show activity against the target (i.e., hits). This can be like searching for a needle in a haystack, so any techniques that can help accelerate and focus this search are of great value. The following techniques are fairly new but are in current use.

1. *Develop sets of compounds to screen for activity against the target.* The first order of business is to construct focused sets of compounds to screen against the target. Structural proteomics and combinatorial chemistry can play major roles at this stage. As mentioned before, an x-ray crystallographic structure for the protein provides the golden standard for three-dimensional structure. Given such structural information, it is possible to develop much more focused compound sets for screening libraries than would otherwise be possible. Combinatorial chemistry can then be used to design the libraries of such compounds. These are numerous small modifications of a basic small molecule or side group that is likely to fit the known binding pocket of the target protein based on the structural information.

Computational chemistry is also adding to this effort. It is being used to generate large *virtual compound libraries* as a part of *structure-based drug design* (see below). Like the combinatorial compound libraries, these are sets of compounds that are likely to fit well with the target's binding site. However, these compounds are all *in silico*, in the computer only.

2. *Screen for compounds that affect the target.* (In most cases, inhibition is aimed for. Most drug targets are enzymes, and most of these are overactive, either by being overabundant or by being stuck in an active state. Therefore, most drug candidates attempt to inhibit or shut off the enzyme by binding to its active site.)

High-throughput laboratory screening can now proceed. This technique uses 96- or 384-well plates to combine the target protein with each of the screening compounds, one per well. In this way, the compounds are individually tested for activity against the target. Once activity is detected, even if it is only moderate, that compound is designated as a hit. Chemists can later attempt to increase its activity by modifying the compound and retesting.

Virtual screening is another, newer technique that is showing some promise. Using the virtual compound libraries, virtual screening uses elaborate computational chemistry techniques to determine *in silico* the fit between each virtual compound model and the binding site of the protein model derived from the x-ray crystallographic structure. This involves computing the chemical affinity, the steric fit, and the energetics of the compound in the binding pocket. When certain thresholds are reached, a hit is declared, and reserved for further optimization (Sneider et al., 2002).

3. *Structure-based drug design.* As mentioned above, virtual screening entails the computational identification of a drug candidate from the ground up. This is called *structure-based* or *rational drug design.* Structural proteomics provides the core information needed to achieve this. There are two main approaches used in structure-based drug design (Wolf et al., 2003):

   a. *Building up an optimized ligand from a known inhibitor molecule.* If structural information is available for a known inhibitor ligand, this can be modeled and used as a starting point. Placed computationally into the target protein's binding site, it can be manipulated on the computer by chemical changes or the addition, moving, or subtraction of chemical groups or even atoms until its fit is considered strong. Some docking programs first decompose the known ligand into fragments before the user docks them appropriately in the binding pocket and begins optimizing.

   b. *De novo ligand design and/or docking into the binding pocket model.* If no known inhibitor exists, or there is no structural information on one, a ligand can be built up from scratch. A base fragment is initially placed in the binding site, then additional fragments or atoms are added according to sets of rules derived from many known protein–ligand structures.

8.9.4 Lead Optimization

At this point, one or more promising drug candidate hits have been identified and are ready to be promoted to lead status. This entails verifying or optimizing the previously mentioned qualities necessary for the drug candidate to be pursued: ease of synthesis, adherence to the Lipinsky rules, target specificity, and efficacy against the disease. Following are two lead optimization endeavors in which proteomics and related computational techniques are making significant inroads.

1. *Optimize the hits rationally (i.e., not by trial and error).* Continuing the theme of rational drug design, let us suppose we now have a model of a lead compound that was generated by one of the methods outlined above. Structural proteomics can continue to be of use in providing precise docking and binding information for the optimization of the lead. Computational chemistry is used to calculate the compound's adherence to the Lipinsky rules. Then structural proteomics and computational chemistry can team up to evaluate *quantitative structure–activity relationships* (QSARs) of the lead candidate. QSAR is the attempt to correlate the structural descriptors of a compound with its activity. The analysis is done computationally, and can help predict not only the strength of the desired activity of the compound, but also activities that might tend toward toxicity or impaired ADME characteristics. The generation and interpretation of QSAR models is complex and can be greatly aided by the application of cheminformatics and data mining techniques (Weaver, 2004). The application of this sort of structural proteomics screening and optimization has been applied successfully in developing HIV protease inhibitors, antimicrobial drugs, and influenza virus neuraminidase inhibitors (Walgren et al., 2004).

Of course, such computational results must be verified in the laboratory. The in vitro analog of QSAR analysis is accomplished by the use of *activity-based probes*. These are chemically reactive compounds that can be used to identify and quantify a specific biological activity of a protein based on its active component (Walgren et al., 2004). Both QSAR analysis and in vitro testing with activity-based probes help to verify target specificity and strength of activity.

2. *Identify a closely related animal model for initial in vivo testing.* Eventually, it is highly desirable to identify and use a valid animal model for in vivo testing of the lead compound to help verify efficacy against the disease in a living organism. This can be a very difficult endeavor, but it can be helped by the application of sequence analysis and comparative genomics. Assuming that we now have a validated target protein, it is possible to take the amino acid or nucleotide sequence of the protein and search for orthologs in animals typically used for models. Hopefully, an ortholog with high sequence similarity can be found, and that animal would then be a candidate for in vivo testing of the lead candidate. Much effort goes into developing a breed of laboratory animal that exhibits the disease phenotype and in which the disease is demonstrably similar to that in humans. If such a model has been developed for this disease, the association between the homologous protein and the disease phenotype can be established. For instance, if that protein is overexpressed in the disease, RNAi can be used to silence the gene for the protein, and the effect on the course

of the disease is observed. Once this association has been established, the animal is now ready for in vivo testing with the lead candidate.

8.9.5 Pharmacology and ADME-Tox

Solid leads have been identified and optimized by this point, and shown to possess the right druglike qualities and efficacy in in vitro studies. However, the large gap between the lab assay and the living being must now be narrowed. Further tests with animal models, plus early human trials, have been and will continue to be the mainstay of this assessment. However, proteomics can play a major role by enabling ADME-tox analyses and screenings *earlier* in the drug discovery process.

1. *Eliminate early on those candidate compounds that have poor "druggability" characteristics (i.e., poor ADME or toxicity profile).* In traditional drug discovery and development, ADME and toxicity assessments were normally done at this stage, with in vitro and, especially, in vivo studies. In fact, this assessment continues on into clinical trials, and it has sometimes been true that ADME-tox properties were unable to be fully understood until late in clinical trials, after tens or hundreds of millions of dollars had been spent. Clearly, failure of a drug candidate at such a late stage is financially very onerous. Therefore, there has been intense interest in the notion of "fail fast": Try to identify and eliminate as early as possible any compound which has ADME-tox properties that are likely to make it a poor or unacceptable drug. Accordingly, drug companies are working to move ADME-tox analysis into increasingly earlier stages of the discovery process.

Many of the analytical methods discussed above can aid in this. Structural proteomics-based QSAR and its relative, *quantitative structure–toxicity relationship* (QSTR), help to identify qualities such as poor absorption or distribution, suboptimal metabolism (e.g., too many metabolic products), too-fast or too-slow elimination, and toxicity. These analyses can be performed earlier, in the lead optimization stage, so that by the time the lead candidate enters clinical trials there is a better understanding of its pharmacology and toxicology. It is even possible to perform preliminary ADME-tox screenings earlier than that, such as in the target validation or screening stages.

2. *Verify specificity; screen lead drug candidates against nontarget proteins in the proteome to ensure minimal cross-reactivity.* Proteomics can be used in another way to build confidence in low side effects and toxicity. High-throughput protein interaction studies can determine whether the lead candidate shows activity against other human proteins that are not the intended target. As the technology matures, functional protein microarrays should prove to be an excellent method for this type of assessment. A chip can be arrayed with a proteome-scale collection of proteins and then incubated with the lead candidate compound, then monitored for off-target protein interactions. In the near term, 96- or 384-well arrays used in high-throughput mode are more feasible for this type of assay, although somewhat lower scale.

There is another interesting use of this approach. This kind of high-throughput assay can also identify *potentially beneficial* off-target interactions. For instance, imatinib mesylate (Gleevec) has been developed as a very successful therapeutic for chronic myeloid leukemia. It works by specifically inhibiting BCR-ABL tyrosine kinase. Recently, it has been shown that Gleevec is also active against other tyrosine kinases and therefore may prove to be useful as a drug for other diseases where other tyrosine kinases are overactive

(Nadal et al., 2004). That kind of beneficial off-target activity of a lead candidate can be identified much earlier, using proteomics approaches such as functional protein chips arrayed, for instance, with a variety of kinases in the case of a kinase-inhibitor such as Gleevec. The drug development process could then split into several parallel tracks, each targeting a different disease and a different kinase.

3. *Enable tissue-specific delivery of a drug candidate within the body.* This is an area of intense interest to the industry. Very promising drug leads have had to be abandoned due to the inability to get them delivered to the target tissue. It is particularly difficult to get drugs delivered with specificity to solid tumor cancer cells. However, new approaches are showing potential to help.

Example In the target identification section, we gave the example of proteomic identification of tissue-specific blood-vessel signature display proteins as potential targets (Oh et al., 2004). The methods used in that study also uncovered a means for enabling tissue-specific drug *delivery*. The study demonstrated the use of proteomic techniques to identify and map display proteins that can help direct drugs to their intended specific tissue target of action. Enabling tissue-specific drug targeting in this way can also help reduce side effects by directing drugs *away* from other areas of the body.

8.9.6 Clinical Trials: Biomarkers and Pharmacogenomics

Even in the clinical trials phase of drug development, proteomics can play several pivotal roles: the identification of biomarkers for diagnostics, for safety, and for measuring therapeutic efficacy. One significant area enabled by biomarkers is *pharmacogenomics*.

Biomarkers Biomarkers are biological indicators found in tissue, serum, urine, and so on, used to determine the following:

- Normal versus disease states
- Pathogenic processes
- Positive and negative responses to therapeutic intervention

In clinical trials, they can play several crucial roles: as diagnostics, to indicate the efficacy of a drug, to help guide dose selection, and to predict or indicate toxicity. Historically, biomarkers have typically been single indicators. However, it has been increasingly recognized that many diseases, such as many forms of cancer, are complex and heterogeneous. In such cases, single-analyte biomarkers simply cannot capture enough information to serve as valid indicators of disease, drug efficacy, and so on. For these cases the much broader-scope approaches of genomics and proteomics are being used increasingly to discover and provide multianalyte biomarker profiles, or panels.

In genomics, this strategy has already proven useful in several cases. For instance, Genomic Health of Redwood City, California, has launched its Oncotype DX diagnostic, which measures expression levels of a panel of 16 genes to predict the course of breast cancer. In this case, that means predicting which tumors are likely to metastasize. Another company, Agendia of Amsterdam, is also marketing a breast cancer profiling test, Mammaprint. This uses a panel of 70 genes found to correlate very highly with future metastasis (Garber, 2004).

These diagnostic panels were developed using a technique that is common to both genomic and proteomic biomarker profile discovery. The technique uses the analysis of large-scale expression patterns to uncover statistically strong associations between over- or underexpression of sets of genes or proteins and the target biological metric, such as disease vs. normal states. This is made possible by a combination of high-throughput technologies such as microarrays along with advanced computational and statistical analysis tools that can mine these large datasets for such associations and determine their statistical significance. One interesting point about this approach is that it is not absolutely necessary to know the function of all the genes or proteins in the panel. Some may even be novel genes or proteins whose functions have never been elucidated. Still, if their expression differences can be strongly and repeatably correlated with the desired biological metric, they can become validated biomarkers.

Genomic biomarker profiles can be very useful, but they also have some inherent drawbacks: They are usually inadequate for determining drug efficacy and toxicity, two key biological metrics to be measured in clinical trials. This is because it is the proteins that are ultimately involved in every disease process, and almost all drugs have proteins as their target. Also, toxicity is usually due to a drug's undesirable metabolic changes or its binding to off-target proteins that negatively alters normal functioning. Therefore, proteomics approaches can play a major part in discovering and developing biomarker profiles that measure drug efficacy and toxicity.

Several proteomics approaches are showing utility in single- and multianalyte biomarker panel discovery. In many cases, they can be used with easily collected biological fluids such as blood or urine. Many of the technologies discussed in Section 8.6 are applicable. 2D-PAGE has been used successfully for identifying differential protein expression, although it has the aforementioned limitations, such as replicability issues and the requirement for relatively high analyte volumes. Protein chips, although not yet a fully mature technology, are a natural fit for highly parallel protein expression profiling leading to biomarker panels (Rocken et al., 2004). For example, an ELISA-based capture chip was used successfully to identify a panel of five circulating angiogenic factor protein levels as biomarkers for gynecological tumors (Huang et al., 2004). Liquid chromatography and mass spectrometry are being used in several ways for biomarker discovery. One is to analyze the low-molecular-weight portion of the proteome to generate new classes of disease biomarker profiles. These have shown significant success in early disease detection (Johann et al., 2004). In another approach, a version of MALDI called *surface-enhanced laser desorbtion ionization* (SELDI) has been used with MS to identify a number of panels of multiple protein biomarkers, each with an expression profile specific to one of several different tumor types, including breast cancer (Li et al., 2002; Shin et al., 2002). One very promising area is the harnessing of known immune response to some early-stage cancers to discover and validate tumor-specific antigens in serum and in tumor lysates as biomarkers for cancer. This has been demonstrated effectively using 2D-PAGE and MS to identify biomarkers for breast cancer, and validated with Western immunoblot analysis (Shin et al., 2002). Another area where proteomics techniques are clearly indicated is the identification of posttranslationally modified forms of proteins. In some cases it is the modification, especially cleavage or glycosylation, that identifies a protein as a biomarker. For example, several techniques, including 1- and 2-D PAGE followed by MS, as well as protein chips, have been used to detect new PTM-specific biomarkers for breast cancer (Shin et al., 2002). These techniques show significant promise in developing biomarkers for diagnostics, for understanding pathogenic process, and for measuring drug efficacy and toxicity in clinical trials and beyond.

Pharmacogenomics Often, significant differences in drug efficacy and/or side effects appear in clinical trials among the test subjects. There has been considerable recent interest in finding correlations between these differences and gene variations or expression patterns of genes or proteins among the test subject population. If such a correlation is found, the variation or expression pattern can become biomarkers that indicate higher likelihood of either efficacy or unwanted side effects (or even toxicity). This area is called *pharmacogenomics*. In November 2003, the FDA ruled for the first time that pharmacogenomic data can be included in drug-approval applications. This is widely believed to be the opening of the door to greater FDA support for *targeted medicine*, that is, therapeutics that are marketed to the subset of the population that is likely to benefit from a particular drug, or that is unlikely to experience serious negative side effects. Biomarkers are the key enabler of this kind of medicine; they are essential in identifying the right patients for targeted therapeutics.

In the genomics world, genotyping is used effectively in determining correlations between a single nucleotide polymorphism (SNP, an inherited gene variation) and the *likelihood* of a person contracting a particular disease. Thus, SNP genotyping is a statistically predictive tool. However, SNP variants are not considered the best biomarkers for either the *presence* of a disease or as a measure of *therapeutic activity* against the disease. There are too many other ways for a gene to get damaged than just a SNP: environmental influences are a major factor. In addition, the SNP gene can be inactivated anyway (e.g., through methylation, which silences genes). Here again, gene and protein expression analysis can detect the activation or inactivation of a gene(s), regardless of the cause. Most of the biomarker discovery techniques mentioned above can be enlisted to identify single- or multianalyte biomarkers for pharmacogenomics. To date, however, the most fully developed biomarkers that are in clinical use are single gene traits that influence drug response in humans.

Example One of the most dramatic examples of pharmacogenomics and targeted therapy is the HER2/Herceptin story. It was discovered that the gene for HER2, a growth factor receptor protein, is overexpressed in 25 to 30% of breast cancers. In those cases, tumor aggressiveness is greatly increased. It was further discovered that a humanized anti-HER2 antibody, Herceptin (trastuzumab), significantly increases effectiveness of chemotherapy against HER2-overexpressing metastatic breast cancer in a combined regimen (Slamon et al., 2001). This combined therapy has significantly improved the outlook in those 25 to 30% of patients for later stages of breast cancer that had previously been much more resistant to treatment. In the Slamon et al. study, the HER2 overexpression biomarker was measured by immunohistochemical analysis. Since that study, a number of other means, both genomic and proteomic, for detecting and measuring this important biomarker have been studied. The proteomic quantitation of circulating HER2 receptor protein in serum has been shown to hold promise in predicting disease outcome and response to therapy (Ross et al., 2003).

8.9.7 Case Study

We have seen how proteomics approaches can influence drug discovery significantly at every stage of the process. The following is a specific case study showing how proteomics can be used successfully in the target identification and validation stages. It is about the development of LAF389, an antitumor drug candidate developed at Novartis (Towbin et al., 2003). It depicts the successful application of 2D-PAGE scaled up as a proteome-wide high-throughput method for protein profiling. The example demonstrates the effective use of proteomic techniques for target identification and validation.

The scientists had a compound, bengamide, which was known to exhibit a significant antitumor effect. However, the mechanism of action of action of bengamide was unknown, and its target was yet to be identified. The main use of a proteomic approach was to apply high-scale 2D gel electrophoresis to narrow down the possible targets for bengamides without bias, then pursue the candidates with more detailed assays.

The scientists used 2D gel electrophoresis at high scale to generate 1500 protein spots before and after treatment with bengamide E, with no presumption of a specific target. Any protein spots that showed reproducible differences following bengamide treatment were then considered candidate bengamide targets and were analyzed further with MALDI-MS to identify them. The spots showed differences in charge, indicating a possible PTM effect.

Isoforms of the protein called 14-3-3 stood out in this set, due to their known role in modulation of signaling, cell cycle control, transcriptional control, and apoptosis, suggesting a role in cancer if impaired. So the scientists tested induction of 14-3-3 proteins with LAF389, a bengamide analog. But which isoform was being induced? The scientists used an analytical blotting method to separate 14-3-3 isoforms, then did Western blotting using antibodies against 14-3-3 proteins. IEF immunoblotting was then used with isoform-specific polyclonal antibodies to identify 14-3-3γ as the induced isoform.

Two different MALDI-MS analyses were used to detect the cause of the induced p*I* difference. These determined that an N-terminal 14-3-3γ peptide sequence was acetylated in the absence of LAF389, but retained the unprocessed N-terminal Met (oxidized) in the LAF389-induced samples. This was verified by using a monoclonal antibody to detect unprocessed 14-3-3γ. These results suggested that LAF389 might be inhibiting MetAp, which normally acetylates 14-3-3γ.

An enzymatic assay verified that LAF389 does inhibit MetAp (1 and 2). This established MetAps as direct targets of LAF389. (Their later in vivo studies showed that LAF389 inhibited MetAp2 only.)

Finally, the scientists validated the target using the following two methods:

1. They used siRNA to knock down MetAp2 level by 75%, after which unprocessed 14-3-3γ was detected, verifying the role of MetAp2 n processing 14-3-3γ.
2. Using x-ray crystallography, they determined the crystal structure of MetAp2 bound with LAF153, another bengamide analog. The structure shows that LAF153 binds in the same manner expected of a polypeptide substrate.

Note that a proteomic approach like this was essential for target identification. Gene expression studies could not have detected this mechanism (and did not, as the scientists report). This is because 14-3-3γ is expressed equally in cultures with and without LAF389; the effect of LAF389 (via inhibition of MetAp) is *strictly on a posttranslational modification* (i.e., the acetylation of 14-3-3γ).

8.10 CONCLUSIONS

In this chapter we have described several recently developed enabling technologies that are being applied to drug discovery and development. There has been much interest in increasing the rate and productivity of drug discovery, as the blockbuster drugs that have sustained the industry recently are rapidly coming off-patent and being replaced by cheaper generics. The use of technologies based on genomics, proteomics, metabolomics, and so on, has led drug discovery and development efforts away from a single target mentality and in the direction of an integrated *systems biology* perspective. With this major shift to larger scale comes a resulting major increase in both speed and quality of leads. The emerging technologies described in this chapter aim to address the many aspects of drug development and to enable more effective understanding of targets and compounds through high-throughput high-content analysis. These emerging high-throughput technologies are beginning to show considerable promise for improving and accelerating drug discovery and development and helping to fill the pipelines of pharmaceutical companies.

Improved clinical outcomes of therapeutic candidates at all stages of development would contribute significantly to reducing development costs and improving the number of candidates that succeed in obtaining approval for commercialization. By using the technologies described herein, companies will find that the quality of their drug leads should be higher and they should be less likely to fail in the clinic. New methods that enable better target identification and validation, a deeper understanding of total systems biology and the implications for any specific drug or target in development, and methods to understand the nature of molecular targets to enable more effective screening and lead identification will be extremely valuable in meeting the needs for cost-effective development of new and useful therapeutics for human disease.

ACKNOWLEDGMENTS

The authors wish to thank Ioannis Moutsatsos, Director of Protein Informatics at Wyeth, whose exceptional graduate course in proteomics at Brandeis University has been an invaluable source of knowledge and inspiration in writing this chapter.

REFERENCES

Angenendt, P., et al., (2004), Cell-free protein expression and functional assay in nanowell chip format, *Anal. Chem.* 76:1844–1849.

Bock, C., et al., (2004), Photoaptamer arrays applied to multiplexed proteomic analysis, *Proteomics* 4:609–618.

Bonneau, R., et al. (2001), Rosetta in CASP4: progress in ab initio protein structure prediction, *Proteins*, Suppl. 5:119–126.

Danner, S., et al. (2001), T7 phage display: a novel genetic selection system for cloning RNA-binding proteins from cDNA libraries, *Proc. Natt. Aced. Sci. USA* 98 (23):12954–12959; Nov. 6.

Drews, J. (2000), Drug discovery: a historical perspective, *Science* 287:1960–1964.

Garber, K. (2004), Gene expression tests foretell breast cancer's future, *Science* 303:1754–1755, Mar. 19.

Gavin, A. C., et al. (2002), Functional organization of the yeast proteome by systematic analysis of protein complexes, 1, *Nature* 415(6868):141–147., Jan. 10.

Gershon, D. (2003), Proteomics technologies: probing the proteome, *Nature* 424:581–587, July 31.

Hanash, S. (2003), Disease proteomics, *Nature*, 422, March 13.

Hannon, G. (2002), RNA interference, *Nature*, 418, July.

Huang, R. et al. (2004), Enhanced protein profiling arrays with ELISA-based amplification for high-throughput molecular changes of tumor patients' plasma, *Clin. Cancer Res.* 10:598–609.

Ideker, T., et al. (2001), Integrated genomic and proteomic analysis of a systematically perturbed metabolic network, *Science* 292:929–934.

Ito, T., et al. (2001), A comprehensive two-hybrid analysis to explore the yeast protein interactome, *Proc. Natl. Acad. Sci. USA* 98(8):4569–4574, Apr. 10.

Johann, D. J. et al. (2004), Clinical proteomics and biomarker discovery, *Ann. N.Y. Acad. Sci.* 1022:295–305, June.

Kopp, J. et al. (2004), Automated protein structure homology modeling: a progress report, *Pharmacogenomics* 5(4):405–416, June.

Lee, J., et al. (2003), Differential protein analysis of spasomolytic polypeptide expressing metaplasia using laser capture microdissection and two-dimensional difference gel electrophoresis, *Appl. Immunohistochem. Mol. Morphol.* 11(2):188–193, June.

Li, J., et al. (2002), Proteomics and bioinformatics approaches for identification of serum biomarkers to detect breast cancer, *Clin. Chem.* 48:1296–1314.

Lopez, M., and Pluskal, M. (2003), Protein micro- and macroarrays: digitizing the proteome, *J. Chromatog. B*, 787:19–27.

Luthra, R., et al. (2004), TaqMan RT-PCR assay coupled with capillary electrophoresis for quantification and identification of bcr-abl transcript type, *Mod. Pathol.* 17(1):96–103, Jan.

MacBeath, G., and Schreiber, S. (2000), Printing proteins as microarrays for high-throughput function determination, *Science*, 289; Sept. 8.

MacNeil, J. (2004), Bring on the protein biochips, *Genome Technol.* 43, May.

Mayer, G., et al. (2003), Aptamers: multifunctional tools for target validation and drug discovery, *DrugPlus Int.*, Nov–Dec.

Nadal, E., et al. (2004), Imatinib mesylate (Gleevec/Glivec) a molecular-targeted therapy for chronic myeloid leukaemia and other malignancies, *Int. J. Clin. Pract.* 58(5):511–516, May.

Oh, P., et al. (2004), Subtractive proteomic mapping of the endothelial surface in lung and solid tumours for tissue-specific therapy, *Nature* 429(6992):629–635, June 10.

Pellois, J., et al. (2002), Individually addressable parallel peptide synthesis on microchips, *Nature Biotechnol.* 20(9):922–926, Sept.

Phizicky, E., et al. (2003), Protein analysis on a proteomic scale, *Nature* 422:208–215, Mar. 13.

Robinson, W., et al. (2003), Protein microarrays guide tolerizing DNA vaccine treatment of autoimmune encephalomyelitis, *Nature Biotechnol.* Sept.

Rocken, C., et al. (2004), Proteomics in pathology, research and practice, *Pathol. Res. Pract.* 200(2):69–82.

Rohl, C. A., et al. (2002), De novo determination of protein backbone structure from residual dipolar couplings using Rosetta, *J. Am. Chem. Soc.* 124(11):2723–2729, Mar. 20.

Ross, J. S., et al. (2003), Targeted therapy for cancer: the HER-2/neu and Herceptin story, *Clin. Leadership Manage. Rev.* 17(6):333–340, Nov–Dec.

Shin, B. K., et al. (2002), Proteomics approaches to uncover the repertoire of circulating biomarkers for breast cancer, *J. Mammary Gland Biol. Neoplasia* 7(4), Oct.

Simon, R., et al. (2003), Tissue microarrays in cancer diagnosis, *Expert Rev. Mol. Diagn.* 3(4): 421–430, July.

Slamon, D. J., et al. (2001), Use of chemotherapy plus a monoclonal antibody against HER2 for metastatic breast cancer that overexpresses HER2, *N. Engl. J. Med.* 344:783–792.

Sneider, G., et al. (2002), Virtual screening and fast automated docking methods, *Drug Discov. Today* 7:64–70.

Taussig, M., et al. (2002), Protein arrays resource page, ESF Programme on Integrated Approaches for Functional Genomics, 2002.

Taussig, M., et al. (2003), Progress in antibody arrays, *Targets* 2(4), Aug.

Towbin, H., et al. (2003), Proteomics-based target identification: bengamides as a new class of methionine aminopeptidase inhibitors, *J. Biol. Chem.* 278(52):52964–52971, Dec. 26.

Uehling, M., (2003), Fishing chips, *Bio-IT World*, Sept.

Uetz, P., et al. (2000), A comprehensive analysis of protein–protein interactions in *Saccharomyces cerevisiae*, *Nature* 403(6770):623–627, Feb. 10.

Walgren, J., et al. (2004), Application of proteomic technologies in the drug development process, *Toxicol. Lett.* 149(1–3):377–385, Apr. 1.

Weaver, D. C. (2004), Applying data mining techniques to library design, lead generation and lead optimization, *Curr. Opin. Chem. Biol.* 8(3):264–270, June.

Wolf, K., et al. (2003), Information-based methods in the development of antiparasitic drugs, *Parasitol. Res.* 90:S91–S96.

Zhu, H., et al. (2000), Analysis of yeast protein kinases using protein chips, *Nature Genet.* 26, Nov.

Zhu, H., et al. (2001), Global analysis of protein activities using proteome chips, *Science* 293, Sept. 14.

APPENDIX: PUBLIC-DOMAIN SOFTWARE TOOLS AND DATABASES

Nucleotides, Genomics, Documentation, General

Tools	Purpose
BLAST	Sequence alignment and homology studies (amino acids and DNA formats) http://www.ncbi.nlm.nih.gov/BLAST/
WebLogo	Consensus sequence logo generator http://weblogo.berkeley.edu/
GenScan, WebGene	Gene prediction (including splice sites, CpG islands, repeat elements, polyA sites, promoter regions) http://genes.mit.edu/GENSCAN.html
RepeatMasker	Annotation and masking of interspersed repeats and low-complexity areas (used in conjunction with GenScan) http://repeatmasker.genome.washington.edu/cgi-bin/RepeatMasker

Databases	Purpose
NCBI Entrez	Gateway page for all the NCBI databases: nucleotide and protein DBs, PubMed, Genome, Structure, OMIM, etc. http://www.ncbi.nlm.nih.gov/Entrez/
GeneCard	Compendium of information related to genes gathered automatically; useful; good collection of links in one place, but slow http://bioinfo.weizmann.ac.il/cards/index.html
UCSC Genome Browser	Graphical genome browser; highly configurable http://genome.ucsc.edu/

(Continued)

Databases	Purpose
Map Viewer	Graphical exploration of sequence data with respect to chromosome loci http://www.ncbi.nlm.nih.gov/mapview/
LocusLink	Chromosome and gene location information http://www.ncbi.nlm.nih.gov/LocusLink/
KEGG	Kyoto Encyclopedia of Genes & Genomes: interaction networks (e.g., metabolic and regulatory pathways) http://www.genome.ad.jp/kegg/
DAVID	DB for Annotation, Visualization, and Integrated Discovery: GoCharts (functional), KeggCharts (metabolic), DomainCharts http://apps1.niaid.nih.gov/david/
Homologene	Homologs in model organisms http://www.ncbi.nlm.nih.gov/entrez/query.fcgi?db=homologene
Entrez SNP	SNP analysis http://www.ncbi.nlm.nih.gov/entrez/query.fcgi?db=snp
Entrez Taxonomy	Taxonomic information for phylogenetic work http://www.ncbi.nlm.nih.gov/entrez/query.fcgi?db=Taxonomy
PubMed	Published information on genes and proteins http://www.ncbi.nlm.nih.gov/entrez/query.fcgi?db=PubMed
OMIM	Information and references (good summaries) http://www.ncbi.nlm.nih.gov/entrez/query.fcgi?db=OMIM

Proteins, Proteomics

Tool	Purpose
Clustal	Software for creating multiple sequence alignment (MSA). The source code for CLUSTAL X and several executable versions for different machines are freely available by anonymous ftp to ftp igbmc.u-strasbg.fr. Hypertext documentation can be viewed at: http://www-igbmc.u-strasbg.fr/BioInfo/clustalx/ The NCBI Vibrant Toolkit is available by anonymous ftp from: http://ncbi.nlm.nih.gov
BioEdit	MSA editing and manipulating http://www.mbio.ncsu.edu/BioEdit/bioedit.html
MEGA	Molecular evolutionary genetics analysis: phylogenetic trees from MSAs http://www.megasoftware.net/
DeepView (spdbv)	Swiss-PDB viewer: for protein 3D structure analysis, homology modeling, and visualization http://www.expasy.org/spdbv/
PeptIdent	Identification of proteins using experimental pI, Mw, and peptide mass fingerprinting data; compares to SwissProt database to find closest matches http://us.expasy.org/tools/peptident.html
Mascot	Uses mass spectrometry data to identify proteins from primary sequence databases http://www.matrixscience.com/
PredictProtein	Predicts secondary protein structure, given an amino acid sequence http://www.embl-heidelberg.de/predictprotein/

Tool	Purpose
ProtScale	Protein profiling tools using a number of scales: hydrophobicity, polarity, α-helix, β-sheet, etc. http://www.expasy.org/cgi-bin/protscale.pl
Rosetta	Ab initio 3D protein fold prediction using David Baker's Rosetta algorithm. Rosetta server available at: http://www.bioinfo.rpi.edu/~bystrc/hmmstr/server.php

Database	Purpose
Expasy–SwissProt	Curated, highly annotated protein database http://us.expasy.org/ http://us.expasy.org/sprot/
InterPro, PFAM	Protein families, domains; graphical domain structure alignments http://www.ebi.ac.uk/interpro/ http://www.sanger.ac.uk/Software/Pfam/index.shtml
InterDom	Database of interacting domains: database of putative interacting protein domains from both experimental and computational sources http://interdom.lit.org.sg/
ProSite	Documentation; consensus patterns http://us.expasy.org/prosite/
CDD	Conserved domain database: more comparative structure and domain information; includes MSAs http://www.ncbi.nlm.nih.gov/entrez/query.fcgi?db=cdd
PDB, PDBsum	Protein database; primarily for 3D structures; crystallographic http://www.rcsb.org/pdb/index.html http://www.biochem.ucl.ac.uk/bsm/pdbsum/index.html
DIP	Database of interacting proteins http://dip.doe-mbi.ucla.edu/
Reactome	Curated database of core human biology reactions and pathways http://www.reactome.org/
BioCarta	Categorization of pathways of all kinds: metabolic, cell signaling, cell cycle regulation, apoptosis, etc http://www.biocarta.com/genes/index.asp
Swiss-2DPage	Two-dimensional polyacrylamide gel electrophoresis database: identification of proteins by mass/charge spots in 2D gel database http://au.expasy.org/ch2d/
SCOP	Structural classification of proteins (e.g., by secondary, tertiary, etc.) http://scop.mrc-lmb.cam.ac.uk/scop/index.html

9

USING DRUG METABOLISM DATABASES DURING DRUG DESIGN AND DEVELOPMENT

Paul W. Erhardt
The University of Toledo College of Pharmacy
Toledo, Ohio

9.1 INTRODUCTION

The type of drug typically pursued during pharmaceutical research and development is one that can be used with a wide margin of safety via oral administration to humans. In this chapter we consider drug metabolism within that context. Thus, it should be appreciated immediately that beyond a drug's specific structural motifs that dictate its distinct interactions with selected sets of human metabolizing enzymes, other structural features that affect a drug's absorption, distribution, and elimination also represent key factors that become equally important toward influencing the overall course of its metabolism. For example, a drug that is metabolized by enzyme A to a greater extent than by enzyme B during a pair of in vitro assays may still be significantly metabolized by B within an in vivo setting if its exposure to B is substantially greater than its exposure to A. The degree of exposure to an enzyme will be determined by the prevalence of that enzyme in various compartments coupled with the drug's distribution into those compartments. Figure 9.1 displays the interplay of these considerations by tracing the path that a central nervous system (CNS) drug will traverse on route to its site of efficacious action after oral administration. Steps highlighted by an asterisk represent compartments that are known to have high levels of metabolic activity. Tables 9.1 and 9.2 indicate how the spectrum, as well as the overall level, of metabolic activity varies across several different tissues and across several classic human phenotypes, respectively.[1] In the end, the overall course of a drug's metabolism will depend upon the drug's initial and continued distribution to the sets of metabolizing enzymes variously displayed within the metabolically active compartments combined with

Drug Discovery and Development, Volume 1: Drug Discovery, Edited by Mukund S. Chorghade

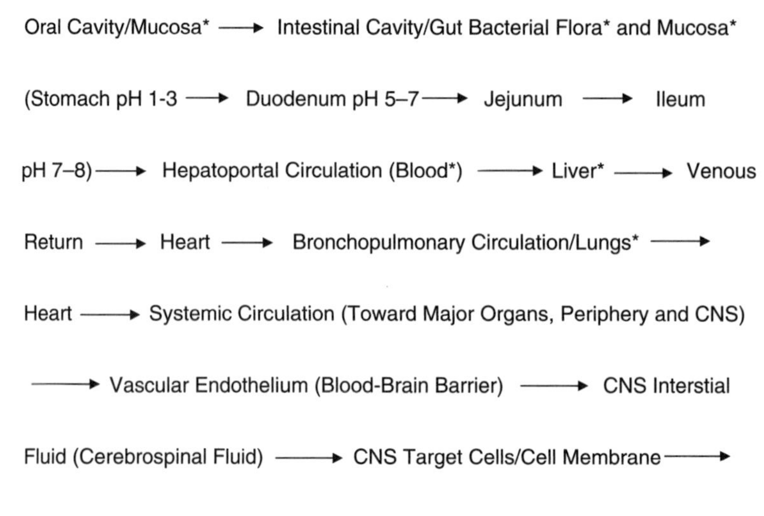

Figure 9.1 Path taken by a CNS drug to its site of action after oral administration. Compartments having high metabolic activity are marked by an asterisk. Although the intestinal mucosa and liver are regarded as the principal organs associated with the *first-pass effect*, the lungs also display considerable xenobiotic metabolizing capability and can be considered to represent the key pseudo-first-pass organs[2] for all routes of administration other than oral or intraperitoneal. The asterisk placed on the blood compartment reflects high levels of specific esterase and amidase activities rather than a high level of overall metabolic capability.

TABLE 9.1 Metabolic and Excretion Capabilities Displayed by Selected Tissues

Tissue	Metabolic/Excretion Activities[a]
Gastrointestinal mucosa	Rich in CYP 3A, but essentially all CYPs are represented, gradually peaking in the duodenum region and then falling off toward the ileum; significant glucuronide and sulfate conjugation pathways; significant monamine oxidase activity.
Blood	Rich in various esterase and amidase enzymes.
Liver	Rich in essentially all CYPs and conjugation pathways. Excrete compounds into bile having MW $\geq$ 500 g, particularly when highly polar, conjugated molecules.
Lungs	Represented by most of the CYPs and conjugation pathways with certain levels being comparable or even higher than levels found in the liver (e.g., CYP 2A13 as recently implicated in the carginogenicity of nicotine).[3]
Vasculature	Neither the BBB, the placenta, or the mammary gland *barriers* exhibit significantly enhanced overall xenobiotic metabolizing capabilities.
Kidneys	Rich in the various protease enzymes; excrete compounds having MW $\leq$ 500 g; highly polar water-soluble compounds are not reabsorbed from the urine.

[a]CYP, cytochrome P450 enzyme[4]; MW, molecular weight; BBB, blood–brain barrier.

TABLE 9.2 Metabolic Capabilities Displayed by Selected Human Phenotypes

Phenotype	Characteristic Metabolic Activities[a]
Debrisoquine[b]	*Extensive* versus *poor metabolizers*. Inherited defect in CYP 2D6 expression: 5 to 10% Caucasian, 2% Oriental, and 1% Arabic populations.
Phenytoin[b]	*Normal* versus *poor metabolizers*. Inherited defect in CYP 2C18: 15 to 20% Oriental and 2 to 6% Caucasian populations.
Aldehyde oxidase[c]	Lack of indicated capability; significant polymorphism among the Oriental population (ca. 50%).
Acetylation	*Fast* and *slow acetylators*. Individual variations in Nat2: 30 to 40% Caucasian, 80 to 90% Oriental, and 100% Eskimo populations are *fast*.
Neonatal	Overall, CYP isoform activities typically lower than children to adult ranges except for the 3A subfamily; blood esterase activity appears to be about 50% at birth[5]; immature UGT pathways fairly common leading to decreased clearance of bilirubin or *gray baby syndrome*.
Elderly	Most studies directed toward enzymatic activity per se suggest that levels remain comparable to younger populations; alternatively, decreased hepatic blood flow can significantly lead to a decrease in the overall metabolism and excretion of xenobiotics.

[a]CYP, cytochrome P450 enzymes[4]; NAT2, *N*-acetyl transferase isoform 2[4]; UGT, uridinediphosphoglucuronosyl transferase.[4]

[b]This phenotype extends to numerous other drugs.

[c]This phenotype is largely characterized by an altered pathway for ethanol metabolism.

the duration of the drug's stay and its substrate suitability for those enzymes, all as a function of drug concentration and as the summation of competing metabolic processes from one compartment to another over time.

Given the complexity of the aforementioned scenario, two approaches can be contemplated toward predicting the metabolism of new drugs being considered for use in humans. The first involves dividing the in vivo setting into simpler parameters that are amenable to either quick experimental or theoretical assessment, followed by reassembling the individual sets of data obtained for a new compound in a manner that accounts for the complexity of the intact human. The second approach involves the query of established databases for structures or structural elements that are similar to the new drug wherein the database's information has been assembled from drug metabolism studies already undertaken within the in vivo setting. Both of these approaches are being deployed by the pharmaceutical industry at very early points in the overall process of drug discovery, the first by adopting high-throughput screening (HTS) methods to assess at least certain of the simplified parameters, and the second by using commercially available drug metabolism databases.[6] The status of the latter approach is reviewed specifically in the remainder of this chapter.

9.2 HISTORICAL PERSPECTIVE

Drug metabolism efforts within the pharmaceutical industry have necessarily tended to focus on the specific issues associated with the development of lead compounds selected for their potential as therapeutic agents and not for their suitability to serve as molecular or mechanistic probes to answer metabolic questions. As a result, the drug metabolism literature has to a large extent become anecdotal, replete with case-by-case examples but having less information pertaining to systematic investigations directed toward the rational development of broader principles useful for predicting the biotransformations of new, biologically interesting compounds in general.[6,7] Nevertheless, as a result of many early workshops, reviews, and textbook treatments (e.g., refs. 8, 9 to 11, and 12 to 14, respectively) as well as to numerous updates in the form of periodic reviews and specialist reports dedicated to the topic of drug metabolism (e.g., refs. 15 to 17), generalizations about the likely metabolic reactions of certain chemical features have gradually come to be accepted with a high degree of confidence, the propensity for ester hydrolysis, O- and N-dealkylations, and for aromatic hydroxylation being exemplary.[18]

As drug metabolism data have continued to accumulate, it has become convenient to store and sort this information using computerized approaches (e.g., refs. 19 to 26). In this regard, two expert systems emerged as the first leaders in this area. Metabol Expert[27–29] and META[30–33] have been constructed by compiling literature intentionally limited to well-established sources such as textbooks and reviews so as to assure the quality of the information contained within the database. By using computer-driven queries across these databases, one can identify sites on a new molecule where metabolic biotransformations are likely to occur. Unfortunately, these early databases have indiscriminately combined metabolism data available from studies that employed a variety of mammalian species, and consequently, their programs tend to predict all of the metabolic possibilities for an exogenous material when the latter is placed in a *theoretical "average" mammal*.[31] Recognizing the need to prioritize the often numerous metabolic possibilities, a priority number based on a scale of 1 (fast biotransformation) to 9 (slow biotransformation) also accompanies each prediction from the META database.

Subsequent to the development of the early expert systems, two databases that represent collections of drug metabolism data have also become available. Metabolite[34,35] is a broad collection of metabolism data that are being accumulated without bias as to literature source. Thus, the strength of this database may eventually lie in its extensive quantity of data rather than in the quality of each of its data entries. Alternatively, the Accelrys Metabolism Database[36,37] represents a computerized version of data taken from the well-respected Biotransformations series edited by Hawkins.[17] Substructure queries across these databases can be used to predict biotransformations for new compounds by finding data for actual compounds within the database that are closest in structure to each query. Finally, METEOR[38,39] is a recently released expert system that, like META, has sought to provide a ranking for the predicted metabolic possibilities arising from a new structural query. These five databases are summarized in Table 9.3.

9.3 PRESENT STATUS

A survey on the use of drug metabolism databases within the industry has recently been assembled as part of a book devoted to this overall topic.[6] While the survey's responses largely reflect experiences that relate to using some of the older databases during the later stages of the drug discovery process, the survey's consensus points become informative toward appreciating the pros and cons of today's attempts to develop and deploy such databases during the early stages of drug discovery. The survey's consensus points are highlighted in Table 9.4.

TABLE 9.3 Commercially Available Drug Metabolism Databases[a]

Database	Vendor	Approx. Number		Search Paradigms	Recommended Platform	Compatible Software	Approx. Cost (thousands of dollars)
		Cmpd.	Biotran.				
MetabolExpert	CompuDrug, Inc.[b] South San Francisco, CA 415-271-8800 www.compudrug.com	—	—	Chemical structure; transformation; test system	PC	PALLAS	10
META	Multicase, Inc. Beachwood, OH 216-831-3740 www.multicase.com	—	—	Chemical structure	VAX; Open VMS; Windows-based PC	—	20
Metabolite	MDL Information Systems, Inc. San Leandro, CA 510-895-1313 www.mdli.com	30,000	100,000	Chemical structure; substructure; similar structure; transformation; data searcher	PC; MAC		40
Metabolism Database	Accelrys, Ltd. Leeds, UK 44 113 224 9788 www.accelrys.com	3,000	25,000	Chemical structure; transformation; test system; author; journal name; keywords	SGI; Windows NT; Sun Solaris; Open VMS	ISIS; REACCS; ACCORD Database Explorer	50
METEOR	LHASA University of Leeds Leeds, UK 44 113 233 6531 www.chem.leeds.ac.uk/luk						

[a]Entries reflect status at end of 2001.

TABLE 9.4 Consensus Points from Industry Survey About Using Drug Metabolism Databases

- Predicting all metabolic possibilities within a theoretical "average" mammal creates a dense forest of information from which it becomes difficult to discern the specific pathways that might be associated with the human response.
- In most cases, many of the suggestions were already suspected from a simple visual inspection of the query molecules, whereas in other cases, biotransformations were missed despite the existence of specific literature precedent.
- The need to convey statistically derived metabolic probabilities rather than a list of possibilities is critical.

Source: Ref. 6.

In addition to the points listed in Table 9.4, two major areas of concern were identified. Both concerns need to be addressed if we are to more advantageously exploit computerized approaches toward the adoption of metabolism considerations into the early stages of drug discovery. The two areas involve initial biological data entry/sorting, followed by chemical structural entry/inquiry. In terms of biologically-related issues, concerns about indiscriminately combining metabolic data from different species have already been alluded to. Although there is some analogy to the drug action literature where simple allometric scaling and physiologically based pharmacokinetic (PBPK) modeling methods can be employed to adjust or better quantitate appropriate dosing of the efficacious levels for a drug to be used in different species,[40] the potential impact that species differences have upon the metabolism pathways are still of concern because the latter can also deviate significantly in a qualitative fashion. For example, consistently distinct differences in phase II metabolism (conjugation pathways) between various species are well established.[41,42] Alternatively, differences in phase I metabolism seem to vary in an inconsistent manner. In most cases where systematic studies have been conducted across different species, qualitative as well as quantitative differences have been observed between the metabolic profiles obtained for a given set of selected xenobiotics, even when the species have been restricted to various mammals.[43] Similarly, as mentioned in the introduction, the potential for obtaining significantly divergent results within the same species when comparing in vitro versus in vivo data should be reiterated. Thus, the metabolic profile for a drug that is distributed away from the liver after intravenous administration is likely to be quite different from that which would be suggested by its in vitro study using liver microsomal fractions. As above, in most cases where systematic studies have been conducted across different tissues within the same species, differences have been observed between the metabolic profiles obtained for any given series of selected xenobiotics[44] (Table 9.1).

Interestingly, the factors pertaining to distribution and pharmacokinetic half-life may actually be larger concerns while attempting to cross-relate data obtained from in vitro versus in vivo studies in the same species than while cross-relating data from in vivo studies obtained between two different species. For example, analyses among our collaborators have shown that there is a statistically significant correlation between the half-lives of a wide range of drugs when determined in two very different species, rat data versus human data.[45,46] The final concern in the biological area has also become encompassed by the new field of pharmacogenetics. It involves the growing appreciation that there can be significant differences in drug metabolism due simply to subtle individual differences in enzymatic phenotype. Here, the metabolism of the same drug studied in the same species using the same method (in vitro or in vivo) can still vary from individual to individual based on classic differences in population phenotypes[47] (Table 9.2), on disease-related differences in phenotype,[48,49] or due to the impact on phenotype resulting from a person's total metabolic history of previous and ongoing exposures to xenobiotics.[50] It should be clear at this point that many of the issues that have complicated earlier systematic investigations within the field of drug metabolism remain and serve to exacerbate the additional challenges associated with today's construction and use of much larger, generalized databases.

It can be imagined that at least some of these issues might be addressed as a series of initial sorting steps when data are being entered. Although this strategy would be cumbersome initially, it would result in a series of biologically intelligent and perhaps more manageable databases. Alternatively, if enough searchable terms pertaining to the concern areas were entered into a single relational database along with the actual metabolism data, appropriately factored searching paradigms, perhaps coupled with rational PBPK considerations throughout, could seemingly be devised to help surmount these concerns on the query end. Ultimately, whether these biological "apples and oranges" are separated as the data are entered, searched, or subsequently rationalized, the desired search query will eventually need to be linked with chemical structure. The latter constitutes a different aspect of constructing databases which has its own set of challenges. An initial consideration of key chemical issues follows.

In terms of chemical-related issues, there are two areas that come to immediate attention. The first involves a fundamental question that is associated with the pursuit of structure metabolism relationships (SMRs) and their use in the effective deployment of metabolism databases. Namely, what is the proper role to expect the entire structure to be playing versus the discrete roles of its displayed organic functionality? After all, it is the latter that actually undergo metabolism, and just as a pharmacophore defines the pattern of structural elements requisite for a drug's interaction with a biologic receptor or enzyme active site, it will be the particular array of functional groups and their immediate molecular environments which dictate what happens once a drug is present at the compartment where a specific metabolic conversion is to take place. Numerous secondary issues also stem from this question, but all of these lead to a common concern about how much detail needs to be included for a given chemical structure data entry or query. While the level of detail that can be applied toward molecular description/searching spans a considerable range, a second chemical issue additionally presents itself at this juncture: namely, the accuracy of the initial chemical depiction. For example, one extreme would be a thorough, three-dimensional electronic surface map across the entire structure as obtained from x-ray analysis and/or rigorous computational treatments, additionally coupled with experimental physicochemical information or descriptors to also address the distribution issues. The other extreme would be a simple two-dimensional figure, perhaps energy minimized with only the aid of an automated drawing program. This particular chemical quandary is addressed further in Section 9.4. For the present, practicality has dictated that the most simple structural data entry possible be used due to the sheer amount of valuable anecdotal information that is already available. However, even in this case it may be expecting too much from a given database to be able to accurately predict the entire metabolic outcome for a new compound using a single structural query. Rather, it may be more appropriate to think in terms of a multistep approach that might first include analyses across a parent database in terms of various metabolic functional groups so as to initially produce a series of specific SMR maps which characterize key *metabophore*[2,7] elements in terms of the probabilities that they might undergo a particular metabolic reaction based on such occurrences relative to other possibilities. The metabophores may then need to be better refined by inputting considerably more precise structural detail. Such a strategy would certainly be in line with today's trend toward smarter libraries as well as with the trends being utilized to integrate various [other bioinformatic and chemoinformatic] systems.[51] Employing this set of species specific atlases, each having its own list of refined metabophore/relative metabolic probability maps, in conjunction with a new drug structure query then becomes a third step in an overall searching paradigm that will probably also need to be continuously guided in a rational manner by appropriately considering distribution and pharmacokinetic issues throughout.

9.4 FUTURE PROSPECTS

From the foregoing discussions, two critical hurdles loom in front of using xenobiobitic metabolism databases to more effectively predict human drug metabolism in the future. The first involves the need for an improved treatment of three-dimensional structure within chemical-related databases in general, and the second involves the need to be able to better correlate the various types of metabolism-related data to the human clinical experience. The chemical structure hurdles are addressed first.

Handling chemical structures and chemical information within the setting of large databases represents a specialized exercise complicated enough to merit its own designation as a new field, that of *chemoinformatics*.[52–55] As mentioned, there is a significant need for improvement in the handling of chemical structures beyond what appears to be occurring within today's database assemblies. For example, that "better correlations are sometimes obtained by using 2D displays of a database's chemical structures than by using 3D displays" only testifies to the fact that we are still not doing a very good job at developing the later.[56] In general, the handling of small molecules and of highly flexible molecular systems[57] is controversial, with the only clear consensus being that treatments of small molecules for use within database collections "have, to date, been extremely inadequate."[58] Certainly, a variety of automated, three-dimensional chemical structure drawing programs are available that can start from simple two-dimensional representations by using Dreiding molecular mechanics or other user-friendly automated molecular mechanics-based algorithms, as well as by using data expressed by a connection table or linear string.[59] Some programs are able to derive three-dimensional structure "from more than 20 different types of import formats."[60] Furthermore, several of these programs can be directly integrated with the latest versions of more sophisticated quantum mechanics packages such as Gaussian 98 MOPAC (with MNDO/d) and extended Hückel.[55,59] Thus, electronic handling of chemical structures and to a certain extent comparing them in three-dimensional formats has already become reasonably well worked out.[52–55,59–62] Table 9.5 is a list of some of the three-dimensional molecular modeling products that have become available during the 1990s.[61]

The fundamental problem that remains, however, is how the three-dimensional structure is derived initially in terms of its chemical correctness, the latter being dependent on what assumptions might have been made during the process of energy minimization. This situation is further complicated by the additional need to understand how a given drug molecule's conformational family behaves during its interactions with each of the biological environments of interest: those associated with all of the compartments traversed in Fig. 9.1, along with those associated with each of the specific metabolizing enzymes that the drug

TABLE 9.5 Three-Dimensional Molecular Modeling Packages that Became Available During the 1990s

Package	Company	Platform	Description
Low-end sophistication			
Nano Vision	ACS Software	MAC	Simple, effective tool for viewing and rotating structures, especially large molecules and proteins
Ball & Stick	Cherwell Scientific	MAC	Model building and visualization; analysis of bond distances, angles
MOBY	Springer-Verlag	IBM (DOS)	Model building and visualization; classical and quantum mechanical computations; large molecules and proteins; PDB files
Nemesis	Oxford Molecular	IMB (Wind) MAC	Quick model building and high-quality visualization; geometric optimization (energy minimization)
CSC Chem. 3D/Chem 3D Plus	Cambridge Scientific	MAC	Easy-to-use building and visualization; geometric optimization; integrated 2D program and word processing
Alchemy III	Tripos Assoc.	IBM (DOS, Wind) MAC	Quick model building; energy minimization; basic calculations; easy integration to high-end systems
PC Model	Serena Software	IBM (DOS) MAC	Low cost with sophisticated calculations; platform flexibility
Mid range sophistication			
CAChe	Tektronic	MAC	Sophisted computation tools; distributed processing
HyperChem	Auto Desk	IBM (Wind) Silicone Graphics	Easy-to-use array of computation tools (classical and semiempirical quantum mechanics)
Lab Vision	Tripos Assoc.	IBM (RISC-6000) Silicone Graphics DEC, VAX	Sophisticated but practical modeling for research
High-end sophistication			
SYBL	Tripos Assoc.	IBM (RISC-6000) Silicone Graphics DEC VAX, Sun 4, Convex	Integrated computation tools for sophisticated structural determination and analysis; database management
CERIUS	Molecular Simulations	Silicone Graphics IBM (RISC-6000) Stardent Titan	Suite of high-performance tools for building and simulating properties

Source: Ref. 61.

will eventually encounter. To track these conformational behaviors in a comprehensive manner, it becomes necessary to consider a drug's multiple conformational possibilities by engaging as many different types of conformational assessment technologies as possible while initially taking an approach that is unbiased by any molded relationship dictated by a specific interacting environment. For example, three common methodological approaches include (1) x-ray; (2) solution spectroscopic methods such as nuclear magnetic resonance (NMR), which can often be done in both polar and nonpolar media; and, (3) computational approaches, which can be done with various levels of solvent and heightened energy content but are limited by the assumptions and approximations that need to be taken in order to simplify the mathematical rigor so as to allow computational solutions to be derived in practical time periods. Analogous to the simple, drawing program starting points, programs are available for conversion of x-ray and NMR data into three-dimensional structures.[63]

An example of how three-dimensional conformation might be addressed is provided by the following description of an ongoing project in our labs that pertains to construction of a human drug metabolism database. Structures are initially considered as closed-shell molecules in their electronic and vibrational ground states with protonated and unprotonated forms, as appropriate, also being entered. If a structure possesses tautomeric options, or if there is evidence for the involvement of internal hydrogen bonding, the tautomeric forms and the hydrogen-bonded forms are additionally considered. Determination of three-dimensional structure is carried out in two steps. Preliminary geometry optimization is affected by using a molecular mechanics method, in our case the gas-phase structure being determined by applying the MacroModel 6.5 modeling package running on a Silicone Graphics Indigo 2 workstation with modified (and extended) AMBER parameters. Multiconformational assessment using systematic rotations about several predefined chemical bonds with selected rotational angles is then conducted to define the low-energy conformers and conformationally flexible regions for each starting structure. In the second step, the initial family of entry structures are subjected to *ab initio* geometry optimizations, which in our case use the Gaussian 98 package running at the T90 machine in the Ohio Supercomputer Center resource. Depending on the size of the molecule, 3-21G* or 6-32G* basis sets[64] are used for conformational and tautomeric assessments. Density functional theory using the B3LYP functional[65] is applied for the consideration of exchange correlation energy while keeping the required computer time at reasonable levels. The highest-level structure determination is performed at the B3LYP/6-31G* level. To ascertain the local energy minimum character of an optimized structure, vibrational frequency analysis is carried out using the harmonic oscillator approximation. Determination of vibrational frequencies also allows for obtaining thermal corrections to the energy calculated at 0 K. Free energies are then calculated at 310 K (human body temperature). From the relative free energies calculated, the gas-phase equilibrium constant and the composition of the equilibrium mixture can be determined. Although these values may not be relevant in polar media such as an aqueous environment or the blood compartment, the calculated conformational distribution is relevant for nonpolar environments that may be encountered when a drug passively traverses membranes or begins to enter the cavity of a nonhydrated receptor/enzyme active site just prior to binding. Repetition of this computational scheme from biased starting structures based on actual knowledge about the interacting biological systems or from x-ray or NMR studies (particularly when the latter have been conducted in polar media), followed by studies of how the various sets of information become interchanged and how they additionally behave when further raised in energy, complete the chemical conformational analyses that are being done for each structure being adopted into our human drug metabolism database.

As mentioned earlier, however, after taking an unbiased structural starting point, structures also need to be considered by ascertaining what their relevant conformations might be during interactions within various biological milieus. It can be imagined that at least within the immediate future, a useful range of such environments to be considered will include aqueous solutions of acidic and neutral pH: namely, at about 2 (stomach) and 7.4 (physiological), respectively; one or more lipophilic settings, such as might be encountered during passive transport through membranes; and finally, specific biological receptors and/or enzyme active site settings that are of particular interest. Importantly, with time this list can then be expected to grow further so as to include several distinct environmental models deemed to be representative for interaction with various transportophore relationships; several distinct environmental models deemed to be relevant for interaction with specific metabophore relationships such as within the active site of a specific cytochrome P450 metabolizing enzyme; and finally, several distinct environmental models deemed to be relevant for interaction with specific toxicophore relationships. If x-ray, NMR, and so on, can be further deployed to assess any one or combination of these types of interactions, a composite approach that deploys as many as possible of these techniques will again represent the most ideal way to approach future conformational considerations within the variously biased settings. Advances toward experimentally studying the nature of complexes where compounds are *docked* into real and model biological environments are proceeding rapidly in all of these areas. In addition to the experimental approaches, computational schemes will probably always be deployed because they can provide the relative energies associated with all of the various species. Furthermore, computational methods can be used to derive energy paths to get from the first set of unbiased structures to a second set of environmentally accommodated conformations in both aqueous media and at biological surfaces. Importantly, these paths and their energy differences can then be compared within database settings along with the direct comparison of the structures themselves, while attempting to uncover and define correlations between chemical structure and some other informational field.

Finally, it should be noted that by using computational paradigms, these same types of comparisons (i.e., among and between distinct families of conformationally related members) can also be done for additional sets of conformations that become accessible at increased energy levels (i.e., at one or more 5 kcal/mol increments of energy) so as to simulate the beneficial losses of energy that might be obtained during favorable binding with receptors or active sites.[66] These types of altered conformations can also become candidates for structural comparisons between databases. The latter represents another important refinement that could become utilized as part of SAR queries that will need to be undertaken in the future. With time, each structural family might be addressed by treating the three-dimensional displays in terms of coordinate point schemes or *graph theory* matrices.[67] This is because these older methods lend themselves to the latest thoughts pertaining to utilizing intentionally *fuzzy coordinates*[68,69] (e.g., $x \pm x'$, $y \pm y'$, *and* $z \pm z'$ for each atomic point within a molecular matrix wherein the specified variations can be derived intelligently from the composite of aforementioned computational and experimental approaches). Alternatively, the fuzzy strategy might become better deployed during the searching routines, or perhaps both knowledgeably fuzzy data entry and knowledgeably fuzzy data searching engines handled, in turn, by fuzzy hardware[70] will ultimately best identify the correlations that are being sought in any given search paradigm of the future. It should be noted, however, that for the fuzzy types of structural treatments, queries will be most effective when the database has become large enough to rid itself statistically of the additional noise that such fuzziness will initially create.

Regardless of how they are evolved exactly, what this section points to is that ultimately, the chemical structural databases of the future will probably have several *tiers*[71] of organized chemical and conformational information available which can be mined distinctly according to the specified needs of a directed searching scheme while still being able to be mixed completely within an overall relational architecture such that undirected *knowledge-generating mining paradigms* can also be undertaken.[72–77] Certainly, simple physicochemical data will need to be included among the parameters for chemical structure storage. Similarly, searching engines will need to allow for discrete substructure queries as well as for assessing overall patterns of *similarity* and *dissimilarity*[78–85] across entire electronic surfaces.

It can be noted that it is probably already feasible to place most of the clinically used drugs into a structural database that could at least begin to approach the low to midtier levels of sophistication because considerable portions of such data and detail are probably already available within the literature, even if it is spread across a variety of technical journals for each drug. On the other hand, it should also be clear that an alternative strategy will be needed to handle the mountains of research compounds associated with a single HTS survey. Taken together, the present discussions suggest that we have a long way to go toward achieving the aforementioned tiers of conformational treatments when dealing with large databases and applying them to the process of drug discovery. Nevertheless, because of the importance of chemoinformatics toward understanding, fully appreciating, and ultimately implementing bioinformatics along the practical avenues of new drug discovery, it can be imagined that future structural fields within databases, including those associated with drug metabolism, may be handled according to the following scenario, as summarized from the ongoing discussion in this section and as also conveyed in Fig. 9.2.

For optimal use in the future, it is suggested that several levels of sophistication will be built into database architectures so that a simple two-dimensional format can be input immediately. Accompanying the simple two-dimensional structure field would be a field for experimentally obtained or calculated physicochemical properties. Although this simple starting point would lend itself to some types of rudimentary structure-related searching paradigms, the same compound would then gradually progress by further conformational study through a series of more sophisticated chemical structure displays. As mentioned earlier, x-ray, NMR, and computational approaches toward considering conformation will be deployed for real compounds, whereas virtual compound libraries and databases will rely on computational approaches or on knowledgeable extrapolation from experimental data derivable by analogy to structures within overlapping similarity space. Eventually, structures would be manipulated to a top tier of structural information. This tier might portray the population ratios within a conformational family for a given structure entry expressed as both distinct member and averaged electrostatic surface potentials wherein the latter can be further expanded so as to display their atomic orientations by fuzzy graph theory for fuzzy three-dimensional coordinate systems. Thus, at this point it might be speculated that an intelligently fuzzy coordinate system could eventually represent the highest level of development for tomorrow's three-dimensional *quantitative SAR*[86,87]-based searching paradigms. Furthermore, it can be imagined that this top tier might actually be developed in triplicate for each compound: that is, one informational field for the environmentally unbiased structural entries, another involving several subsets associated with known or suspected interactions with the biological realm, and a third for tracking conformational families when raised by about 5 to 10 kcal/mol in energy. Finally, conformational and energetic considerations pertaining to a compound's movement between its various displays can also be expected to be further refined so as ultimately to allow future characterization

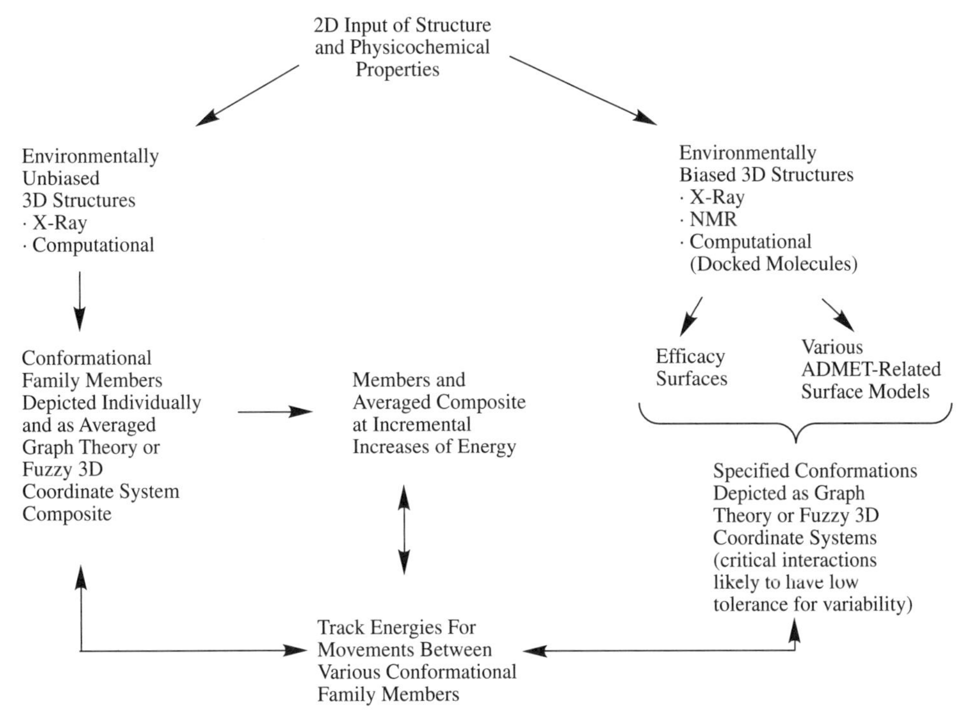

Figure 9.2 Handling chemical structures within databases of the future. This figure depicts the quick entry and gradual maturation of structures. Structure entry would be initiated by a simple two-dimensional depiction that is gradually matured in conformational sophistication through experimental and computational studies. Note that structures would be evolved in both an unbiased and in several environmentally biased formats. The highest structural tier represents tracking and searching the energies required for various conformational movements that members would take when going from one family to another. Search engines, in turn, would also provide for a variety of flexible query paradigms involving physical properties with both full and partial (sub)structure searching capabilities using pattern overlap/recognition, similarity–dissimilarity, CoMFA, and so on.

and searching of the dynamic chemical events that occur at the drug–biological interface (e.g., modes and energies of docking trajectories and their associated molecular motions relative to both ligand and receptor/active site). This top tier is extremely valuable for fully understanding the interactions of interest to drug metabolism, a situation made apparent by the large amount of effort already going on today in this area.[88–93]

Similarly, chemical structure search engines of the future will probably be set up so that they can be undertaken at several tiers of sophistication, the more sophisticated requiring more expert-based inquiries and longer search times for the correlations to be assessed. A reasonable hierarchy for search capability relative to the structural portion of any query might become (1) simple two-dimensional structure with and without physicochemical properties; (2) three-dimensional structure at incremented levels of refinement; (3) two-dimensional and three-dimensional substructures; (4) molecular similarity–dissimilarity indices; (5) fuzzy coordinate matrices; (6) docked systems from either the drug's or the receptor or active site's view at various levels of specifiable precision; and finally, in the more distant future, (7) energy paths for a drug's movement across various biological milieu, including the trajectories and molecular motions associated with drug-receptor/active site docking scenarios. Emphasizing informatics flexibility, this type of approach, where data entry can occur rapidly for starting structure displays and then be gradually matured to more sophisticated displays as conformational details are accurately accrued, coupled with the ability to query at different levels of chemical complexity and visual displays[94] at any point during database maturation, should allow for chemically creative database mining strategies to be effected in the near term as well as into the more distant future.

The second major hurdle toward more effectively deploying drug metabolism databases involves the critical need to better define the correlation of preclinical data with what can then be expected to occur in humans. In other words, a ready mechanism or protocol needs to be available that can be used to provide for validation of a given HTS model relative to the model's contribution toward projecting the eventual clinically observed composite of variously parameterized HTS events. In this regard, it can be emphasized that the amount of metabolism data produced at the clinical level is actually quite small compared to the rather large amount of metabolism data available from preclinical studies. This situation has prompted an effort by our laboratories to establish a specific human drug metabolism database.[95] This relational database coupled with substructure searching capability should be derived solely from human clinical results that are continually being contributed by all interested practitioners. In turn, the database should be available on the WWW via a nonprofit mechanism. Thus, the operation and utility of this metabolism database might be imagined to be somewhat similar to that of the Cambridge x-ray collection, the protein databank, or to some of the newer gene-related informational resources that have been made available over the WWW on a nonprofit basis.

The sheer size of such a common database can overcome the anecdotal nature of the numerous smaller collections presently being held individually by the big pharma members of the pharmaceutical enterprise. Importantly, the database's growing size will eventually allow it to be utilized to develop more accurate and meaningful human SMRs. Selected aspects of the overall SMRs, in turn, can still be applied by individuals in a proprietary fashion to better predict the metabolic fate of their own, specific structural motifs. Similarly, a specific human metabolism database would support rather than compete with the ongoing activities of the existing metabolism database vendors. The latter have already collected data from numerous species and various testing paradigms, all of which will still be very much required as critical road maps during new drug development for quite some time. Finally, and perhaps most important, the assembly of this type of database may be the only way to assess and validate the actual utility of the ongoing explosion of biochemical and in vitro metabolism data and HTS techniques presently being directed toward resolving metabolism issues at the earliest possible stages of drug discovery. The benefits of such a database are summarized in Table 9.6.

TABLE 9.6 Utility of a Human Xenobiotic Metabolism Database

- Will be available on the WWW via a nonprofit format
- Will allow explicit structure searching of standards selected to validate proprietary drug metabolism screens
- Will allow substructure searching to identify analogous metabolic occurrences within humans relative to proprietary compounds undergoing drug development
- Will have large number of biotransformation entries so that statistically derived probability assessments can be made about all metabolic possibilities

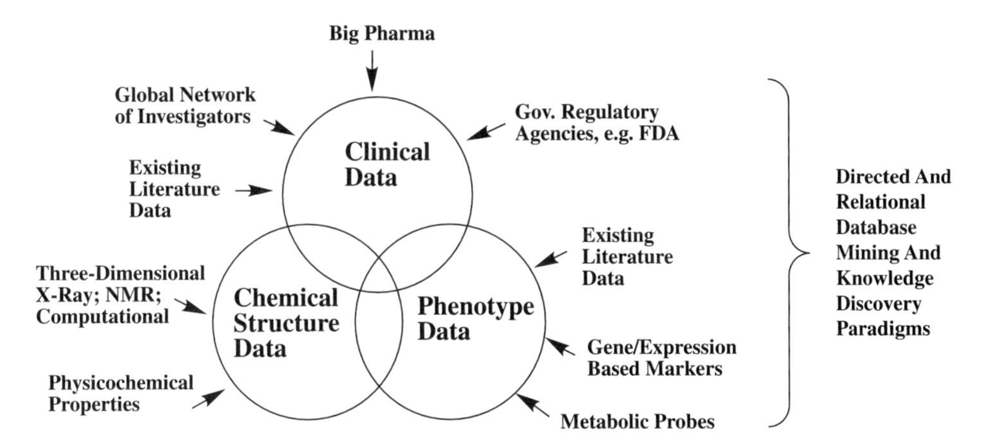

Figure 9.3 Informational fields to be included in the human drug metabolism database. This database should be available on the WWW via a nonprofit format. A ready mechanism should be made available to continually receive human (clinical) metabolism data from any source. However, actual data entry will have to be monitored for quality control via the database's maintenance organization.

The informational fields being constructed within this database are shown in Fig. 9.3, which also depicts the database's overall architecture. The treatment of three-dimensional chemical structures has already been outlined in Fig. 9.2, which depicts the progression for maturing structural entries according to the prior discussion.

9.5 SUMMARY

The gradual accumulation of drug metabolism studies has afforded a vast field of data which offers the potential to be used for predicting the metabolic outcomes of new drug candidates. Five major expeditions have ventured into this field to provide maps of varying detail which are available commercially as expert systems or databases with chemical structure searching capabilities. An analysis of several case studies indicates that attempts to predict metabolic outcomes for new structures by using these types of databases have been only marginally successful. The reasons for this shortcoming include both biological and chemical factors. Some of the biological issues include species and phenotypic variation, as well as how to properly interrelate PBPK types of parameterization and HTS screening results so as to better reflect the intact human situation. The major chemical issues include a fundamental question about how much of a structure should be included within a metabophore during SMR assessments, and the long-standing issue of how to accurately derive three-dimensional structure. The discussions within this chapter have considered these issues and have suggested some new approaches that could enhance the utility of future chemical structure–biological information databases in general. In particular, the suggestions may also be useful for the specific assembly of drug metabolism databases that might partner in a synergistic manner with HTS metabolism data acquisition methodologies.

For the biological issues it is imperative that a human drug metabolism database be made available as a standard so that all types of preclinical sets of data from whatever type of theoretical model, in vivo or in vitro experimental model, or parameterized HTS assay can then be monitored for their predictabilities in statistical terms if not actually validated within a more traditional format for their direct utilities. Similarly, in terms of chemical issues, it has been suggested that although two-dimensional representations may constitute a practical starting point for the input of structures, it is imperative that methods be evolved to mature these displays into accurate representation of the relevant three-dimensional conformational families. Specific approaches toward the construction of a commonly held human drug metabolism database and for the handling of three-dimensional chemical structure have been elaborated herein. Both approaches are presently being explored within our laboratories.

REFERENCES AND NOTES

1. Tables 9.1 and 9.2 represent a compilation of information taken largely from the drug metabolism sections of three long-standing textbooks in the areas of medicinal chemistry, pharmacology, and toxicology, respectively: (a) D. Williams. Drug metabolism, in *Medicinal Chemistry*, 4th ed., W. Foye, T. Lemke, and D. Williams (Eds.), Williams & Wilkins, Baltimore, pp. 83–140 (1995). (b) L. Z. Benet (Ed.). General principles, Sec. I (5 chapters), in *Goodman & Gilman's The Pharamacological Basis of Therapeutics*, 9th ed., J. G. Hardman, L. E. Lombard, P. B. Molino, R. W. Radon and A. G. Gilman (Eds.), McGraw-Hill, New York, pp. 1–101 (1996). (c) K. K. Rozman and C. D. Klaassen; A. Parkinson; and M. A. Medinsky and C. D. Klaassen, contributing authors. Disposition of toxicant, Unit 2 (3 chapters), in *Casarett & Doull's Toxicology: The Basic Science of Poisons*, 5th ed., C. D. Klaassen, M. O. Amdur, and J. Doull (Eds.), McGraw-Hill, New York, pp. 89–198 (1996).
2. This phrase is being considered for adoption by the International Union of Pure and Applied Chemistry's nomenclature working party on drug metabolism terms chaired by P. W. Erhardt.
3. T. Su, Z. Bao, Q-Y. Zhang, T. Smith, J-Y. Hong, and X. Ding. Human cytochrome P450 CYP2A13: predominant expression in the respiratory tract and its high efficiency metabolic activation of a tobacco-specific carcinogen, 4-(methylnitrosamino)-1-)3-pyridyl)-1-butanone. *Cancer Res.*, **60**, 5074–5079 (2000).
4. Nomenclature for selected metabolizing enzymes. (a) Cytochrome P450s (CYPs): family, a subset of the CYP superfamily sharing at last 40% amino acid sequence homology denoted by an Arabic numeral after the uppercase letters CYP, (e.g., CYP3); subfamily, a subset of the CYP superfamily sharing at lest 55% homology denoted by an Arabic uppercase letter after the family designation (e.g., CYP3A); isoform, an individual form of CYP denoted by adding a second Arabic numeral after the subfamily designation (e.g., CYP3A4). (b) *N*-Acetyl transferases (NATs) are divided into two major families (i. e., 1 and 2), for which there are also many individually displayed polymorphic versions. (c) Uridinedisphosphoglucuronosyl transferases (UGSs) are divided into two major families (i. e., 1 and 2), based on having at least 50% amino acid sequence homology followed by additional subfamily designations such as A and B, all of which also have many individually displayed additional polymorphic versions (e.g., UGT1A1).
5. Recent results from this laboratory: e.g., J. Liao. Development of Soft Drugs for the neonatal population, M. S. thesis, Univ. Toledo (2000).
6. *Drug Metabolism: Databases and High Throughput Testing During Drug Design and Development*, IUPAC, Chapel Hill, NC (1999).
7. P. Erhardt. Drug metabolism data: past and present status, *Med. Chem. Res.*, **8**, 400–421 (1998).
8. National Academy of Sciences (U.S.) Committee. Problems of drug safety, *Workshops* (1966–1968).

9. E. Boyland and J. Booth. The metabolic fate and excretion of drugs, *Annu. Rev. Pharmacol.*, **2**, 129–142 (1962).
10. R. T. Williams. Detoxication mechanisms in man, *Clin. Pharm. Ther.*, **4**, 234–254 (1963).
11. J. R. Gillette. Metabolism of drugs and other foreign compounds by enzymatic mechanisms, *Prog. Drug Res.*, **6**, 11–73 (1963).
12. R. T. Williams, *Detoxication Mechanisms: The Metabolism and Detoxication of Drugs, Toxic Substances and Other Organic Compounds*, Wiley, New York (1959).
13. T. B. Binns. *Absorption and Distribution of Drugs*, Williams & Wilkins, Baltimore (1964).
14. B. N. LaDu, H. G. Mandel, and E. L. Way. *Fundamentals of Drug Metabolism and Drug Disposition*, Williams & Wilkins, Baltimore (1972).
15. J. W. Bridges and L. F. Chasseaud. *Progress in Drug Metabolism*, Vol. 1, Wiley, New York (1976).
16. The Chemical Society/The Royal Society of Chemistry (London), *Foreign Compound Metabolism in Mammals*, Vols. 4–6, *Specialist Periodical Reports* (1977, 1979, and 1981).
17. D. R. Hawkins. *Biotransformations: A Survey of the Biotransformations of Drugs and Chemicals in Animals*, Vols. 1–8, Royal Society of Chemistry, London (1989–1995).
18. R. V. Smith, P. W. Erhardt, and S. W. Leslie. Microsomal *O*-demethylation, *N*-demethylation and aromatic hydroxylation in the presence of bisulfite and dithiothreitol, *Res. Commun. Chem. Pathol. Pharmacol.*, **12**, 181–184 (1975).
19. D. E. Atkinson. *Cellular Energy Metabolism and Its Regulation*, Academic Press, New York (1977).
20. M. L. Spann, K. C. Chu, W. T. Wipke, and G. Ouchi. Use of computerized methods to predict metabolic pathways and metabolites, *J. Environ. Pathol. Toxicol.*, **2**, 123–130 (1978).
21. T. H. Varkony, D. H. Smith, and C. Djerassi. Computer-assisted structure manipulation: studies in the biosynthesis of natural products, *Tetrahedron*, **34**, 841–852 (1978).
22. G. I. Ouchi. Computer-assisted prediction of plausible metabolites of xenobiotic compounds, *Univ. Micro. Int.*, Univ. California, Santa Cruz, CA (1978).
23. L. Goldberg. *Structure–Activity Correlation as a Predictive Tool in Toxicology*, Hemisphere, New York (1983).
24. J. F. Tinker and H. Gelernter. Computer simulation of metabolic transformation, *J. Comput. Chem.*, **7**, 657–665 (1986).
25. M. L. Mavrovounitis and G. Stephanopoulous. Computer-aided synthesis of biochemical pathways, *Biotechnol. Bioeng.*, **36**, 1119–1132 (1990).
26. S. Barcza, L. A. Kelly, and C. D. Lenz. Computerized retrieval of information on biosynthesis and metabolic pathways, *J. Chem. Inf. Comput. Sci.*, **30**, 243–251 (1990).
27. F. Darvas. Metabol expert: an expert system for predicting the metabolism of substances, in *QSAR in Environmental Toxicology II*, K. L. Kaiser (Ed.) Reidel Co., Dordrecht, The Netherlands (1987), pp. 71–81.
28. F. Darvas. Predicting metabolic pathways by logic programming, *J. Mol. Graph.*, **6**, 80–86 (1988).
29. Commercially available from ComGenex, Inc.
30. G. Klopman, M. Dimayuga, and J. Talafous. META. 1. A program for the evaluation of metabolic transformation of chemicals, *J. Chem. Inf. Comput. Sci.*, **34**, 1320–1325 (1994).
31. J. Talafous, L. M. Sayre, J. J. Mieyal, and G. Klopman. META. 2. A dictionary model of mammalian xenobiotic metabolism, *J. Chem. Inf. Comput. Sci.*, **34**, 1326–1333 (1994).
32. G. Klopman and M. Tu. Meta. 3. A genetic algorithm for metabolic transform priorities optimization, *J. Chem. Inf. Comput. Sci.*, **37**, 329–339 (1997).
33. Commercially available from Multicase, Inc.

34. E. M. Gifford, M. A., Johnson, and C.-C. Tsai. Using the xenobiotic metabolism database to rank metabolic reaction occurrences, *Mol. Connect. MDL Newsl. Commun. Customers*, **13**, 12–13 (1994).

35. *The metabolite* database is available from MDL Information Systems, Inc.

36. J. Hayward. Synopsys metabolism database, in *Drug Metabolism: Databases and High Throughput Testing During Drug Design and Development*, P. Erhardt (Ed.), IUPAC, Chapel Hill, NC, pp. 281–288 (1999).

37. Commercially available from Accelrys, Ltd.

38. N. Greene. Knowledge based expert systems for toxicity and metabolism prediction, in *Drug Metabolism: Databases and High Throughput Testing During Drug Design and Development*, P. Erhardt (Ed.), IUPAC, Chapel Hill, NC, pp. 289–296 (1999).

39. Commercially available from Lhasa, UK.

40. K. A. Bachmann and R. Ghosh. The use of in vitro methods to predict in vivo pharmacokinetics and drug interactions, *Curr. Drug Metabol.*, **2**, 299–314 (2001).

41. J. N. Smith. Comparative Biochemistry of Detoxification, in *Comparative Biochemistry*, M. Florkin and H. S. Mason (Eds.), Academic Press, New York, pp. 403–457 (1964).

42. G. Mulder. *Conjugation Reactions in Drug Metabolism*, Taylor & Francis, London (1990).

43. K. W. DiBiasio. In vitro testicular xenobiotic metabolism of model substrates in rats, mice, monkeys, and humans, *Univ. Micro. Int.*, Univ. California. Davis, CA (1989).

44. J. Baron, J. M. Voigt, T. B. Whitter, T. T. Kawabata, S. A. Knapp, F. P. Guengerich, and W. B. Jakoby. Identification of intratissue sites for xenobiotic activation and detoxication, *Adv. Exp. Med. Biol.*, **197**, 119–144 (1986).

45. K. Bachmann, D. Pardoe, and D. White. Scaling basic toxicokinetic parameters from rat to man, *Environ. Health Perspect.*, **104**, 400–407 (1996).

46. J. Sarver, D. White, P. Erhardt, and K. Bachmann. Estimating xenobiotic half-lives in humans from rat data: influence of log *P*, *Environ. Health Perspect.*, **105**, 1204–1209 (1997).

47. S. Hayashi, J. Watanabe, and K. Kawajiri. Genetic polymorphisms in the 5′-flanking region change transcriptional regulation of the human cytochrome P450IIE 1 gene, *J. Biochem. (Tokyo)*, **110**, 559–565 (1991).

48. H. Sugimura, N. H. Caporaso, G. L. Shaw, R. V. Modali, F. J. Gonzalez, R. N. Hoover, J. H. Resau, B. F. Trump, A. Weston, and C. C. Harris. Human debrisoquine hydroxylase gene polymorphisms in cancer patients and controls, *Carcinogenesis*, **11**, 1527–1530 (1990).

49. S. Kato, M. Onda, N. Matsukura, A. Tokunaga, T. Tajiri, D. Y. Kim, H. Tsuruta, N. Matsuda, K. Yamashita, and P. G. Shields. Cytochrome P4502E1 (CYP2E1) genetic polymorphism in a case–control study of gastric cancer and liver diseases, *Pharmacogenetics*, **5**, S141–S144 (1995).

50. K. Tominaga, Y. Koyama, M. Sasagawa, M. Hiroki, and M. Nagai. A case–control study of stomach cancer and its genesis in relation to alcohol consumption, smoking and familial cancer history, *Jpn. J. Cancer Res.*, **82**, 974–979 (1991).

51. S. Borman. Combinatorial chemistry. *C & E News*, Feb. 24, 43–62 (1997).

52. F. K. Brown. Chemoinformatics: what is it and how does it impact drug discovery, *Annu. Rep. Med. Chem.*, **33**, 375–384 (1998).

53. M. Hann and R. Green. Chemoinformatics: a new name for an old problem, *Curr. Opin. Chem. Biol.*, **3**, 379–383 (1999).

54. P. Ertl, W. Miltz, B. Rohde, and P. Selzer. Web-based chemoinformatics for bench chemists, *Drug Discov. World*, **1**(2), 45–50 (2000).

55. ChemFinder at www.chemfinder.com; Chem Navigator at www.chemnavigator.com; Info Chem GmbH at www.infochem.com; Institute for Scientific Info. at www.isinet.com; MDL Information Systems at www.mdli.com; MSI at www.msi.com; Oxford Molecular at www.oxmol.com;

Pharmacopeia at www.pharmacopeia.com; QsarIS at www.scivision. com/qsaris.html; Synopsys Scientific Systems at www.synopsys.co.uk; Tripos at www.tripos.com.

56. Personal communications while attending several recent workshops directed toward expediting drug development.
57. G. Karet. Multiplicity, *Drug Discov. Dev.*, Jan., 28–32, (2001).
58. Personal communications while attending several recent scientific conferences in the area of medicinal chemistry.
59. ChemPen 3D at http://home.ici.net/~hfevans/chempen.htm; CS Chem Draw and CS Chem Draw 3D at www.camsoft.com; Corina at www.mol-net.de.
60. A. Nezlin. Chem Draw Ultra 5.0: an indispensable everyday tool for an industrial organic chemist, *Chem. News. Commun.*. **9**, 24–25 (1999).
61. M. Endres. High-resolution graphics combined with computational methodology make 3D molecular modeling exciting, colorful and downright sexy, *Today's Chem. at Work*, Oct., 30–44 (1992).
62. G. Karet. Software promotes targeted design. *Drug Discov. Dev.*, Aug./Sept., 42–48 (2000).
63. QUANTA (system for determining three-dimensional protein structure from x-ray data) and FELIX (system for determining three-dimensional protein structure from NMR data) available from MSI-Pharmacopeia (www.msi.com).
64. W. J. Hehre, L. Radom, P. v. R. Schleyer, and J. A. Pople. *Ab Initio Molecular Orbital Theory*, Wiley, New York (1986).
65. A. D. Becke. Density-functional thermochemistry. III. The role of exact exchange, *J. Chem. Phys.*, **98**, 5648–5652 (1993).
66. (a) B. C. Oostenbrink, J. W. Pitera, M. van Lipzig, J. H. Meerman, and W. F. van Gunsteren. Simulations of the estrogen receptor ligand-binding domain: affinity of natural ligands and xenoestrogens, *J. Med. Chem.*, **43**, 4594–4605 (2000). (b) N. Wu, Y. Mo, J. Gao, and E. F. Pai. Electrostatic stress in catalysis: structure and mechanism of the enzyme orotidine monophosphate decarboxylase, *Proc. Natl. Acad. Sci. USA*, **97**, 2017–2022 (2000). (c) A. Warshel, M. Strajbl, J. Villa, and J. Florian. Remarkable rate enhancement of orotidine 5′-monophosphate decarboxylase is due to transition-state stabilization rather than to ground-state destabilization, *Biochemistry*, **39**, 14728–14738 (2000).
67. J. Devillers and A. T. Balaban (Eds.). Topological Indices and Related Descriptors in QSAR and QSPR, *Gordon and Breach Science*, Amsterdam (1999).
68. C. Wrotnowski. Counting on computational intelligence, *Mod. Drug Discov.*, Nov./Dec., 46–55 (1999).
69. C. M. Henry. Protein structure by mass spec, *Chem. Eng. News*, Nov. 27, 22–26 (2000).
70. E. K. Wilson. Quantum computers, *Chem. Eng. News*, Nov. 6, 35–39 (2000).
71. B. Ladd. Intuitive data analysis: the next generation Rx. *Mod. Drug Discov.*, Jan./Feb., 46–52 (2000).
72. V. Venkatsubramanian. Computer-aided molecular design using neural networks and genetic algorithms, in *Genetic Algorithms in Molecular Modeling*, J. Devillers (Ed.), Academic Press, New York (1996).
73. A. Globus. Automatic molecular design using evolutionary techniques, *Nanotechnology*, **10**(3), 290–299 (1999).
74. T. Studt. Data mining tools keep you ahead of the flood, *Drug Discov. Dev.*, Aug./Sept., 30–36 (2000).
75. R. Resnick. Simplified datamining, *Drug Discov. Dev.*, Oct., 51–52 (2000).
76. J. Jaen-Oltra, T. Salabert-Salvador, F. J. Garcia-March, F. Perez-Gimenez, and F. Tohmas-Vert. Artificial neural network applied to prediction of fluoroquinolone antibacterial activity by topological methods, *J. Med. Chem.*, **43**, 1143–1148 (2000).

77. Bio Reason at www.bioreason.com; Columbus Molecular Software (Lead Scope) at www.columbus-molecular.com; Daylight Chemical Information Systems at www.daylight.com; IBM (Intelligent Miner) at www.ibm.com; Incyte (Life Tools and Life Prot) at www.incyte.com; Molecular Applications (Gene Mine 3.5.1) at www.mag.com; Molecular Simulations at www.msi.com; Oxford Molecular (DIVA 1.1) at www.oxmol.com; Pangea Systems (Gene World 3.5) at www.pangeasystems.com; Pharsight at www.pharsight.com; SAS Institute (Enterprise Miner) at www.sas.com; SGI (Mine Set 3.0) at www.sgi.com; Spotfire (Spotfire Pro 4.0 and Leads Discover) at www.spotfire.com; SPSS (Clementine) at www.spss.com; Tripos at www.tripos.com.
78. M. A. Johnson, E. Gifford, and C.-C. Tsai. In *Concepts and Applications of Molecular Similarity*, M. A. Johnson (Ed.), Wiley, New York, pp. 289–320 (1990).
79. E. Gifford, M. Johnson, and C.-C. Tsai. A graph-theoretic approach to modeling metabolic pathways. *J. Comput.-Aided Mol. Des.*, **5**, 303–322 (1991).
80. E. M. Gifford, M. A. Johnson, D. G. Kaiser, and C.-C. Tsai. Visualizing relative occurrences in metabolic transformations of xenobiotics using structure–activity maps, *J. Chem. Inf. Comput. Sci.*, **32**, 591–599 (1992).
81. E. M. Gifford, M. A. Johnson, D. G. Kaiser and C.-C. Tsai. Modeling the Relative metabolic occurrence of alkyl-nitrogen bond cleavage using structure–reactivity maps, *Xenobiotica*, **25**, 125–146 (1995).
82. E. M. Gifford. Applications of molecular similarity methods to visualize xenobiotic metabolism structure–reactivity relationships, *Ph.D. thesis, Univ. Toledo* (1996).
83. K. M. Andrews and R. D. Cramer. Toward general methods of targeted library design: topomer shape similarity searching with diverse structures as queries, *J. Med. Chem.*, **43**, 1723–1740 (2000).
84. T. Borowski, M. Krol, E. Broclawik, T. C. Baranowski, L. Strekowski, and M. J. Mokrosz. Application of similarity matrices and genetic neural networks in quantitative structure–activity relationships of 2- or 4-(4-methylpiperazino) pyrimidines: 5-HT_{2A} receptor antagonists, *J. Med. Chem.*, **43**, 1901–1909 (2000).
85. E. A. Wintner and C. C. Moallemi. Quantized surface complementarity diversity (QSCD): a model based On small molecule–target complementarity, *J. Med. Chem.*, **43**, 1933–2006 (2000).
86. M. Pastor, G. Cruciani, I. McLay, S. Pickett, and S. Clementi. Grid-independent descriptors (GRIND): a novel class of alignment-independent three-dimensional molecular descriptors, *J. Med. Chem.*, **43**, 3233–3243 (2000).
87. C. Gnerre, M. Catto, F. Leonetti, P. Weber, P.-A. Carrupt, C. Altomare, A. Carotti, and B. Testa. Inhibition of monoamine oxidases by functionalized coumarin derivatives: biological activities, QSARS, and 3D-QSARS, *J. Med. Chem.*, **43**, 4747–4758 (2000).
88. H. A. Carlson, K. M. Masukawa, K. Rubins, F. D. Bushman, W. L. Jorgensen, R. D. Lins, J. M. Briggs, and J. A. McCammon. Developing a dynamic pharmacophore model for HIV-1 integrase, *J. Med. Chem.*, **43**, 2100–2114 (2000).
89. E. K. Bradley, P. Beroza, J. E. Penzotti, P. D. Grootenhuis, D. C. Spellmeyer, and J.L. Miller. A rapid computational method for lead evolution: description and application to α_1-adrenergic antagonists, *J. Med. Chem.*, **43**, 2770–2774 (2000).
90. M. Graffner-Nordberg, J. Marelius, S. Ohlsson, A. Persson, G. Swedberg, P. Andersson, S. E. Andersson, J. Aqvist, and A. Hallberg. Computational predictions of binding affinities to dihydrofolate reductase: synthesis and biological evaluation of methotrexate analogues, *J. Med. Chem.*, **43**, 3852–3861 (2000).
91. R. Garcia-Nieto, I. Manzanares, C. Cuevas, and F. Gago. Increased DNA binding specificity for antitumor ecteinascidin 743 through protein–DNA interactions? *J. Med. Chem.*, **43**, 4367–4369 (2000).
92. A. Vedani, H. Briem, M. Dobler, H. Dollinger, and D. R. McMasters. Multiple-conformation and protonation-state representation in 4D-QSAR: the neurokinin-1 receptor system, *J. Med. Chem.*, **43**, 4416–4427 (2000).

93. M. L. Lopez-Rodriguez, M. J. Morcillo, M. Fernandez, M. L. Rosada, L. Pardo, and K.-J. Schaper. Synthesis and structure–activity relationships of a new model of arylpiperazines. 6. Study of the 5-HT_{1A}/α_1-adrenergic receptor affinity by classical Hansch analysis, artificial neural networks, and computational simulation of ligand recognition, *J. Med. Chem.*, **44**, 198–207 (2001).

94. R. Wedin. Visual data mining speeds drug discovery, *Mod. Drug Discov.*, Sept./Oct., 39–47 (1999).

95. P. Erhardt. Epilogue, in *Drug Metabolism: Databases and High-Throughput Testing During Drug Design and Development*, P. Erhardt (Ed.), IUPAC, Chapel Hill, NC, p. 320 (1999).

10

DISCOVERY OF THE ANTIULCER DRUG TAGAMET

C. Robin Ganellin
University College London
London, UK

10.1 HISTORICAL BACKGROUND

10.1.1 Prologue

Tagamet has the nonproprietary name (INN) *cimetidine*. At the time that it was discovered, cimetidine represented a novel type of drug action that revolutionized the treatment of peptic ulcer disease. It was the product of a rational approach to drug design, involving a very close collaboration between pharmacologists and medicinal chemists. The pharmacology involved a detailed analysis that ran counter to the general wisdom at the time, and the chemistry was unusual in that there was no known lead molecule with the required properties to build upon (neither natural product nor synthetic chemical compound). The chemistry was also unusual in that it relied on a detailed application of physical–organic chemistry to structure–activity analysis. The subject has become a classic textbook example. To understand and appreciate why this discovery was unusual it is necessary to consider the level of knowledge (or lack of it) prevailing at the time (in 1964–1972) the research was carried out. Much of the time was spent in the usual fog that accompanies research into the unknown where there is no precedent but only analogy on which to base ideas. Now it all seems obvious and very logical, but at the time it was bedevilled with controversy and uncertainty. Indeed, the work was conducted at the SmithKline & French Research Institute in Welwyn Garden City, UK, and the lack of early progress led the top American management in Philadelphia to order the closure of the research program after the first three years. Fortunately, the UK scientists did not accede to this order.

Drug Discovery and Development, Volume 1: Drug Discovery, Edited by Mukund S. Chorghade

10.1.2 Pharmacological Receptors

Natural transmitter substances such as the biogenic amines have provided a rich source for the development of new drugs. They act at specific sites on cells in the body that we call receptors, and in so doing they stimulate the cells to produce a specific response. Usually, this involves a sequence of biochemical events that modulate the cell's function. The biogenic amines therefore act as chemical messengers to affect what the body does. The remarkable thing is that the biogenic amines have a chemical individuality that is specifically recognized at their own receptor sites; they do not act at other receptor sites because the sites discriminate among different messengers. This makes for a highly selective process. It follows that the design of compounds to act in competition with the biogenic amines for occupation of these sites is a potential way of attaining specificity of drug action. It also appears that the receptor sites for a particular amine are not homogeneous. Drugs have been discovered that differentiate among sites for the same amine, suggesting different populations of amine receptors. This allows further scope for introducing selectivity of drug action.

Now, in the twenty-first century we know that these receptor sites are protein molecules on the cell membrane that are either ion channels or are coupled to enzymes involved in phosphorylation which amplify the original stimulus. In the first half of the twentieth century there was very little proof of the nature of the amine receptors. Indeed, for many years they were the figment of pharmacologists′ imagination. Their presence was only inferred from the nature of dose–response relationships obtained by relating the concentration of biogenic amine applied and its effect on the function of a tissue or organ. By 1950, acetylcholine was considered to act at two types of receptor named after natural products (nicotine and muscarine), adrenaline and noradrenaline had just been suggested to act at α and β receptors, and histamine had recently been proposed to possibly act at a second type of receptor. The pharmacological characterization of histamine receptors had been established by the seminal work of Heinz Schild, who had used histamine and antihistamines as the means to provide the general basis for classifying competitive reversible antagonism in the decade 1945–1955 (recall the now well-known Schild plot). Subsequently (1966), he classified the actions of histamine that were blocked by the antihistamine drugs as being mediated by histamine type 1 (H_1) receptors.

10.1.3 Peptic Ulcer Disease

Peptic ulceration is the most common disease of the gastrointestinal tract. It produces considerable illness and pain, and it used to result in great economic loss to the patients and their communities; it could even be fatal. It comprises duodenal and gastric ulcers and affects large numbers of people who are otherwise relatively fit. Duodenal and gastric ulcers are localized erosions of the mucous membrane of the duodenum or stomach, respectively, which expose the underlying layers of the gut wall to the acid secretions of the stomach and to the proteolytic enzyme pepsin. In the 1960s the cause of acute peptic ulcer was not properly understood, but for many years the main medical treatment had been aimed at reducing acid production, based on the hope that neutralizing gastric acid would reduce its irritating effects and also reduce the efficacy of pepsin, and so allow ulcers to heal.

Secretion of gastric acid by the parietal cells of the stomach is initiated by the thought, sight, smell, or taste of food and is mediated by the autonomic nervous, system via the

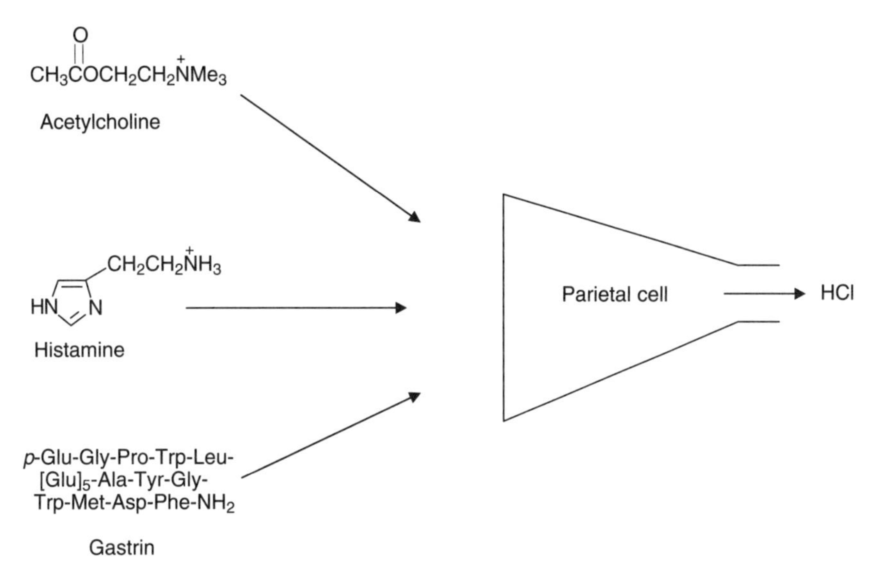

Scheme 10.1 Three chemical messengers stimulate the production of hydrochloric acid from the gastric parietal cell. The formulas show acetylcholine cation, histamine monocation (the most prevalent species at physiological pH 7.4), and human gastrin I.

vagus nerves, which provide parasympathetic innervation to the stomach and small intestine; the neurotransmitter released by stimulation of the vagus is acetylcholine. Branches of the vagus, innervating the antral region of the stomach, stimulate the release of the peptide hormone gastrin from special gastrin-producing G cells. The presence of food in the stomach further stimulates release of gastrin, which passes into the bloodstream and is carried to the parietal cells, where it acts to stimulate them to secrete hydrochloric acid. In addition to acetylcholine and gastrin, a third chemical secretagog, histamine, was known to be involved (Scheme 10.1).

The relationship between the three secretagogues acetylcholine, gastrin, and histamine has been a source of considerable controversy among physiologists for many years. When it was found (in the late 1940s and early 1950s) that antihistamine drugs did not reduce acid secretion, the role of histamine as a secretagog was placed in serious question. By 1964, when gastrin had been isolated (Gregory and Tracy, 1961) and sequenced (Gregory et al., 1964) at Liverpool University, most gastric physiologists were convinced that gastrin played the key role in the physiological control of gastric acid and considered histamine to be unimportant (Johnson, 1971).

For many years the main medical treatment for peptic ulcers relied on the use of antacids to neutralize the gastric acid, but when taken in sufficient quantities they may cause unpleasant side effects. Anticholinergic drugs (to block the acetylcholine transmission) can decrease gastric acid secretion, but their use in the treatment of peptic ulceration was limited by side effects such as dryness of the mouth, urinary retention, and blurred vision. The alternative to drug treatment is surgery. This aims to cut out part of the acid secretory and gastrin-producing regions of the stomach (e.g., partial gastrectomy) or to selectively cut the branches of the vagal nerve (e.g., selective vagotomy) that supply the acid-secretory region. This was a difficult and sometimes dangerous operation.

10.1.4 Search for New Antiulcer Drugs

With the background to treatment described above it is not surprising that most major pharmaceutical companies set up research programs aimed at discovering antiulcer agents. For many years the main approach was to induce ulcer formation in rats, or to stimulate production of gastric acid in rats, and to screen compounds for their ability to protect against ulcer formation or to inhibit acid production. The record of success was very poor. In the 1960s a more scientific and rational approach was beginning to take shape. With increasing understanding of the physiology of gastric acid secretion, several companies established research programs to discover specific inhibitors of the action of the then known chemical messengers. In the UK, ICI Pharmaceuticals initiated a search for a gastrin antagonist, and Pfizer and SmithKline & French independently sought a histamine antagonist.

The two secretagogs, gastrin and histamine, therefore presented alternative targets for inhibiting acid production. If one of them has a controlling influence on acid secretion, blocking it might succeed in controlling acidity, but since blocking one site still leaves two others to act, it is by no means certain that such an approach will be successful. Furthermore, as with acetylcholine, gastrin and histamine have other actions, and an agent that blocks their effects on acid stimulation may also block other sites, leading to unacceptable side effects (as with the anticholinergics). Thus, it can be appreciated that such an approach to drug discovery is highly speculative, and one certainly has no reason to presume that it will be successful in providing a new therapy. On the other hand, it does have a scientific rationale founded on working hypotheses that are capable of investigation and being put to the test. An important part of this analysis is provided by the pharmacologist's view of drug receptors.

10.2 SEARCH FOR AN H_2-RECEPTOR HISTAMINE ANTAGONIST

10.2.1 Histamine Receptors

Histamine was discovered at the beginning of the twentieth century, and in 1910 Dale and Laidlaw published their seminal work on the actions of histamine on smooth muscle and blood pressure. Subsequent studies led to a view that histamine was a principal mediator of inflammation and shock; eventually, Bovet (1950) and colleagues discovered antihistamines, and in the 1940s these new drugs were introduced for the treatment of allergic conditions such as urticaria and hay fever. Pharmacological studies with antihistamines such as mepyramine (**1**, Scheme 10.2) led to the view that they were reversible competitive antagonists and that antagonism was surmountable. Schild identified pharmacological criteria so that the antihistamines could be viewed as acting in competition with histamine for occupation of its specific receptors sites. By 1948, studies on vascular tissues had shown that the antihistamines could only reduce the intensity of the action of large doses of histamine but did not totally abolish its vasodilator effects, and this led Folkow et al. (1948) to suggest that there may be two types of histamine receptor, only one of which could be blocked by antihistamines such as diphenhydramine (**2**, Scheme 10.2). The suggestion appeared to lay dormant in the literature for many years.

The antihistamines were used by pharmacologists to explore the various actions of histamine in different tissues. Several actions of histamine had been noted that could not be specifically antagonized by these drugs: for example, stimulation of gastric acid secretion in the rat, cat, or dog; simulation of isolated atria of the guinea pig; inhibition of rat uterus contractions. Some pointers to the differentiation of histamine receptors had been obtained

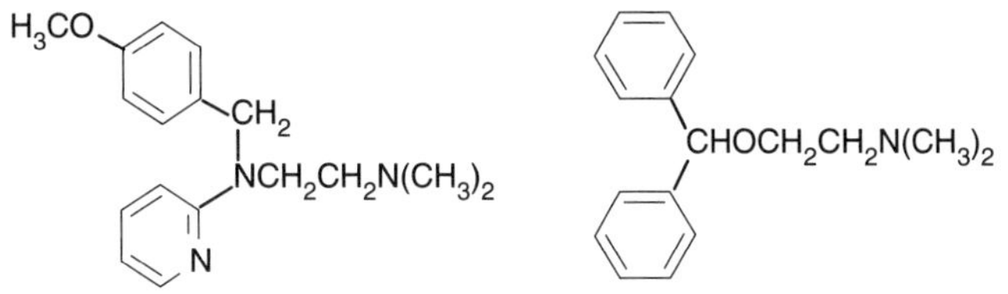

(**1**) Mepyramine (**2**) Diphenhydramine

Scheme 10.2

by considering the selectivity of action of agonists on these different tissue systems, and the results led Ash and Schild to propose in 1966 that the actions of histamine blocked by the antihistamine drugs characterized one type of histamine receptor, which they named the *H_1 receptor*. They suggested that other actions of histamine not specifically antagonized were probably mediated by other histamine receptors, but the characterization of these receptors awaited the discovery of specific antagonists; in the meantime they were to be regarded as non-H_1 receptors.

10.2.2 Biological Approach to a Histamine Antagonist at Non-H_1 Receptors

The inability of the antihistamine drugs (H_1-receptor antagonists) to inhibit histamine-stimulated gastric acid secretion had been known for many years and there have been a few published reports of concerted efforts to discover a specific antagonist to this action of histamine. In collaboration with the Eli Lilly Company in the United States, Grossman et al. (1952) reported on an extensive study of compounds, chemically related to histamine, that were examined for their action on acid secretion and also tested as possible inhibitors of histamine stimulation, but Grossman did not uncover a histamine antagonist.

A similar analysis by Sir James Black led him to establish at the SmithKline & French research Laboratories in Welwyn Garden City, in 1964, the test procedures needed to detect antagonists of these other effects of histamine. It was hoped that the work would lead to a new type of pharmacological agent with possible clinical utility: namely, a possible means for selective pharmacological control of gastric acid secretion with the potential for treating peptic ulcer disease. Compounds were tested for their ability to inhibit histamine-stimulated gastric acid secretion in anaesthetized rats using a refined Ghosh and Schild (1958) preparation. Other researchers (Ash and Schild, 1966; Grossman et al., 1952; Lin et al., 1962; Van den Brink, 1969) also examined close analogs of histamine for possible antagonism of histamine-stimulated gastric acid secretion. None of these studies established a histamine antagonist.

Since other types of inhibitors of gastric secretion could also act in this test, compounds found to be active were also tested on isolated tissue systems to provide additional criteria for specific antagonism to histamine. Two in vitro test systems involving two different animal species were set up: histamine-induced stimulation of guinea-pig right atrium (which continues to beat spontaneously in vitro because it contains the pacemaker and histamine increases the rate of beating) and inhibition by histamine of evoked contractions of rat uterus.

It is worth noting that the atmosphere prevailing in gastroenterological science at that time was strongly against the search for a histamine antagonist as a means of controlling gastric acid secretion. In the 1960s many researchers turned their attention to seeking specific inhibitors of gastrin-induced acid secretion. The import of histamine was difficult to prove and there was a widely held view that histamine had no place in the physiological maintenance of gastric acid secretion. The unsuccessful effort by researchers at Eli Lilly in the 1950s to find an antagonist of histamine-stimulated acid secretion added further to the general feeling that the approach was "played out."

10.2.3 Chemical Approach to an Antagonist: Generating a Lead

Once the problem of obtaining a competitive antagonist was posed in biological terms it was necessary to consider how to approach it chemically. How could one obtain such a compound? Where should one start, given no obvious lead compound? Nothing was known chemically about the physiological site of action of histamine. Returning to first principles, the structure of histamine was used as a chemical starting point. The view was taken that since the search was for a molecule that would compete with histamine for its receptor site, such a molecule would have to be recognized by the receptor and then bind more strongly than histamine, but not trigger the usual response. It therefore seemed worthwhile to retain in potential antagonist structures some chemical features of histamine to aid receptor recognition, and to include chemical groups that might assist the binding.

In this type of research one has to work by analogy. Considering other areas where other antagonists had been developed by chemical means to assist binding (e.g., antimetabolites, enzyme inhibitors, and other receptor systems, such as anticholinergic and antiadrenergic drugs). The structure of histamine was therefore modified to alter its chemical properties deliberately, while retaining some definite aspect of its structure or chemistry. Some examples have been discussed elsewhere by Ganellin et al. (1976).

Many compounds were made, based on the structure of histamine. In the first four years some 200 compounds were synthesized and tested, without providing a blocking drug. The problem for the chemist is that there are too many possible compounds for synthesis. Even small modifications of the natural stimulant histamine introduce many variables. Two hundred compounds may sound to be a rather low number by today's standards of high-throughput screening and parallel synthesis. It must be remembered, however, that most of the compounds were selected to explore a possible mode of interaction. They were not selected because there were suitable intermediates available commercially (i.e., ease of synthesis was not allowed to dictate the medicinal chemistry), although when all other things were equal, chemical accessibility was, of course, the determinant. Furthermore, imidazole chemistry can be demanding and compounds had to be purified by ion-exchange chromatography in aqueous solution.

During this period there developed considerable uncertainty about whether the receptors for histamine in gastric acid secretion might not be accessible. Also during this period, all seven isomers of monomethyl histamine had been synthesized and tested to explore where methyl groups could be accommodated in the histamine molecule without loss of affinity. There was considerable relief when it was found that 4-methylhistamine was selective for stimulating acid secretion, in comparison with its effect in stimulating the ileum. Conversely, 2-methylhistamine showed some selectivity for the ileum (in comparison with histamine). Here was evidence for the existence of at least two types of histamine receptors.

Toward the end of this time many doubts were expressed about whether it really would prove possible to block the action of histamine on gastric acid secretion, and indeed, there was considerable pressure within the company to abandon the project (see above). The scientists involved in the project were, however, firmly resolved to continue, and the test

TABLE 10.1 Structure and Antagonist Activities of Some Simple Imidazolylalkylisothioureas, Imidazoylalkylguanidines, and Imidazoylalkylcarboxamidines

NH, $(CH_2)_nX$—C, NH_2, HN, N

NH, $(CH_2)_nNH$—C, Z, HN, N

Structure	n	Substituent	Activity[a]
3	2	X = NH	+
4	2	X = S	++
5	3	X = NH	+++
6	3	X = S	±
7	2	Z = SMe	±
8	2	Z = Me	±
9	3	Z = SMe	+++
10	3	Z = Me	+++

[a]Tested for inhibition of histamine-stimulated gastric acid secretion in the lumen-perfused anaesthetized rat. Results represented semiquantitively as ±, detectable; +, $ID_{50} > 500\,\mu mol/kg$; ++, $ID_{50} \sim 200\,\mu mol/kg$; +++, $ID_{50} = 100$ to $50\,\mu mol/kg$. ID_{50} is the intravenous dose that reduces a near-maximal secretion to 50%.

system was refined. It is very important to conduct research in such a way as to learn from negative results. Even a list of inactive compounds is informative if they have been selected for particular reasons. Having tested many compounds with lipophilic substituents without seeing antagonism, the pharmacologists reexamined some of the early hydrophilic compounds. One of these polar hydrophilic compounds showed some blocking activity. It was very weak but it provided the vital lead. It was missed originally because this compound also acted as a stimulant; in fact, it is a partial agonist. The compound is a histamine derivative in which a guanidine group replaces the amino group in the side chain: namely, N^{α}-guanylhistamine (**3**) (Table 10.1).

10.2.4 Lead Optimization

The lead compound (**3**) was very weakly active, but within a few days an analogous compound was retested and found to be more active: (*S*)-[2-(imidazol-4-yl)ethyl]isothiourea (**4**) (Table 10.1). Still, a much more active compound was required.

An immediate question to be answered was whether activity was due to the presence of the guanidine or isothiourea groups (amidines) per se or to the structural resemblance to histamine. Structure–activity studies suggested that for these structures the imidazole ring was important; antagonism did not appear to be a property of amidines in general. It was also necessary to identify the particular chemical properties that conferred antagonist activity in order to make analogs of increased potency.

The amidine groups are strong bases and are protonated and positively charged at physiological pH. Thus, the molecules resemble histamine monocation but also differ in several ways; the amidinium group is planar (whereas the ammonium group of histamine is tetrahedral) and the positive charge is distributed over three heteroatoms. It was noted that the distance between ring and terminal nitrogen is potentially greater than in histamine, and that there are several nitrogen sites for potential interactions instead of one, thereby affording more opportunities for intermolecular hydrogen bonding.

It was envisaged that amidines might act as antagonists through additional binding being contributed by the amidine group. A type of bidentate hydrogen bonding between ion pairs can occur with an amidinium cation and an oxyacid anion (e.g., carboxylate, phosphate, or sulfate), which might be part of the receptor protein. Structural variables therefore identified for study initially were the amidine groups, amidino N-substituents (to search for hydrophobic interactions as a contribution to affinity), side-chain length, and alternatives to imidazole.

Many analogs of compounds **3** and **4** were made, but most turned out to be less active; at that time, unsubstituted imidazole appeared to be the best ring; alkyl substitution on the amidine N gave inconsistent results, and lengthening the side chain gave another breakthrough but threw up an apparent contradiction. For the guanidine structure, increasing the chain length led to a compound (**5**) showing an increase in antagonist activity. However, for the isothiourea, the reverse result was obtained; that is, increasing the chain length gave a compound (**6**) of reduced antagonist activity (Ganellin, 1981).

Thus, although the first two compounds discovered [guanidine (**3**) and isothiourea (**4**)] appeared to be closely related in structure in simply being isosteric nitrogen and sulfur analogs, the results with the homologs suggested that the situation was more complex. In an attempt to rationalize these differences, various related amidines were examined (Scheme 10.3). It was found that the reversed isothioureas (**7**, **9**) (side chain on N instead of S) resembled the guanidines in chain-length requirements, as did carboxamidines (e.g., **8** and **10**).

The apparent nonadditivity between structural change and biological effect posed a typical problem familiar to all practicing medicinal chemists: With so many structural variables

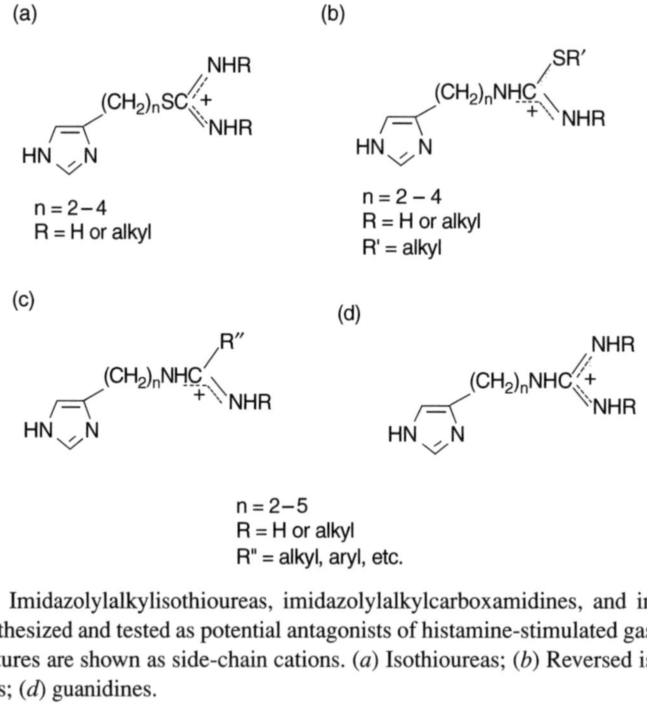

(a) $(CH_2)_nSC(NHR)_2^+$, HN N
n = 2–4
R = H or alkyl

(b) $(CH_2)_nNHC(SR')(NHR)^+$, HN N
n = 2 – 4
R = H or alkyl
R' = alkyl

(c) $(CH_2)_nNHC(R'')(NHR)^+$, HN N

(d) $(CH_2)_nNHC(NHR)_2^+$, HN N

n = 2–5
R = H or alkyl
R" = alkyl, aryl, etc.

Scheme 10.3 Imidazolylalkylisothioureas, imidazolylalkylcarboxamidines, and imidazolylalkylguanidines synthesized and tested as potential antagonists of histamine-stimulated gastric acid secretion. The structures are shown as side-chain cations. (*a*) Isothioureas; (*b*) Reversed isothioureas; (*c*) carboxamidines; (*d*) guanidines.

to study (e.g., ring, side-chain length, amidine system, amidine substituents), there are many millions of structures incorporating different combinations of these variables, and one cannot make and test them all. What, then, should govern the selection?

An essential feature of the discipline in medicinal chemistry is to find logical reasons for defining the boundary conditions for the selection of structures for synthesis. In the case under study there was a continuous search for useful physicochemical models for studying the chemistry of these compounds, and the inconsistencies in the structure–activity pattern were used to challenge the model or to reexamine the meaning of the biological test results. This dialogue, a search for self-consistency between the chemistry and biology, is vital to new drug research where no precedent exists.

To explore structure–activity relationships further, it became desirable to increase the side-chain length still more, but problems of chemical synthesis were experienced and new synthetic routes were required. Exploration of amidines and substituents continued but progress became very slow. The problem was that the compounds had mixed activities. In the main they acted as both agonists and antagonists; that is, they appeared to be partial agonists. This meant that although the compounds antagonized the action of histamine, they were not sufficiently effective inhibitors of gastric acid secretion because of interference through their inherent stimulatory activity. This appeared to impose a limitation on the potential of this type of structure for providing antagonists.

10.2.5 Validating the Research Program

Thus, a critical stage was reached in the need for selectivity: to achieve a separation between agonist and antagonist activities. It seemed that these compounds might act as agonists by mimicking histamine chemically, since like histamine, they have an imidazole ring, and being basic amidines, the side chain at physiological pH is protonated and carries a positive charge. It also seemed likely that these features would permit receptor recognition and provide binding for a competitive antagonist. This posed a considerable dilemma because the chemical groups that appeared to be required for antagonist (blocking) activity were the same groups that seemed to confer the agonist (stimulant) effect.

To separate these activities, the strongly basic guanidine group was replaced by nonbasic groups that although polar, would not be charged. Such an approach furnished analogs that indeed were not active as agonists; however, the first examples were also not active as antagonists. Eventually, one example, the thiourea derivative (**11**, SK&F 91581) (Scheme 10.4), that did not act as a partial agonist exhibited weak activity as an antagonist. Thioureas are essentially neutral in water because of the electron-withdrawing thiocarbonyl group. Conjugation forces the nitrogen atoms into a planar form and limits the availability of the nitrogen lone electron pairs, as in amides.

$(CH_2)_nNHC(=S)NHR^2$ (imidazol-4-yl; HN, N ring)

(**11**)	SK&F 91581	n = 3	R^2 = H
(**12**)	SK&F 91863	n = 4	R^2 = H
(**13**)	Burimamide	n = 4	R^2 = Me

Scheme 10.4 Imidazolylalkylthioureas.

TABLE 10.2 Structures and H_2-Receptor Histamine Antagonist Activities of Burimamide, Metiamide, Cimetidine, and Isosteres

Structure: imidazole ring (HN, N) bearing R^5 and $CH_2 X CH_2CH_2NHC(=V)NHMe$

Compound		Structure			H_2-Receptor Activities: In Vitro, Atrium[a] K_B (95% limits) $\times 10^{-6} M$		H_2-Receptor Activities: In Vitro, Uterus[b] K_B (95% limits) $\times 10^{-6} M$		In Vivo, Acid Secretion[c] ID_{50} (μmol/kg)
Number	Trivial Name	R^5	X	V					
13	Burimamide (thiourea)	H	CH_2	S	7.8	(6.4–8.6)	6.6	(4.9–8.3)	6.1
14	Thiaburimamide	H	S	S	3.2	(2.5–4.5)	3.2	(2.5–4.5)	5
15	Metiamide (thiourea)	Me	S	S	0.92	(0.74–1.15)	0.75	(0.40–1.36)	1.6
16	Urea isostere	Me	S	O	22	(8.9–65)	7.1	(1.6–30)	27
17	Guanidine isostere	Me	S	NH_2^+	16	(8.1–32)	5.5	(2.8–13)	12
18	Nitroguanidine (isostere)	Me	S	N–NO_2	1.4	(0.79–2.8)	1.4	(0.72–3.2)	2.1
19	Cimetidine (cyanoguanidine)	Me	S	N–CN	0.79	(0.68–0.92)	0.81	(0.54–1.2)	1.4

[a] Activities determined against histamine simulation of guinea-pig right atrium in vitro. The dissociation constant (K_B) was calculated from the equation $K_B = B/(x-1)$, where x is the respective ratio of concentrations of histamine needed to produce half-maximal responses in the presence and absence of different concentrations (B) of antagonist.

[b] Activities determined against histamine inhibition of electrically evoked concentrations of rat uterus in vitro.

[c] Activities as antagonists of histamine-simulated gastric acid secretion in the anaesthetized rat as indicated in footnote *a* of Table 10.1.

Problems of synthesizing higher homologous amines were solved by this time. The amine with the four-carbon atom chain length was synthesized and further exploration revealed that with this type of structure, extension of the alkylene side chain resulted in a marked increase in antagonist potency. It was not until the side chain had been lengthened that the significance of the result with SK&F 91581 (**11**) became clear and the desired aim was achieved, that is, a pure competitive antagonist without agonist effects. This compound (**12**, SK&F 91863) paved the way for an exploration of alkyl groups seeking a potential hydrophobic interaction and led to the *N*-methyl analog, which was given the name *burimamide* (**13**).

Burimamide (**13**) was an extremely important compound. It was highly selective, showed no agonist activity, and antagonized the action of histamine in a competitive manner on the two in vitro non-H_1 systems, guinea pig atrium and rat uterus (Table 10.2). It fulfilled the criteria required for characterizing the existence of another set of histamine receptors, the H_2 receptors. Thus, it allowed these tissue systems to be defined as H_2-receptor systems, so burimamide was defined as an H_2-receptor antagonist. This discovery was announced in *Nature* in 1972; the work had taken six years (Black et al., 1972). It provided the first definition of histamine H_2 receptors.

Burimamide also antagonized the action of histamine as a stimulant of gastric acid secretion in the rat, cat, and dog, and it was the first H_2-receptor antagonist to be investigated in humans. Given intravenously, it blocked the action of histamine as a stimulant of gastric acid secretion in humans, thereby confirming that burimamide behaves in humans as it does in animals. Thus, burimamide validated the research program. However, its one drawback was that it was not sufficiently active to be given orally. Thus, although burimamide was selective enough to define H_2 receptors, it was not active enough to permit proper drug development.

10.3 DEVELOPMENT OF A CLINICAL CANDIDATE DRUG

10.3.1 Dynamic Structure–Activity Analysis

Various ways to alter the structure of burimamide were examined in an attempt to increase potency. One approach that proved successful resulted from two lines of exploration that merged. Attempts were being made to overcome the problem of synthesizing the side chains by inserting a thioether link. Meanwhile, a study was being made of the pK_a characteristics of burimamide since it was realized that burimamide in aqueous solution is a mixture of many chemical species in equilibrium. At physiological pH there are three main forms of the imidazole ring, three planar configurations of the thioureido group (a fourth is theoretically possible but is disfavored by internal steric hindrance), and various trans and gauche rotamer combinations of the side chain CH_2–CH_2 bonds (Scheme 10.5). This means that at a given instant only a small proportion of the drug molecules would be in a particular form.

The existence of a mixture of species leads one to question which may be biologically active and whether altering drug structure to favor a particular species would alter drug potency. This is a process of dynamic structure–activity analysis (DSAA) (Ganellin, 1981). There are substantial energy barriers to interconversion between the species of burimamide, so it is quite likely that a drug molecule presenting itself to the receptor in a form unfavorable for drug–receptor interaction might diffuse away again before having time to rearrange into a more favorable form. The relative population of favorable forms might therefore determine the amount of drug required for a given effect.

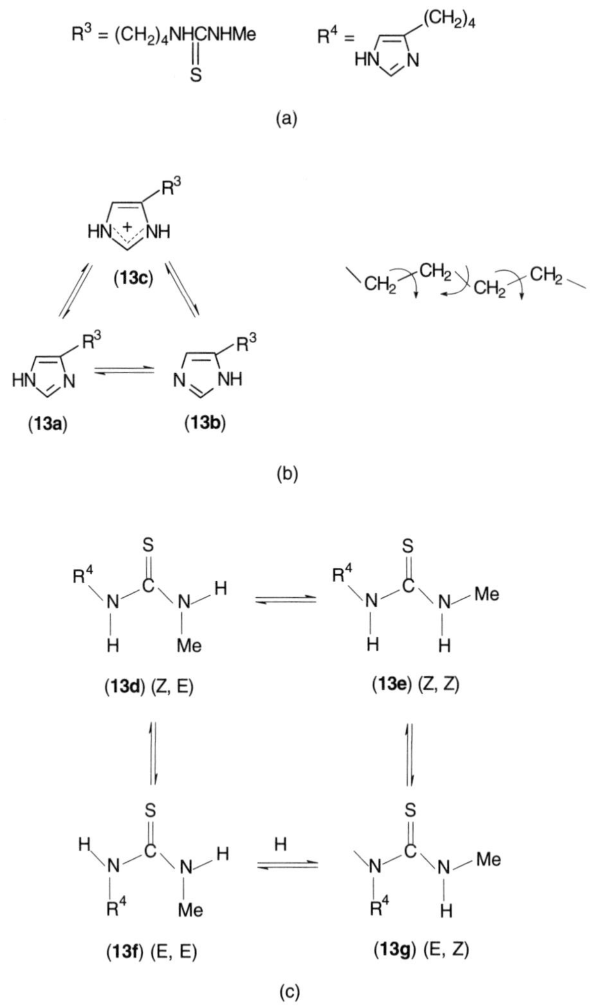

Scheme 10.5 Burimamide species equilibria in solution: (*a*) imidazole ring (ionization and tautomerism); (*b*) alkane chain (C—C bond rotation gives trans and gauche conformers); (*c*) thiourea group (configurational isomerism).

The various species of burimamide do not inconvert instantaneously, but whereas the rotamers of the side chain and thioureido groups are interconverted simply by rotation of a C–C or C–N bond, interconversion of the ring forms probably involves a water-mediated proton transfer. It was argued that if a molecule presents itself to the receptor with the ring in an unfavorable form, it might not readjust unless there were suitably oriented water molecules (or other hydrogen donor–acceptors) present.

10.3.2 Imidazole Tautomerism and Sulfur Methylene Isosterism

The arguments above led to a study of imidazole tautomerism and the population of imidazole species. At physiological pH the main species are (Scheme 10.5) the cation (**13c**) and two uncharged tautomers (**13a** and **13b**), and their populations were estimated qualitatively from the electronic influence of the side chain using pK_a data and the Hammett equation (Charton, 1965b):

$$pK_{a(R)} = pK_{a(H)} + \rho\sigma_m$$

where ρ is the Hammett reaction constant and σ_m the Hammett substituent constant. For burimamide, the ring pK_a (7.25 at 37°C) is greater than that of unsubstituted imidazole (6.80), indicating that the side chain is mildly electron releasing. In contrast, for histamine the ammonium ethyl side chain was seen to be electron withdrawing since it lowered the pK_a of the imidazole ring (pK_a = 5.90). Thus, although both histamine and burimamide are monosubstituted imidazoles, the structural similarity is misleading in that the electronic properties of the respective imidazole rings are different.

If the active form of burimamide were tautomer **13a**, the form most preferred for histamine, increasing its relative population might increase activity; for example, incorporating an electronegative atom into the antagonist side chain should convert it into an electron-withdrawing group and favor species **13a**. This would not be the only requirement for activity, and it would be necessary to minimize disturbance to other biologically important molecular properties such as stereochemistry and lipid–water interactions. For reason of synthesis, the first substitution to be made was the replacement of a methylene group ($-CH_2-$) by the isosteric thioether linkage ($-S-$) at the carbon atom next but one to the ring, to afford thiaburimamide (**14**) (Table 10.2), which was found to be more active as an antagonist.

It was argued that further stabilization of tautomer **13a** might be obtained by incorporating an electron-releasing substituent in the vacant 4(5) position of the imidazole ring. A methyl group was selected since it was thought that it should not interfere with receptor interaction, 4-methylhistamine having been shown to be an effective H_2-receptor agonist. This approach was successful, and introduction of a methyl group into the ring of the antagonist furnished a more potent drug, which was named *metiamide* (**15**) (Black et al., 1974) (Table 10.2). Metiamide represented a major improvement, being 10 times more potent than burimamide in vitro and a potent inhibitor of stimulated acid secretion in humans. It was investigated in patients and shown to produce a significant increase in the healing rate of duodenal ulcers and marked symptomatic relief. However, of 700 patients treated, there were a few cases of granulocytopenia (causing a reduction in the number of circulating white cells in the blood and leaving patients open to infection). Although reversible, this severely limited the amount of clinical work, so that another compound was required for clinical development.

10.3.3 Isosteres of Thiourea and the Discovery of Cimetidine

There now arose a truly critical issue for drug development. The question had to be faced whether the granulocytopenia was due to the pharmacological properties of an H_2-receptor histamine antagonist. Remember, this was the first time that such a pharmacological agent had been tested in the clinic. If H_2 receptors were involved in the generation of white blood cells, this would clearly limit the usefulness of an H_2-antagonist drug. Alternatively, the problem might be due to a toxic effect of the drug structure, in which case it would be a chemical problem, and therefore, in principle, it should be capable of a solution. One possibility explored was that the granulocytopenia associated with metiamide was caused by the thiourea group in the molecule. Exploration had continued with other possible structures, in particular with alternatives to the thiourea group. Isosteric replacement of the thiourea sulfur atom (=S) of metiamide by carbonyl oxygen (=O) gave the urea analog (**16**), but this was much less active. Isosteric replacement by imino nitrogen (−N) afforded the guanidine (**17**), which interestingly, though charged, was not a partial agonist but a fairly active antagonist. However, in vitro, the urea and guanidine isosteres were both approximately 20 times less potent than metiamide; therefore, other ways were investigated for removing the positive charge on the guanidine derivative.

Guanidine basicity is markedly reduced by electron-withdrawing substituents, and Charton (1965a) had demonstrated a high correlation between the inductive substituent constant σ_I and pK_a for a series of monosubstituted guanidines. The cyano and nitro groups are sufficiently electron withdrawing to reduce the pK_a by over 14 units, to values <0. The nitroguanidine (**18**) and cyanoguanidine (**19**) analogs of metiamide were synthesized and found to be active antagonists comparable with metiamide (Durant et al. 1977). Of these two compounds, the cyanoguanidine (**19**) is slightly more potent and was selected for development (Brimblecombe et al., 1975, 1978), being given the nonproprietary name *cimetidine*.

Cyanoguanidines exist predominantly in the cyanoimino form, and in cimetidine the cyanoimino group (=NCN) replaces the thione (=S) sulfur atom of metiamide. The similar behavior of cimetidine and metiamide as histamine H_2-receptor antagonists and the close similarity in physicochemical characteristics of thiourea and cyanoguanidine permit the description of thiourea and cyanoguanidine groups in the present context as bioisosteres (Durant et al., 1977). Nitroguanidine may also be considered to be a bioisostere of thiourea in this series of structures.

10.3.4 Cimetidine: A Breakthrough in the Treatment of Peptic Ulcer Disease

Cimetidine was shown to be a specific competitive antagonist of histamine at H_2 receptors in vitro and to be effective in vivo at inhibiting histamine-stimulated gastric acid secretion in the rat, cat, and dog. The ID_{50} values determined in the rat, cat, and dog were not significantly different from each other. Cimetidine was also shown to be active when administered orally in the dog.

Cimetidine was also found to be an effective inhibitor of pentagastrin-stimulated acid secretion. (Pentagastrin is a synthetic biologically active analog of gastrin which contains the terminal four amino acid residues of gastrin: *N*-*t*-BOC-β-Ala-Trp-Met-Asp-Phe-NH_2.) The ID_{50} values indicated that the potency of cimetidine against pentagastrin-stimulated secretion is very similar to its potency against histamine-stimulated secretion. This finding demonstrated that gastrin and histamine are somehow linked in the gastric secretory process, and the results firmly established that histamine has a physiological role in gastric acid secretion.

Cimetidine given orally at 1 to 1.2 g/day was shown to relieve symptoms and promote healing of lesions in a majority of patients with peptic ulcer disease. Fortunately, cimetidine did not cause granulocytopenia, thereby demonstrating the lack of any link with H_2-receptor antagonism. Cimetidine was marketed first in the United Kingdom in November 1976 and was marketed in the United States in November 1977; by 1979 it was sold in over 100 countries under the trademark Tagamet, representing all the major markets with one important exception—it was not granted approval for use in Japan until 1982. Cimetidine changed the medical management of peptic ulcer disease and became a very successful product (Freston, 1982; Winship, 1978). In 1983 its annual worldwide sales reached nearly $1 billion, and in several countries it was the leading prescription product in sales; indeed, it was the first of the modern "blockbuster" medicines.

10.4 SUMMARY AND FURTHER OBSERVATIONS

In this discovery, the medicinal chemistry was very dependent on applications of physical–organic chemistry principles, at a time when this was rather unusual. In particular, critically important contributions were made by considering pK_a values and applications of the Hammett equation. The design process also involved several examples of the use of isosteric replacement. Retrospective structure–activity analysis has demonstrated the existence of correlations with log P, dipole moments, and conformational analysis. The clear message is that in medicinal chemistry it is necessary to think as an organic chemist but not only about synthesis.

The long-term nature of pharmaceutical research and development and the need for tenacity to continue in the face of considerable difficulty and disappointment is well illustrated by this case history of drug discovery (see Fig. 10.1). The research project was initiated in 1964, and it took six years (to 1970) to obtain burimamide, which was used to characterize pharmacologically histamine H_2 receptors, to verify the basic concept, and to be studied in human volunteers. The next drug, metiamide, a more potent and orally active compound, was investigated clinically, but its use was severely restricted by a toxic effect. Cimetidine followed from metiamide: it was first synthesized in 1972 and made generally available in 1977. An overall time of 13 years had elapsed!

It is self-evident that any account of drug discovery must be incomplete, and certainly only a small proportion of the total studies made have been described in this chapter. Many avenues examined during structure–activity analysis turned out to be ineffective, and since in the main, these are not mentioned here, the net effect may be to make the work appear to be more rational and more perceptive than is warranted. To limit the scale of the problem for lead generation and optimization, the early work concentrated on imidazole derivatives.

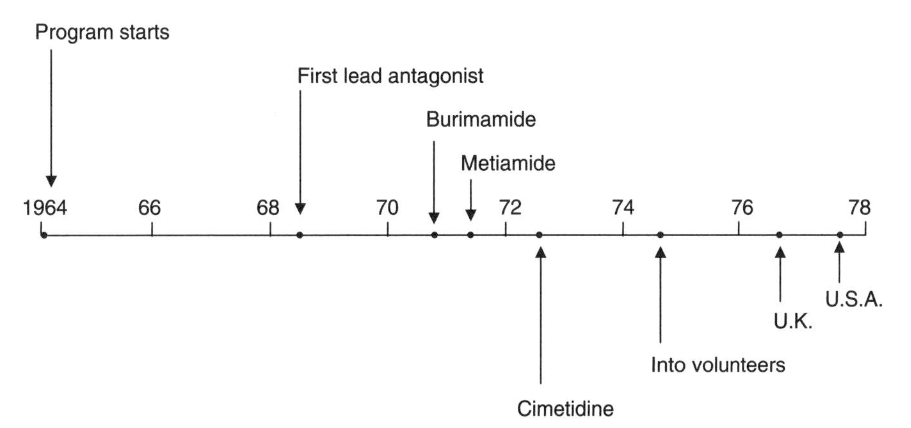

Figure 10.1 Thirteen years to discover and develop cimetidine and make it generally available for therapeutic use.

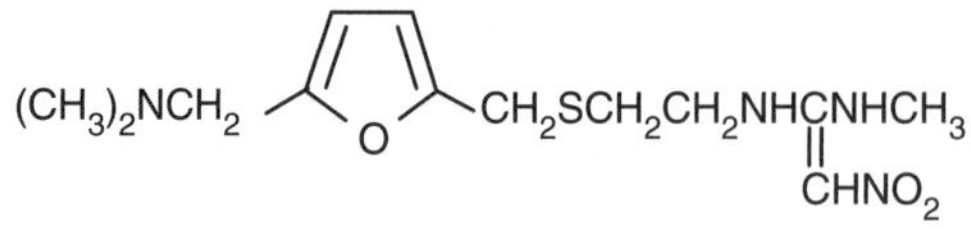

$(CH_3)_2NCH_2$–[furan]–$CH_2SCH_2CH_2NHC(=CHNO_2)NHCH_3$

(**20**) Ranitidine

Scheme 10.6

However, it was soon demonstrated that other heterocycles such as pyridine and thiazole could be used effectively in place of imidazole. Later, researchers at Glaxo demonstrated that it was not even necessary to have a nitrogen heterocycle, and they synthesized a furan derivative, ranitidine (**20**), which became the second histamine H_2-receptor antagonist to be introduced clinically (under the trademark Zantac) for treatment of peptic ulcer disease (Bradshaw et al., 1982) and eventually became the second modern blockbuster medicine (Scheme 10.6). This was followed by three other H_2-receptor antagonist products: nizatidine from Eli Lilly (Lin et al., 1983), famotidine from Yamanouchi (Takeda et al., 1982), and roxatidine from Teikoku.

Post Scriptum The treatment of diseases is developing continuously. The last 40 years have seen notable changes in gastroenterology. Until Tagamet became available, the medicinal management of peptic ulcer disease and related problems of hypersecretion of acid such as dyspepsia were not really effective. Surgery had become the mainstay for treatment. Tagamet changed this situation dramatically and offered physicians a medicine that acted pharmacologically to "turn off the tap" of acid secretion. Blocking histamine H_2 receptors gave physiologists a tool for studying the physiology of gastric acid secretion. This led to the development of omeprazole, the first of the H^+/K^+ ATPase inhibitors, which represented a new class of antiulcer product, the proton-pump inhibitors. This still left a mystery problem: Why did patients relapse after treatment had permitted their ulcers to heal? Then came another breakthrough: the discovery that ulcers were caused by an infection of *Helicobacter pylori*. So treatment has developed into a regimen that aims first to reduce acidity and so relieve patients of pain and then to give a combination of antibiotic and antibacterial agents to eradicate the infection.

At the time that it was discovered, burimamide appeared to be a selective drug and was used to characterize H_2 receptors. There were, however, some anomalous pharmacological effects with its use; we now know that, ironically, it is actually much more potent as an antagonist at histamine H_3 receptors than at H_2 receptors. Thus, from the early struggle in the 1960s to prove the existence of more than one type of histamine receptor, we have now had the contribution of molecular biological techniques that inform us that there are at least four classes of histamine receptor, designated as H_1, H_2, H_3, and H_4. It remains to be seen whether new drug therapies will be developed as a consequence of these later discoveries.

REFERENCES

1. Ash, A. S. F., and H. O. Schild (1966), *Br. J. Pharmacol. Chemother.* **27**, 427–439.
2. Black, J. W., W. A. M. Duncan, G. J. Durant, C. R. Ganellin, and M. E. Parsons (1972), *Nature* **236**, 385–390.
3. Black, J. W., G. J. Durant, J. C. Emmett, and C. R. Ganellin (1974), *Nature* **248**, 65–67.

4. Bovet, D. (1950), *Ann. N. Y. Acad. Sci.* **50**, Art. 9, 1089–1126.
5. Bradshaw, J., M. E. Butcher, J. W. Clitherow, M. D. Dowle, R. Hayes, D. B. Judd, J. M. McKinnon, and B. J. Price (1982), in A. M. Creighton and S. Turner, eds., *Chemical Regulation of Biological Mechanisms*, Special Publication 42, Royal Society of Chemistry, London, pp. 45–57.
6. Brimblecombe, R. W., W. A. M. Duncan, G. J. Durant, J. C. Emmett, C. R. Ganellin, and M. E. Parsons (1975), *J. Int. Med. Res.* **3**, 86–92.
7. Brimblecombe, R. W., W. A. M. Duncan, G. J. Durant, J. C. Emmett, C. R. Ganellin, G. B. Leslie, and M. E. Parsons (1978), *Gastroenterology* **74**, 339–347.
8. Charton, M. (1965a), *J. Org. Chem.* **30**, 3346–3350.
9. Charton, M. (1965b), *J. Org. Chem.* **30**, 969–973.
10. Dale, H. H., and P. P. Laidlaw (1910), *J. Physiol. (London)*, **41**, 318–344.
11. Durant, G. J., J. C. Emmett, C. R. Ganellin, P. D. Miles, H. D. Prain, M. E. Parsons, and G. R. White (1977), *J. Med. Chem.* **20**, 901–906.
12. Folkow, B., K. Haeger, and G. Kahlson (1948), *Acta Physiol. Scand.* **15**, 264–278.
13. Freston, J. W. (1982), *Ann. Intern. Med.* **97**, 573–580.
14. Ganellin, C. R. (1981), *J. Med. Chem.* **24**, 913–920.
15. Ganellin, C. R., G. J. Durant, and J. C. Emmett (1976), *Fed. Proc. Fed. Am. Soc. Exp. Biol.* **35**, 1924–1930.
16. Ghosh, M. M., and H. O. Schild (1958), *Br. J. Pharmacol. Chemother.* **13**, 54–61.
17. Gregory, H., R. C. Sheppard, D. S. Jones, P. M. Hardy and G. W. Kenner (1964), *Nature (London)* **204**, 931–933.
18. Gregory, R. A., and H. J. Tracy (1961), *J. Physiol. (London)* **156**, 523–543.
19. Grossman, M. I., C. Robertson, and C. E. Rosiere (1952), *J. Pharmacol. Exp. Ther.* **104**, 277–283.
20. Johnson, L. R. (1971), *Gastroenterology* **61**, 106–118.
21. Lin, T. M., R. S. Alphin, F. G. Henderson, D. N. Benslay, and K. K. Chen (1962), *Ann. N. Y. Acad. Sci.* **99**, 30–44.
22. Lin, T. M., D. C. Evans, M. W. Warrick, R. P. Pioch, and R. R. Ruffalo (1983), *Gastroenterology* **84**, 1231.
23. Takeda, M., T. Tagaki, Y. Yashima, and H. Maeno (1982), *Arzneim. Forsch.* **32**(ii), 734–737.
24. Van den Brink, F. G. (1969), in *Histamine and Antihistamines, Molecular Pharmacology, Structure–Activity Relations, Gastric Acid Secretion*, Drukkerij Gebr. Janssen, Nijmegen, The Netherlands, p. 179.
25. Winship, D. H. (1978), *Gastroenterology* **74**, 402–406.

11

DISCOVERY OF POTENT NONPEPTIDE VASOPRESSIN RECEPTOR ANTAGONISTS

BRUCE E. MARYANOFF
Johnson & Johnson Pharmaceutical Research and Development
Spring House, Pennsylvania

11.1 INTRODUCTION

The practice of drug discovery has undergone a paradigm shift over the past 30 years or so.[1] Around 1975, there were fairly limited options available to the "drug hunter" keen on nabbing the ultimate prey: a new marketed product. Drug discovery at that time was usually centered on the synthesis of sizable samples (2 to 5 g) of new chemical compounds and broad pharmacological screening in animal models that represented a disease state. For the most part, researchers did not conduct in vitro testing with enzymes and receptors. If one were seeking a novel, patentable, biologically active compound in the "old days," one would prepare interesting compounds with structures like those of known drugs, bioactive natural products, or endogenous mediators (e.g., norepinephrine and acetylcholine), but with interesting chemical twists and turns, so to speak. Drug discovery scientists have now become very partial to the pursuit of discrete molecular targets such as receptors, enzymes, and ion channels, with an emphasis on finding compounds that act directly, with good selectivity. In addition, the abundance of detailed structures for pharmaceutically interesting macromolecules has led to very "structure-based" approaches to drug discovery. It would be wonderful if all of the technological advances, such as genomics/proteomics, molecular biology, vast chemical libraries, high-throughput screening, protein structures, and computational methods, could be applied to a worthy research goal and actually deliver a clinical candidate that would sail all the way to the marketplace. Unfortunately, the experience of the last 10 to 15 years has revealed a disquieting paradox. Despite major investments of capital and human resources in high-tech drug discovery, the output of pharmaceutical products has actually diminished.[2]

Drug Discovery and Development, Volume 1: Drug Discovery, Edited by Mukund S. Chorghade

Although we have numerous exciting modern tools and approaches at our disposal, we probably should not lose sight of the old-fashioned low-tech approaches. It is my sensibility to embrace *all* possible avenues of attack in trying to find drug candidates with staying power, candidates that possess the attributes to enter human clinical trials and to progress to the market.

My discovery of TOPAMAX topiramate, which is marketed worldwide for the treatment of epilepsy and migraine, occurred under the old paradigm.[3] A synthetic intermediate from an enzyme inhibitor project, McN-4853, was screened in a panel of pharmacological assays, one of which was the maximal electroshock seizure test in mice. The interesting anticonvulsant activity that we found was confirmed in an NIH laboratory (National Institute of Neurological Diseases and Stroke), and a high level of enthusiasm was generated. Topiramate was placed on the development pathway and ultimately led to a commercial product with annual sales in excess of $1 billion. It is doubtful that this valuable molecule would ever have surfaced through a process of *direct intent*.

Since 1990, we have been pursuing serine protease inhibitors by the process of structure-based drug discovery. Nearly eight years of direct intent transpired before we delivered a compound into development that advanced to human clinical trials.[1] Our first drug candidate from this project was the tryptase inhibitor RWJ-56423, for the treatment of asthma and allergic rhinitis.[1,4] Subsequently, on related enzyme targets, we delivered a first-in-class, dual inhibitor of cathepsin G and chymase, RWJ-355871, for the treatment of airway inflammatory diseases.[1,5] This novel protease inhibitor emanated from the auspicious confluence of high-throughput screening and structure-based drug design. It can be viewed as encouraging that these high-tech approaches yielded such favorable outcomes, but the time frame was protracted and a marketed product is not even on the horizon.

In another fundamental design project, we sought compounds to block certain integrins, which are cell-surface adhesion molecules. We wanted to mimic key aspects of the KQAGD sequence in the fibrinogen γ-chain to generate oral fibrinogen receptor (glycoprotein IIb/IIIa) antagonists for treating thrombotic disorders. By applying high-throughput solid-phase parallel synthesis in the lead optimization stage, we were able to identify elarofiban (RWJ-53308) and propel it into human clinical trials in a reasonably short time frame.[6] Although our oral integrin antagonist elarofiban progressed successfully through phase IIa, the therapeutic modality of oral GPIIb/IIIa antagonists was unfortunately drawn into serious question by troublesome results in the clinical trials of certain oral agents from other companies.[7] As a consequence, this promising antithrombotic therapy virtually disintegrated overnight.

Another opportunity to score a drug product entailed a low-tech rather old-fashioned approach: the chemical modification of competitor structures to attain the desired biological activity and suitable patentability. This "follow-on" regime may sound less risky in that a validated starting point exists and the characteristics of the clinical candidate can be envisioned more easily. However, the risk is just moved downstream in the pathway in the sense that the clinical candidate must meet many additional requirements, unlike an agent that is first-in-class. It is necessary to impart significant improvements over key existing compounds from competitors, and in many cases, the requirements can be a very tall order, indeed. Since it may not be readily apparent how to engineer the desired molecular attributes, the acquisitive researchers must resort to Edisonian methods. We came to apply this modus operandi to the discovery of potent, nonpeptide vasopressin receptor antagonists.

11.2 GENESIS OF THE VASOPRESSIN RECEPTOR ANTAGONIST PROJECT

In 1997, the drug discovery team that I was leading in concert with biologist Dr. Patricia Andrade-Gordon was working on a number of interesting molecular targets. However, these targets were largely viewed as high risk in that their therapeutic potential had not been defined by existing compounds with that mechanism of action in a clinical environment. We wanted to balance the project portfolio of the Vascular Research Team by initiating work on some follow-on, or *fast-second*, targets. After reviewing various possibilities with key team members, we whittled the original list of 10 targets down to two, one of which was vasopressin receptor antagonists. The idea of entering this new area did not capture our imagination immediately. Rather, it was our attendance at the annual meeting of the American Heart Association (AHA) in November of that year that provided the motivation. After studying a poster presented by a scientist from Lederle Laboratories (now part of Wyeth Pharmaceuticals), we became more intrigued with the potential for quickly delivering a useful clinical entity. However, it was a luncheon engagement at the AHA meeting that brought us to a crisp decision.

Our team was involved in a multiyear collaboration with COR Therapeutics, Inc. (now part of Millennium Pharmaceuticals) in the pursuit of thrombin receptor (PAR-1) antagonists, a cutting-edge class of antithrombotic agents.[8] We happened to encounter Dr. Charles Homcy, the senior vice president of R&D at COR, at the AHA meeting and went to lunch with him in Orlando's Peabody Hotel. Since Charles had been at Lederle, we mentioned the exciting poster on vasopressin receptor antagonists, whereupon he indicated that the project was initiated by him and driven forward during his tenure. Charles's enthusiasm was so infectious that Patricia and I contracted the disease. Simultaneously, she and I felt a surge of resolution, and we were pleased to pay the entire bill. Our determination to enter this field was communicated to key team members on our return to Spring House.

A target proposal was written and submitted to Dr. Per Peterson, our senior vice president of drug discovery, for review and approval. Interestingly, another team, one based in Raritan, New Jersey, was also planning to embark on this target as a fast-second project. The Endocrine Therapeutics Team viewed vasopressin from the perspective of its being a hypophyseal hormone. How would this conflict be settled? Per's philosophy was clear on this issue: We compete vigorously with other drug companies but not within our own. Therefore, our two teams combined forces and drew up a plan for a cross-site collaboration early in 1998. Most of the biology and pharmacology would be conducted in Raritan, and the chemistry would be split between the two sites, with different novel structures being designed and synthesized at each site. This collaboration has been exceptionally fruitful, with five compounds entering preclinical development, one of which progressed successfully through phase IIa clinical trials. In this chapter I intend to address the chemical series that were pursued in my group at Spring House.

11.3 VASOPRESSIN, ITS RECEPTORS, AND DISEASE

Arginine vasopressin (AVP) is a nonapeptide hormone that is mainly secreted from the posterior pituitary gland and that exerts multiple biological actions, as a hormone and a neurotransmitter.[9] AVP is synthesized in magnocellular neurons within the hypothalamus and released into the circulation from the posterior pituitary. It can have multiple peripheral

TABLE 11.1 Vasopressin Receptor Subtypes and Biological Actions

Subtype	Location	Physiological Actions	Signal Transduction[a]
V_{1a}	Smooth muscle	Stimulates smooth muscle contraction	G_q coupled:
	Liver	Stimulates hepatic glycogenolysis	IP formation
	CNS	Memory; behavior	Ca^{2+} mobilization
V_{1b}	Anterior pituitary	Stimulates ACTH release from pituitary	G_q coupled: IP formation Ca^{2+} mobilization
V_2	Kidney	Stimulates renal reabsorption of water	G_s coupled: adenylate cyclase

[a]IP, inositol phosphate.

actions, including contraction of uterine, bladder, and vascular smooth muscle, stimulation of hepatic glycogenolysis, induction of platelet aggregation, release of corticotropin from the anterior pituitary, and the stimulation of renal water reabsorption. As a neurotransmitter within the central nervous system (CNS), AVP can affect various fundamental behaviors, the stress response, and the memory process. These diverse actions of AVP are mediated through specific G-protein-coupled receptors (GPCRs), which have been classified into three subtypes: V_{1a}, V_{1b}, and V_2 (Table 11.1). The V_{1a} and V_{1b} receptors signal through phosphatidylinositol hydrolysis to mobilize intracellular calcium (Ca^{2+}), whereas V_2 receptors signal through cyclic adenosine monophosphate (cAMP). The V_{1a} receptors mediate the contraction of smooth muscle, hepatic glycogenolysis, and the CNS effects of vasopressin; the V_{1b} receptors stimulate corticotropin release from the pituitary gland; and the V_2 receptors, which are found only in the kidney, are responsible for the antidiuretic actions of AVP. Within the renal collecting ducts, AVP regulates the water-selective channel aquaporin-2, inducing the translocation of aquaporin-2 to the plasma membrane to increase water permeability.[10]

From a cardiorenal standpoint, we were especially interested in the V_{1a} and V_2 receptor subtypes. Vasopressin-induced antidiuresis maintains normal plasma osmolarity and blood volume via renal epithelial V_2 receptors, while vasoconstriction is mediated by V_{1a} receptors. In pathological states, plasma AVP levels may be unusually high at a given plasma osmolarity, resulting in water retention and hyponatremia (low plasma Na^+ concentrations), an electrolyte abnormality that occurs in 1% of hospitalized patients.[11] Hyponatremia is associated with edema in conditions such as hepatic disease/cirrhosis, congestive heart failure (CHF), and renal failure. Presently, no drug is available that will selectively cause water excretion, or *aquaresis*.

There is substantial evidence to support a pathophysiological role of AVP in CHF.[12] In a rat model of CHF, resulting from ischemic cardiomyopathy, a peptide-based V_2 receptor antagonist increased cardiac output, decreased peripheral resistance, and increased urine output four- to ten-fold.[13] The first nonpeptide V_2-receptor antagonist, OPC-31260, induced marked aquaresis in conscious dogs with CHF. A combined V_2/V_{1a} receptor antagonist could have particular benefit in heart failure via hemodynamic and renal mechanisms.[14] In pivotal clinical trials, three nonpeptide vasopressin receptor antagonists improved the fluid status, osmotic balance, and hemodynamics of patients with CHF; however, long-term, adequately populated studies are needed to prove the value of such drugs relative to clinical outcomes and quality of life.[12d]

Nonosmotic release of AVP is well documented in liver cirrhosis,[15] which poses a serious unmet medical need. The failure to suppress AVP activity after central volume expansion may be one of the early mechanisms responsible for water–electrolyte imbalance in liver cirrhosis in children.[16] Administration of a V_2-receptor antagonist in rats with experimental cirrhosis has produced significant improvement.[17] Vasopressin is implicated in experimental brain edema, and a V_2-receptor mechanism has been postulated.[18] In rats, a V_2 antagonist increased plasma osmolarity and reversed the increase in brain water caused by bilateral occlusion of the carotid arteries; it also prevented cerebral edema and increases in brain sodium while normalizing diuresis induced by subarachnoid hemorrhage.[19]

Researchers at several pharmaceutical companies, recognizing the potential therapeutic utility of vasopressin receptor antagonists, have been inspired to identify clinical candidates.[20] In early studies, marked species differences were found in the effects of peptide-based vasopressin antagonists.[21] Although SKF-101926 was a V_2-receptor antagonist in rats, dogs, and monkeys in vivo, and human renal tissue in vitro, it showed agonist activity in human clinical trials. The variability between rats, dogs, monkeys, and humans suggested critical structural differences between vasopressin receptor proteins. A similar observation of interspecies divergence was made with backup compound SKF-105494. During the 1980s, SmithKline & French scientists sought a pure water diuretic, partly in support of their diuretic franchise involving the treatment of hypertension, and thus advanced these interesting peptide molecules. However, the problematic species dependency hampered their development plans and suitably alerted subsequent practitioners.

A major step forward was made with the discovery of the first nonpeptide vasopressin antagonists, OPC-21268 (V_1-selective)[22] and OPC-31260 (V_2-selective).[23] Prior to this milestone, potent vasopressin antagonists were peptide analogs of AVP, with the expected liabilities of poor oral bioavailability and short duration of action.[24] OPC-31260 (**1**; mozavaptan) is an orally active V_2-selective antagonist in rats and humans,[25] and numerous structure–activity studies have been performed around this chemical class (Scheme 11.1).[23b,26] This work at Otsuka had a great impact on the field because it established a key structural motif, or pharmacophore, for nonpeptide vasopressin receptor antagonists: the 1-(4-benzamidobenzoyl)benzazepine template. Further strides on this central theme were made in the late 1990s by researchers at Lederle/Wyeth[27] and Yamanouchi.[28] At this point in time, various benzazepine-type series have been disclosed in the scientific literature,[29] including several from Johnson & Johnson.[30] Some compounds of particular interest are the V_2-selective antagonists tolvaptan (**2**; OPC-41061)[26,31] and lixivaptan (**3**; VPA-985),[27,32] and the V_{1a}/V_2 antagonist conivaptan (**4**; YM-087),[28a,33] all of which have advanced to phase II or phase III clinical trials (Scheme 11.1). Other important contributions have emanated from researchers at Sanofi–Synthelabo, who have reported V_{1a}-, V_{1b}-, and V_2-receptor antagonists of clinical interest with very different structural motifs: indoline prolinamides, such as V_{1a}-selective SR-49059 (relcovaptan) and V_{1b}-selective SSR-149415, and spiroindolinones, such as V_2-selective SR-121463A.[34] Our drug discovery effort, which has focused on the (benzamidobenzoyl)benzazepine class, is described in the remainder of this chapter.

11.4 THE GAME PLAN

When initiating a new drug discovery project, it helps to have a well-defined goal, a vision for success, some novel chemical ideas, and a workable plan for moving compounds forward. In the absence of a sound plan with clear-cut decision-making steps (the *critical*

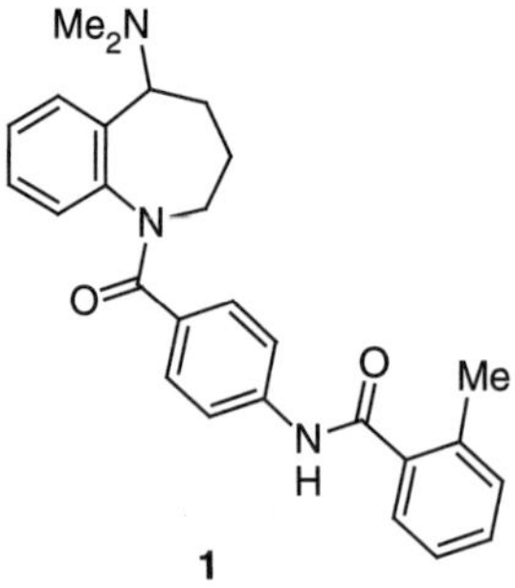

Scheme 11.1

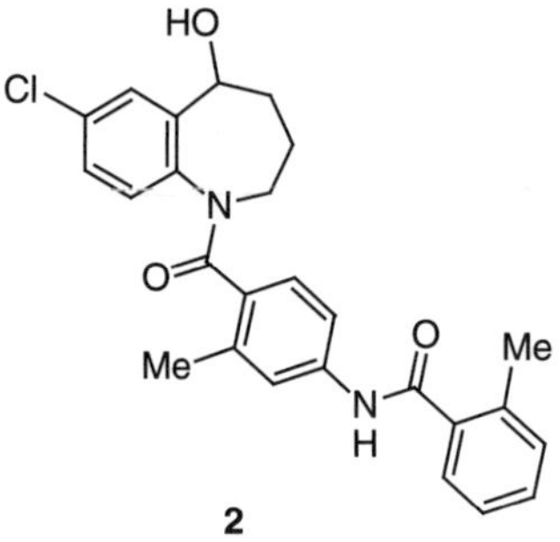

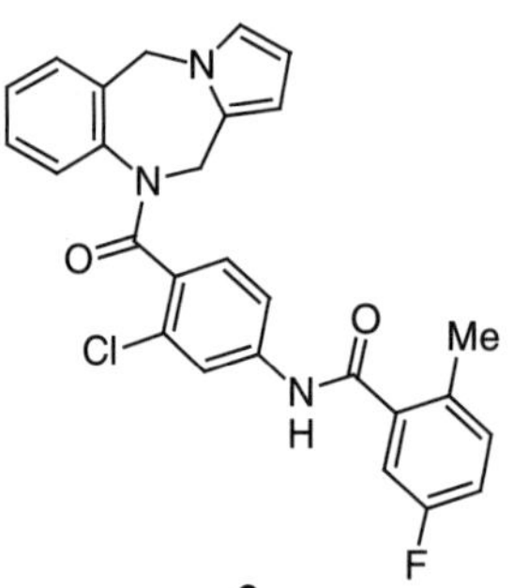

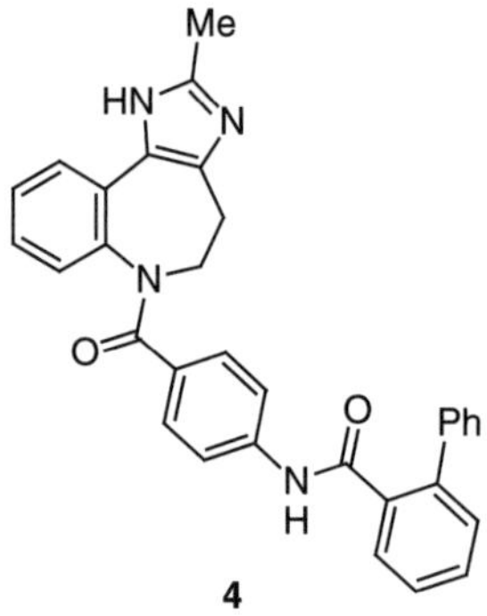

path), there is no reason at all to design and synthesize target molecules. Importantly, the early bioassays in the critical path need to be established beforehand, and suitable pharmacology models need to be available. At the outset we were going to rely on human receptor binding assays coupled with human receptor cell-based assays.

The 418-amino acid human V_{1a} receptor has a 72% sequence identity with the rat V_{1a} receptor, 36/37% identity with the human/rat V_2 receptors, and 45% identity with the human oxytocin receptor; the cloned 371-amino acid human V_2 receptor has 50% identity with the rat V_2 receptor and 45% identity with the human oxytocin receptor.[35] A crucial starting point for our project was establishing and validating binding assays based on cloned and expressed human V_{1a} and V_2 receptors. We used recombinant human V_{1a} or V_2 receptor preparations derived from the cell membranes of transfected HEK-293 cells, with test compounds being evaluated for their ability to displace [^{3}H]AVP. Cellular functional assays[36] were also put in place to ascertain whether a compound that binds is a receptor agonist or antagonist. The inhibition of receptor activation caused by AVP was quantified in HEK-293 cells expressing either human V_{1a} or V_2 receptors, and changes in intracellular concentrations of either Ca^{2+} (for V_{1a}) or cAMP (for V_2) were measured. For follow-up work, we had cellular functional assays based on rat V_{1a} and V_2 receptors to address any issues that might arise in the rat pharmacology work. Also, a human oxytocin receptor binding assay was available to rule out that undesirable action. Compounds of interest for in vivo evaluation would be tested for their pharmacokinetic properties, such as metabolic stability to rat and human liver microsomes, oral bioavailability and duration, and binding to plasma proteins.

Obviously, if a compound with good in vitro activity were metabolically unstable or poorly absorbed orally, it would not proceed to the next stage.

Our early in vivo assessment would entail hypertension (V_{1a} action) and aquaresis (V_2 action) experiments in rats. A V_{1a} receptor antagonist should reverse the hypertension induced by AVP (saline solution infused at 30 ng/kg per minute intravenously) in pentobarbital-anesthetized rats. The degree of reduction in mean arterial blood pressure from the AVP-induced level of +50 to +60 mmHg would be recorded, and ED_{50} values would be calculated from the linear portion of the dose–response curve. The acute diuretic effect of a V_2-receptor antagonist would be determined in hydrated, conscious rats given single oral doses of test compounds. Spontaneously voided urine would be collected over 4 hours, and the urine volume, osmolality, and electrolyte concentrations would be measured. Advanced leads would be tested acutely for aquaresis in dogs and monkeys, then multidose studies would be performed in rats and monkeys. Thus, a workable game plan was established.

11.5 NOVEL CHEMOTYPES: VARIATIONS ON A THEME

My experience at the Orlando AHA meeting in 1997 brought the (benzamidobenzoyl)benzazepine class of vasopressin antagonists into the forefront. Lixivaptan (**3**) had entered human clinical trials[20a,32] and analog CL-385004 (**5**) also looked like an interesting compound (Scheme 11.2).[27c] Such structures appeared to offer a suitable platform for branching off to a novel chemotype. We conceived of compounds with general formula **6**,[30c,d] and an examination of the patent literature in 1998 indicated that no compounds of this variety had been disclosed. Since these are fairly sizable molecules that could present challenges for oral drug development due to unfavorable physical properties, such as high molecular weight (450 to 600 Da), high hydrophobicity (log $P > 4$), and limited aqueous solubility (<1 mg/mL),[37] we thought that the presence of one, or even two, basic amine centers in targets **6** could be a positive influence. It was hoped that these novel chemical entities would give rise to receptor antagonists with good in vitro and in vivo potency and favorable oral bioavailability. Even though we became highly enamored with this chemical approach, we decided to diversify by pursuing parallel paths to enhance the probability of success. Such a parallel-processing strategy can be applied in the early stage of a new project, but ultimately needs to be narrowed down by the assessment of critical data. Thus, we also conceived of novel indole-fused compounds with generic structure **7**, which are somewhat akin to **1** and **4**,[30a] and novel bridged bicyclic derivatives with general structure **8**,[30b] which are related to targets **6** (Scheme 11.2).

11.5.1 Azepinoindoles

An assortment of indolobenzodiazepine derivatives, **7a–m**, were prepared, mainly according to the route in Scheme 11.3, and tested (Table 11.2).[30a] Intermediate azepino[4,3,2-*cd*]indole (**11**) was obtained in seven steps from 4-nitroindole (**10**),[38] which was prepared from 2-methyl-3-nitroaniline (**9**) in two steps.[39] Acylation of **11** with a 4-nitroaroyl chloride ($RNO_2PhCOCl$) provided **12**, which was reduced with zinc and ammonium chloride to provide aniline **13**. Acylation of **13** with aroyl chlorides ($R_1PhCOCl$) provided target compounds **14** (encompassing **7a–g**). In general, this series of compounds (Table 11.2) did not bind effectively to the V_{1a} receptor, although three compounds exhibited modest V_{1a} binding of 30 to 60% at 1 μ*M*: **7b**, 58%; **7d**, 33%; **7h**, 49%. Mozavaptan congener **7a** was

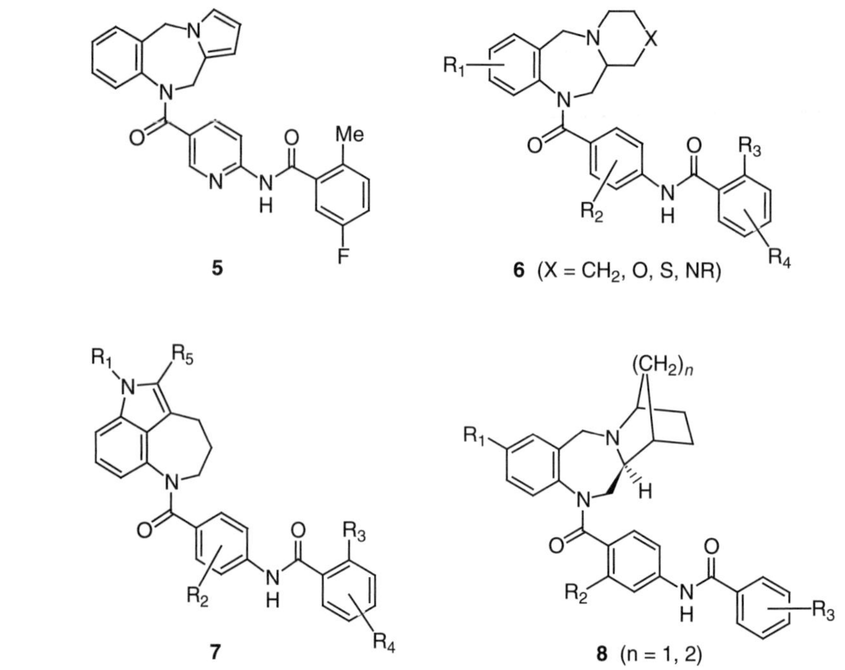

Scheme 11.2 (*S*,*S*,*S*) or (*S*) enantiomer shown.

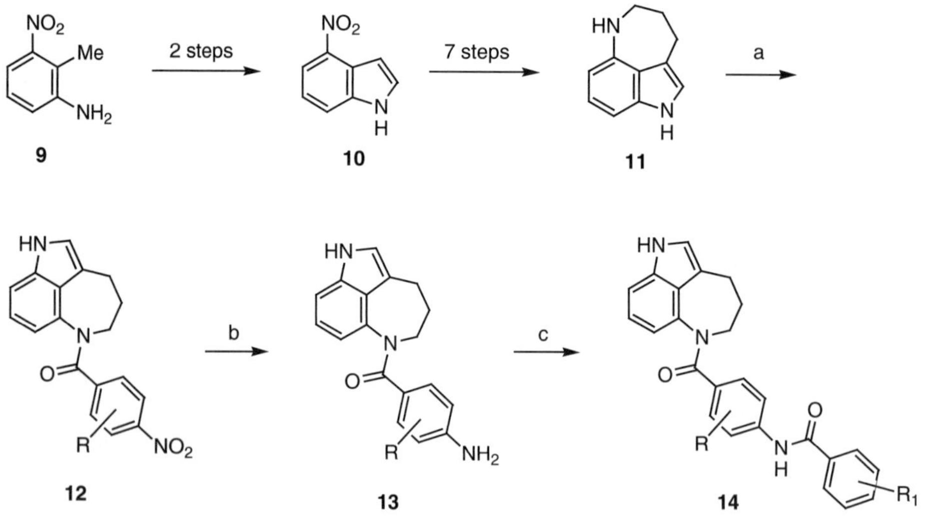

Scheme 11.3 Synthesis of **14**. Reagents and conditions: a, 4-NO_2ArCOCl, Et_3N, CH_2Cl_2, 0°C; b, Zn dust, NH_4Cl, MeOH, reflux; c, Ar′COCl, Et_3N, CH_2Cl_2, 0°C.

TABLE 11.2 Indoloazepine Derivatives with Their Vasopressin Receptor Effects

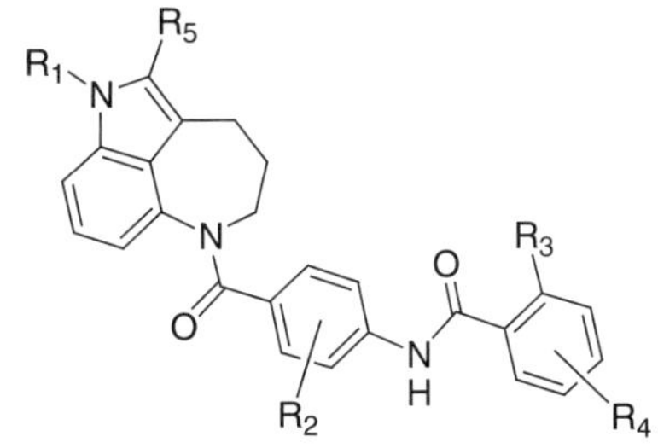

Compound[a]	R_2[b]	R_3	R_4	R_1	R_5	V_2 Binding K_i (nM)[c]	V_2 Funct. K_i (μM)[d]
7a	H	2-Me	H	H	H	IA	
7b	H	2-Me	5-F	H	H	90	
7c	H	2-Ph	H	H	H	9.0	0.070
7d	H	2-(4-MePh)	H	H	H	5.0	0.070
7e	*o*-Cl	2-Ph	H	H	H	7.0	
7f	*m*-Cl	2-Ph	H	H	H	20	0.33
7g	H	2-Me	3-F	H	H	83	
7h	H	2-(4-MePh)	H	Me	H	10	
7i	H	2-(4-MePh)	H	C(O)Me	H	80	
7j	H	2-(4-MePh)	H	Pr	H	IA	
7k	H	2-Ph	H	H	Cl	20	
7l	H	2-Ph	H	H	SO_3H	6.0	
7m	*o*-Cl	2-Ph	H	H	SO_3H	6.0	
VPA-985[e]						2.3	0.023

[a]Purified by reverse-phase semiprep. HPLC; >95% pure by reverse-phase HPLC/MS (215/254 nm); characterized by 300-MHz ^{1}H NMR and MS.

[b]Position of substitution is ortho or meta to the carboxyl.

[c]Binding to human V_2 receptors ($N = 3$ to 6). IA = inactive, i.e., <30% inhibition of radioligand binding at 100 nM. All analogs had >100-fold selectivity for V_2 over V_{1a}.

[d]Inhibition of functional human V_2-receptor activity ($N = 3$ to 5).

[e]Reference standard. V_{1a} binding $K_i = 44$ nM.

inactive in V_2 binding, and lixivaptan congener **7b** was moderately potent (V_2 $K_i = 90$ nM) but much weaker than lixivaptan (V_2 $K_i = 2.3$ nM). However, it was the 2-phenylbenzoyl group, which is contained in conivaptan (**4**), that made a big difference. For example, **7c–e** had V_2 K_i values in the range 5 to 10 nM, and substitution of the indole 2-position with a sulfonic acid group surprisingly did not diminish the binding potency (i.e., **7l** and **7m**). Compounds **7c** and **7d** exhibited reasonable functional antagonism, as their V_2 K_i values of 17 nM compared favorably with that of lixivaptan (VPA-985; V_2 $K_i = 23$ nM). Compound **7c** was evaluated orally in the rat aquaresis assay at a dose of 10 mg/kg, but it lacked activity. Besides oral bioavailability being a problem with this series, the compounds tended to have low aqueous solubility, which is why we tried the sulfonic acid group.

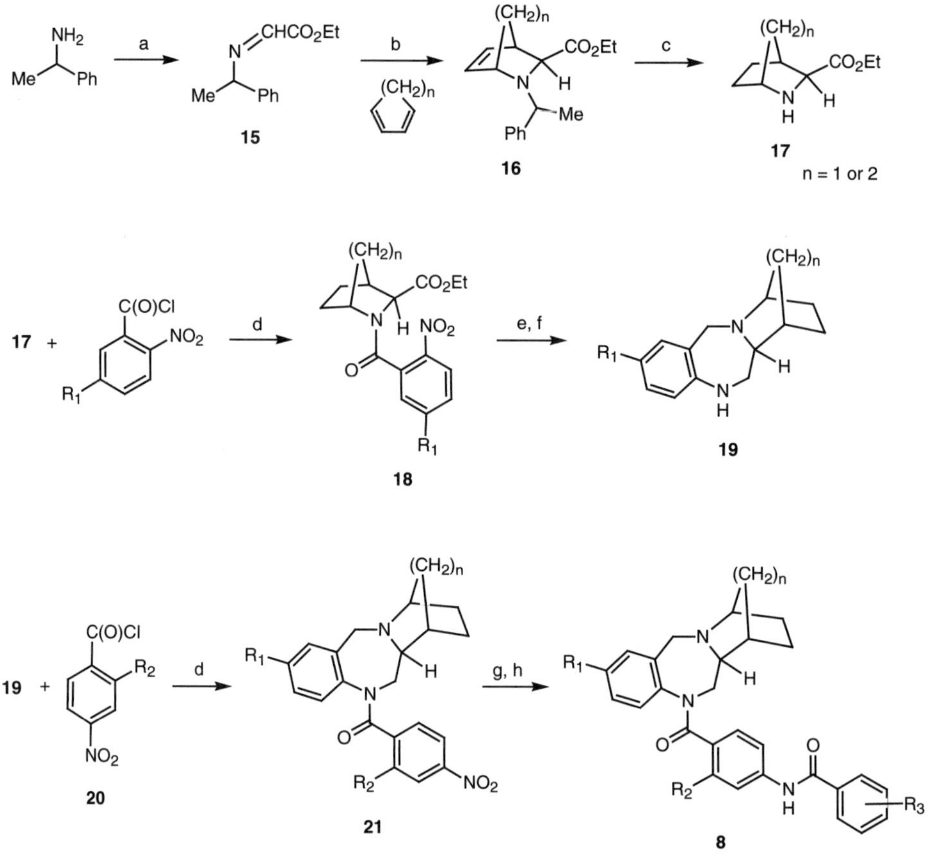

Scheme 11.4 Synthesis of **8**. Reagents and conditions: a, EtO_2CCHO, toluene, reflux ($-H_2O$); b, CF_3CO_2H ($n = 1$) or $CF_3CO_2H/BF_3 \cdot Et_2O$ ($n = 2$), CH_2Cl_2, $-20°C$; c, H_2, Pd/C, EtOH; d, Et_3N, CH_2Cl_2, 0°C; e, Fe, AcOH, reflux; f, $LiAlH_4$, THF, 0°C; g, Zn dust, NH_4Cl, MeOH, reflux; h, ArCOCl, Et_3N, CH_2Cl_2, 0°C.

11.5.2 Bridged Bicyclic Derivatives

Bridged bicyclic benzodiazepines, represented by **8a–l**, were synthesized (Scheme 11.4) and evaluated (Table 11.3).[30b,40] Fortunately, the chemistry for obtaining the key bridged bicyclic amino acids in enantiomerically enriched form had already been described.[41] Hetero Diels–Alder reaction of cyclopentadiene ($n = 1$) or 1,3 cyclohexadiene ($n = 2$) with activated imine **15** (from ethyl glyoxalate and α-methylbenzylamine) yielded exo-substituted cycloadducts **16** with high ($n = 1$) to moderate ($n = 2$) diastereoselectivity. The (*S,S,S or S*)-cycloadduct was produced predominantly from the (*R*)-(+)-amine, and vice versa. The diastereomers were separated by flash-column chromatography, and the one isomer (shown) was hydrogenated to reduce the alkene and deprotect the nitrogen to furnish bicyclic amino esters **17** in good yields (>95% e.e.). 2-Nitrobenzamides **18** were reduced and cyclized to intermediate benzodiazepinediones (not shown), which were reduced with $LiAlH_4$ to give the key benzodiazepine core structure, **19** (50 to 65% overall yield based on **17**).[42] Importantly, these transformations occurred without any loss of stereochemical

TABLE 11.3 Bridged Bicyclic Derivatives with Their Vasopressin Receptor Effects

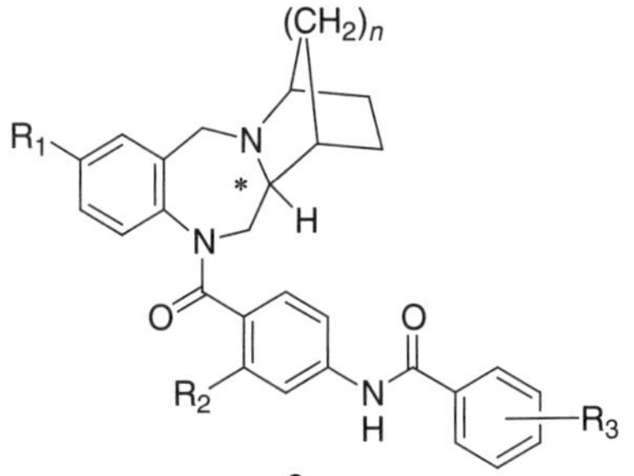

8

Compound[a]	R_1	R_2	R_3	n	Config.[b]	V_{1a} Binding K_i (nM)[c]	V_2 Binding K_i (nM)[c]	V_{1a} Funct. K_i (μM)[d]	V_2 Funct. K_i (μM)[d]
8a	H	Cl	2-Ph	1	*S*	7.0	1.8	0.13	0.009
8b	H	Cl	2-Ph	1	*R*	7%[e]	7.0		0.03
8c	Cl	H	2-(4-MePh)	1	*S*	7.0	2.3	0.023	0.013
8d	Cl	Cl	2-(4-MePh)	1	*S*	10	1.4		
8e	Cl	H	2-F	1	*S*	20	12	2.6	0.180
8f	Cl	H	2-Cl	1	*S*	1.5	4.8	0.13	0.032
8g	Cl	H	2-CF_3	1	*S*	1.2	3.2	0.04	0.009
8h	H	H	2-Ph	2	*R/S*	~30	2.8		
8i	H	Cl	2-Ph	2	*R/S*	30%[e]	1.8		
8j	H	Cl	3-F-5-Me	2	*S*	15%[e]	9.0		
8k	Cl	H	2-Cl	2	*S*	10	6.0	0.62	0.52
8l	Cl	H	2-(4-MePh)	2	*S*	17	33		
VPA-985[f]						44	2.3	6.0	0.023
OPC-31260[f]						70	13	0.42	0.04

[a]Purified by chromatography and crystallized as HCl salts; >95% pure by reverse-phase HPLC/MS (215/254 nm); characterized by 300-MHz ^{1}H NMR and MS. Elemental analyses for compounds tested in vivo. Where relevant, >99% e.e. by chiral HPLC.

[b]Absolute configuration for the stereocenter with the asterisk. For $n = 1$, (*S,S,S*) or (*R,R,R*) enantiomers; for $n = 2$, (*S*) and/or (*R*) enantiomers.

[c]Binding to human V_{1a} and V_2 receptors ($N = 3$ to 6).

[d]Inhibition of functional human V_{1a} or V_2 activity ($N = 3$ to 5).

[e]Percent inhibition at 100 nM.

[f]Reference standard.

integrity at the originally established stereogenic center. Acylation of **19** with 4-nitrobenzoyl chlorides (**20**) yielded precursors **21**, which were reduced and acylated to provide the desired target molecules, **8**.

This series of compounds (Table 11.3) afforded very potent V_2 antagonists, possessing single-digit nanomolar K_i values in receptor binding (i.e., **8a**, **8c**, **8d**, **8g**, **8h**, and **8i**). Several compounds with the (*S*) configuration were dual V_{1a}/V_2 antagonists in the binding assays. Compound **8g** is a notable example, with V_{1a} and V_2 K_i values of 3.2 and 1.2 nM. This compound and **8c** also showed well in both functional assays: V_2 and V_{1a} K_i values of 9 and

TABLE 11.4 In Vivo Diuretic Effects of 8a–c in Rats

Oral Dose (mg/kg)	N	Urine Vol. (mL)[a]	Urine Osmol. (mOs/kg)[a]	Oral Dose (mg/kg)	N	Urine Vol. (mL)[a]	Urine Osmol. (mOs/kg)[a]
		8a				**8b**	
Vehicle	10	1.1 ± 0.2	663 ± 62	Vehicle	10	0.6 ± 0.1	892 ± 70
0.3	9	2.5 ± 0.3	444 ± 79	0.3	9	1.4 ± 0.1	408 ± 37
1	10	6.0 ± 0.7	299 ± 34	1	10	3.4 ± 0.4	303 ± 29
3	10	15.1 ± 1.4	180 ± 9	3	10	9.4 ± 1.1	238 ± 19
10	10	29.9 ± 2.0	138 ± 11	10	10	19.5 ± 1.4	172 ± 9
		8c				*OPC-31260*	
Vehicle	10	2.3 ± 0.3	441 ± 30	Vehicle	16	2.1 ± 0.4	608 ± 50
1	10	2.9 ± 0.2	387 ± 15	1	8	2.3 ± 0.5	530 ± 42
3	10	5.8 ± 0.5	226 ± 9	3	10	5.4 ± 0.7	372 ± 39
10	10	17.8 ± 1.3	159 ± 9	10	9	12.2 ± 1.0	238 ± 18
				30	8	21.5 ± 3.0	165 ± 17

[a]Mean ± SE.

40 nM for **8g** and 13 and 23 nM for **8c**. The (R)-enantiomer (**8b**) is a potent V_2-selective antagonist with V_2 K_i values of 7 nM in the binding assay and 30 nM in the functional assay.

We tested potent derivatives **8a** (10-fold V_2-selective), **8b** (V_2-selective), and **8c** (balanced V_{1a}/V_2) in vivo for their aquaretic activity in rats (Table 11.4). Compounds **8a** and **8b** exhibited reasonable activity at a low oral dose of 0.3 mg/kg, whereas **8c** was considerably less potent (3 mg/kg). Compounds **8a–c** had modest oral bioavailability in rats ($F = 14\%$, 10%, and 7%, respectively), with **8c** being the lowest; the oral half-lives were useful, ranging from 2.5 to 5 h. The weaker potency of **8c** may be related to its unimpressive oral bioavailability. We studied the ability of the VPA-985, **8a**, **8c**, and **8g** to reverse AVP-induced hypertension in rats (V_{1a} related). VPA-985 had an ED_{50} value of 1120 ± 520 μg/kg (mean ± SE, $N = 5$), whereas **8a**, **8c**, and **8g** showed greater potency with ED_{50} values of 220 ± 43 μg/kg ($N = 6$), 170 ± 120 μg/kg ($N = 8$), and 380 ± 290 μg/kg ($N = 5$), respectively, which relates to their V_{1a} affinities. The potent compounds in this series tended to have low aqueous solubility and some significant inhibition of cytochrome P450 enzymes, which are involved in the oxidative metabolism of drug molecules (e.g., inhibition of Cyp 3A4).

11.5.3 Thiazino-, Oxazino-, and Pyrazinobenzodiazepines

Our best prospects for clinical candidates were realized by investigating compounds in series **6**.[30c,d] Initially, we carried out some work with **6** ($X = CH_2$) and its pyrrolidine congener, but the patent landscape surrounding this type of compound became unattractive as a result of information disclosed by Ohtake and co-workers.[29f,43] Thus, the compounds with X = S, O, and NR became much more attractive. Most of the target compounds were obtained by assembling the requisite tricyclic intermediates **25**, as shown in Scheme 11.5, and then converting them by the standard protocol mentioned earlier to target molecules, which were subjected to biological assays (Tables 11.5 to 11.7).[30c,d,44] In certain cases the racemates of the tricyclic intermediates were resolved into individual enantiomers, which were converted to single-enantiomer targets.[30c,d,44] Herein, we discuss the following representative target molecules: thiazino compounds (**26–36**), oxazino compounds (**37–50**),

Scheme 11.5 Synthesis and resolution of tricyclic intermediates **25**.

and pyrazino compounds (**51–64**). A synthetic example involving **54** serves to illustrate the procedure for introducing a 2,2,2-trifluoroethyl group and for executing the end game (Scheme 11.6).

The thiazino series, **6** (X = S), was explored first with numerous racemic targets being prepared from **25a**, some of which are presented in Table 11.5 (**26–37**). The direct analog of VPA-985, **26**, had much weaker binding to the V_{1a} and V_2 receptors than the reference,

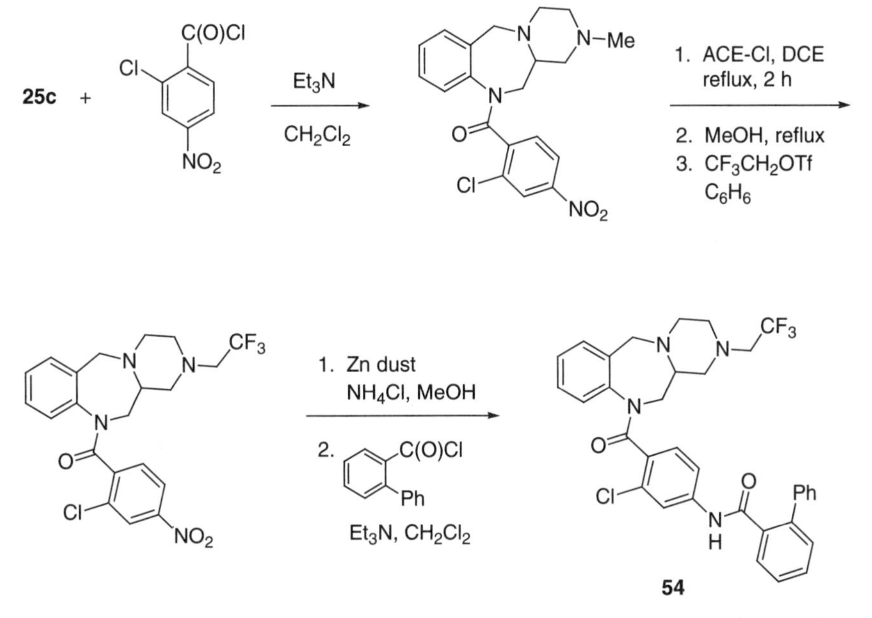

Scheme 11.6 Synthesis of target **54**. ACE, α-chloroethoxycarbonyl; DCE, 1,2-dichloroethane; Tf, trifluoromethanesulfonyl (triflyl).

TABLE 11.5 Vasopressin V_{1a} and V_2 Binding and Functional Data for Thiazinobenzodiazepines

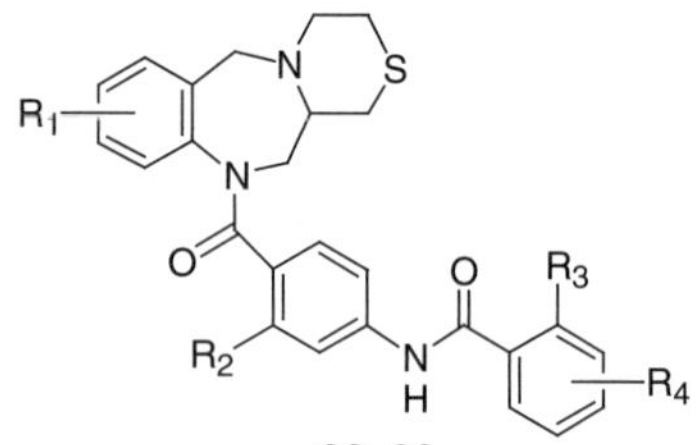

26–36

Compound[a]	R_1	R_2	R_3	R_4	V_{1a} Binding[b] K_i (nM)	V_2 Binding[b] K_i (nM)	V_{1a} Funct.[c] K_i (μM)	V_2 Funct.[c] K_i (μM)
26	H	Cl	Me	5-F	IA	60	14	0.3
27	H	Cl	Ph	5-F	IA	37	14	0.07
28	H	H	Ph	H	49	5.0	8.0	0.02
(+)-**28**[d,e]	H	H	Ph	H	84	3.2	1.4	0.015
(−)-**28**[e,f]	H	H	Ph	H	290	25	1.3	0.04
29	H	Cl	Ph	H	490	11		
(+)-**29**[d,e]	H	Cl	Ph	H	250	3.7	14	0.017
30	H	Cl	Ph	4-F	410	3.7		
31	H	F	Ph	H	IA	10		
32	H	Me	Ph	H	IA	10		
33	H	OMe	Ph	H	IA	15		
34	9-Cl	H	Ph	H	IA	10		
35	8-Me	H	Ph	H	18	8.0	14	0.023
36	8-F	H	Ph	H	IA	9.0		
VPA-985[g]					44	2.3	6.0	0.023

[a]Purified by reverse-phase semi-prep. HPLC and isolated as trifluoroacetate salts unless noted otherwise; >95% pure by reverse-phase HPLC/MS (215/254 nm); characterized by ESI-MS, with selected compounds analyzed by 300-MHz ^{1}H NMR. Compounds are racemates unless noted otherwise.

[b]Binding to human V_{1a} and V_2 receptors ($N = 3$ to 6). IA = inactive (<30% inhibition at 100nM).

[c]Inhibition of functional human V_{1a} or V_2 activity ($N = 3$ to 5).

[d](S)-(+) enantiomer.

[e]HCl salt.

[f](R)-(−) enantiomer.

[g]Reference standard.

although the V_2 K_i value of 60 nM is still quite respectable. The reduced V_2 affinity of **26** is consistent with its potency in the V_2 functional assay ($K_i = 300$ nM). Introduction of an *o*-phenyl group, as in **27**, enhanced the V_2 receptor affinity ($K_i = 37$ nM), which was reflected in the V_2 functional assay ($K_i = 70$ nM). Compound **27** showed good V_2 selectivity, especially as depicted from the functional assays, with a V_{1a}/V_2 ratio of 200. Thus, we adopted the *o*-phenyl substituent for other analogs. Des-halogen parent **28** had similar V_2 affinity ($K_i = 5.0$ nM) and selectivity ($V_{1a}/V_2 = 10$). Its two enantiomers, (S)-(+)-**28** and (R)-(−)-**28**, were examined and each had significant V_2 receptor affinity ($K_i = 3.2$ and 25 nM, respectively). The mere eightfold difference in V_2 affinity for the two enantiomers of **28** suggests that the geometry around this portion of the ligand is not particularly critical for V_2 receptor binding interactions.

The excellent V_2 affinity of (*S*)-(+)-**28**, coupled with the 26-fold selectivity for V_2 over V_{1a}, was encouraging. Subsequent variation of R_2 on the 4-aminobenzamide ring provided potent, reasonably V_2-selective analogs (cf. **28** with **29** and **30–33**). Overall, notable V_2 affinities were realized for (*S*)-(+)-**29** ($K_i = 3.7\,nM$), (*S*)-(+)-**28** ($K_i = 3.2\,nM$), and **30** ($K_i = 3.7\,nM$). The results for **34–36** indicate that R_1 substitution is fairly well tolerated, although this modification can strengthen V_{1a} affinity, as observed for **35** (V_{1a} $K_i = 18\,nM$).

To follow up on these results, we studied several compounds in cell-based functional assays. Compounds (+)-**29**, (*S*)-(+)-**28**, (*R*)-(−)-**28**, **35**, and VPA-985 potently antagonized the effects of AVP on human V_2 receptors ($K_i = 0.015$ to $0.04\,\mu M$), whereas they had just weak potency against human V_{1a} receptors ($K_i = 1.3$ to $15\,\mu M$). Enantiomer (*S*)-(+)-**28** gave V_{1a} and V_2 K_i values of 1400 and 15 n*M*, for about 90-fold V_2 selectivity, and (*S*)-(+)-**29** gave V_{1a} and V_2 K_i values of 14,000 and 17 n*M*, for an impressive functional V_2 selectivity of about 820-fold. The potency of each enantiomer of **28** in the V_2-receptor functional assay ($K_i = 15$ to $40\,nM$) is reasonably consistent with the binding data.

For the oxazino series, **6** (X = O), we mostly studied the (*S*)-(+) enantiomers because they have better V_2 receptor affinity (Table 11.6; **37–50**). In comparing **40** and **41**, each enantiomer has a significant V_2 affinity ($K_i = 0.9$ and $13\,nM$, respectively), but there is a 15-fold preference for **40**. Interestingly, **40** also has a high affinity for the V_{1a} receptor ($K_i = 24\,nM$), which accounts for a 25-fold V_2 binding selectivity. Direct VPA-985 analog **37** exhibited excellent V_2 receptor affinity as well as modest V_{1a} affinity, in contrast to the related thiazinobenzodiazepine (**26**). In general, the *o*-phenyl group resulted in good-to-excellent V_2 affinity. Indeed, several oxazino analogs had single-digit nanomolar V_2 receptor binding with good selectivity versus V_{1a}, such as **39**, **40**, **42–44**, **46**, **47**, and **49**. The most potent compounds possessed the 2-phenyl (**40**, V_2 $K_i = 0.9\,nM$), 2-phenyl-4-fluoro (**46**, V_2 $K_i = 1.9\,nM$), and 2-phenyl-4-hydroxy (**49**, V_2 $K_i = 1.4\,nM$) groups. In the V_2 functional assay, **37**, **38**, **40**, **42–47**, and **49** were very potent in antagonizing the effects of AVP ($K_i = 0.002$ to $0.02\,\mu M$), whereas they were weak in the V_{1a} functional assay ($K_i = 0.7$ to $11\,\mu M$). In the cell-based assays, **40** exhibited K_i values of 420 and 3 n*M* for V_{1a} and V_2, reflecting 140-fold V_2 selectivity. The (*R*)-(−) enantiomer **41** showed less functional potency ($K_i > 15{,}000$ and $170\,nM$) relative to **40**, in contrast to a comparison of (*S*)-(+)-**28** and (*R*)-(−)-**28**.

Oral administration of (*S*)-(+)-**28** to rats elicited a dose-dependent aquaretic effect. At a dose of 10 mg/kg p.o., urine output ($N = 8$) was increased 300% over untreated controls ($N = 8$), with a reduction in urine osmolality of 70%. Oral administration of **40** to rats produced a dose-dependent aquaretic effect with remarkable potency. An oral dose of just 1 mg/kg caused a 700% increase of urine output ($N = 10$) over untreated controls ($N = 18$), with a 60% reduction of urine osmolality. For comparison, at oral doses of 10 and 1 mg/kg, OPC-31260 (**1**) is reported to modify urine output/osmolality by +500%/–65% and 0%/–10%, respectively;[23b] VPA-985 (**3**) is reported to modify urine output/osmolality by +450%/−70% and +200%/−50%, respectively.[27a]

Compound **40** showed excellent pharmacokinetics in rats and dogs. In rats, the oral bioavailability was very favorable ($F = 68\%$), with an oral $t_{1/2}$ value of 3.7 h. The physical properties of **40** (MW = 538 Da) appeared to be better than those for reference compound lixivaptan (**3**; MW = 474 Da): for **40**, log $P = 3.63$; log D (pH3) = 2.01 (p$K_a = 4.75/12.7$); for lixivaptan: calculated log $P = 6.1$ (calculated p$K_a = 11.9$).[45] The log P value for **40** of 3.63 is within a range (essentially 0 to 4) that is favorable for membrane transport.

Scheme 11.7 Solid-state structure of **40**·TsOH [(*S*)-(+) isomer], showing the cationic subunit with its atom-numbering scheme (standard atom color code; H, cyan).

The structure of **40**, as a tosylate salt, was determined by x-ray diffraction. The tricycle adopts a chairlike conformation for the seven-membered ring, with the pendant amide carbonyl (C-13) in an axial orientation. The carbonyl oxygen of this amide, O-2, is *anti* to the fused benzene ring (Scheme 11.7). This arrangement is analogous to that observed for related *N*-acyltetrahydrobenzazepines in solution.[46] We carried out a Monte Carlo conformational search on protonated **40** with the OPLS–AA force field and GB/SA water model.[30c] The global energy minimum contains a *trans*-fused oxazinobenzodiazepine with a chairlike seven-membered ring, an axial amide, and an amide carbonyl anti to the fused benzene ring (Scheme 11.8). The tricyclic nucleus in this structure is closely superimposable on

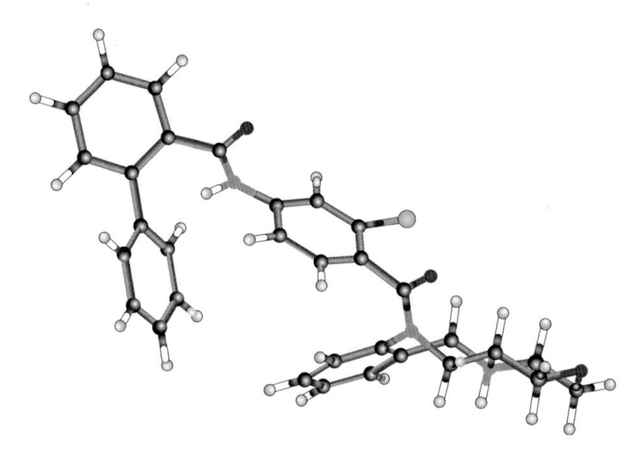

Scheme 11.8 Structure of the global minimum-energy conformation of protonated **40** (standard atom color code).

that of the x-ray structure. The next-higher-energy structure is 1.7 kcal/mol (7.1 kJ/mol) less stable than the global minimum. It has a *cis*-fused tricycle, resulting from inversion of the ammonium center (NH^+), a chairlike seven-membered ring, an axial amide, and an amide carbonyl *anti* to the fused benzene.[30c] This structural information could be helpful in developing a pharmacophore model for V_2-receptor binding, and some modeling studies in this regard have been reported.[47]

From our studies with the thiazino- and oxazinobenzazepine series, we identified potent nonpeptide vasopressin receptor antagonists, some of which have low-nanomolar V_2-receptor affinity and at least 20-fold selectivity for V_2 over V_{1a} receptors. In the thiazino class, (*S*)-(+)-**28** has excellent V_2 affinity ($K_i = 3.2\,nM$), moderate binding selectivity ($V_{1a}/V_2 = 26$), good functional selectivity ($V_{1a}/V_2 = 93$), and oral efficacy as an aquaretic agent in rats. Also, (*S*)-(+)-**29** has excellent V_2 affinity ($K_i = 3.7\,nM$) and notable V_2 selectivity (binding $V_{1a}/V_2 = 68$; functional $V_{1a}/V_2 = 820$). In the oxazino class, (*S*)-(+) enantiomer **40** has excellent V_2-receptor affinity ($K_i = 0.9\,nM$), moderate binding selectivity ($V_{1a}/V_2 = 27$), good functional selectivity ($V_{1a}/V_2 = 140$), and impressive oral potency as an aquaretic agent in rats. Although some compounds had reasonably high affinity for the human V_{1a} receptor, such as **35**, **40**, and, **49**, significant activity was not manifested in the V_{1a} functional assay (Tables 11.5 and 11.6); this behavior was also observed with lixivaptan (**3**; VPA-985). In general, the thiazino- and oxazinobenzodiazepine classes were not conducive to providing potent functional V_{1a}-receptor antagonists. On the basis of an extensive array of preclinical data, oxazinobenzodiazepine **40** (JNJ-17048434; RWJ-351647) was advanced into human clinical studies. On oral administration to humans this V_2-selective antagonist was found to be effective as an aquaretic agent, with remarkable potency. Thus, **40** appears to have potential for the treatment of edematous conditions in patients.

Finally, the pyrazino series, **6** (X = NR), was examined with an eye toward identifying a suitable backup to **40**.[30d] In this case we hoped that the presence of two basic amine centers might offer advantages relative to the physical properties or, at least, offer somewhat different and useful physical properties. The pyrazinobenzodiazepines were evaluated for binding to human V_{1a} and V_2 receptors and for function activity on cells with human V_2 receptors. The results for representative compounds, **51**–**64**, are given in Table 11.7. For **51** and its two enantiomers, there was very little difference in affinity, which is consistent with the results for the corresponding sulfur and oxygen systems (*vide supra*), although the lack of a distinction in this case is more emphatic. (*R*)-(+)-**51** has a V_2 K_i value of 12 n*M*, which is 13-fold less potent than that for oxygen analog **40** ($K_i = 0.9\,nM$).[48] The potency of (*R*)-(+)-**51** in the V_2 functional assay is also less than that of **40**, by a factor of about 9, but it is in the same range as VPA-985 (**3**). In general, the pyrazino series exhibited very weak V_{1a} affinity, except for **56**. The nitrogen substituent, R_3, plays an important role in determining V_2 affinity. When R_3 is Me (**51**) or H (**52**), the V_2 K_i value is 16 or 21 n*M*; however, when R_3 is *i*-Pr (**53**) or CF_3CH_2 (**54**) the V_2 affinity is very weak. Another aspect relates to the R_1 substituent on the 4-aminobenzoyl unit, compare the V_2 affinity of **51** with that of **55**–**60**. The *o*-chloro (**51**) and *o*-methyl (**56**) groups gave reasonably potent V_2 affinity, whereas the *o*-methoxy (**57**) and *o*-trifluoromethyl (**58**) groups did not. As for R_2, a 4-methyl group (**63**) was quite good, but a 4-methoxy group (**64**) was not. Substantial potency in the V_2 functional assay ($K_i < 250\,nM$) was obtained for **51**, (*R*)-(+)-**51**, (*S*)-(−)-**51**, **56**, **62**, and **63**. Compound **61**, a direct analog of VPA-985 (**3**), showed 20-fold lower V_2 affinity; however, the affinity was improved to some degree (factor of 3) by using a 2-phenyl group for R_2 (i.e., **62**). (*R*)-(+)-**51** turned out to be very selective for V_2-receptor action versus V_{1a}, as it had a V_{1a} binding $K_i > 2900\,nM$ and a V_{1a} functional $K_i > 28{,}600\,nM$.

TABLE 11.6 Vasopressin V_{1a} and V_2 Binding and Functional Data for Oxazinobenzodiazepines

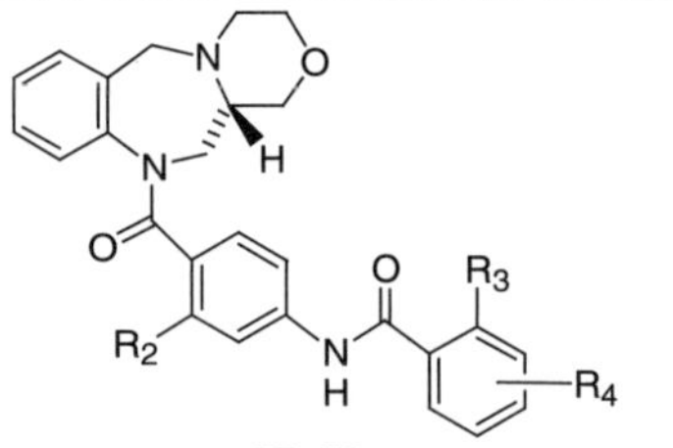

37–50

Compound[a]	R_2	R_3	R_4	V_{1a} Binding[b] K_i (nM)	V_2 Binding[b] K_i (nM)	V_{1a} Funct.[c] K_i (μM)	V_2 Funct.[c] K_i (μM)
37	Cl	Me	5-F	100	2.8	2.0	0.012
38	Cl	Ph	5-F	ca. 300[d]	11	6.4	0.012
39[e]	H	Ph	H	ca. 30[f]	3.7		0.016
40[g,h]	Cl	Ph	H	24±7	0.9±0	0.42±0.05	0.003±0
41[h,i]	Cl	Ph	H	640±150	13±1	>15	0.17±0.02
42	Cl	4-MeOPh	H	IA	2.8	1.9	0.003
43	Cl	3-MeOPh	H	IA	3.7	3.4	0.002
44	Cl	4-OH-Ph	H	IA	4.2	4.2	0.011
45	Cl	3-OH-Ph	H	IA	5.0	11	0.02
46	Cl	Ph	4-F	IA	1.9	0.7	0.002
47	Cl	Ph	4-OMe	IA	4.2	1.3	0.005
48	Cl	Ph	5-OMe	IA	17		
49	Cl	Ph	4-OH	ca. 30	1.4	1.6	0.006
50	Cl	Ph	5-OH	IA	13	1.9	0.07
VPA-985[j]				44	2.3	6.0	0.023

[a]Same as for Table 11.5, except compounds are (S)-(+) enantiomers (as depicted in the formula) unless noted otherwise.

[b]Binding to human V_{1a} and V_2 receptors ($N = 3$ to 6). IA = inactive (<30% inhibition at 100nM).

[c]Inhibition of functional human V_{1a} or V_2 activity ($N = 3$ to 5).

[d]62% inhibition at 1000 nM.

[e]Racemic mixture.

[f]69% inhibition at 100 nM.

[g]$C_{32}H_{28}ClN_3O_3$•HCl•1.3H_2O (analyzed correctly for C/H/N/H_2O); 98.7% enantiomeric purity by chiral HPLC.

[h]K_i data (V_{1a}/V_2) for racemic mixture (**40** + **41**): 75/2 nM; 1.0/0.01 μM.

[i]HCl salt; (R)-(−) enantiomer related to **40**.

[j]Reference standard.

Additional studies were conducted with (R)-(+)-**51**. Its functional responses, in terms of K_i values, with human and rat V_2/V_{1a} receptors were 47±5/>28,600 nM and 16±1/>3900 nM, respectively. There was no inhibition of AVP-induced activation of human V_{1b} or oxytocin receptors (K_i > 17,000 nM).[49] In rats, (R)-(+)-**51** had excellent oral bioavailability ($F = 81\%$) with an oral plasma half-life of 3.8 h; its oral bioavailability was also noteworthy in beagle dogs ($F = 49\%$) and cynomolgus monkeys ($F = 62\%$).

TABLE 11.7 Vasopressin V_{1a} and V_2 Binding and Functional Data for Pyrazinobenzodiazepines

51–64

Compound[a]	R_1[b]	R_2	R_3	V_{1a} Binding[c] K_i (nM)	V_2 Binding[c] K_i (nM)	V_2 Funct.[d] K_i (μM)
51	*o*-Cl	2-Ph	Me	IA	16	0.099
(+)-**51**[e,f]	*o*-Cl	2-Ph	Me	>2900	12 ± 2	0.047 ± 0.005
(−)-**51**[g]	*o*-Cl	2-Ph	Me	IA	20	0.017
52	*o*-Cl	2-Ph	H	IA	21	0.129
53	*o*-Cl	2-Ph	*i*-Pr	IA	IA	
54	*o*-Cl	2-Ph	CH_2CF_3	IA	IA	
55	H	2-Ph	Me	IA	IA	
56	*o*-Me	2-Ph	Me	ca. 30	ca. 50	0.077
57	*o*-OMe	2-Ph	Me	IA	IA	
58	*o*-CF_3	2-Ph	Me	IA	IA	
59	*m*-Me	2-Ph	Me	IA	IA	
60	*m*-OMe	2-Ph	Me	IA	ca. 50	0.32
61	*o*-Cl	2-Me,5-F	Me	IA	ca. 50	
62	*o*-Cl	2-Ph,5-F	Me	IA	17	0.094
63	*o*-Cl	2-(4-MeC_6H_4)	Me	IA	15	0.062
64	*o*-Cl	2-(4-$MeOC_6H_4$)	Me	IA	IA	
40				24	0.9	0.004
VPA-985[h]				43	2.3	0.023

[a]Target compounds were purified by reverse-phase semi-prep. HPLC and isolated as bis-trifluoroacetate salts, unless noted otherwise; >95% pure by reverse-phase HPLC/MS (215/254 nm); characterized by ESI-MS, with selected compounds analyzed by 300-MHz ^{1}H NMR. Compounds are racemates unless noted otherwise.

[b]The position of substitution is *ortho* or *meta* to the carboxy group.

[c]Binding to human V_{1a} and V_2 receptors (N = 2 to 6). IA = inactive (i.e., <30% inhibition of radioligand binding at 100 nM).

[d]Inhibition of functional human V_2 activity (N = 2 to 5).

[e](R)-(+) enantiomer; >99% enantiomeric purity by chiral HPLC.

[f]Di-HCl salt.

[g]98.3% enantiomeric purity by chiral HPLC.

[h]Reference standard.

Oral administration of (R)-(+)-**51** to hydrated conscious rats produced a dose-dependent aquaretic effect (Table 11.8). For example, at a oral dose of 10 mg/kg it caused a 1030% increase of urine output over untreated controls, with a 23% reduction of urine osmolality. For comparison, at an oral dose of 10 mg/kg, OPC-31260 is reported to alter

TABLE 11.8 Effect of (*R*)-(+)-51 on Urine Volume and Osmolality in Rats[a]

Oral Dose	Urine Volume		Urine Osmolality	
(mg/kg)	mL	% of Vehicle	mOs/kg	% of Vehicle
Vehicle	2.3±0.2		692±189	
0.3	3.3±0.4	143	368±37[b]	53
1	4.5±0.8	196	334±34[b]	48
3	9.4±1.2[b]	409	239±23[b]	35
10	23.7±1.9[b]	1030	156±12[b]	23

[a](*R*)-(+)-**51** was administered orally to conscious hydrated male rats at the dose specified. Each value represents the mean±SE ($N = 7$ to 8). Values representing the percent of vehicle (vehicle = 100%) are also provided.
[b]$P < 0.05$ versus vehicle values.

urine output/osmolality by +500%/−65%,[23b] and lixivaptan is reported to alter urine output/osmolality by +450%/−70%.[27a] (*R*)-(+)-**51** was also a very efficacious aquaretic agent in beagle dogs: at an oral dose of 1 mg/kg, the values for urine output/osmolality were +1016%/−81%.

Our investigation of the pyrazinobenzodiazepine series was fruitful in that it furnished (*R*)-(+)-**51** (JNJ-16240198; RWJ-659834), a highly selective V_2-receptor antagonist that inhibits vasopressin-induced renal water resorption without inducing electrolyte loss in rats and dogs. On the basis of an assortment of preclinical data, (*R*)-(+)-**51** was advanced into development as a backup to oxazinobenzodiazepine **40**. Vasopressin V_2-selective antagonists of this type could be useful for several clinical indications for which marketed diuretics are used, such as congestive heart failure, hypertension, and edema, with the added benefit of aquaresis without electrolyte loss. Particular areas for therapeutic application would be hyponatremia and liver cirrhosis.

11.6 EPILOGUE

Our drug discovery efforts on vasopressin receptor antagonists were initiated in late 1997. To propel this project, we assembled a cross-site collaboration between my research team in Spring House, Pennsylvania, and a research team in Raritan, New Jersey. Together, a great drug discovery plan of action was formulated and executed. Thus, this bilateral relationship turned out to be highly productive, perhaps even beyond our original dreams, with five compounds having entered preclinical development. One of these, **40**, progressed successfully through phase IIa clinical trials and another (not discussed here) advanced to human clinical studies as a mixed V_{1a}/V_2-receptor antagonist.[30f,g] In this chapter I adopted a personal perspective and concentrated on the medicinal chemistry work that was performed by my group in Spring House.[30a–d,44] In a similar vein, the excellent medicinal chemistry work of the scientists in Raritan was also very fruitful.[30e–g] In fact, the combined operations of our two chemistry groups served to greatly expand the range of structural diversity under exploration and proved to be very complementary as well. When a chemotype from one side of the Delaware River entered preclinical development and then fell out (for unforeseen reasons that will go unmentioned), a chemotype from the other side filled the vacuum. This alternation of contributed clinical candidates from each group in the partnership was a powerful formula for success. In retrospect, the results from our vasopressin-receptor antagonist research exceeded our original expectations. We applied direct intent, and it seems that clinical candidates kept coming our way. Point and click—well, perhaps not that easy. At the end of the day, it has been a very rewarding experience to see our well-considered research plan come together in an excellent collaborative effort, to "deliver the goods."

ACKNOWLEDGMENTS

I am indebted to numerous colleagues at Johnson & Johnson PRD who have played important roles in our vasopressin antagonist studies; their names appear in the articles cited in footnote 30. Relative to the work presented here, I want to recognize especially the outstanding scientific leadership afforded by Drs. Patricia Andrade-Gordon, Keith Demarest, Joseph Gunnet, Dennis Hlasta, and William Hoekstra. Dr. Gunnet was particularly instrumental in driving the biology of this project forward. I also thank Jay Matthews and Alexey Dyatkin for their excellent medicinal chemistry efforts, and Lawrence de Garavilla for his pharmacology contributions. Jay Matthews' chemical leadership was critical to the success of our vasopressin V_2 antagonist series.

REFERENCES AND NOTES

1. Maryanoff, B. E. *J. Med. Chem.* 2004, *47*, 769–787.
2. Thayer, A. M. *Chem. Eng. News* 2003, *81*, 33–37 (July 28). Also see Service, R. F. *Science* 2004, *303*, 1796–1799.
3. (a) For information on TOPAMAX topiramate, refer to: www.topamax.com. (b) Also see Maryanoff, B. E.; Nortey, S. O.; Gardocki, J. F.; Shank, R. P.; Dodgson, S. P. *J. Med. Chem.* 1987, *3*, 880–887. Shank, R. P.; Gardocki, J. F.; Vaught, J. L.; Davis, C. B.; Schupsky, J. J.; Raffa, R. B.; Dodgson, S. J.; Nortey, S. O.; Maryanoff, B. E. *Epilepsia* 1994, *35*, 450–460. Maryanoff, B. E.; Costanzo, M. J.; Nortey, S. O.; Greco, M. N.; Shank, R. P.; Schupsky, J. J.; Ortegon, M. P.; Vaught, J. L. *J. Med. Chem.* 1998, *41*, 1315–1343. Shank, R. P.; Gardocki, J. F.; Streeter, A. J.; Maryanoff, B. E. *Epilepsia* 2000, *41(Suppl. 1)*, S3–S9. Garnett, W. R. *Epilepsia* 2000, *41* (Suppl. 1), S61–S65. Ben-Menachem, E. *Rev. Contemp. Pharmacother.* 1999, *10*, 163–184. Tomson, T. *Curr. Sci.* 2002, *82*, 698–706. Faught, E. *Drugs Today* 1999, *35*, 49–57. Michelucci, R.; Passarelli, D.; Riguzzi, P.; Volpi, L.; Tassinari, C. A. *CNS Drug Rev.* 1998, *4*, 165–186. Brandes, J. L.; Saper, J. R.; Diamond, M.; Couch, J. R.; Lewis, D. W.; Schmitt, J.; Neto, W.; Schwabe, S.; Jacobs, D. *J. Am. Med. Assoc.* 2004, *291*, 965–973.
4. Costanzo, M. J.; Yabut, S. C.; Almond, H. R., Jr.; Andrade-Gordon, P.; Corcoran, T. W.; de Garavilla, L.; Kauffman, J. A.; Abraham, W. M.; Recacha, R.; Chattopadhyay, D.; Maryanoff, B. E. *J. Med. Chem.* 2003, *46*, 3865–3876.
5. Greco, M. N.; Hawkins, M. J.; Powell, E. T.; Almond, H. R., Jr.; Corcoran, T. W.; de Garavilla, L.; Kauffman, J. A.; Recacha, R.; Chattopadhyay, D.; Andrade-Gordon, P.; Maryanoff, B. E. *J. Am. Chem. Soc.* 2002, *124*, 3810–3811. Greco, M. N.; Almond, H. R., Jr.; de Garavilla, L.; Hawkins, M. J.; Maryanoff, B. E.; Qian, Y.; Walker, D. G.; Cesco-Cancian, S.; Nilsen, C. N.; Patel, M. N.; Humora, M. J. PCT Intl. Patent Appl. WO 0335654, 2003. A Novel, Potent Dual Inhibitor of the Leukocyte Proteases Cathepsin G and Chymase: molecular mechanisms and anti-inflammatory

activity in vivo. de Garavilla, Lawrence; Greco, Michael N.; Sukumar, Narayanasami; Chen, Zhi-Wei; Pineda, Agustin O.; Mathews, F. Scott; Di Cera, Enrico; Giardino, Edward C.; Wells, Grace I.; Haertlein, Barbara J.; Kauffman, Jack A.; Corcoran, Thomas W.; Derian, Claudia K.; Eckardt, Annette J.; Damiano, Bruce P.; Andrade-Gordon, Patricia; Maryanoff, Bruce E. Drug Discovery, School of Medicine, Johnson & Johnson Pharmaceutical Research and Development, Spring House, PA, USA. Journal of Biological Chemistry (2005), 280(18), 18001-18007. Publisher: American Society for Biochemistry and Molecular Biology, CODEN: JBCHA3 ISSN: 0021-9258. Journal written in English. CAN 143:19271 AN 2005:379217 CAPLUS (Copyright © 2006 ACS on SciFinder (R)).

6. Hoekstra, W. J.; Maryanoff, B. E.; Andrade-Gordon, P.; Cohen, J. H.; Costanzo, M. J.; Damiano, B. P.; Haertlein, B.; Harris, B. D.; Kauffman, J. A.; Keane, P. M.; McComsey, D. F.; Villani, F. J., Jr.; Yabut, S. B. *Bioorg. Med. Chem. Lett.* 1996, *6*, 2371–2376. Hoekstra, W. J.; Poulter, B. L. *Curr. Med. Chem.* 1998, *5*, 195–204. Hoekstra, W. J.; Maryanoff, B. E.; Damiano, B. P.; Andrade-Gordon, P.; Cohen, J. H.; Costanzo, M. J.; Haertlein, B. J.; Hecker, L. R.; Hulshizer, B. L.; Kauffman, J. A.; Keane, P.; McComsey, D. F.; Mitchell, J. A.; Scott, L.; Shah, R. D.; Yabut, S. C. *J. Med. Chem.* 1999, *42*, 5254–5265. Damiano, B. P.; Mitchell, J. A.; Giardino, E.; Corcoran, T.; Haertlein, B. J.; de Garavilla, L.; Kauffman, J. A.; Hoekstra, W. J.; Maryanoff, B. E.; Andrade-Gordon, P. *Thromb. Res.*, 2001, *104*, 113–126.
7. Quinn, M. J.; Byzova, T. V.; Qin, J.; Topol, E. J.; Plow, E. F. *Arterioscler. Thromb. Vasc. Biol.* 2003, *23*, 945–952. Salam, A. M.; Al Suwaidi, J. *Expert Opin. Invest. Drugs* 2002, *11*, 1645–1658. Quinn, M. J.; Plow, E. F.; Topol, E. J. *Circulation* 2002, *106*, 379–385. Chew, D. P.; Bhatt, D. L.; Topol, E. J. *Am. J. Cardiovasc. Drugs* 2001, *1*, 421–428. Chew, D. P.; Bhatt, D. L.; Sapp, S.; Topol, E. J. *Circulation* 2001, *103*, 201–206.
8. Derian, C. K.; Maryanoff, B. E.; Zhang, H.-C.; Andrade-Gordon, P. *Drug Dev. Res.* 2003, *59*, 355–366. Maryanoff, B. E.; Zhang, H.-C.; Andrade-Gordon, P.; Derian, C. K. *Curr. Med. Chem.–Cardiovasc. Hematol. Agents* 2003, *1*, 13–36. Derian, C. K.; Maryanoff, B. E.; Zhang, H.-C.; Andrade-Gordon, P. *Expert Opin. Invest. Drugs* 2003, *12*, 209–221.
9. Reeves, W. B.; Andreoli, T. E. In *Williams Textbook of Endocrinology*, 8th ed., Wilson, J. D.; Foster D. W. (Eds.), W.B. Saunders, Philadelphia, 1992, pp. 311–356.
10. Nielson, S.; Chou C. L.; Marples, D.; Christensen, E. I.; Kishore, B. K.; Knepper, M. A. *Proc. Natl. Acad. Sci. USA* 1995, *332*, 1540–1545.
11. Anderson, R. J.; Chung, H.-M.; Kluge, R.; Schrier, R. W. *Ann. Intern. Med.* 1985, *102*, 164–168. Chung, H. M.; Kluge, R.; Schrier, R. W.; Anderson, R. J. *Arch. Intern. Med.* 1986, *146*, 333–336. Sane, T.; Rantakari, K.; Poranen, A.; Taehtelae, R.; Vaelimaeki, M.; Pelkonen, R. *J. Clin. Endocrinol. Metab.* 1994, *79*, 1395–1398.
12. (a) Kortas, C.; Bichet, D. G.; Rouleau, J. L.; Schreier, R. W. *J. Cardiovasc. Pharmacol.* 1986, *8* (suppl. 7), S107–S110. (b) Johnston, C. I.; Arnolda, L. F.; Tsunoda, K.; Phillips, P. A.; Hodsman, G. P. *Can. J. Physiol.* 1987, *65*, 1706–1711. (c) Van Zwieten, P. A. *Prog. Pharmacol. Clin. Pharmacol.* 1990, *7*, 49–66. (d) Thibonnier, M. *Curr. Opin. Pharmacol.* 2003, *3*, 683–687. (e) Thibonnier, M.; Coles, P.; Thibonnier, A.; Shoham, M. *Annu. Rev. Pharmacol. Toxicol.* 2001, *41*, 175–202.
13. Mulinari, R. A.; Gavras, I.; Wang, Y. X.; Franko, R.; Gavras, H. *Circulation* 1990, *81*, 308–311.
14. Naitoh, M.; Suzuki, H.; Murakamj, M.; Matsumoto, A.; Arakawa, K.; Ichihara, A.; Nakamoto, H.; Oka, K.; Yamamura, Y.; Saruta, T. *Am. J. Physiol.* 1994, *267*, H2245–H2254. Lee, C. R.; Watkins, M. L.; Patterson, J. H.; Gattis, W.; O'Connor, C. M.; Gheorghiade, M.; Adams, K. F., Jr. *Am. Heart J.* 2003, *146*, 9–18.
15. Martin, P-Y.; Schrier, R. W. *Kidney Int.* 1997, *59*, S43–S49.
16. Winnicki, C.; Warnawin, K.; Januszewicz, P.; Socha, J. *Pol. J. Pediatr. Gastroenterol. Nutr.* 1994, *19*, 425–430.
17. Tsuboi, Y.; Ishikawa, S.-E.; Fujisawa, G.; Okada, K.; Saito, T. *Kidney Int.* 1994, *46*, 237–244.

18. Laszlo, F. A.; Csati, S.; Balaspiri, L. *Acta Endocrinol.* 1984, *106*, 56–60. Reeder, R. F.; Nattie, E. E.; North, W. G. *J. Neurosurg.* 1986, *64*, 941–950. Rosenberg, G. A.; Kyner, W. T.; Estrada, E.; Patlak, C. S. *Ann. NY Acad. Sci.* 1986, *481*, 383–389.

19. Tang, A. H.; Ho, P. M. *Life Sci.* 1988, *43*, 399–403. Laszlo, F. A.; Varga, C.; Balaspiri, L. *Ann. N.Y. Acad. Sci.* 1993, *689*, 627–629.

20. (a) Thibonnier, M. *Expert Opin. Invest. Drugs* 1998, *7*, 729–740. (b) Trybulski, E. J. *Annu. Rep. Med. Chem.* 2001, *36*, 159–168. (c) Albright, J. D.; Chan, P. S. *Curr. Pharm. Des.* 1997, *3*, 615–632. (d) Paranjape, S. B.; Thibonnier, M. *Expert Opin. Invest. Drugs* 2001, *10*, 825–834.

21. Ruffolo, R. R.; Brooks, D. P.; Huffman, W. F.; Poste, G. *Drug News Perspect.* 1991, *4*, 217–222. Kinter, L. B.; Caltabiano, S.; Huffman, W. F. *Biochem. Pharmacol.* 1993, *45*, 1731–1737.

22. (a) Yamamura, Y.; Ogawa, H.; Chihara, T.; Kondo, K.; Onogawa, T.; Nakamura, S.; Mori, T.; Tominaga, M.; Yabuuchi, Y. *Science* 1991, *252*, 572–574. (b) Ogawa, H.; Yamamura, Y.; Miyamoto, H.; Kondo, K.; Yamashita, H.; Nakaya, K.; Chihara, T.; Mori, T.; Tominaga, M.; Yabuuchi, Y. *J. Med. Chem.* 1993, *36*, 2011–2017.

23. (a) Yamamura, Y.; Ogawa, H.; Yamashita, H.; Chihara, T.; Miyamoto, H.; Nakamura, S.; Onogawa, T.; Yamashita, H.; Hosokawa, T.; Mori, T.; Tominaga, M.; Yabuuchi, Y. *Br. J. Pharmacol.* 1992, *105*, 787–791. (b) Ogawa, H.; Yamashita, H.; Kondo, K.; Yamamura, Y.; Miyamoto, H.; Kan, K.; Kitano, K.; Tanaka, M.; Nakaya, K.; Nakamura, S.; Mori, T.; Onogawa, T.; Tominaga, M.; Yabuuchi, Y. *J. Med. Chem.* 1996, *39*, 3547–3555.

24. Manning, M.; Sawyer, W. H. J. *Receptor Res.* 1993, *13*, 195–214. Laszlo, F. A.; Laszlo, F., Jr.; De Weid, D. *Pharmacol. Rev.* 1991, *43*, 73–108. Manning, M.; Chan, W. Y.; Sawyer, W. H. *Regul. Peptides* 1993, *45*, 279–283.

25. Fujisawa, G.; Ishikawa, S.; Tsuboi, Y.; Okada, K.; Saito, T. *Kidney Int.* 1993, *44*, 19–23. Ohnishi, A.; Orita, Y.; Takagi, N.; Fujita, T.; Toyoki, T.; Ihara, Y.; Yamamura, Y.; Inoue, T.; Tanaka, T. *J. Pharmacol. Exp. Ther.* 1995, *272*, 546–551.

26. Kondo, K.; Ogawa, H.; Yamashita, H.; Miyamoto, H.; Tanaka, M.; Nakaya, K.; Kitano, K.; Yamamura, Y.; Nakamura, S.; Onogawa, T.; Mori, T.; Tominaga, M. *Bioorg. Med. Chem.* 1999, *7*, 1743–1757.

27. (a) Albright, J. D.; Reich, M. F.; Delos Santos, E. G.; Dusza, J. P., Sum, F.-W.; Venkatesan, A. M.; Coupet, J.; Chan, P. S.; Ru, X.; Mazandarani, H., Bailey, T. *J. Med. Chem.* 1998, *41*, 2442–2444. (b) Aranapakam, V.; Albright, J. D.; Grosu, G. T.; Chan, P. S.; Coupet, J.; Saunders, T.; Ru, X.; Mazandarani, H. *Bioorg. Med. Chem. Lett.* 1999, *9*, 1733–1736. (c) Aranapakam, V.; Albright, J. D.; Grosu, G. T.; Delos Santos, E. G.; Chan, P. S.; Coupet, J.; Ru, X.; Saunders, T.; Mazandarani, H. *Bioorg. Med. Chem. Lett.* 1999, *9*, 1737–1740. (d) Albright, J. D.; Delos Santos, E. F.; Dusza, J. P.; Chan, P. S.; Coupet, J., Ru, X.; Mazandarani, H. *Bioorg. Med. Chem. Lett.* 2000, *10*, 695–698. (e) Ashwell, M. A.; Bagli, J. F.; Caggiano, T. J.; Chan, P. S.; Molinari, A. J.; Palka, C.; Park, C. H.; Rogers, J. F.; Sherman, M.; Trybulski, E. J.; Williams, D. K. *Bioorg. Med. Chem. Lett.* 2000, *10*, 783–786.

28. (a) Norman, P.; Leeson, P. A.; Rabasseda, X.; Castaner, J.; Castaner, R. M. *Drugs Future* 2000, *25*, 1121–1130. Matsuhisa, A.; Taniguchi, N.; Koshio, H.; Yatsu, T.; Tanaka, A. *Chem. Pharm. Bull.* 2000, *48*, 21–31. (b) Shimada, Y.; Taniguchi, N.; Matsuhisa, A.; Sakamoto, K.; Yatsu, T.; Tanaka, A. *Chem. Pharm. Bull.* 2000, *48*, 1644–1651. Matsuhisa, A.; Kikuchi, K.; Sakamoto, K.; Yatsu, T.; Tanaka, A. *Chem. Pharm. Bull.* 1999, *47*, 329–339.

29. (a) Cho, H.; Murakami, K.; Nakanishi, H.; Fujisawa, A.; Isoshima, H.; Niwa, M.; Hayakawa, K.; Hase, Y.; Uchida, I.; Watanabe, H.; Wakitani, K.; Aisaka, K. *J. Med. Chem.* 2004, *47*, 101–109. (b) Shimada, Y.; Taniguchi, N.; Matsuhisa, A.; Yatsu, T.; Tahara, A.; Tanaka, A. *Chem. Pharm. Bull.* 2003, *51*, 1075–1080. (c) Sum, F.-W.; Dusza, J.; Delos Santos, E.; Grosu, G.; Reich, M.; Du, X.; Albright, J. D.; Chan, P.; Coupet, J.; Ru, X.; Mazandarani, H.; Saunders, T. *Bioorg. Med. Chem. Lett.* 2003, *13*, 2195–2198. (d) Tsukada, J.; Tahara, A.; Tomura, Y.; Wada, K.; Kusayama, T.; Ishii, N.; Aoki, M.; Yatsu, T.; Uchida, W.; Taniguchi, N.; Tanaka, A. *Eur. J. Pharmacol.* 2002,

446, 129–138. (e) Kakefuda, A.; Suzuki, T.; Tobe, T.; Tsukada, J.; Tahara, A.; Sakamoto, S.; Tsukamoto, S. *J. Med. Chem.* 2002, *45*, 2589–2598. (f) Ohtake, Y.; Naito, A.; Hasegawa, H.; Kawano, K.; Morizono, D.; Taniguchi, M.; Tanaka, Y.; Matsukawa, H.; Naito, K.; Oguma, T.; Ezure, Y.; Tsuriya, Y. *Bioorg. Med. Chem.* 1999, *7*, 1247–1254.

30. (a) Matthews, J. M.; Greco, M. N.; Hecker, L. R.; Hoekstra, W. J.; Andrade-Gordon, P.; de Garavilla, L.; Demarest, K. T.; Ericson, E.; Gunnet, J. W.; Hageman, W.; Look, R.; Moore, J. B.; Maryanoff, B. E. *Bioorg. Med. Chem. Lett.* 2003, *13*, 753–756. (b) Dyatkin, A. B.; Hoekstra, W. J.; Hlasta, D. J.; Andrade-Gordon, P.; de Garavilla, L.; Demarest, K. T.; Gunnet, J. W.; Hageman, W.; Look, R.; Maryanoff, B. E. *Bioorg. Med. Chem. Lett.* 2002, *12*, 3081–3084. [Corrigendum: *Bioorg. Med. Chem. Lett.* 2004, *14*, 3363] (c) Matthews, J. M.; Hoekstra, W. J.; Dyatkin, A. B.; Hecker, L. R.; Hlasta, D. J.; Poulter, B. L.; Andrade-Gordon, P.; de Garavilla, L.; Demarest, K. T.; Ericson, E.; Gunnet, J. W.; Hageman, W.; Look, R.; Moore, J. B.; Reynolds, C. H.; Maryanoff, B. E. *Bioorg. Med. Chem. Lett.* 2004, *14*, 2747–2752. (d) Matthews, J. M.; Hlasta, D. J.; Andrade-Gordon, P.; Demarest, K. T.; Ericson, E.; Gunnet, J. W.; Hageman, W.; Look, R.; Moore, J. B.; Maryanoff, B. E. *Lett. Drug Design Discov.* 2005, *2*, 219–223. (e) Urbanski, M. J.; Chen, R. H.; Demarest, K. T.; Gunnet, J.; Look, R.; Ericson, E.; Murray, W. V.; Rybczynski, P. J.; Zhang, X. *Bioorg. Med. Chem. Lett.* 2003, *13*, 4031–4034. (f) Xiang, M. A.; Chen, R. H.; Demarest, K. T.; Gunnet, J.; Look, R.; Hageman, W.; Murray, W. V.; Combs, D. W.; Patel, M. *Bioorg. Med. Chem. Lett.* 2004, *14*, 2987–2989. (g) Xiang, M. A.; Chen, R. H.; Demarest, K. T.; Gunnet, J.; Look, R.; Hageman, W.; Murray, W. V.; Combs, D. W.; Rybczynski, P. J.; Patel, M. *Bioorg. Med. Chem. Lett.* 2004, *14*, 3143–3146. (h) RWJ-351647 and RWJ-659834 are the subject of an issued U.S. patent (Mar. 30, 2004): Dyatkin, A. B.; Hoekstra, W. J.; Maryanoff, B. E.; Matthews, J. M. U.S. patent 6,713,475 (2004) [WO 0043398; EP 1147115].

31. Sorbera, L. A.; Castaner, J.; Bayes, M.; Silvestre, J. *Drugs Future* 2002, *27*, 350–357. Gheorghiade, M.; Niazi, I.; Ouyang, J.; Czerwiec, F.; Kambayashi, J.; Zampino, M.; Orlandi, C. *Circulation* 2003, *107*, 2690–2696.

32. Martinez-Castelao, A. *Curr. Opin. Invest. Drugs* 2001, *2*, 525–530. Chan, P. S.; Coupet, J.; Park, H. C.; Lai, F.; Hartupee, D.; Cervoni, P.; Dusza, J. P.; Albright, J. D.; Ru, X.; Mazandarani, H.; Tanikella, T.; Shepherd, C.; Ochalski, L.; Bailey, T.; Lock, T. Yeung W.; Ning, X.; Taylor, J. R.; Spinelli, W. *Adv. Exp. Med. Biol.* 1998, *449*, 439–443.

33. Martinez-Castelao, A. *Curr. Opin. Invest. Drugs* 2002, *3*, 89–95. Tahara, A.; Tomura, Y.; Wada, K.; Kusayama, T.; Tsukada, J.; Takanashi, M.; Yatsu, T.; Uchida, W.; Tanaka, A. *J. Pharmacol. Exp. Ther.* 1997, *282*, 301–308. Gerbes, A. L.; Guelberg, V.; Gines, P.; Decaux, G.; Gross, P.; Gandjini, H.; Djian, J. *Gastroenterology* 2003, *124*, 933–939.

34. Martinez-Castelao, A. *Curr. Opin. Invest. Drugs* 2001, *2*, 1423–1427. Serradeil-Le Gal, C. *Cardiovasc. Drug Rev.* 2001, *19*, 201–214. Jimenez, W.; Serradeil-Le Gal, C.; Ros, J.; Cano, C.; Cejudo, P.; Morales-Ruiz, M.; Arroyo, V.; Pascal, M.; Rivera, F.; Maffrand, J.-P.; Rodes, J. *J. Pharmacol. Exp. Ther.* 2000, *295*, 83–90. Serradeil-Le Gal, C.; Wagnon, J.; Garcia, C.; Lacour, C.; Guiraudou, P.; Christophe, B.; Villanova, G.; Nisato, D.; Maffrand, J. P.; Le Fur, G.; Guillon, G.; Cantau, B.; Barberis, C.; Trueba, M.; Ala, Y.; Jard, S. *J. Clin. Invest.* 1993, *92*, 224–231. Serradeil-Le Gal, C.; Lacour, C.; Valette, G.; Garcia, G.; Foulon, L.; Galindo, G.; Bankir, L.; Pouzet, B.; Guillon, G.; Barberis, C.; Chicot, D.; Jard, S.; Vilain, P.; Garcia, C.; Marty, E.; Raufaste, D.; Brossard, G.; Nisato, D.; Maffrand, J. P.; Le Fur, G. *J. Clin. Invest.* 1996, *98*, 2729–2738. Griebel, G.; Simiand, J.; Serradeil-Le Gal, C.; Wagnon, J.; Pascal, M.; Scatton, B.; Maffrand, J.-P.; Soubrie, P. *Proc. Natl. Acad. Sci. USA* 2002, *99*, 6370–6375. Gard, P. *Curr. Opin. Oncol. Endocrinol. Metabol. Invest. Drugs* 2000, *2*, 265–272.

35. (a) Barberis, C.; Morin, D.; Durroux, T.; Mouillac, B.; Guillon, G.; Seyer, R.; Hibert, M.; Tribollet, E.; Manning, M. *Drug News Perspect.* 1999, *12*, 279–292. (b) Barberis, C., Seibold, A., Ishido, M., Rosenthal, W., Birnbaumer, M. *Regul. Peptides* 1993, *45*, 61–69. (c) Thibonnier, M.; Auzan, C.; Madhun, Z; Wilkins, P.; Berti-Mattera, L.; Clauser, E. *J. Biol. Chem.* 1994, *269*, 3304–3310. (d) Hirasawa, A.; Shibata, K.; Kotosai, K.; Tsujimoto, G. *Biochem. Biophys. Res.*

Commun. 1994, *203*, 72–79. (e) Morel, A.; Lolait, S. J.; Brownstein, M. J. *Regul. Peptides* 1993, *45*, 53–59. (f) Morel, A.; O'Carroll, A. M.; Brownstein, M. J.; Lolait, S. J. *Nature* 1992, *356*, 523–526. (g) Lolait, S. J.; O'Carroll, A. M.; McBride, O. W.; Konig, M.; Morel, A.; Brownstein, M. J. *Nature* 1992, *357*, 336–339.

36. Tahara, A.; Saito, M.; Sugimoto, T.; Tomura, Y.; Wada, K.; Kusayama, T.; Tsukada, J.; Ishii, N.; Yatsu, T.; Uchida, W.; Tanaka, A. *Br. J. Pharmacol.* 1998, *125*, 1463–1470.
37. Oprea, T. I. *J. Comput.-Aided Mol. Design* 2000, *14*, 251–264. Walters, W. Patrick; A.; Murcko, M. A. *Curr. Opin. Chem. Biol.* 1999, *3*, 384–387. Lipinski, C. A.; Lombardo, F.; Dominy, B. W.; Feeney, P. J. *Adv. Drug Deliv. Rev.* 1997, *23*, 3–25.
38. Hester, J. B., Jr. *J. Org. Chem.* 1964, *29*, 1158–1160. Hester, J. B., Jr. *J. Org. Chem.* 1967, *32*, 4095–4098. Nagasaka, T.; Ohki, S. *Chem. Pharm. Bull.* 1977, *25*, 3023–3033.
39. Bergman, J.; Sand, P. *Org. Synth.* 1987, *65*, 146–149. Quick, J.; Saha, B. *Tetrahedron Lett.* 1994, *35*, 8553–8556.
40. There are some errors involving the interchange of compound numbers **4** and **5** in ref 30b, which are reported in the noted Corrigendum.
41. Abraham, H.; Stella, L. *Tetrahedron* 1992, *48*, 9707–9718. Bertilsson, S. K.; Ekegren, J. K.; Modin, S. A.; Andersson, P. G. *Tetrahedron* 2001, *57*, 6399–6406.
42. Carabateas, P. M. U.S. patent 3,860,600 (1975); *Chem. Abstr.* 1975, *83*, 58892.
43. Ohtake, Y.; Naito, A.; Naito, K.; Matsukawa, H.; Saito, Y.; Toyofuku, H. *PCT Int. Appl.* 1998, 1–156 (WO 9843976); *Chem. Abstr.* 1998, *129*, 290149.
44. (a) Dyatkin, A. B.; Freedman, T. B.; Cao, X.; Dukor, R. K.; Maryanoff, B. E.; Maryanoff, C. A.; Matthews, J. M.; Shah, R. D.; Nafie, L. A. *Chirality* 2002, *14*, 215–219. (b) Matthews, J. M.; Dyatkin, A. B.; Evangelisto, M.; Gauthier, D. A.; Hecker, L. R.; Hoekstra, W. J.; Poulter, B. L.; Maryanoff, B. E. *Tetrahedron: Asymmetry* 2004, *15*, 1259–1267.
45. SciFinder, Advanced chemistry development software, Solaris v4.67.
46. Hassner, A.; Amit, B.; Marks, V.; Gottlieb, H. E. *J. Org. Chem.* 2003, *68*, 6853–6858.
47. Gieldon, A.; Kazmierkiewicz, R.; Slusarz, R.; Ciarkowski, J. *J. Comput.-Aided Mol. Design* 2001, *15*, 1085–1104.
48. (*R*)-(+)-**51** has the same spatial arrangement at the stereogenic center as does **40**; however, the stereochemical descriptor for the absolute configuration of (+)-**51** ["*R*"] differs from that of **40** ["*S*"] because of the Cahn–Ingold–Prelog priority rules: Cahn, R. S.; Ingold, C.; Prelog, V. *Angew. Chem. Int. Ed.* 1966, *5*, 385–415.
49. Inhibition of AVP activation of human V_{1b} or oxytocin receptors in transfected HEK-293 cells.

12

DISCOVERY AND DEVELOPMENT OF THE ULTRASHORT-ACTING ANALGESIC REMIFENTANIL

PAUL L. FELDMAN

GlaxoSmithKline Research and Development
Research Triangle Park, North Carolina

12.1 INTRODUCTION

To the layman general anesthesia is archetypically viewed as a patient gently falling asleep following the doctor or nurse placing a mask over the patient's nose and mouth. The patient loses consciousness and the surgeons begin their work. To the practitioner general anesthesia is a state induced by a combination of agents that induce hypnosis, the sleep component, along with amnesia, analgesia, and muscle relaxation (Fig. 12.1). The combinations of agents that produce these effects facilitate surgical or procedural intervention at minimal risk to the patient while providing for optimal recovery.

Hypnotics, such as propofol, induce sleep and a lack of awareness. Analgesics such as fentanyl and remifentanil synergize with hypnotics to provide profound hypnosis as well as to blunt the severe pain a patient may realize as a result of the surgical procedure. Amnestics such as midazolam are anxiolytics and also provide for the absence of recall of surgical events. It is most often the experience of patients undergoing a surgical event that they cannot recall anything that took place after being placed under anesthesia. Muscle relaxants such as *cis*-atracurium and succinylcholine disrupt the nerve to muscle signal transmission and are used clinically as adjuncts to anesthesia to facilitate intubations and relax muscles. The muscle relaxation, or paralysis, caused by these agents allow a surgeon to more easily manipulate a patient undergoing a surgical procedure. It is the skillful combination of these agents administered to patients by anesthesiologists that provides for humane and tolerable surgical interventions.[1]

Drug Discovery and Development, Volume 1: Drug Discovery, Edited by Mukund S. Chorghade

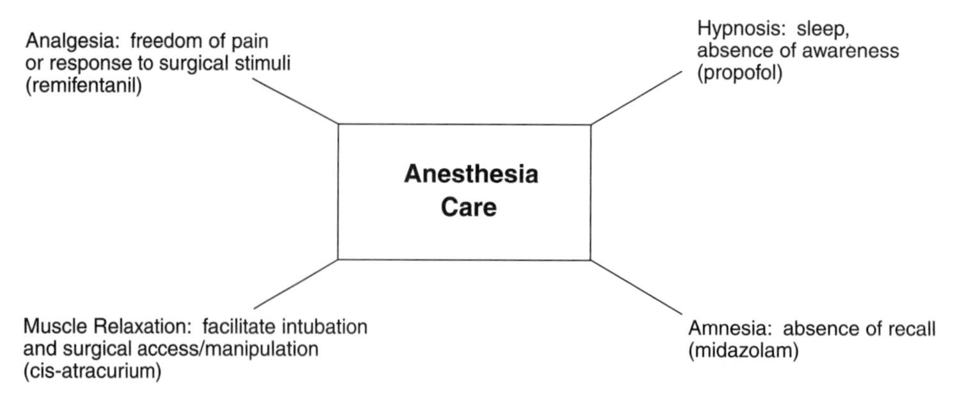

Figure 12.1 Anesthesia quadrad.

Anesthesiologists require agents that are extremely safe and predictable because they service a large and varied patient population. They must deliver agents that provide the benefits of anesthetic care while ensuring that the anesthetic agents themselves will manifest minimal risk. Typically, the anesthesiologist titrates the anesthetic drugs throughout a surgical procedure depending on the manipulation the surgeon is performing on the patient. For example, the anesthesiologist may need to deepen the level of hypnotic and analgesic when a patient is undergoing a particularly painful part of the surgical procedure. Once that part of the procedure is completed the anesthesiologist may lighten the anesthesia so that at the end of the procedure they can provide for a rapid recovery of the patient. If a deep anesthetic state is maintained throughout a surgical event, it may compromise a rapid recovery and leave lingering effects of the anesthesia long after the patient has left the operating room.

One of the limitations of many anesthetic drugs is that they accumulate during a procedure and this prolongs a patient's recovery from the surgery and anesthesia. A goal of the pharmaceutical industry has been to provide new anesthetic agents that have an ultrashort duration of action so that there is a rapid and predictable recovery of patients. To date, there are no clinically used ultrashort-acting hypnotics or amnestics.[2] The most popular hypnotics and amnestics are short-acting, but if given over a prolonged period of time, they will accumulate and prolong patient recoveries. Succinlycholine is an ultrashort-acting muscle relaxant that is used routinely given its favorable pharmacokinetic properties; however, it suffers some serious pharmacodynamic limitations that preclude it being the ideal ultrashort-acting muscle relaxant.[3] The introduction of remifentanil, Ultiva, provides anesthesiologists with the analgesic component of anesthetic care that delivers a strong analgesic with a rapid onset of action, ultrashort and predictable duration of action, and no cumulative effects.[4] Hopefully, over the next several years, safe, effective, and ultrashort-acting hypnotics, muscle relaxants, and amnestics will be discovered and progressed to provide anesthesiologists with the other agents they need to complement remifentanil in the anesthesia quadrad.

12.2 DISCOVERY OF REMIFENTANIL

Based on the premise that the clinically used analgesic opiate anesthetic agents were highly efficacious yet suffered from pharmacokinetic liabilities, we began a research program

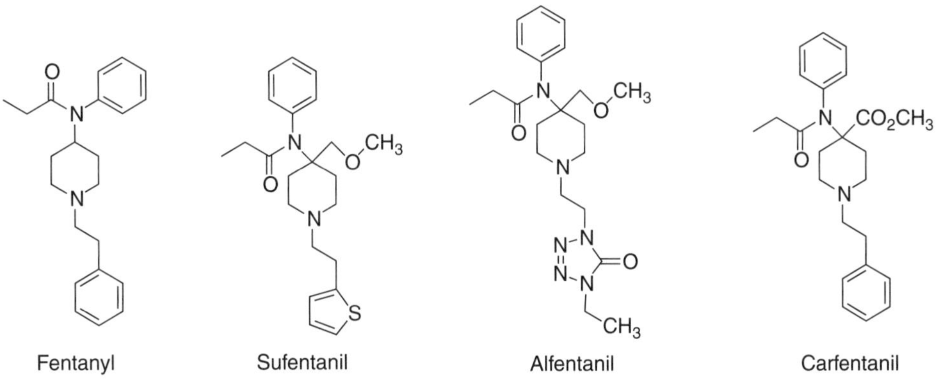

Fentanyl Sufentanil Alfentanil Carfentanil

Scheme 12.1 Known opiate analgesics.

to discover an analgesic opiate with an ultrashort duration of action for use as an adjunct to anesthesia.[5] The popular opiate analgesics used as adjuncts to anesthesia are fentanyl, sufentanil, and alfentanil (Scheme 12.1). All of the analgesics are potent μ-opioid agonists that were discovered by Paul Janssen and co-workers at Janssen Pharmaceutica in the 1960s and 1970s. Our goal was to discover a drug with μ-opioid agonist properties, such as fentanyl and its congeners, but which would undergo extremely rapid metabolism and elimination. The rapid metabolism of the drug would be independent of liver and kidney function and the rate of metabolism would not change with prolonged or high-dose administration. The importance of discovering a drug that undergoes metabolism independent of liver or kidney function is directly related to the predictability of drug half-life. Anesthetics that rely on the liver or kidney for clearance may have varied durations of action across patient populations because the functions of these organs may differ depending on the state of health of a patient. We sought to discover a drug whose metabolism would be absolutely predictable regardless of the health status of the patient who received the agent.

The design of potent μ-opioid analgesics that are ultrashort acting was based on the hypothesis that an ester incorporated into a fentanyl motif would retain the potent μ-opioid agonist properties. In addition, the ester functional group would rapidly be metabolized to an acid by ubiquitous blood and tissue esterases and would render the new molecule inactive at the μ-opioid receptor. Inspecting the structure of fentanyl and its analogs, we noted that the most potent μ-opioid agonists in this structural class possessed a two-carbon chain attached to a lipophilic moiety pendant to the piperidine nitrogen. We reasoned that the lipophilic group, a phenyl group in fentanyl, docked into a lipophilic portion of the μ-opioid receptor. Replacement of the aryl group with an ester function, also a lipophilic group, would provide for potent μ agonists. However, once the ester group was hydrolized to an acid by blood or tissue esterases, the polar and charged acid function would lose significant affinity for the μ receptor due to unfavorable interactions. One attractive feature to this design strategy is the use of ubiquitous esterases to metabolize and inactivate the analgesic versus reliance on the liver or kidney to inactivate or clear the drug. Given the broad distribution of esterases in humans, it was hypothesized that the proposed mechanism of metabolism would be operative in all patient populations.

The design we proposed had precedent in the medicinal chemistry literature (Scheme 12.2). In the early 1980s Erhardt and co-workers published their successful

Esterases

Highly Potent Inactive

Scheme 12.2 Medicinal chemistry approach.

approach to generating ultra short-acting beta blockers.[6] They replaced one of the aryl rings of propanolol with a ethyl propionate moiety. Esmolol was found to be a potent and effective ultrashort-acting beta blocker that was commercialized and used in the hospital setting. Just after we had begun our work, two additional reports were published describing similar approaches to generate ultrashort-acting antiarrythmics and opioid analgesics.[7,8] Unfortunately, the approach published using the fentanyl scaffold to generate ultrashort-acting analgesics did not succeed. A possible explanation for the failure of this approach is that the esters were hydrolized too slowly given the sterically hindered nature of the ester moiety. In the successful approaches using this strategy, the esters are unhindered and thus readily accepted as substrates for esterase enzymes. This general approach to discovering drugs, active agents that are predictably metabolized to inactive metabolites, has been coined *soft drugs* and has been applied to several different therapeutic areas.[9]

The testing scheme we used to evaluate our opioid analgesics is shown in Fig. 12.2. Initially, molecules were tested for their μ-opioid affinity and efficacy in the well-characterized and standard in vitro guinea pig ileum assay. Compounds that were shown to be potent μ-opioid agonists were then tested in the rat tail withdrawal reflex model. In this in vivo model compounds were tested for their analgesic efficacy as well as their duration of action. Ideally, we wished to discover compounds as potent as fentanyl and its congeners both in vitro and in vivo, but with durations of action in the rat significantly less than fentanyl. For those compounds with the desired profile in the primary assays, the in vitro

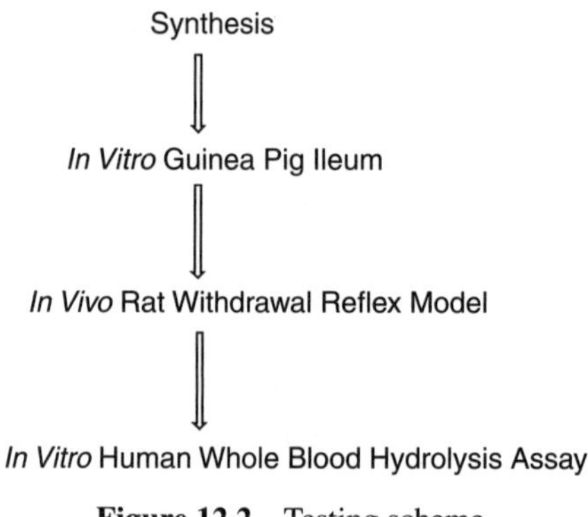

Figure 12.2 Testing scheme.

half-lives in human whole blood were determined to assess the ability of the compounds to be metabolized by esterases in human blood.[10]

Our initial attempts to generate potent μ-opioid agonists were disappointing. When the phenethyl group of fentanyl was replaced by acetate, propionate, and butyrate esters, the compounds were approximately 1000-fold less active than fentanyl in the guinea pig ileum assay. Although these compounds were weak μ-opioid agonists, we were encouraged to pursue this approach given that these esters possessed the desired duration of action in the rat. Compared to fentanyl, these compounds were approximately four to six times shorter acting. Therefore, to discover the drug candidate with the desired profile, we needed to retain the structural features of these initial compounds, which gave analgesics with an ultrashort duration of action, but we needed to increase their μ-opioid agonist potency.

Janssen and co-workers had shown that the hydrogen at the 4-position of the piperidine portion of fentanyl could be substituted with a methyl ester (cf. carfentanil) or methoxy methyl group (cf. sufentanil and alfentanil) to obtain very potent μ-opioid agonists. Indeed, we found that replacement of the hydrogen with a methyl ester had a profound affect on the μ-opioid agonist properties in our series of molecules. Replacement of the phenethyl group of carfentanil with a methyl propionate provided an extremely potent μ-opioid agonist with an ultrashort duration of action. This compound, remifentanil, was approximately twofold less potent than fentanyl in vitro (EC_{50} 3.5 nM versus 1.8 nM in the guinea pig ileum assay), equipotent to fentanyl in the rat tail withdrawal reflex model (ED_{50} 4.5 μg/kg for remifentanil versus 4.6 μg/kg for fentanyl), and was four times shorter acting (15-min versus 60-min duration) in rat tail withdrawal reflex model (Table 12.1).

We hypothesized that the ultrashort duration of action that remifentanil possessed in the rat is due to its hydrolysis to an inactive acid metabolite. Indeed, in subsequent biometabolism studies we demonstrated that remifentanil is very rapidly and nearly quantitatively metabolized to its acid metabolite in the rat. In addition to the ultrashort duration of action in the rat, we also discovered that remifentanil is rapidly hydrolized to its acid metabolite

TABLE 12.1 Structure–Activity Relationships for Ultrashort-Acting Analgesics

N R O N n CO_2Me

n	R = H	R = CO_2Me
1	$1.27 \pm 0.57 \times 10^{-5} M^{a}$	$5.39 \pm 1.02 \times 10^{-6} M$
2	$1.66 \pm 0.59 \times 10^{-6} M$ $(3.2/10 - 15)^{b}$	$3.55 \pm 0.23 \times 10^{-9} M$ (0.0044/15)
3	$3.60 \pm 0.30 \times 10^{-6} M$	$1.50 \pm 0.98 \times 10^{-8} M$
4	$5.44 \pm 1.51 \times 10^{-6} M$	$1.02 \pm 0.63 \times 10^{-8} M$
Fentanyl	$1.76 \pm 0.36 \times 10^{-9} M$ (0.0046/60)	

[a]EC_{50} values for inhibition of electrically evoked contraction in guinea pig ileum.

[b]ED_{50} values (mg/kg) for analgesic effects in rat tail withdrawal assay/duration of in vivo effect in minutes.

in vitro in human whole blood. The acid metabolite was synthesized and tested for its μ-opioid in vitro and in vivo and efficacy. Gratifyingly, the acid metabolite of remifentanil was shown to be more than 500-fold less active than remifentanil in the in vitro guinea pig ileum assay and greater than 350-fold less potent in the rat tail withdrawal reflex model. From these data we surmised that the acid has a significantly lower affinity for the μ-opioid receptor due to the polar and ionized functional group being unable to bind effectively into a lipophilic pocket in the μ-opioid receptor.

It was essential to demonstrate in animals that remifentanil has a significantly shorter duration of action than fentanyl and its congeners after a single dose. In addition, it was just as significant to demonstrate that upon prolonged infusion of remifentanil the recovery of animals would be just as rapid as it was when given a single dose. To test whether remifentanil would accumulate upon prolonged administration, rats were infused with high doses of remifentail for 1 hour and the time it took to recover was measured. Indeed, even after the 1-hour infusion the animals recovered from the analgesic affects of remifentanil in the same time frame as if given a single bolus dose. Also, an experiment was conducted where rats were given multiple successive bolus doses of remifentanil to see if the recovery from its effects would be prolonged when treated successively with the drug. Again, no prolongation of duration was noted, suggesting that the mechanism for metabolism of remifentanil was not saturated and accumulation of the drug was not significant. All of these preclinical data suggested that we had discovered a potent μ-opioid agonist that had an ultrashort duration of action based on a liver- and kidney-independent metabolic pathway to a significantly less active metabolite.

12.3 CHEMICAL DEVELOPMENT OF REMIFENTANIL

The synthetic route we employed to generate remifentanil and its analogs is patterned after the route developed by Janssen and co-workers (Scheme 12.3).[11] The route began with the Strecker reaction performed on *N*-benzyl-4-piperidinone to generate **2**. Conversion of the nitrile to the amide proved to be capricious, due to the reversion of the amino nitrile to *N*-benzyl-4-piperidinone and decomposition depending on acid strength, reaction time, and temperature. Carefully controlled hydrolysis of the nitrile to amide could be accomplished with 85% sulfuric acid. Conversion of the primary amide to the acid was accomplished under harsh basic conditions. Following conversion of the potassium salt to the sodium salt the methyl ester was made by alkylating the sodium carboxylate with methyl iodide. In addition to the desired product, this reaction also produced very polar material believed to be the quaternary ammonium salt. This undesired by-product could easily be separated from the desired ester by extractive workup; however, the ester was isolated as a viscous oil. Acylation of the aniline nitrogen was accomplished by heating **5** in neat propionic anhydride at 167°C followed by crystallization of the product as its oxalic acid salt. The benzyl group was removed using high-pressure hydrogenation, and the secondary amine was immediately subjected to a Michael reaction with methyl acrylate to form the free base of remifentanil. The white crystalline hydrochloride salt of remifentanil was formed with HCl in MeOH and ether.

Although the medicinal chemistry route to remifentanil is relatively short, there were a number of problems that prevented the use of this route on a large scale (Scheme 12.4). In the first stage the use of KCN in acetic acid generates HCN, which is a safety issue when this reaction is performed on a large scale. In the hydrolysis of the nitrile to the primary

KCN, $PhNH_2$

HOAc, H_2O
20–25 °C, 2h
(71%)

85% H_2SO_4

45 °C, 24h
(66%)

1. 4M KOH
ethylene glycol
150 °C, 8h
2. conc. HCl
3. 10M NaOH
(93%)

$(MeO)_2SO_2$
DMF
50 °C, 0.75h
(72%)

1. $(EtCO)_2O$
167 °C, 3h
2. $(HO_2C)_2$
MeOH
(80%)

H_2 (50 psi)
10% Pd-C
MeOH/HOAc
20–25 °C, 24h
(99%)

1. $CH_2CHCO_2CH_3$
MeOH, 20–25 °C, 16h
2. HCl, $MeOH/Et_2O$
(89%)

1, 2, 3, 4, 5, 6, 7, **Remifentanil-HCl**

Scheme 12.3 Medicinal chemistry route to remifentanil.

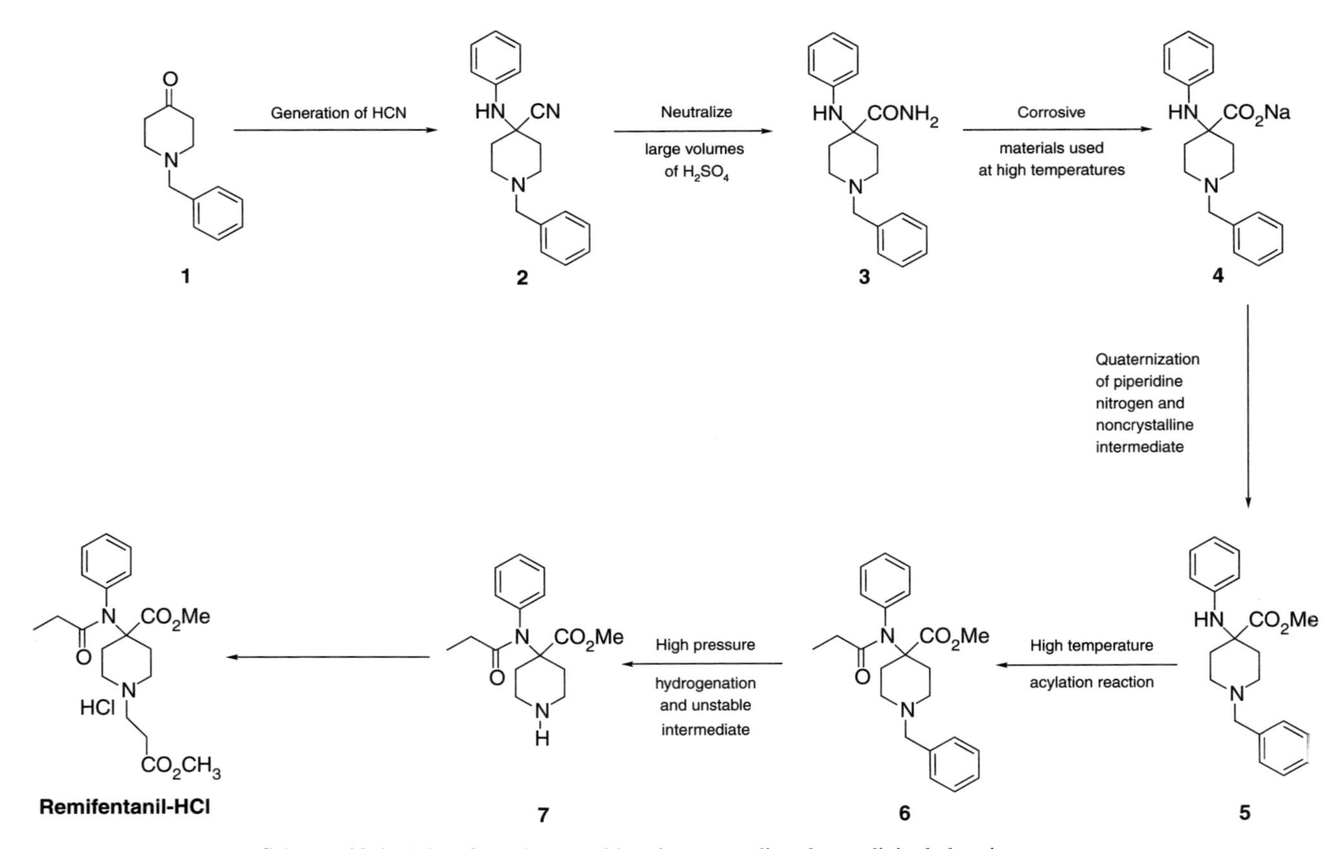

Scheme 12.4 Manufacturing considerations regarding the medicinal chemistry route.

amide the product was isolated as an oil after neutralizing large volumes of sulfuric acid. This method of isolation proved to be impractical on a large scale. Subsequent hydrolysis of the amide to acid **4** using KOH in ethylene glycol at 150°C needed to be avoided given the corrosive nature of these reagents at high temperature, which would be difficult to operate on a large scale. Conversion of the acid to the methyl ester also needed to be modified to avoid both isolation of the noncrystalline ester **5** and quaternization of the piperidine nitrogen. Milder conditions were also sought for the acylation of the ester **5** to give **6** as well as the removal of the benzyl group to give **7**. Avoiding the high-pressure hydrogenation to remove the benzyl group would significantly simplify the manufacture of remifentanil. All of these issues needed to be addressed to accomplish a cost-effective, simplified, robust, and safe route to remifentanil.

The ultimate manufacturing route to remifentanil is depicted (in Scheme 12.5). Except for stage 4, the general synthetic route to remifentanil is similar to the medicinal chemistry route. However, a number of significant improvements were made which enhanced the safety, robustness, and scaling of the reactions. In the first stage (the Strecker reaction), by using acetone cyanohydrin, the cyanide is slowly released in a controlled fashion and thus avoids the expulsion of gaseous HCN from the reaction. To avoid a complex workup of a toxic mixture, a solvent from which the product would crystallize directly was sought. Use of the organic reagent cyanide (acetone cyanohydrin) avoided the presence of metal salts, and by using industrial mineral spirits the crystallization occurred directly, thus driving the equilibrium in favor of the desired product (**2**), which crystallizes from the reaction,

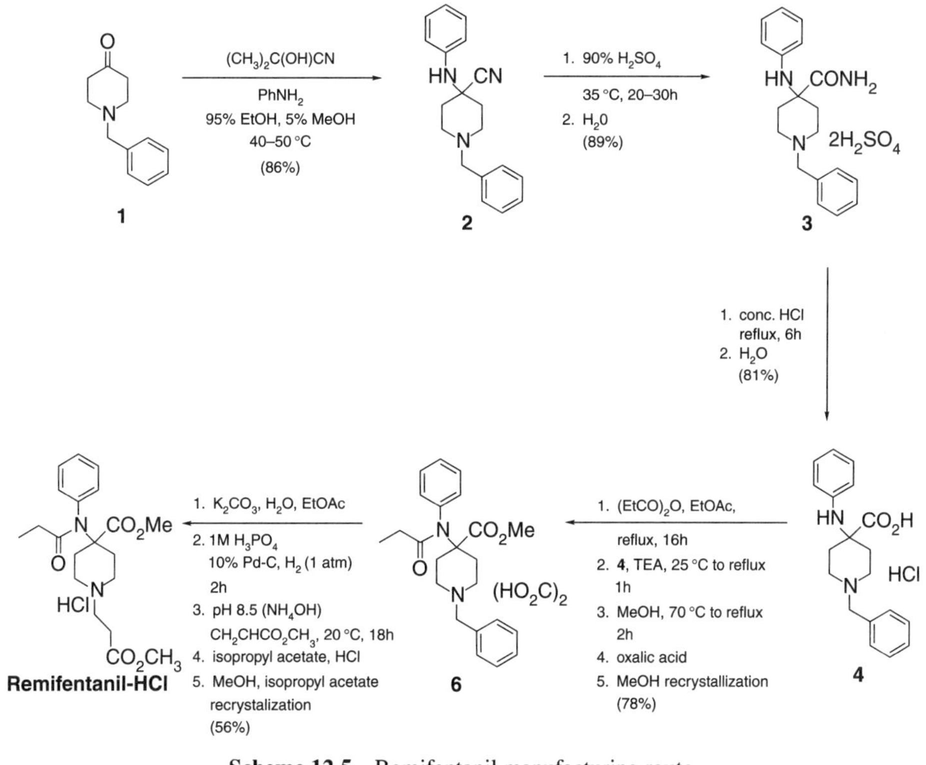

Scheme 12.5 Remifentanil manufacturing route.

mixture in very high purity. Hydrolysis of the nitrile to the primary amide (**3**) was accomplished with sulfuric acid, as had been done in the medicinal chemistry route, and the product was crystallized from the reaction mixture by the careful addition of water to the reaction once the hydrolysis had been completed. Isolation of the bis-sulfate salt simplified this reaction workup considerably versus the tedious neutralization of sulfuric acid that had been standard in the medicinal chemistry process. Hydrolysis of the amide to the acid was accomplished using concentrated HCl versus the strongly basic conditions used in the medicinal chemistry scheme. Following completion of the hydrolysis, the crude product is filtered, then slurried with water, and the monohydrochloride salt was isolated as a crystalline solid.

The most significant and elegant change to the medicinal chemistry route is the one-stage conversion of acid **4** to ester **6**[12] (Scheme 12.6). In this step the acid is esterified and the aniline nitrogen is acylated under mild conditions. By using this procedure we have obviated the need to isolate the oily ester **5** and avoid using the harsh conditions to acylate **5** to generate **6**. The reaction is conducted by initially refluxing ethyl acetate with propionic anhydride to remove any alcohols that may be contaminating the ethyl acetate solvent. Once this part of the process is completed, the reaction is cooled to room temperature, and **4**, along with triethyl amine, is added and the reaction is heated to reflux for 1 hour. Following completion of the 1-hour reflux, the reaction is cooled to 70°C, methanol is added, and then the reaction is again refluxed for 2 hours. Upon completion of the reaction the product is isolated as its crystalline oxalate salt (**6**). This salt is recrystallized from methanol to yield **6** in an overall yield of 78% from **4**.

It is hypothesized that the initial step in the production of **6** is formation of the mixed anhydride of the 4-anilidopiperidine acid and propionic acid. Ring closure onto the anilino nitrogen generates intermediate **8**, which can either open to form acid **9** or lose water to generate iminium ion (**10**). Acid **9** can also be converted to **10** via formation of another

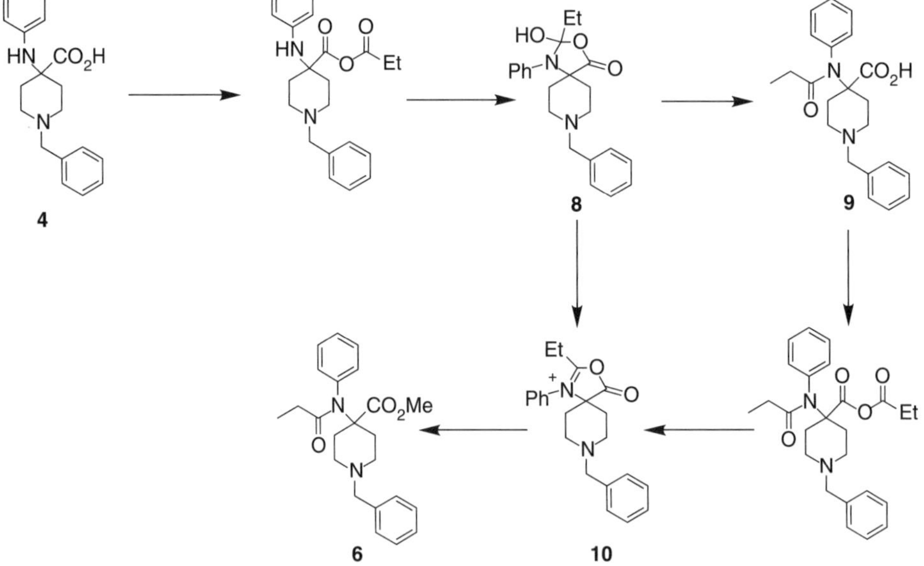

Scheme 12.6 Mechanism proposal for amide–ester (**6**) formation.

mixed anhydride, followed by ring closure. The electrophilic carbonyl of **10** is susceptible to nucleophilic attack and in the presence of methanol will generate the desired product **6**.

The mechanism of this reaction has precedent in the literature. Formation of α-amido esters or amides from α-amido acids can be accomplished by treating the acids with an anhydride to generate azlactones, which are then opened by alcohols or amines with catalytic acid to generate the desired products.[13] In the synthesis of **6** the α-amino nitrogen of **4** is a secondary amine, so that when the azlactone-like intermediate (**10**) is formed it is activated toward nucleophilic attack without the need for an acid catalyst. The generality of this reaction has been demonstrated by the use of various anhydrides and alcohol nucleophiles.[12]

The completion of the synthesis is accomplished by hydrogenolysis of the benzyl group and subsequent Michael addition of methyl acrylate onto the piperidine nitrogen to yield remifentanil. This same sequence was used in the medicinal chemistry route, however, in the process route different conditions were used to avoid the use of high-pressure hydrogenation and the isolation of the oily intermediate secondary piperidine. After the free base of **6** is made using K_2CO_3 in water–methanol, the piperidine is extracted into dilute phosphoric acid and hydrogenated at 1 atm over Pd–C. The pH of the solution containing the secondary piperidine is adjusted to 8.5 using ammonium hydroxide, and methyl acrylate is added to the aqueous solution to effect the Michael addition. The pH of the solution is extremely important because if it is too acidic, the nitrogen is protonated and the Michael addition does not take place. Alternatively, if the solution is too basic, the unhindered methyl ester of remifentanil is rapidly hydrolized. The isolation of remifentanil is accomplished by extracting the free base into isopropyl acetate, acidifying with HCl, and then crystallizing remifentanil–HCl from a mixture of MeOH and isopropyl acetate.

The manufacturing process to generate remifentanil is five stages, all isolated intermediates are solids, and the synthesis proceeds in approximately 25% overall yield. The process used to generate remifentanil provides a safe, robust, inexpensive, and elegant manufacturing route.

12.4 HUMAN CLINICAL TRIALS WITH REMIFENTANIL

The early clinical development of remifentanil demonstrated that it produced a dose-dependent increase in analgesia, it had a rapid onset of action, and it had a significantly more rapid offset of activity versus alfentanil (5.4 versus 54 minutes).[4] As predicted by the preclinical work, the rapid inactivation of remifentanil in humans was shown to be primarily the result of hydrolysis of the unhindered ester by nonspecific plasma and tissue esterases to the significantly less potent carboxylic acid. As a consequence of this mechanism of inactivation it was not surprising, but gratifying, to demonstrate that the pharmacokinetics of remifentanil in patients who had impaired hepatic or renal functions were unaltered compared to healthy volunteers.

Remfentanil is indicated for use as an analgesic during induction and maintenance of general anesthesia and for postoperative analgesia under close supervision in postanesthesia or intensive care unit setting. Based on the data gathered from human trials, the recommended dose guidelines for remifentanil's use in general anesthesia and as an analgesic in the immediate postoperative period were established. Induction of anesthesia is typically done with an infusion of 0.5 to 1 μg/kg per minute of remifentanil along with a hypnotic agent. Maintenance of anesthesia can be conducted with an infusion of remifentanil along with volatile agents, such as N_2O or isoflurane, or an intravenous hypnotic, such as propofol. Following completion of the surgical procedure, remifentanil can be infused at a rate of 0.025 to 2 μg/kg per minute in the immediate postoperative arena to continue analgesic relief. Given the ultrashort duration of action of remifentanil, careful management of a patient's pain in the postoperative period must be considered. This can be achieved by slowly reducing the infusion of remifentanil in the postoperative setting or administering a long-acting analgesic just prior to cessation of the surgical procedure. Therefore, as the analgesic effects of remifentanil rapidly dissipate, the patient's pain will continue to be managed with the longer-acting analgesic.

Remifentanil is also approved for use during monitored anesthesia care. Monitored anesthesia care is typically done for patients who are undergoing outpatient procedures. These procedures include very short surgical procedures, such as a breast biopsy, or short surgical procedures that utilize a nerve block. In a monitored anesthesia setting, remifentanil may be administered either by using a single dose or by the continuous infusion method. In either method remifentanil may be administered in combination with the anxiolytic midazolam followed by a local or regional anesthetic block.

In humans, remifentanil demonstrates the typical pharmacology of a μ-opioid agonist. In addition to its beneficial efficacy as an analgesic and hypnotic, remifentanil may cause respiratory depression, bradycardia, hypotension, and skeletal muscle rigidity during procedures in which it is used for induction and/or maintenance of general anesthesia. Due to the pharmacokinetic profile of remifentanil, the duration of these adverse side effects can be controlled by discontinuing or decreasing the rate of the remifentanil infusion. In addition, the incidence and magnitude of these effects are dependent on the type and dose of other anesthetic agents coadministered with remifentanil.

Remifentanil is unique among the clinically used opioid anesthetic agents as a result of its pharmacokinetic profile. Its rapid and predictable pharmacokinetic profile provides anesthesiologists with an analgesic that is optimal for different patient types and surgical procedures ranging from short procedures performed in an outpatient setting to protracted surgical cases that require rapid intraoperative titration and/or rapid recovery. The discovery and development of remifentanil is a result of careful consideration of the needs of the anesthesia community, the thoughtful design and execution of the chemistry and pharmocology, the creative efforts in development to provide drug products suitable and safe for clinical use, and the diligent efforts during the clinical phase to provide a product that is differentiated and realizes its potential.

ACKNOWLEDGMENTS

Many talented drug discoverers are involved in any program that progresses successfully from the lab to the market. The discovery and development of remifentanil is no exception to this rule. I want to thank the many talented people who worked on remifentanil during all phases of its progression. There are too many people who played significant roles to mention all by name; however, Dr. Michael James, who served as the biology program leader for the discovery of remifentanil, is gratefully acknowledged for his seminal contribution.

REFERENCES

1. For leading references and a general review of research on anesthetic agents, see Rees, D. C.; Hill, D. R. *Annu. Rep. Med. Chem.* 1996, *31*, 41.

2. Recent disclosures have described classes of ultrashort-acting benzodiazepines that may find use as ultrashort-acting amnestics. (a) Stafford, J. A.; Pacofsky, G. J.; Cox, R. F.; Cowan, J. R.; Dorsey, G. F., Jr.; Gonzales, S. S.; Jung, D. K.; Koszalka, G. W.; McIntyre, M. S.; Tidwell, J. H.; Wiard, R. P.; Feldman, P. L. *Bioorg. Med. Chem. Lett.* 2002, *12*, 3215. (b) Pacofsky, G. J.; Stafford, J. A.; Cox, R. R.; Cowan, J. R.; Dorsey, G. F., Jr.; Gonzales, S. S.; Kaldor, I.; Koszalka, G. W.; Lovell, G. G.; McIntyre, M. S.; Tidwell, J. H.; Todd, D.; Whitesell, G.; Wiard, R. P.; Feldman, P. L. *Bioorg. Med. Chem. Lett.* 2002, *12*, 3219.
3. Recent reports have described a new ultrashort-acting nondepolarizing neuromuscular blocking agent (muscle relaxant) that is undergoing human clinical trials. (a) Boros, E. E.; Bigham, E. C.; Mook, R. A., Jr.; Patel, S. S.; Savarese, J. J.; Ray, J. A.; Thompson, J. B.; Hashim, M. A.; Wisowaty, J. C.; Feldman, P. L.; Samano, V. *J. Med. Chem.* 1999, *42*, 206. (b) Samano, V.; Ray, J. A.; Thompson, J. B.; Mook, R. A., Jr.; Jung, D. K.; Koble, C. S.; Martin, M. T.; Bigham, E. C.; Regitz, C. S.; Feldman, P. L.; Boros, E. E. *Org. Lett.* 1999, *1*, 1993. (c) Belmont, M. R.; Lien, C. A.; Savarese, J. J.; Patel, S.; Fischer, G.; Mook, R. A., Jr. *Br. J. Anaesth.* 1999, *82*, A419. (d) Boros, E. E.; Samano, V.; Ray, J. A.; Thompson, J. B.; Jung, D. K.; Kaldor, I.; Koble, C. J.; Martin, M. T.; Styles, V. L.; Mook, R. A., Jr.; Feldman, P. L.; Savarese, J. J.; Belmont, M. R.; Bigham, E. C.; Boswell, S. E.; Hashim, M. A.; Patel, S. S.; Wisouaty, J. C.; Bowers, G. D.; Moseley, C. L.; Walsh, J. S.; Reese, M. J.; Rutowske, R. D.; Sefler, A. M.; Sptizer, T. D. *J. Med. Chem.* 2003, *46*, 2502.
4. A summary of the clinical profile of remifentanil is described in Patel, S. S.; Spencer, C. M. *Drugs*, 1996, *52*, 417.
5. The primary references for the medicinal chemistry and pharmacology leading to the discovery of remifentanil are: (a) Feldman, P. L.; James, M. K.; Brackeen, M. F.; Bilotta, J. M.; Schuster, S. V.; Lahey, A. P.; Lutz, M. W.; Johnson, M. R.; Leighton, H. J. *J. Med. Chem.* 1991, *34*, 2202. (b) James, M. K.; Feldman, P. L.; Schuster, S. V.; Bilotta, J. M.; Brackeen, M. F.; Leighton, H. J. *J. Pharmacol. Exp. Ther.* 1991, *259*, 712.
6. Erhardt, P. W.; Woo, C. M.; Anderson, W. G.; Gorczynski, R. J. *J. Med. Chem.* 1982, *25*, 1408.
7. Stout, D. M.; Black, L. A.; Barcelon-Yang, C.; Matier, W. L.; Brown, B. S.; Quon, C. Y.; Stampfili, H. F. *J. Med. Chem.* 1989, *32*, 1910.
8. Colapret, J. A.; Diamantidis, G.; Spencer, H. K.; Spaulding, T. C.; Rudo, F. G. *J. Med. Chem.* 1989, *32*, 663.
9. Lutz, M. W.; Morgan, P. H.; James, M. K.; Feldman, P. L.; Brackeen, M. F.; Lahey, A. P.; James, S. V.; Bilotta, J. M.; Pressley, J. C. *J. Pharmacol. Exp. Ther.* 1994, *271*, 795.
10. A review of therapeutic agents that were discovered using the soft drug approach can be found in Bodor, N.; Buchwald, P. *Med. Res. Rev.* 2000, *20*, 58.
11. Van Daele, P. G. H.; DeBruyn, M. F. L.; Boey, J. M.; Sanczuk, S.; Agten, J. T. M.; Janssen, P. A. J. *Arzneim.-Forsch.* 1976, *26*, 1521.
12. Coleman, M. J.; Goodyear, M. D.; Latham, D. W. S.; Whitehead, A. J. *Synlett* 1999, *12*, 1923.
13. Carter, H. B. *Org. React.* 1946, *3*, 198.

13

DISCOVERY AND DEVELOPMENT OF NEVIRAPINE

KARL GROZINGER, JOHN PROUDFOOT, AND KARL HARGRAVE
Boehringer-Ingelheim Pharmaceuticals
Ridgefield, Connecticut

13.1 INTRODUCTION

In October 1991 the Center for Disease Control in the United States declared the acquired immune deficiency syndrome (AIDS) to be an epidemic in that country. From 151 cases and 128 deaths reported in 1981, deaths from AIDS in the United States passed the 100,000 mark by the end of 1990. At the end of 2002, the World Health Organization estimated that there were 42 million people living with HIV/AIDS worldwide and that about 3 million people had died from AIDS in 2002 alone.[1] A massive scientific effort aimed at understanding the etioliogy of this debilitating and deadly disease led to the recognition that a retrovirus, human immune deficiency virus type 1 (HIV-1), was the causative agent.[2] The virus invades and destroys $CD4^+$ T cells of the immune system, eventually compromising the immune system to such an extent that normally innocuous pathogens become life-threatening. The first drug approved for the treatment of AIDS, zidovudine (AZT), reached the market in 1987 and was followed by several others. In this chapter we describe the discovery and development of one of these drugs, nevirapine (Scheme 13.1).

The HIV life cycle, depicted schematically in Fig. 13.1, shows the targets for potential therapeutic intervention. Virus replication requires entry into the cell, incorporation of the virus genetic material into the host genome, and subsequent transcription and translation to generate new virus particles. Transformation of the viral genetic material, single-stranded RNA, into double-stranded proviral DNA is accomplished by the enzyme HIV-1 reverse transcriptase (HIV-1 RT). The proviral DNA is subsequently integrated into the host genome by HIV-1 integrase. It remains dormant until activated, when transcription leads to the production of viral RNA, some of which serves as genetic material for viral prog-

Drug Discovery and Development, Volume 1: Drug Discovery, Edited by Mukund S. Chorghade

Nevirapine

Scheme 13.1

eny and some of which acts as a template for viral polyprotein production. A third virus enzyme, HIV-1 protease, is essential for processing this material into individual proteins that along with the viral RNA assemble to produce a new generation of infectious virus particles. The budding or escape of these new virus particles is lethal to the host cell. Therapies based on vaccines or inhibition of the integrase enzyme have so far not progressed beyond clinical evaluation. The first therapeutic agent for prevention of virus entry into the cell was approved in 2003. Inhibition of the RT and protease enzymes has proven to be most successful approach to effective therapies, and nearly all the approved AIDS drugs target one of these two enzymes.

As our interest in developing a new drug for the treatment of AIDS took root, one agent under evaluation in the clinic, AZT,[3] showed efficacy in slowing the progression of the disease. AZT targets the HIV-1 RT enzyme by mimicking a natural substrate of this enzyme, thymidine. Because it mimics a nucleoside naturally present in the cell, AZT can also interfere with the activity of endogenous human enzymes. *It was thought that some of the side effects seen with AZT treatment were due to this lack of selectivity, and it was clear that there were opportunities for additional, more selective inhibitors of HIV-1 RT.* HIV-1 RT is a heterodimeric enzyme with subunits of molecular weight 66 kDa (p66) and 51 kDa (p51). The enzyme possesses three distinct catalytic activities. It can function as an RNA-dependent DNA polymerase that transforms the single-stranded RNA viral genome

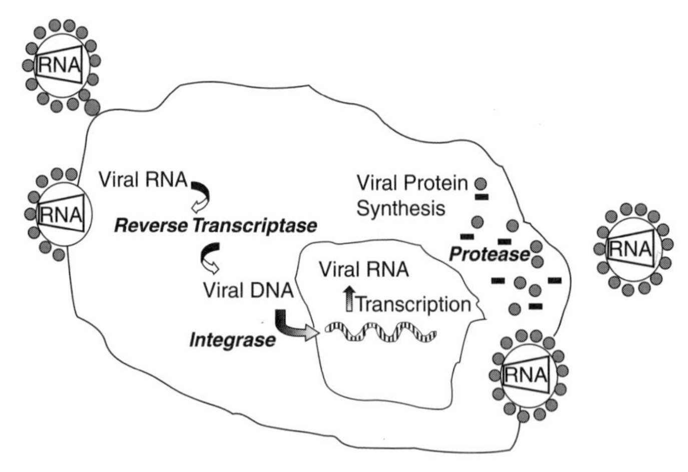

Figure 13.1 HIV life cycle.

to an RNA–DNA hybrid. An RNAse H activity cleaves the original RNA strand, and a DNA-dependent DNA polymerase activity synthesizes a complementary DNA providing double-stranded proviral DNA. Since RT enzymatic activity is unique to retroviruses and no corresponding activity is found in the normal human cell, this enzyme is a particularly attractive target for drug discovery and holds the promise that a selective inhibitor of RT could be a particularly safe and effective agent. Our goal, therefore, was to generate a drug structure that did not resemble the natural nucleoside substrates of the enzyme.

13.2 LEAD DISCOVERY AND OPTIMIZATION

The decision to search for nonnucleoside inhibitors of HIV-1 RT required that we initiate a screening program to identify a structural lead. This was based not only on the preference to develop a novel structural class but also on the pragmatic basis that no suitable lead structures were known. The screening capabilities at that time (<100 samples per week) would now be considered (very) low-throughput screening compared to the high-throughput screening (HTS) or ultrahigh-throughput screening (UHTS) available with today's robotic systems, which can screen thousands of compounds per hour. Thus, it was fortuitous that after screening only approximately 600 random compounds from the company sample collection that a pyrido[2,3-*b*][1,4]benzodiazepinone (**1**) was found that was weakly active (6 μ*M*) in the HIV-1 RT enzyme assay.[4] The tricyclic backbone in this structure is a common feature in a large series of compounds in our company collection that center around the M_1-selective antimuscarinic agent pirenzepine (Scheme 13.2).[5] Although there were many pirenzepine-analog pyrido[2,3-*b*][1,4]benzodiazepinones (**1**) and isomeric pyrido[2,3-*b*][1,5]benzodiazepinones (**2**) available for testing, there were very few

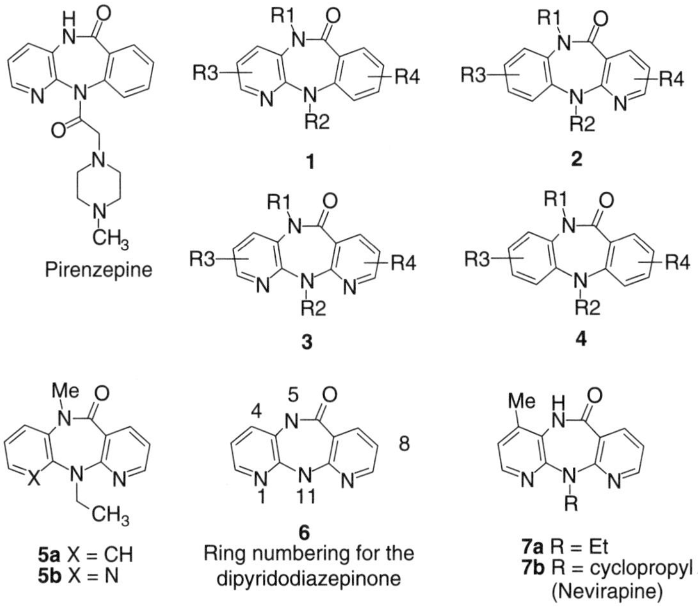

Pirenzepine

1

2

3

4

5a X = CH
5b X = N

6
Ring numbering for the dipyridodiazepinone

7a R = Et
7b R = cyclopropyl (Nevirapine)

Scheme 13.2

dipyrido[3,2-*b*:2′,3′-*e*][1,4]diazepinones (**3**) and dibenzo[*b,e*][1,4]diazepinones (**4**). A focused screening of the available analogs from these four series of related compounds suggested that both the pyrido[2,3-*b*][1,5]benzodiazepinones (**2**) and dipyrido[3,2-*b*:2′,3′-*e*][1,4]diazepinones (**3**) showed the most promise as lead structures, and at this point there was no biological or physicochemical basis on which to choose one series over the other. Accordingly, the synthetically more accessible pyrido[2,3-*b*][1,5]benzodiazepinone (**2**) series was initially selected as the primary series for further studies, with the derivative 6,11-dihydro-11-ethyl-6-methyl-5H-pyrido[2,3-*b*][1,5]benzodiazepinone (**5a**) chosen as a lead structure. This compound was active in the HIV-1 RT enzyme screening assay with an IC_{50} value of 350 n*M*.

The lead optimization process requires that a set of criteria be established that are deemed essential for the selection of compounds (for preclinical and clinical development) with the characteristics necessary to achieve the therapeutic objective, desired route of administration, and so on. For this program the HIV-1 RT enzymatic assay was the primary screen. Active compounds ($IC_{50} < 1\ \mu M$) in this assay were then tested for their ability to block HIV-1 proliferation in cell culture, and a direct correlation was quickly found between the enzymatic and cellular potencies. Specificity was determined by testing selected compounds against reverse transcriptase enzymes from HIV-2, SIV (simian immunodeficiency virus), and feline and murine leukemia viruses. In addition, the compounds were tested for specificity against human DNA polymerases α, β, γ, and δ, and against calf thymus DNA polymerase.[6] Additional criteria, not uncommon to most other drug candidates, required that the compound(s) selected be nontoxic at the intended doses, that they have reasonable aqueous solubility and be orally bioavailable, and that they be sufficiently stable to metabolism to exhibit a reasonable half-life in vivo. Finally, the compound(s) selected needed to have the ability to pass the blood–brain barrier. This was required by the fact that the HIV-1 virus has the ability to infect the central nervous system, sometimes leading to the potentially fatal AIDS dementia complex.

Although, as stated above, most of the initial synthetic efforts focused on the pyrido[2,3-*b*][1,5]benzodiazepinone (**2**) series, a smaller synthetic effort was directed to the dipyrido[3,2-*b*:2′,3′-*e*][1,4]diazepinones (**3**), with the corresponding methyl–ethyl analog **5b** (IC_{50} = 125 n*M*) used as the lead for that series. It quickly became apparent that, in general, structure–activity relationships developed for **2** also applied to **3**. Also, the dipyrido compounds are generally as potent as or more potent than the corresponding monopyrido **2**, and these SAR findings were confirmed as additional compounds were synthesized and tested.

Relatively early in the project, in vivo metabolism studies showed that N,N′-dealkylation occurred fairly rapidly in the compounds with no aromatic ring substitution, although the rate of N,N′-dealkylation was slower in the dipyrido compound (**5b**) than was the corresponding pyridobenzo substance **5a**. Subsequent confirmation with other analogs demonstrated that in general the rate of metabolism of the dipyrido compounds in different animal species was significantly slower than the corresponding pyridobenzo analogs. In addition to this important finding, the dipyrido compounds were also found to have greater aqueous solubility and were less cytotoxic than the corresponding pyridobenzo analogs. Accordingly, the focus of the synthetic program shifted exclusively to the dipyrido series.

Since a primary goal was to develop an orally available drug, emphasis was placed on measuring the rates of metabolism for selected analogs. It was known from the structure-activity relationships that were being developed for this series that N(5)-dealkylation and N(5),N(11)-dedialkylation resulted in substantial reductions in potency. On the other hand, by attaching a methyl group at the 4-position (see **6**) on the pyridyl ring rather than the amide nitrogen (position-5, **6**), three beneficial results were obtained. First, N(5)-dealkylation from the amide nitrogen was no longer possible. Second, rather than observing an increased rate of N-dealkylation at position 11, N-dealkylation at that position actually decreased significantly, with the primary metabolism (hydroxylation) taking place at the 4-methyl group. The rate of this oxidative metabolism was much slower than the N,N-dealkylation observed with **5b**. A third important finding was an increased potency of the 4-methyl derivative compared to the corresponding 5-methyl compound (**5b**). Thus, by one simple change in the structure, the rate of metabolism was lowered significantly and the potency was increased appreciably.

Lead optimization of the series based on the dipyridodiazepinone structure (**3**) provided structure–activity relationships (SAR) that supported the selection of nevirapine (**7b**) as a preclinical candidate. These SAR studies indicated that a methyl group at position 4 resulted in compounds with the highest potency and best overall characteristics. Derivatives with small alkyl or acyl substituents at the lactam nitrogen (position 5) were also potent but suffered from relatively rapid metabolism as noted above. Moreover, small alkyl groups (ethyl and cyclopropyl) at the N-11-position were required for optimum enzyme inhibition since both smaller functional groups (methyl and hydrogen) and larger alkyl groups (including propyl and *i*-propyl) conferred significantly lower potency. Aromatic ring substitution other than at position 4 generally lowered potency and reduced aqueous solubility. The most potent compounds had a 4-methyl substituent provided that there was no substitution (i.e., $R_1 = H$) at the adjacent 5-position.

On the basis of the results above, the N-11 ethyl (**7a**) and cyclopropyl (**7b**) derivatives were selected as preclinical candidates. These two compounds were evaluated in a wide range of tests and generally exhibited similar biological and physicochemical characteristics, although the ethyl analog was more potent than the cyclopropyl analog against the HIV-1 RT enzyme, and it was equipotent in its ability to block virus proliferation in the cellular assay. The determining factor in the selection of the cyclopropyl analog (**7b**, nevirapine) over the ethyl analog for clinical evaluation was the significantly higher oral bioavailability of nevirapine in monkeys and rats. This enabled the attainment of much higher blood levels at a given oral dose. High oral bioavailability (93%) and a relatively long half-life (25 hours) were subsequently measured in human clinical trials.[7]

13.3 CHEMICAL DEVELOPMENT AND PROCESS RESEARCH

The initial synthesis[4] of nevirapine (**7b**) developed by the medicinal chemistry group employed 2-chloro-3-nitro-4-methylpyridine (**8**) as a key intermediate (Scheme 13.3). Catalytic reduction to 2-chloro-3-amino-4-pyridine (**9**) and condensation with 2-chloronicotinoylchloride (**10**) gave the 2,2′-dihaloamide (**11**). Treatment of **11** with four equivalents of cyclopropylamine (**12**) in xylene at 120 to 140°C under autogenous pressure produced the 2′-alkylamino adduct (**13**). Subsequent ring closure with sodium hydride in pyridine at 80 to 100°C gave nevirapine.

A significant concern during early drug development was the limited availability of intermediate **8**, which could be obtained commercially only in small quantities. It is derived, via **16**, by nitration of **14** (Scheme 13.3), and a yield of 82% has been reported for this reaction.[8] However, the method is not scalable, due to a rapid uncontrollable exotherm when the reaction is conducted on a large scale. In addition, the nitration method gives a mixture of regioisomers **15** and **16**, and separation of these isomers proved to be industrially

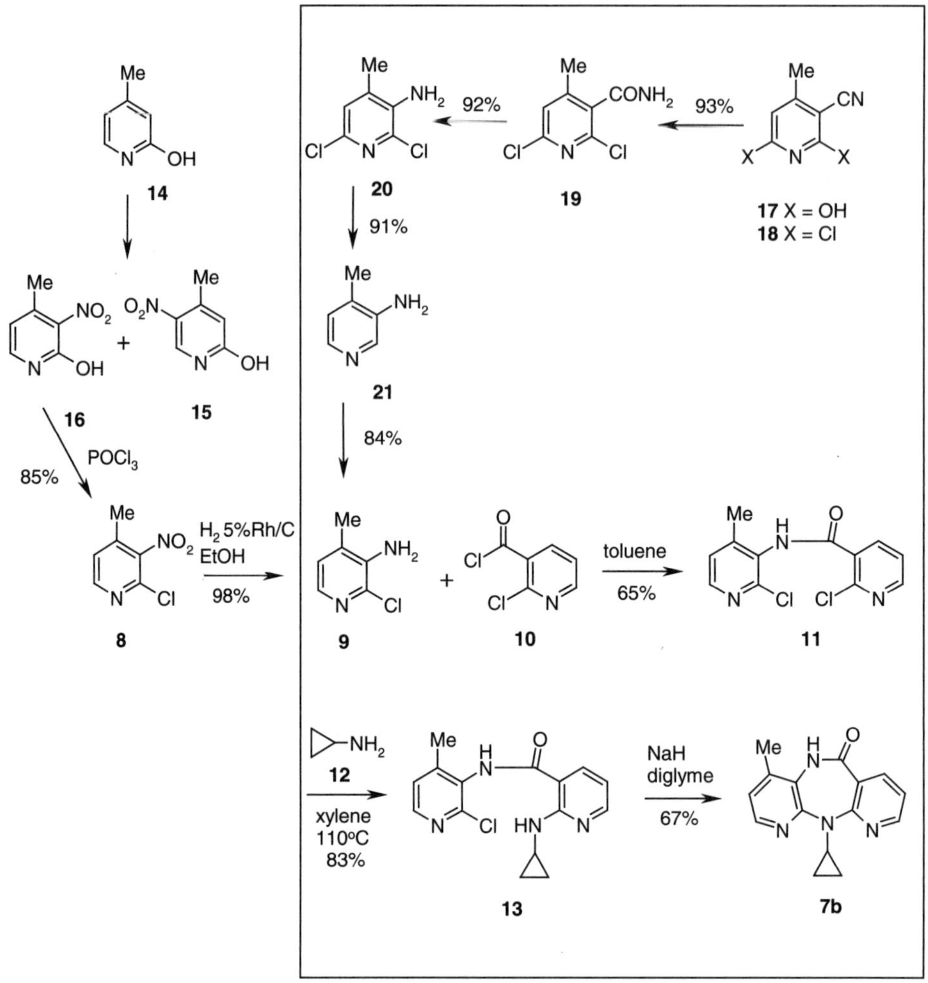

Scheme 13.3

impractical. Based on the information obtained from these experiences, the use of **8** as a precursor was abandoned and chemical and process development activities were directed toward development of an alternative approach to 3-amino-2-chloro-4-methylpyridine (**9**) that could be scaled up to produce pilot plant quantities.

Although 2,3,4-trisubstituted pyridines are rather difficult to access, 2,3,4,6-tetrasubstituted pyridines, particularly those with substituent patterns similar to **17** (Scheme 13.3), are relatively easy to produce on a large scale,[9] and in fact, **17** is commercially available in multiton quantities. As part of the SAR efforts around the nevirapine structure, we had experience in transforming this material (**17**) into dichloroaminopyridine (**20**). Treatment with phosphorous oxychloride gives **18**, which is followed by acid hydrolysis of the 3-cyano substituent and conversion to the amine under Hofmann rearrangement conditions. Efforts to selectively remove the 6-chloro substituent from intermediate **18**, **19**, or **20** were unsuccessful. However, removal of both chlorine atoms by catalytic dechlorination followed by selective rechlorination in the 2-position gave the required product (**9**).[10] Although this

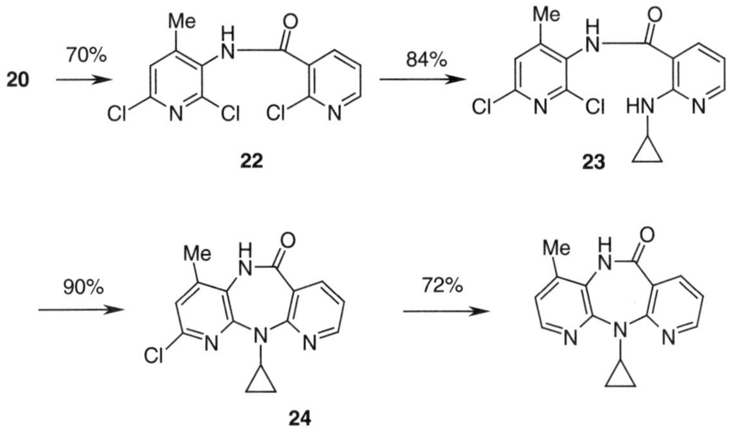

20 —70%→ 22 —84%→ 23 —90%→ 24 —72%→

Scheme 13.4

synthetic approach lacks atom economy with respect to the removal and addition of chlorine atoms, it provided the opportunity to meet the short-term active pharmaceutical ingredient (API) supply needs and established a synthetic strategy upon which further development activities could be founded. It should also be noted that all process steps from the commercially available raw material 2,6-dihydroxy-4-methyl-3-cyanopyridine (**17**) avoid the use of halogenated organic solvents, making it environmentally attractive.

Intermediate **20** can also be transformed into nevirapine via **22**, **23**, and **24** in an alternative synthetic sequence that eliminates the dechlorination and rechlorination process steps (Scheme 13.4).[11] In this approach, **24** is dechlorinated with palladium on carbon and hydrogen to give nevirapine in high yield.[12] Although this option presented the opportunity to eliminate one process step, all the intermediates from this method differ in chemical composition from the original process (Scheme 13.3). For this reason, reevaluation of the API impurity profile, toxicology, and other pharmacological and regulatory issues would be required. Because this option was identified late in the chemical development process, it was decided that the potential process benefits were more than offset by the additional time and effort required to requalify this process, and this option for API production was abandoned.

The nevirapine process scheme used during chemical development provided the basis on which to begin process development studies, with the objective of defining reaction conditions that would allow this process to be carried out on a routine commercial basis. In this process, the basic elements of the molecule are introduced at the step involving the condensation of **9** and **10**. Using the U.S. Food and Drug Administration (FDA) guidelines[13] for defining the starting point in the synthesis for regulatory purposes, 3-amino-2-chloro-4-methylpyridine (**9**) was considered a raw material in the synthesis. This provided the opportunity to implement further process improvements in the preparation of this molecule after the product launch, with limited regulatory impact.

The use of cyclopropylamine (**12**) presented a significant process optimization opportunity. In the initial chemical development pilot studies, the conversion of **11** to **13** required 4 molar equivalents of cyclopropylamine in the reaction medium. Although **12** is a simple building block, it is rather expensive on a per kilogram basis. In this reaction, 1 mol of **12** is used to absorb the by-product HCl from the reaction. Calcium oxide was found to be a

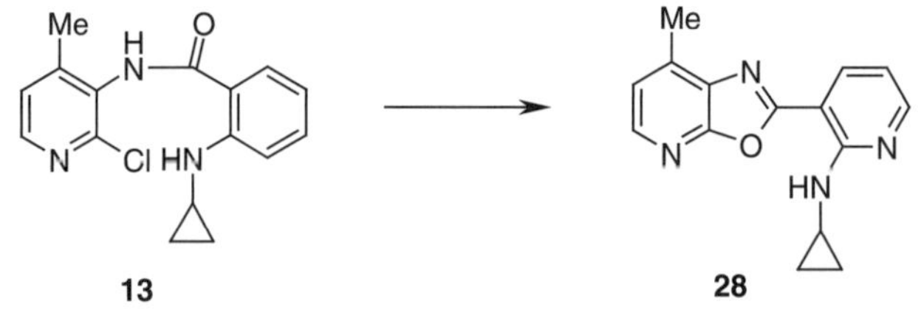

Me H O N N Cl HN 13 → Me N N O N HN 28

Scheme 13.5

much more cost-effective neutralizing agent. However, even with calcium oxide present, a 2.5 *M* excess of cyclopropylamine is required to carry the reaction to completion. Efforts to combine this reaction with the subsequent cyclization step were successful, and **13** was treated as a nonisolated intermediate in process development pilot runs as well as on a commercial scale.

Particular attention was also paid during development activities to the specific reaction conditions employed in the final cyclization step. The medicinal chemistry procedure as well as the initial chemical development efforts used either pyridine or diglyme as solvent and sodium hydride as the base. Sodium hydride in 2.8 *M* excess is required to carry the cyclization reaction to completion. The first mole of sodium hydride is consumed with the deprotonation of the more acidic amide proton. In the event that a base of insufficient strength is used to remove the amino proton of **13**, ring closure to the oxazolo[5,4]pyridine (**28**, Scheme 13.5) from the displacement of the chlorine atom by the amide carbonyl oxygen occurs. The same undesired product was observed when sodium carbonate or other base systems were used. No industrially practical substitute for sodium hydride as a reagent base was identified. It was recognized from solvent screening studies that the reaction pathway for the ring closure was also very solvent dependent. When dimethylformamide (DMF) was used as the solvent, oxazole is the exclusive reaction product.

One problem with the use of sodium hydride as a reagent in pilot and commercial operations relates to the storage and handling requirements for this material. Sodium hydride is received from the supplier in 10-kg bags as a 60% amalgam in mineral oil to stabilize the reagent. In the nevirapine process, the mineral oil tends to agglomerate with the product upon precipitation from the reaction mixture. An intermediate purification step was developed using DMF as a crystallization medium. The crude product was dissolved in hot DMF followed by charcoal treatment to remove the residual mineral oil. The charcoal was then removed by filtration followed by crystallization of the product after addition of water. The method outlined in the box in Scheme 13.3 was used to produce the nevirapine-API requirements through phase III clinical trials, commercial launch, and production.

13.4 MECHANISM OF ACTION

As nevirapine progressed through development into the clinic, the details of its mode of action were gradually uncovered. Enzyme kinetics experiments revealed that nevirapine and related molecules were noncompetitive inhibitors of HIV-1 RT[14] and bind at a site distinct from the catalytic, or active, site. Nevirapine was also found to display a remarkable selectivity profile. It did not inhibit human DNA polymerases, which gave hope for an improved safety profile over nucleoside inhibitors in the clinic. Very surprisingly, nevirapine was

TABLE 13.1 Nature of Amino Acid Residues at Positions 181 to 188

	181	182	183	184	185	186	187	188	Nevirapine IC_{50} (n*M*)
HIV-1 sequence	Y	Q	Y	M	D	D	L	Y	40
HIV-2 sequence	I	Q	Y	M	D	D	I	L	>10,000
HIV-1 (Y188L)	Y	Q	Y	M	D	D	L	L	>10,000
HIV-1 (Y181C)	C	Q	Y	M	D	D	L	Y	3000

also inactive against the reverse transcriptase (HIV-2 RT) from the closely related virus, HIV-2. Initially, the origin of the selectivity toward HIV-2 RT was difficult to understand, since nucleoside analogs typically display comparable potency against both HIV-1 RT and HIV-2 RT. These data, in combination with the noncompetitive enzyme kinetics, prompted experiments designed to identify the location of the binding site for nevirapine on RT. Ultimately, a combination of biophysical[15] and crystallographic[16] studies located the drug-binding site in the larger p66 subunit of the enzyme adjacent to tyrosine residues at positions 181 and 188. These amino acids are close in sequence to aspartic acids 185 and 186, which constitute part of the enzyme active site. The corresponding residues in HIV-2 reverse transcriptase are isoleucine 181 and leucine 188. The data in Table 13.1 illustrate the importance of the nature of the amino acid residues at these two positions for the binding of nevirapine to RT. Replacement of even one of the tyrosine residues in the native HIV-1 sequence, for example with leucine as shown in Table 13.1, renders the mutant enzyme insensitive to nevirapine.[17] Concurrent in vitro experiments in which virus was grown in the presence of increasing concentrations of the drug yielded virus that was resistant to nevirapine. Analysis of the RT enzyme from the resistant virus showed that it had a mutation encoding for the replacement of tyrosine 181 with a cysteine residue.[18] This single amino acid change decreased the sensitivity of the enzyme to the drug by about 100-fold and alerted us to the possibility of virus resistance to the drug in a clinical setting.

13.5 CLINICAL STUDIES

As nevirapine progressed into the clinic in 1991, the initial expectations were that it could be used successfully as a monotherapeutic agent. Pharmacokinetic studies showed that the drug was well absorbed, with high bioavailability and a long half-life allowing twice-daily dosing.[19,20] Exposure form the 200- and 400-mg/day dosing regimens gave drug plasma levels many times above the in vitro IC_{50} value for inhibition of virus replication, and at these doses the drug caused an initial rapid decrease in viral load in treated patients. However, over a period of weeks to months there was a rebound in virus back to original levels.[21] The rebound is due to the emergence of resistant virus[22] and is due mainly to the virus containing the mutant Y181C RT enzyme discussed above.

The focus changed to determining the utility of the drug in combination with other anti-AIDS agents. In vitro studies had provided evidence that combinations of anti-HIV drugs could be more effective than single agents.[23] Evidence supporting combination therapy in the clinic was also emerging; for example, AZT in combination with didanosine or zalcitabine was more effective than AZT alone.[24] Clinical studies showed that nevirapine in combination with a variety of other nucleoside and protease inhibitors is effective in suppressing virus replication for prolonged periods of time.[25] Nevirapine was approved by the FDA in 1996 and was the first member of the nonnucleoside reverse transcriptase inhibitor (NNRTI) class to reach the market.

ACKNOWLEDGMENTS

The successful discovery and development campaign described above required the creative input and hard work of many individual scientists whose scientific contributions are captured in the references cited. We would also like to particularly acknowledge the leadership and commitment of Jay Merluzzi, Alan Rosenthal, Julian Adams, Robert Eckner, and Peter Grob.

REFERENCES

1. http://www.who.int/hiv/en/.
2. Curran, J. W.; Morgan, W. M.; Hardy, A. M.; Jaffe, H. W.; Darrow, W. W.; Dowdle, W. R. The epidemiology of AIDS: current status and future prospects. *Science* 1985, 229, 1352.
3. Mitsuya, H.; Weinhold, K. J.; Furman, P. A.; St. Clair, M. H.; Nusinof Lehrman, S.; Gallo, R. C.; Bolognesi, D.; Barry, D. W.; Broder, S. 3′-Azido-3′-deoxythymidine: an antiviral agent that inhibits the infectivity and cytopathic effect of human T cell lymphotropic virus type III/lymphadenopathy-associated virus in vitro. *Proc. Natl. Acad. Sci. USA*, 1985, 82, 7096.
4. Hargrave, K. D.; Proudfoot, J. R.; Grozinger, K. G.; Cullen, E.; Kapadia, S. R.; Patel, U. R.; Fuchs, V. U.; Mauldin, S. C.; Vitous, J.; Behnke, M. L.; Klunder, J. M.; Pal, K.; Skiles, J. W.; McNeil, D. W.; Rose, J. M.; Chow, G. C.; Skoog, M. T.; Wu, J. C.; Schmidt, G.; Engel, W. W.; Eberlein, W. G.; Saboe, T. D.; Campbell, S. J.; Rosenthal, A. S.; Adams, J. Novel non-nucleoside inhibitors of HIV-1 reverse transcriptase. 1. Tricyclic pyridobenzo- and dipyridodiazepinones. *J. Med. Chem.*, 1991, 34, 2231–2241.
5. Carmine, A. A.; Brogden, R. N. *Drugs*, 1985, 30, 85.
6. Merluzzi, V. J.; Hargrave, K. D.; Labadia, M.; Grozinger, K.; Skoog, M.; Wu, J. C.; Shih, C. K.; Eckner, K.; Hattox, S.; Adams, J.; Rosenthal, A. S.; Faanes, R.; Eckner, R. J.; Koup, R. A.; Sullivan, J. L. Inhibition of HIV-1 replication by a nonnucleoside reverse transcriptase inhibitor. *Science*, 1990, 250, 1411–1413.
7. (a) Lamson, M. J.; Sabo, J. P.; MacGregor, T. R.; Pav, J. W.; Rowland, L.; Hawi, A.; Cappola, M.; Robinson, P. Single dose pharmacokinetics and bioavailability of nevirapine in healthy volunteers. *Biopharm. Drug Dispos.*, 1999, 20, 285–291. (b) Cheeseman, S. H.; Hattox, S. E.; McLaughlin, M. M.; Koup, R. A.; Andrews, C; Bova, C. A.; Pav, J. W.; Roy, T.; Sullivan, J. L.; Keirns, J. J. Pharmacokinetics of nevirapine: initial single-rising-dose study in humans. *Antimicrob. Agents Chemother.*, 1993, 37, 178–182.
8. Burton, A. G.; Halls, P. J.; Katritzky, A. R. *Tetrahedron Lett.*, 1971, 24, 20211.
9. Bobbitt, M.; Scala, D. A. *J. Org. Chem.*, 1960, 25, 560.
10. Grozinger, K. G.; Hargrave, K. D.; Adams, J. U.S. patent 5,668,287, 1997.
11. Grozinger, K. G.; Fuchs, V.; Hargrave, K. D.; Mauldin, S.; Vitous, J.; Campbell, S.; Adams, J. Synthesis of nevirapine and its major metabolite. *J. Heterocycl. Chem.*, 1995, 32, 259–263.
12. Grozinger, K. G.; Hargrave, K. D.; Adams, J. U.S. patent 5,571,912, 1996.
13. Guidance for Industry BACPAC I: *Intermediates in Drug Substance Synthesis Bulk Active Post-approval Changes: Chemistry, Manufacturing, and Controls Documentation*, U.S. Department

of Health and Human Services Food and Drug Administration Center for Drug Evaluation and Research (CDER), Center for Veterinary Medicine (CVM), February 2001.

14. (a) Wu, J. C.; Warren, T. C.; Adams, J.; Proudfoot, J.; Skiles, J.; Raghavan, P.; Perry, C.; Farina, P. F.; Grob, P. M. A novel dipyridodiazepine inhibitor of HIV-1 reverse transcriptase acts through a nonsubstrate binding site. *Biochemistry*, 1991, 30, 2022–2026. (b) Spence, R. A.; Kati, W. M.; Anderson, K. S.; Johnson, K. A. Mechanism of inhibition of HIV-1 reverse transcriptase by non-nucleoside inhibitors. *Science*, 1995, 267, 988–993.
15. Cohen, K. A.; Hopkins, J.; Ingraham, R. H.; Pargellis, C.; Wu, J. C.; Palladino, D. E. H.; Kincade, P.; Warren, T. C.; Rogers, S.; Adams, J.; Farina, P. F.; Grob, P. M. Characterization of the binding site for nevirapine (BI-RG-587) a nonnucleoside inhibitor of human immunodeficiency virus type-1 reverse transcriptase. *J. Biol. Chem.*, 1991, 266, 14670–14674.
16. Kohlstaedt, L. A.; Wang, J.; Friedman, J. M.; Rice, P. A.; Steitz, T. A. Crystal structure at 3.5A resolution of HIV-1 reverse transcriptase complexed with an inhibitor. *Science*, 1992, 256, 1783–1790.
17. Shih, C. K.; Rose, J. M.; Hansen, G. L.; Wu, J. C.; Bacolla, A.; Griffin, J. A. Chimeric human immunodeficiency virus type 1/type 2 reverse transcriptases display reversed sensitivity to non-nucleoside analog inhibitors. *Proc. Natl. Acad. Sci. USA*, 1991, 88, 9878–9882.
18. Richman, D.; Shih, C. K.; Lowy, I.; Rose, J.; Prodanovich, P.; Goff, S.; Griffin, J. Human immunodeficiency virus type 1 mutants resistant to nonnucleoside inhibitors of reverse transcriptase arise in tissue culture. *Proc. Natl. Acad. Sci. USA*, 1991, 88, 11241–11245.
19. Lamson, M. J.; Sabo, J. P.; MacGregor, T. R.; Pav, J. W.; Rowland, L.; Hawi, A.; Cappola, M.; Robinson, P. Single dose pharmacokinetics and bioavailability of nevirapine in healthy volunteers. *Biopharm. Drug Dispos.*, 1999, 20, 285–291.
20. Pollard, R. B.; Robinson, P.; Dransfield, K. Safety profile of nevirapine, a nonnucleoside reverse transcriptase inhibitor for the treatment of human immunodeficiency virus infection. *Clin. Ther.*, 1998, 20(6), 1071–1092.
21. De Jong, M. D.; Vella, S.; Carr, A.; Boucher, C. A. B.; Imrie, A.; French, M.; Hoy, J.; Sorice, S.; Pauluzzi, S.; et al. High-dose nevirapine in previously untreated human immunodeficiency virus type 1-infected persons does not result in sustained suppression of viral replication. *J. Infect. Dis.*, 1997, 175, 966–970.
22. Richman, D. D.; Havlir, D.; Corbeil, J.; Looney, D.; Ignacio, C.; Spector, S. A.; Sullivan, J.; Cheeseman, S.; Barringer, K.; et al. Nevirapine resistance mutations of human immunodeficiency virus type 1 selected during therapy. *J. Virol.*, 1994, 68, 1660–1666.
23. Harmenberg, J.; Aakesson-Johansson, A.; Vrang, L.; Cox, S. Synergistic inhibition of human immunodeficiency virus replication in vitro by combinations of 3′-azido-3′-deoxythymidine and 3′-fluoro-3′-deoxythymidine. *AIDS Res. Hum. Retroviruses*, 1990, 6, 1197–1202.
24. Jablonowski, H. Studies of zidovudine in combination with didanosine and zalcitabine. *J. Acquir. Immune Defic., Syndr. Hum. Retroviruses*, 1995, 10(Suppl. 1), S52–S56.
25. (a) Reliquet, V.; Ferre, V.; Hascoet, C.; Besnier, J. M.; Bellein, V.; Arvieux, C.; Molina, J. M.; Breux, J. P.; Zucman, D.; Rozenbaum, W.; Allavena, C.; Raffi, F. Stavudine, didanosine and nevirapine in antiretroviral-naive HIV-1-infected patients. *Antiviral Ther.*, 1999, 4, 83–84. (b) D'Aquila, R. T.; Hughes, M. D.; Johnson, V. A.; Fischl, M. A.; Sommadossi, J.-P.; Liou, S.-H.; Timpone, J.; Myers, M.; Basgoz, N.; et al. Nevirapine, zidovudine, and didanosine compared with zidovudine and didanosine in patients with HIV-1 infection: a randomized, double-blind, placebo-controlled trial. *Ann. Intern. Med.*, 1996, 124, 1019–1030.

14

APPLICATIONS OF NUCLEAR IMAGING IN DRUG DISCOVERY AND DEVELOPMENT

JOHN W. BABICH
Molecular Insight Pharmaceuticals, Inc.
Cambridge, Massachusetts

WILLIAM C. ECKELMAN
Molecular Tracer, LLC
Bethesda, Maryland

14.1 INTRODUCTION

14.1.1 Process and Challenges of Drug Development

The process of preclinical drug discovery has changed fundamentally over the past decade, providing pharmaceutical companies with a large number of novel targets as well as novel tools designed to accelerate the discovery process. Many of the new molecular target structures, however, have not been thoroughly validated, which increases the risk of unforeseeable side effects or lack of efficacy. In addition, novel drug discovery technologies, including high-throughput screening involving libraries of thousands of compounds, have increased the number of molecules being considered for lead optimization with high-speed chemistry and ultimately for in vivo investigation in a wide array of suitable animal systems. So far, these developments have added additional costs to the drug development process without necessarily increasing its efficacy or improving its cost-effectiveness (Lehmann Brothers and McKinsey, 2001).

The process of early clinical drug development, in contrast, has changed little over the past 20 years. Chances for a drug candidate that has emerged successfully from preclinical studies to advance from a phase I trial to an approved product average around 22%, and even for

Drug Discovery and Development, Volume 1: Drug Discovery, Edited by Mukund S. Chorghade

compounds in phase III, the failure rate is still 57% (Lawrence, 2004). A significant proportion of these failures are associated with inappropriate drug metabolism and pharmacokinetics of candidate molecules. At the same time, costs associated with successful drug development are now estimated to exceed $800 million, with the bulk of these costs accruing in advanced clinical trials (Dickson and Gagnon, 2004). This situation defines the urgent need for drug development companies to recognize the success and failure potential early in the R&D process and has spurred interest for reliable cost-effective methods to generate pharmacokinetic, toxicology, proof of concept, and efficacy data with high predictability in preclinical animal studies and early clinical trials that provide a good rationale for go/no-go decisions.

14.1.2 Role and Contribution of Positron Emission Tomography

Positron emission tomography (PET) and single photon emission tomogrpahy (SPECT) are compelling quantitative imaging techniques to monitor and assess intra- and extracellular events such as biochemical pathways, molecular interactions, drug pharmacokinetics, and pharmacodynamics in the living organism. Endogenous molecules, receptor and enzyme ligands, drugs, or biomolecules are labeled with radionuclides and injected mostly intravenously as molecular probes. In the living organism they participate in biochemical processes while emitting gamma rays, which permit visualization of their location and concentration.

Only recently, PET and SPECT scanners were refined and redesigned such that they also became capable of providing high-resolution images of small animals, such as mice and rats (Cherry et al., 1997; Cherry, 2001). This innovation has catapulted the use of PET and SPECT from patient management or clinical trials to the preclinical stages of drug development. It has also largely promoted its use to validate molecular target structures for drug discovery in vivo, investigate and redesign experimental drugs, study the distribution of a drug, monitor its interaction with its target across organs and tissue in the body, and link that information to its biological function and clinical efficacy in the same organism.

Since PET and SPECT capture and visualize in vivo processes at the molecular level and since new drug candidates increasingly target specific molecular structures and events, it may not be surprising to note that both disciplines, drug discovery and PET/SPECT, become increasingly intertwined. PET and SPECT identify and validate in vivo molecular structures that are proposed but not yet proven to present good drug development targets, while on the other hand, pharmacogenomics and functional proteomics increasingly provide new probes for PET imaging (Eckelman, 2003).

14.2 PRINCIPLES AND EVOLUTION OF TECHNOLOGY

14.2.1 Introduction to PET Principles

PET does not provide a direct chemical analysis of reaction products; instead, the labeled molecules function as tracers to depict individual steps in a biochemical cascade. The labeled molecule emits positrons, which travel only a very short distance and then combine with an electron. Annihilation occurs; the masses of positron and electron convert into their energy equivalent through emission of two 511-keV photons that are about 180° apart. Those emitted photons are detected as a coincidence event once they strike opposing detectors simultaneously. Between 6 and 70 million detection pairs are recorded from many different angles and allow for reconstruction of an image. The kinetic information is used to calculate the concentration of binding sites over time. Tissue concentration of the probe in conjunction with the time course of the plasma concentration permit calculating the transport and reaction process the probe undergoes in the living organism. As a result, one obtains an image related to the rate of the process under investigation.

The radiolabeled probes are made in miniaturized self-shielded low-energy cyclotrons, which bombard stable nucleon with protons or deuterons to generate proton-rich nuclei. As a result of the proton–neutron imbalance, the nucleus is stabilized by conversion of a proton to a neutron. This conversion occurs by electron capture whereby the charge of a proton is "neutralized" by an orbital electron, or by emission of a positive electron or positron whereby a proton gives up its charge resulting in conversion to a neutron. Those minicyclotrons are combined in automated units that provide, usually under the control of an integrated computer, solvents, reagent additions, solution transfer, column separation, and other steps to produce labeled imaging probes, creating mobile and versatile PET radiopharmacies.

Positron-emitting radionuclides are available for the key elements found in almost all biomolecules and drugs, such as carbon (^{11}C), nitrogen (^{13}N), and oxygen (^{15}O). In addition, fluorine, although not normally a constituent of biomolecules, can act as a bioisostere of hydrogen or the hydroxyl group and is commonly encountered in a variety of drugs. Simple molecules such as $^{13}NH_3$, $^{11}CO_2$, or $H^{18}F$ are most commonly used as precursors to the radiolabeled tracer. They substitute for a stable carbon, nitrogen, or fluorine to yield labeled molecules or drugs that are chemically and biologically indistinguishable from the stable or nonradioactive counterpart. With these building blocks a rich array of physiologically active molecules and drugs can be produced. Usually, compounds are administered in picomole or nanomole amounts to provide dynamic and real-time information on in vivo behavior of radiolabeled molecules. The direct in vivo imaging of molecular interaction of biological processes provides insights into physiological and disease processes at the molecular level that had not previously been possible and have tremendous value to accelerate, advance, and improve drug discovery and drug development processes.

14.2.2 Suitable Targets

The majority, if not all, novel compounds in drug development target in a highly specific and selective fashion cellular or nuclear receptors, cellular enzymes, or transporters or interfere with protein–protein interaction. The tracer technology delivers high sensitivity for imaging those low-density molecular systems in vivo, even though at a cost of lower spatial resolution than for MRI or CT. The high specific activity of the radioligands facilitates detection of low-density sites. Radiolabeled ligands or antibodies are instrumental to map receptor expression across anatomical structures such as different regions of the brain, and to understand in vivo binding kinetics and associated biological functions of drugs and antibodies binding to cell-surface receptors. Radiolabeled ligands, for example, permit quantification of receptor occupancy achieved by doses of a therapeutic agent targeting a specific receptor or receptor subtype.

14.2.3 Suitable Animal Models

Advances in PET imaging technology led in the late 1990s to the evolution of compact low-cost dedicated small-animal PET scanners with high spatial resolution capability that extended the use of PET as a research tool in clinical applications to preclinical drug discovery research by making laboratory animals such as mice and rats, which are so important in the early stages of drug discovery, accessible to in vivo imaging technology (Cherry et al., 1997; Cherry, 2001; Myers and Hume, 2002) and deliver dynamic, real-time information. Small animal imaging technology specifically offers high volumetric resolution at an axial field of view of 1.8 cm. Three-dimensional collections of emission data in combination with Bayesian reconstruction techniques generate high-definition images of the entire mouse body (Qi et al., 1998). These advancements have greatly accelerated PET studies in a variety of animals, including those that are good model systems of human diseases or carry defined genetic alterations, such as gene knock-out animals.

Knock-mice have more recently been instrumental in investigating receptor-subtype selectivity for muscarinic receptor ligands toward each of the four receptor subtypes M1, M2, M3, and M4 in mice where either one of these subtypes was absent (Jagoda et al., 2003). This particular study identified the M2 subtype receptor as the main target of the radioligand tested in distinct regions of the brain. More recently, protein–protein interactions subject to a range of physiological and pathophysiological conditions have been investigated in vivo using PET. The yeast/two-hybrid technology was key in allowing scientists to find interaction partners for any given protein and further map down to the single-amino-acid level the requirements for functional protein–protein interactions. With the help of PET, these protein–protein interactions can now be examined and quantified not just in vitro under highly artificial nonphysiologic conditions but in the living animal. This is achieved by linking the two-hybrid construct to a reporter gene whose expression in turn is visualized by radiolabeled ligands (Luker et al., 2003), allowing researchers to monitor intrinsic binding specificities of proteins and gene regulation during development, disease progression, and under the influence of pharmacologic agents. Given the still prevailing shortcomings in the efficacy of gene transfer, the strength of this approach lies in the preclinical animal studies.

14.3 ROLE IN DRUG DISCOVERY

PET is primarily a functional imaging technology designed for rapid and repeated noninvasive in vivo assessment of molecular and ultimately biological processes. Molecular imaging examines disease biology as the disease evolves and proceeds through its natural history. This capability makes PET an ideal tool to accompany all stages of drug development: from the identification of molecular structures involved in the genesis and evolution of the disease, through the design of drugs to effectively modify those molecular structures and their biological or pathological function in vivo, and ultimately to monitor therapeutic efficacy and response in the patient. Products of modern drug discovery are monoclonal antibodies, vaccines, therapeutic peptides or proteins, and enzyme inhibitors. Usually, there are several iterations in the chemical design of these compounds before a lead with the desired properties emerges. At this stage of the drug development process, radionuclide-based methods are helpful to assess binding characteristics of antibodies or ligands or to measure new transport or enzyme functions in vivo.

14.3.1 Target Validation and Drug Design

Target Identification: Linking Markers to Disease The most significant contribution of PET in finding or confirming molecular structures as targets for therapeutic interventions has occurred over the past decade in neurodegenerative and psychiatric disorders which, by their very nature, are not easily accessible for investigation in animal models. PET visualizes brain neurochemistry in the living brain by using radiotracers that monitor neurotransmitter synthesis, metabolism, enzymes, transporters, and receptors. From those studies, insights into the biological role of these proteins and their involvement in psychiatric or neurodegenerative disorders, including schizophrenia, anxiety disorders, depression, or various forms of dementia, have emerged that in turn identify molecular target structures suitable for pharmacological modulation (Smith et al., 2003). PET has, for example, been instrumental in identifying altered serotonin pathway activities in the genesis and evolution of several eating disorders such as anorexia nervosa and bulimia nervosa, providing a first rational for therapeutic interventions strategies (Barbarich et al., 2003).

Similarly, PET studies with radiolabeled ligand of peripheral-type benzodiazepine receptors, which were believed to be expressed exclusively by peripheral blood cells and linked to peripheral inflammatory responses led to the discovery of benzodiazepine receptor expression associated with neuroinflammatory processes in the brain. Specifically, PET identified benzodiazepine receptors in activated microglia cells as they accompany neuroinflammatory diseases such as multiple sclerosis or neurodegerative diseases such as Alzheimer's. Now the expression of this peripheral benzodiazepine binding sites on microglia is viewed as a key indicator for their transition from a normal resting state to the activated state (Cagnin et al., 2001). Those receptors have subsequently been shown to display all the functional characteristics associated with the initiation of an inflammatory response, such as release of nitrite oxide and tumor necrosis factor-α (Wilms et al., 2003). Taken together, these PET observations led to the identification of a molecular target structure and therapeutic rational to attenuate the neuroinflammatory process associated with Alzheimer's disease or multiple sclerosis.

In Vivo Analysis of Receptors, Enzymes, and Signal Transduction in Normal and Diseased Cells and Animals The therapeutic value of a novel drug that acts through binding a distinct molecular structure in a specific organ or cell population is guided by specificity, saturation kinetics, and selectivity. Each of these parameters can be assessed in vivo using PET. 16α[^{18}F]fluoro-17β-estradiol, a selective estrogen receptor ligand, has, for example, been used to quantify the distribution of the tracer in discrete cerebral areas and to quantify in vivo estrogen receptor binding parameters in the brain (Dehdashti et al., 1999). Cell–cell interactions also govern host–pathogen interactions and can be visualized to monitor pathogen tropism, pathogen life cycle, signal transduction, host response, and cell trafficking in living animals (Piwnica-Worms et al., 2004). Radiolabeled rolipram, a drug that inhibits phosphodiesterase 4, a key intracellular signal transduction molecule that is implied in signaling pathways used by inflammatory signals as well as neurotransmitters, has been used to monitor PDE4 activity in vivo in distinct regions of the brain (DaSilva et al., 2002). Similarly, phosphoinositide turnover accompanies a wide variety of intracellular signal transduction processes, including those initiated by cytokines or neurotransmitters. In the brain its turnover is closely linked to synaptic functionality and to the production of second messengers. Radiolabeled diacylglycerol as a marker for phosphoinositide turnover is suitable to assess postsynaptic biological responses in healthy subjects and under various disease conditions (Imahori et al., 2002).

Intracellular signaling is also visualized and quantified in vivo by examining gene regulation with the help of gene reporter constructs. The regulatory element of interest is combined with reporter genes, such as the secreted alkaline phosphatase, that are easily visualized by PET (Haberkorn et al., 2004). Hereby, gene expression throughout the body is quantified in the presence or absence of a pharmacological agent (Iyer et al., 2001). For example, human T-cells were engineered to express a reporter gene whose product is visualized by PET and whose promoter is regulated by an intracellular transcription factor. The T-cell receptor-dependent nuclear factor of activated T-cells (NFAT) have been used for molecular imaging of T-cell activation in vivo (Ponomarev et al., 2001). This tool has the potential to provide in vivo insights into the course of normal and pathologic immune responses as well as temporal dynamics and immune regulation at different stages of disease and following therapy, such as adoptive immunotherapy for cancer, vaccination, or immunosuppressive drugs.

In Vivo Characterization of Novel Targets: Defining In Vivo Gene Function, Protein–Protein Interaction, Signaling Pathways, and Gene Regulation Protein–protein interaction or receptor occupancy to complex cellular and biological responses can only be observed in the living being using PET. The dopamine D_2 receptor, for example, is implied in some of the behavioral and psychological symptoms of dementia (BPSD), including aggressiveness, wandering, and sleep disturbance. PET using a specific probe for the dopamine D_2 receptor in patients treated with risperidone, an approved drug for Alzheimer disease, was instrumental in documenting the fact that risperidone enhanced the binding potential of the dopamine D_2 receptor without interacting with it directly, thus elucidating a molecular mechanism of action that was linked to some of its clinical effects (Meguro et al., 2004).

Similarly, radiolabeled enzyme inhibitors or substrate are used to quantify in vivo enzyme activity. Failure of cells to repair endogenous DNA continuously has, for example, been identified as an initial step in the evolution of many human malignancies and has also been related to therapy response and outcome. One of the key enzymes ensuring continuous and faithful DNA repair is alkylguanine-DNA alkyltransferase (AGT), which is often faulty in breast cancer. A radiolabeled substrate for AGT has been developed that monitors in vivo: for example, in a rat model of breast cancer, the activity of this enzyme in vivo (Zheng et al., 2003). This sets the stage to investigate, for example, whether novel anticancer compounds may be modulated in their efficacy by the presence or functional activity of AGT, which may be distributed heterogeneously throughout the tumor.

PET advances the study of gene regulation, and signal transduction cascades from extracellular or in vitro experimental systems to the in vivo situation. It monitors temporal and spatial dynamics of expression and function of specific genes and intrinsic binding specificities of proteins that govern gene regulation during disease progression or under the influence of a pharmacological agent. The study of gene regulation of human breast cancer cells in vivo, for example, is suitable to gain more insights into the molecular events accompanying disease progression (Berger and Gambhir, 2000). With the help of small animal imaging, specific binding of the tumor suppressor gene p53 to its viral target, the large T-antigen, is detected and quantified in living mice (Luker et al., 2002). Transgenes brought into living animals are instrumental in investigating the transcriptional activation of endogenous genes by PET. For example, a reporter gene is under the transcriptional control of regulatory elements specific for the p53 tumor suppressor gene functions as an in vivo readout system to measure the p53-governed signal transduction cascade and activation of p53-dependent genes in murine xenograft models of human tumors (Doubrovin et al., 2001). These examples illustrate that PET advances functional proteomics from the extracellular and the in vitro scenario to the in vivo situation and aids drug development through direct interrogation of molecular targets within intact animals.

14.3.2 Preclinical Studies

Small animal PET scans transform established animal disease models to functionally validated visual models for drug development where the biological role of a given molecular target structure is observed directly and the effect of a compound targeting that structure can be visualized and assessed simultaneously in a functional readout system.

Organ Distribution, Pharmacokinetics, and Toxicity of Leads or Candidates PET quantifies, for example, uptake and distribution of novel neuropharmacological agents in experimental animals and links their binding to their respective target structure to physiological and biochemical processes, such as glucose metabolism or blood flow, which can then be used to extrapolate those data to correlate physiological and pharmacological effects to optimize drug design and ultimately, clinical treatment (Halldin et al., 2001). Salazar and Fischman (1999) evaluated BMS 181101, a drug with agonist and antagonist activity at various sites in the serotonin system. The ^{11}C-labeled form of the drug was used to show that the residence time in the brain was short and, as a result, specific binding could not be determined by external imaging. These studies on receptor occupancy showed that the drug may have a narrow therapeutic index and may not be suitable for once- or twice-a-day dosage.

Monoamine oxidases (MAOs) A and B are flavoproteins residing in the outer mitochondrial membrane. They participate in the processing of various neurotransmitters; MAO-A predominantly oxidizes serotonin, norepinephrine, and dopamine. Selective MAO-A inhibitors are used to treat patients with Alzheimer's disease, Parkinson's disease, or depression and other psychiatric disorders. A radiolabeled reversible and specific inhibitor of MAO-A was instrumental in quantifying and evaluating its uptake into the brain of baboons, its tissue distribution across the brain, and its binding kinetic with MAO-A (Bottlaender et al., 2003), creating a tool to evaluate in vivo pharmacokinetics, substrate-binding affinity, and functional efficacy of new MAO-A inhibitors.

Acetylcholinesterase (AChE) is an important target of several marketed drugs treating neurodegenerative diseases. Novel radiotracers, such as [^{11}C]physostigmine, depict regional distribution of AChE activity, as it is known from histological studies on postmortem brains and allow quantifying the occupancy of binding sites on AChE by AChE inhibitors. This facilitates pharmakokinetic studies across brain regions at the molecular level. As a complementary approach, radiolabeled acetylcholine analog substrates measure AChE activity and quantify the efficacy of AChE inhibitors. For example, in patients with Alzheimer's disease, endogenous AChE activity is known to be reduced. Here, the ability of current drugs such as donepezil and rivastigmine to inhibit the activity of AChE allows therapeutic monitoring of these drugs and in the design of improved versions of current drugs by using PET imaging as an indicator for target inhibition in vivo (Shinotoh et al., 2004).

Achieving the right bioavailability and organ distribution of novel compounds which have passed all the in vitro tests and possessed the desired mechanism of action in cell cultures can be a challenging endeavor. For example, promising cancer compounds with a dual-action mechanism that inhibit both topoisomerase I and II may suffer from poor extravascular distribution, which limits their therapeutic efficacy even for compounds with high in vitro cytotoxic profile. PET has been instrumental in studying biodistribution of lead candidates in mice bearing human tumor xenografts and selecting candidate with the best biodistribution, clearance, metabolic stability, and ultimately, efficacy profile against human tumors in a mouse model (Osman et al., 2001).

In Vivo Mechanism of Action of Leads or Candidates in Cells and Animals Small animal PET characterizes exact drug–target interactions and elucidates the mechanism of action of pharmacologic agents that could not otherwise be observed. A dual PET strategy, for example, has been instrumental in defining the effect of methamphetamine and scopolamine in the brain. Researchers investigated simultaneously the binding of these agents to the dopamine D_1 receptor in the striatum of conscious monkeys as well their ability to functionally activate signaling cascades known to be associated with the D_1 receptor (i.e., the D_1 receptor-coupled cAMP messenger determined by phosphodiesterase IV activity as a readout system) (Tsukada et al., 2001). It was shown that neither compound interfered with the binding capacity of the D_1 receptor, but both compounds enhanced PDE4 activity. Dual PET analysis thus examined simultaneously receptor occupancy and second messenger systems and thereby delivered in vivo functional insights on the molecular mode of action of motion-sickness drugs.

Radiolabeled ligands targeting the imidazoline receptor (Hudson et al., 2003) not only facilitated visualization of their uptake kinetics and biodistribution but also linked their interaction with their respective receptor to a biochemical signaling cascade and ultimately to a biological response. PET shows that receptor activation is associated with amelioration of depression in an established rat model of this disease. This experimental setup now sets the stage to investigate simultaneously in vivo bioavailability, receptor binding, and saturation as well as functional efficacy (i.e., behavioral changes of the rat model) of novel compounds targeting these receptors.

Similarly, radiolabeled inhibitors of the 11β-hydroxylase, an enzyme involved in the biosynthesis of cortisol and aldosterone and also known to function as anesthetic drug through the GABAergic system, are useful not only for in vivo imaging of the adrenocortex but also to investigate the function, dynamics, and kinetics of narcotic drugs with PET that are designed to work by targeting the GABAA receptors (Mitterhauser et al., 2003).

Preclinical Proof of the Principle of Novel Treatment Paradigms Applying PET to small animals not only visualizes molecular and biochemical processes in the in vivo situation compared to extracellular or in vitro studies but also provides a unique opportunity to monitor cell behavior and processes that were not accessible for direct investigation and hence proof of principle before. This includes cell trafficking in vivo as it occurs during tumor metastasis or during immune responses. PET detects, in vivo and in real time, tumor cell migrating in mouse models, visualizes directly their organ preference, and investigates how the metastatic potential can be modulated by pharmacological agents that alter the cell-surface profile of the tumor cells, such as their expression pattern of adhesion molecules (Koike et al., 1995).

Adoptive immune therapy using bioengineered T-cells to target tumor cells specifically and systemically is an emerging treatment paradigm in oncology. These T-cells can be equipped with a marker gene that permits their tracking in vivo without compromising their ability to identify and kill distant metastatic tumor cells, as shown recently in a murine model of a human tumor. This technique for imaging the migration of ex vivo-transduced antigen-specific T-cells in vivo is informative, nontoxic, and potentially applicable to humans (Koehne et al., 2003). As PET permits monitoring and quantifying of in vivo expression of transgenes, it assess the tissue specificity, quality, and sustainability of foreign gene expression in the context of preclinical gene therapy studies involving, for example, adenovirus-mediated gene transfer techniques (Mayer-Kuckuk et al., 2003; Groot-Wassink et al., 2004). Further, xenograft models of human tumors have been used to investigate in vivo the feasibility of suicide-gene transfer for the treatment of prostate cancer (Pantuck et al., 2002).

14.3.3 Clinical Studies

Patient Selection Through Noninvasive Identification of Molecular Targets With the arrival of novel drugs designed to interfere with specific molecular processes, the need to monitor in vivo the proposed mechanism of action at the molecular level and to identify prospectively patients who express the molecular target structure in sufficient amounts or do not express target structures that may predispose them for side effects has emerged. PET offers the unique opportunity to monitor patients in vivo in a noninvasive fashion for both (Fishman et al., 1997a,b; Solomon et al., 2003). Good examples are experimental drugs such as Erbitux, which targets the epidermal growth factor receptor, a cell-surface receptor tyrosine kinase whose expression and activity is linked to the genesis of several cancers. Efficacy of this drug in the individual patient is probably correlated with the expression level of the EGF receptor as well as its functional activity. A PET biomarker recognizing specifically the functional active EGF-R tyrosine kinase was developed that also acts as an inhibitor of this tumor-associated target (Ben-David et al., 2003) and should permit both identification of patients suitable for treatment and the monitoring of therapeutic efficacy.

Similarly, very recently the first antiangiogenesis drug, Avastin, was approved for cancer treatment. Avastin targets the vascular–endothelial growth factor (VEGF) receptor on endothelial cells and is designed to shut off tumor vascularization and thereby to kill the cancer. It would be helpful both to monitor treatment efficacy and to identify patients who are likely to benefit from this treatment by visualizing the expression of VEGF. A radiolabeled probe to this end has been developed and has been shown in small animal models of human tumors to recognize VEGF with high affinity and high specificity in both the primary tumor and all metastatic sites, preparing the way for similar studies in patients receiving this experimental drug (Collingridge et al., 2002).

Therapeutic inhibitors of intracellular signaling cascades are thought to constitute more specific and less toxic anticancer compounds. However, appropriate tools to assess in vivo precisely the molecular mechanism by which a novel compound is believed to act have been lacking, and PET may well fill this gap: for example, an experimental compound, 17-allylaminogeldanamycin (17-AAG), designed to inhibit the heat-shock protein Hsp90. 17-AAG causes degradation of the oncogene Her2 as well as other proteins and blocks tumor growth in preclinical animal experiments. Her2 is implemented in the pathogenesis of breast cancer and other solid tumors, and 17-AAG is the first experimental drug targeting this pathway to enter clinical trials. PET permits in vivo imaging of Her2 and the kinetics of its degradation under the influence of 17-AAG in animal tumors (Smith-Jones et al., 2004). This approach allows noninvasive imaging of the pharmacodynamics of a targeted drug and will facilitate the rational design of combination therapy based on target inhibition. Early identification of patients who are because of their pharmacogenomic profile likely to be more susceptible to certain side effects of novel drugs is of pivotal importance to smoothen the drug development process. PET has been instrumental in identifying patients with a distinct pattern of histamine H_1-receptor expression in the brain who are most likely to suffer more severely from subjective sleepiness and objective sedation under the influence of histamine H_1-receptor antagonists (Tashiro and Yanai, 2003). Histamine H_1-receptor occupancy, which is then used to assess the risk for sedation in the individual patient. This experimental setup will also be useful in evaluating the overall potential of novel antihistamines to elicit this side effect during preclinical and early stage clinical studies.

Similarly, PET is helpful in determining whether apparent unresponsiveness to treatment is caused by failure to deliver the drug to the site of disease or by other means of pharmacodynamic resistance. For example, ^{11}C-labeled phenytion ([^{11}C]DPH) was instrumental in quantifying the regional brain concentration of this drug in patients with medically resistant epilepsy. It was shown that the epileptogenic focus accumulated amounts of the radiolabeled drug similar to those in normal brain areas in the same patient, pointing to a pharmacodynamic rather than a pharmacokinetic cause for the resistance observed (Baron et al., 1983). More recently, others have provided PET data to suggest that the cortical GABA-A receptor plasticity differs in a normal brain area, and epileptogenic focus may play a role in causing drug-resistant epilepsy (Marrosu et al., 2003).

In Vivo Biodistribution Radiolabeled drugs of interest (DOIs) enable the investigator to determine the biodistribution or pharmacokinetics and target tissue in normal organs as a function of time and dosage. This is even more important, as innovative drugs often times target very specifically distinct molecular processes and biochemical pathways that play out differently in different organs. Understanding the in vivo distribution and observing in vivo drug effects across organs and tissue can be crucial in defining the appropriate dose, understanding the risks of potential side effects, and obtaining some insight as to how the disease itself, potential co-morbidities, or even unrelated disease states may interfere with the kinetic, distribution, and site-specific effects of a compound. Pathophysiological changes in tissues, for example, may alter the delivery of drugs and make extrapolations from normal tissue data unreliable. Radiolabeled DOI permit comparative pharmacokinetic analysis under normal and pathological conditions: for example, to investigate tissue distribution of antibiotics and determine actual concentrations at the sites of infection (Fischman et al., 1997a–c). Inappropriate drug metabolism and pharmacokinetics is estimated to cause 40% of clinical drug development failures (Lappin and Garner, 2003). Microdosing combines PET with accelerator mass spectrometry to investigate human metabolism by providing both pharmacodynamic and pharmakokinetic information of minute amounts of drugs. It permits safe human studies of experimental drugs in the very early development process, thus reducing sunk costs associated with later-stage clinical trail failure as well as limiting the need for animal experiments (Bergstrom et al., 2003).

PET has been instrumental in defining the biodistribution and concentration range of antibiotics such as trovofloxacin in human volunteers. It was determined that concentrations sufficient to kill most members of enterobacteriacaeae and anaerobes were achieved in all organs, including the brain (Babich et al., 1996; Fischman et al., 1997c, 1998). PET also assists in optimizing formula or administration route of existing drugs. Two studies were carried out to evaluate drug distribution in the gut and in the lung. Producing a true tracer situation is important to these studies. In the analysis of modified-release formulations, drugs such as diltiazem are formulated with small amounts of ^{152}Sm samarium oxide. The tablets can then be activated by neutron activation to produce radioactive ^{153}Sm and can be followed in vivo using planar imaging. This phase I study allowed quantitative distribution of the tablet in the gastrointestinal tract (Maziere et al., 1992) In another example, the lung distribution of the corticosteroid triamcinolone acetonide was studied using the dispenser, Azmacort, a pressurized aerosol metered-dose inhaler formulation. The steroid was radiolabeled with ^{11}C and introduced into the dispenser. The time–activity curve obtained using PET showed a significant increase in the steroid in the lung and a significant decrease in the mouth. These data were submitted to the U.S. Food and Drug Administration as supportive evidence of the Azmacort inhaler's superiority (Berridge and Heald, 1999). The pharmacokinetic profiles of BCNU, an anticancer compound, was compared after intravenous administration and intraarterial administration using [^{11}C]BCNU in patients with recurrent gliomas. Using intraarterial administration, BCNU levels in tumors average 50-fold greater than comparable intravenous administration. The authors suggest that within this small group of patients ($N = 10$), the degree of metabolic trapping of BCNU in tumors correlated with the clinical response to this agent (Tyler et al., 1986).

It is also possible to determine the extent of interaction of competing drugs on the pharmacokinetics and tissue concentration of the DOI. In this case, the DOI is radiolabeled and the organ distribution and kinetics are determined before and after the administration of a nonradioactive drug with a particular pharmacological profile. In this manner it may be possible to determine the receptor affinity of a novel neuroleptic (DOI) in the presence of a well-characterized drug having similar pharmacology.

Pharmacodynamic Evaluation of Biological Parameters to Select Dose and Assess Therapy One of the most important and difficult steps in the drug development process is defining the dose–response relationship. Much can be learned about a candidate drug by measuring its effect on physiological parameters and biochemical processes. For example, several fluoroquinolone antibiotics have been evaluated as to their ability to alter cerebral blood flow, glucose metabolism, and oxygen consumption (Bednarczyk et al., 1990; Green et al., 1991). Other studies aim at establishing in vivo a relationship between plasma drug concentration and drug–target interactions in distinct organs such as the brain. If the new drug acts by binding specifically to a cell-surface or nuclear receptor, PET determines receptor binding parameters (affinity and capacity) in vivo either preclinically in animals or in early-stage trials in humans in binding-site experiments, whereby high-specific-activity radiolabeled drug is administered with increasing amounts of unlabeled drug (Morris et al., 1996). These data should be useful in the determination of dosing, particularly for neuroleptic drugs. If maximal therapeutic effect can be demonstrated to occur at a known level of receptor occupancy and that degree of occupancy can be related to a given dose level, increasing doses will probably result in increased incidence of side effects without further therapeutic benefit. Similarly, Aprepitant, a highly selective substance P antagonist, is an experimental drug designed to ameliorate chemotherapy-induced nausea and emesis. Its efficacy is dictated by its ability to bind to the NK(1) receptor in the brain. PET has been instrumental in a phase I study to assess NK(1) brain occupancy as a function of aprepitant dose in healthy volunteers; these data now allow to predict NK(1) occupancy from plasma drug concentrations and can be used to guide dose selection for clinical trials of NK(1) receptor antagonists in central therapeutic indications (Bergstrom et al., 2004).

The putative antipsychotic drug M100907 was studied indirectly using ^{11}C-labeled spiperone, which binds to both the dopamine D_2 receptor and the 5-HT_{2A} serotonin receptor (Offord et al., 1999). The therapeutic index of M100907 was defined in phase I single- and multiple-dose tolerability studies. PET was then used to confirm the mechanism of action of M100907 in humans. The resulting data also helped investigators to define an appropriate dose range and regimen for subsequent clinical studies in schizophrenia patients. (Offord et al., 1999). ^{11}C-labeled raclopride, a dopamine receptor antagonist, has been used to measure increases in the neurotransmitter dopamine indirectly as a function of the pharmacologic action of an amphetamine-like potential drug candidate under investigation for schizophrenia, building on the insight that elevated synaptic dopamine concentrations are associated with this disease (Breier et al., 1997).

PET has long enjoyed a pivotal position in the evaluation of tumor response to chemotherapy and hence patient management. This includes, foremost, the use of FDG to quantify tumor metabolism as an indicator of tumor response to treatment with chemotherapeutic agents (Eckelman et al., 2000; Lammertsma et al., 2001). In addition, biochemical indicators of cellular proliferation, such as the incorporation of thymidine in actively dividing cells, can be exploited to investigate in vivo in humans the effect of investigational cancer drugs designed to stop the proliferation of tumor cells and to correlate the biological readout with in vivo plasma concentration and biodistribution data (Wells et al., 2003). Similarly, anticancer drugs have been proposed to exert their effect by causing programmed cell death of tumor cells. The molecular marker of apoptosis Annexin V has been instrumental in quantifying in cancer patients undergoing chemotherapy the extent of tumor cell apoptosis. The investigators were further able, despite the small number of study patients, to relate overall survival and progression free survival to the uptake of the apoptosis tracer, suggesting that the radiolabeled apoptosis marker Annexin may be used as a surrogate marker to assess therapeutic efficacy both for approved drugs in the clinical management of cancer patients and in early clinical trials to monitor the efficacy of novel investigational cancer compounds (Belhocine et al., 2002). In this context it is of note that cancer drug development programs are specifically burdened by long observation times required before clinical efficacy is satisfactorily assessed and higher failure rate in the more expensive late-stage clinical studies.

Similarly, the experimental cancer drug Comprestatin A4 phosphate is designed to disrupt tumor blood supply by binding to endothelial cell tubulin, thereby causing morphological changes of endothelials cells that lead to their disruption. This mechanism of action has been confirmed in animal studies. PET was instrumental in visualizing this effect in patients as part of a phase I clinical trial and to assess cooperation and synergetic effects of Comprestatin A4 phosphate when combined with cisplatin and to aid dose selection for a subsequent phase II trial (West and Price, 2004). In a phase II study with the experimental drug razoxane, also designed to interfere with tumor vascularization, PET-guided visualization of the vascular physiology using $[^{15}O]H_2$ and $[^{15}O]C$ as tracers monitored over time the effect of razoxane on renal tumor perfusion compared to normal tissue perfusion. PET provided valuable information on the in vivo biology of angiogenesis in patients and was instrumental in assessing the effects of antiangiogenic therapy (Anderson et al., 2003).

14.4 SUMMARY AND OUTLOOK

The challenges for the pharmaceutical industry in the drug discovery and development process range from the evaluation of potential new drug candidates, the determination of drug pharmacokinetics and pharmacodynamics, the measurement of receptor occupancy as a determinant of drug efficacy, and the pharmacological characterization of mechanisms of action. Among the tools with significant potential to reduce overall costs and improve the reliable identification of promising new compounds is PET. Often, the mechanism of action of new compounds is well defined based on an array of extracellular and in vitro studies. Early recognition that the proposed mechanism of action also holds true in an in vivo scenario such as a wild-type animal or an animal model representing the target disease is of significant value for the drug-development process. Although PET may appear to be expensive, by providing the basis for early go/no-go decisions, it can actually be cost-effective. With the help of PET, the drug candidate is being tested in an intact species. Imaging is especially important in paradigms where animals are studied before and after treatment with the drug candidate. Because paired statistics can be carried out in the same animal, fewer animals are needed. Therefore, in vivo imaging has an advantage over the autoradiographic approaches that are carried out in vivo but require the sacrifice of the animal after each study.

Clinical-stage development programs often could be accelerated if reliable and predictive surrogate markers were available that could be monitored noninvasively. This applies, for example, to cancer trials, where PET could establish almost instantaneous whether a novel compound is capable of affecting tumor cell proliferation. For example, FDG has been instrumental in quantifying tumor response to chemotherapy (Lammertsma, 2001). Fast, efficient, and cost-effective use of PET in the drug discovery and development process relies on the availability of well-characterized binding sites and suitable radioligands. If these are not readily obtainable, well-established radiopharmaceuticals such as O-15 water for blood flow and F-18 FDG for glucose metabolism or well-established radiolabeled ligands that can measure the effect of drugs indirectly may provide valuable alternatives.

Before PET can be firmly integrated in the clinical evaluation of experimental drugs in multicenter international trials, standards defining the visual and analytical methods to be used for quantification and reporting of PET data need to be established. Both the European Organization for Research and Treatment of Cancer and the National Cancer Institute in the United States have made recommendations along these lines for the use of PET in assessing tumor response to chemotherapy (Eckelman et al., 2000). In phase I and II human studies, classic PK measurements can then be coupled with imaging measurements to define an optimal dosing schedule, help formulate the design of phase III studies, and thus contribute to improving their success rates. It is to be expected that pharmacogenomics will identify new surrogate markers for therapy monitoring which may represent potential new tracers for imaging suitable to accelerate and improve the clinical drug development process and ultimately to guide treatment decisions.

REFERENCES

Anderson, H., Yap, J. T., Wells, P., Miller, M. P., Propper, D., Price, P., Harris, A. L. Measurement of renal tumour and normal tissue perfusion using positron emission tomography in a phase II clinical trail of razoxane. *Br. J. Cancer*, July 21, 2003, 89(2):262–267.

Babich, J. W., Rubin, R. H., Graham, W. A., Wilkinson, R. A., Vincent, J., Fishman, A. J. 18F-labeling and biodistribution of the novel fluoro-quinolone antimicrobial agent, trovafloxacin (CP 99,219). *Nucl. Med. Biol.*, Nov. 1996, 23(8):995–998.

Barbarich, N. C. Kaye, W. H., Jimerson, D. Neurotransmitter and imaging studies in anorexia nervosa: new targets for treatment. *Curr. Drug Target CNS Neurol. Disord.* Feb. 2003, 2(1):61–72.

Baron, J. C., Roeda, D., Munari, C., Crouzel, C., Chodkiewicz, J. P., Comar, D. Brain regional pharmacokinetics of ^{11}C-labeled diphenylhydantoin: positron emission tomography in humans. *Neurology* (Cleveland), 1983, 33:580–585.

Bednarczyk, E. M., Green, J. A., Nelson, A. D., et al. Comparison of the effect of temafloxacin, ciprofloxacin, or placebo on cerebral blood flow, glucose and oxygen metabolism in healthy subjects by means of positron emission tomography. *Clin. Pharmacol. Ther.*, 1990, 50:165–171.

Belhocine, T., Steinmetz, N., Hustinx, R., Bartsch, P., Jerusalem, G., Seidel, L., Rigo, P., Green, A. Increased uptake of the apoptosis-imaging agent (99m)Tc recombinant human Annexin V in human tumors after one course of chemotherapy as a predictor of tumor response and patient prognosis. *Clin. Cancer Res.*, Sept. 2002, 8(9):2766–2774.

Bench, C. J., Lammertsma, A. A. Dolan, R. J. Grasby, P. M., Warrington, S. J., Gunn, K., et al., Dose dependent occupancy of central dopamine D2 receptors by the novel neuroleptic CP-88,059-01: a study using positron emission tomography and ^{11}C-raclopride. *Psychopharmacology*, 1993, 112(2–3):308–314.

Ben-David, I., Rozen, Y., Ortu, G., Mishani, E. Radiosynthesis of ML03, a novel positron emission tomography biomarker for targeting epidermal growth factor receptor via the labelling synthon: [^{11}C]acryloyl chloride. *Appl. Radiat. Isot.*, Feb. 2003, 58(2):209–217.

Berger, F., Gambhir, S. S. Recent advances in imaging endogenous or transferred gene expression utilizing redionuclide technologies in living subjects: applications to breast cancer. *Breast Cancer Res.*, 2001, 3(1):28–35.

Bergstrom, M., Grahnen, A., Langstrom, B. Positron emission tomography microdosing: a new concept with application in tracer and early clinical drug development. *Eur. J. Clin. Pharmacol.*, Sept. 2003, 59(5–6):357–366.

Bergstrom, M., Hargreaves, R. J., Burns, H. D., Goldberg, M. R., Sciberras, D., Reines, S. A., Petty, K. J., Orgen, M., Antoni, G., Langstrom, B., Eskola, O., Schenin, M., Solin, O., Majumdar, A. K., Constanzer, M. L., Battisti, W. P., Bradstreet, T. E., Gargano, C., Hietala, J. Human positron emission tomography studies of brain neurokinin 1 receptor occupancy by aprepitant. *Biol. Psychiatry*, May 15, 2004, 55(10):1007–1012.

Berridge, M. S., Heald, D. L. In vivo characterization of inhaled pharmaceuticals using quantitative positron emission tomography. *J. Clin. Pharmacol.*, 1999, Suppl., 25S–29S.

Bottlaender, M., Dolle, F., Guenther, I., Roumenov, D., Fuseau, C., Bramoulle, Y., Curet, O., Jegham, J., Pinquier, J. L., George, P., Valette, H. Mapping the cerebral monoamine oxidase type A: positron emission tomography characterization of the reversible selective inhibitor [^{11}C]befloxatone. *J. Pharmacol. Exp. Ther.*, May 2003, 305(2):467–473.

Breier, A., Su, T.-P., Saunders, R., et al. Schizophrenia is associated with elevated amphetamine induced synaptic dopamine concentrations: evidence from a novel positron emission tomography method. *Proc. Natl. Acad. Sci. USA*, 1997, 94, 2569–2574.

Cagnin, A., Brooks, D. J. Kennedy, A. M., Gunn, R. N., Myers, R., Turkheimer, F. E., Jones, T., Banati, R. B. In vivo measurement of activated microglia in dementia. *Lancet*, Aug. 11 2001, 358(9280):461–467.

Cherry, S. R. Fundamentals of positron emission tomography and applications in preclinical drug development. *J. Clin. Pharmacol.*, May 2001, 41(5):482–491.

Cherry, S. R., Shao, Y., Silverman, R. W., Meadors, K., Siegel, S., Chatziioannou, A., Yound, J. W., Jones, W. F., Moyers, J. C., Newport, D., Boutefnouchel, A., Farquhar, T. H. Andreaco, M., Pulus, M. J., Binkley, D. M., Nutt, R., Phelps, M. E. Micro PET: a high resolution PET scanner for imaging small animals. *IEEE Trans. Nucl. Sci.*, 1997, 44:1161–1166.

Collingridge, D. R., Carroll, V. A., Glaser, M., Aboagye, E. O., Osman, S., Hutchinson, O. C., Barthel, H., Luthra, S. K., Brady, F., Bicknell, R., Price, P., Harris, A. L. The development of [(124)I]iodinated-VG76e: a novel tracer for imaging vasucular endothelial growth factor in vivo using positron emission tomography. *Cancer Res.*, Oct. 15, 2002, 62(20):5912–5919.

DaSilva, J. N., Lourenco, C. M., Meyer, J. H., Hussey, D., Potter, W. Z., Houle, S. Imaging cAMP-specific phosphodiesterase-4 in human brain with R-[^{11}C]rolipram and positron emission tomography. *Eur. J. Nucl. Med. Mol. Imaging*, Dec. 2002, 29(12):1680–1683.

Dehdashti, F., Flanagam, F. L., Mortimer, J. E., Katzenellenbogen, J. A., Welch, M. J., Siegel, B. A. Positron emission tomographic assessment of "metabolic flare" to predict response of metastatic breast cancer to antiestrogen therapy. *Eur. J. Nucl. Med.*, Jan. 1999, 26(1):51–56.

Dickson, M., Gagnon, J. P. Key factors in the rising cost of new drug discovery and development. *Nature Drug Discov.*, May 2004.

Doubrovin, M., Ponomarev, V., Beresten, T., Balatoni, J., Bornmann, W., Finn, R., Humm, J., Larson, S., Sadelain, M., Blasberg, R., Gelovani, Tjuvajev, J. Imaging transcriptional regulation of p53-dependent genes with positron emission tomography in vivo. *Proc. Natl. Acad. Sci. USA.*, July 31, 2001, 98(16):9300–9305.

Eckelman, W. C. Labeled agents as tracers and carriers: theory and practice. In Feinendegen, L. E., Shreeve, W. W., Eckelman, W. C., Bahk, Y. W., Wagner, H. N., Jr; (Eds.), *Molecular Nuclear Medicine: The Challenges of Genomics and Proteomics to Clinical Practice*. Springer-Verlag, New York, 2003, pp. 85–98.

Eckleman, W. C. Tatum, J. L., Kurdziel, K. A., et al. Quantitative analysis of tumor biochemistry using PET and SPECT(1). *Nucl. Med. Biol.*, 2000, 27, 633–691.

Fischman, A. J., Babich, J. W., Alpert, N. M., Vincent, J., Wilkinson, R. A., Callahan, R. J., Correia, J. A., Rubin, R. H. Pharmacokinetics of ^{18}F-labeled trovafloxacin in normal and *Escherichia coli*–infected rats and rabbits studied with positron emission tomography. *Clin. Microbiol. Infect.*, June 1997a, 3(3):379.

Fischman, A. J., Babich, J. W., Alpert, N. M., Vincent, J., Wilkinson, R. A., Callahan, R. J., Correia, J. A., Rubin, R. H. Pharmacokinetics of ^{18}F-labeled trovafloxacin in normal and *Escherichia coli*–infected rats and rabbits studied with positron emission tomography. *Clin. Microbiol. Infect.*, Feb. 1997b, 3(1):63–72.

Fischman, A. J., Alpert, N. M., Babich, J. W., Rubin, R. H. The role of positron emission tomography in pharmacokinetic analysis. *Drug Metab. Rev.*, Nov 1997c, 29(4):923–956.

Fischman, A. J., Babich, J. W., Bonab, A. A., Alpert, N. M., Vincent, J., Callahan, R. J., Correia, J. A., Rubin, R. H. Pharmacokinetics of [^{18}F]trovafloxacin in healthy human subjects studied with positron emission tomography. *Antimicrob. Agents Chemother.*, Aug 1998, 42(8):2048–2054.

Green, J. A., Bednarczyk, E. M., Miraldi, F., et al. Reduction in cerebral glucose metabolism caused by quinoline administration in healthy subjects. *Nucl. Med. Biol. Int. J. Appl. Radiat. Isot. P. B*, 1991, 19:126–129.

Groot-Wassink, T., Aboagye, E. O., Wang, Y., Lemoine, N. R., Reader, A. J., Vassaux, G. Quantitative imaging of Na/I symporter transgene expression using positron emission tomography in the living animal. *Mol. Ther*, Mar. 2004, 9(3):436–442.

Haberkorn, U., Altmann, A., mier, W., Eisenhut, M. Impact of functional genomics and proteomics on radionuclide imaging. *Semin. Nucl. Med.*, Jan. 204, 34(1):4–22.

Halldin, C., Gulyas, B., Farde, L. PET studies with carbon-11 radioligands in neuropsychopharmacological drug development. *Curr. Pharm. Des.*, Dec. 2001, 7(18):1907–1929.

Herbst, R. S., Mullani, N. A., Davis, D. W., Hess, K. R., McConkey, D. J., Charnsangavej, C., O'Reilly, M. S., Kim, H. W., Baker, C., Roach, J., Ellis, L. M., Rashid, A., Pluda, J., Bucana, C., Madden, T. L., Tran, H. T., Abbruzzese, J. L. Development of biologic markers of response and assessment of antiangiogenic activity in a clinical trial of human recombinant endostatin. *J. Clin. Oncol.*, Sept. 15, 2002, 20(18):3804–3814.

Hudson, A. L., Tyacke, R. J., Lalies, M. D., Davies, N., Finn, D. P., Marti, O., Robinson, E., Husbands, S., Minchin, M. C., Kimura, A., Nutt, D. J. Novel ligands for the investigation of imidazoline receptors and their binding proteins. *Ann. N. Y. Acad.*, Dec. 2003, 1009:302–308.

Imahori, Y., Fujji, R., Tujino, H., Kimura, M., Mineura, K. Postitron emission tomography: measurement of the activity of second messenger systems. *Methods*, July 2002, 27(3):251–262.

Iyer, M., Wu, L., Carey, M., Wang, Y., Smallwood, A., Gambhir, S. S. Two-step transcriptional amplification as a method for imaging reporter gene expression using weak promoters. *Proc. Natl. Acad. Sci. USA*, Dec. 4 2001, 98(25):14595–14600.

Jagoda, E. M., Kiesewetter, D. O., Shimoji, K., Ravasi, L., Yamada, M., Gomeza, J., Wess, J., Eckelman, W. C. Regional brain uptake of the muscarinic ligand, [^{18}F]FP-TZTP, is greatly decreased in M2 receptor knockout mice but not in M1, M3 and M4 receptor knockout mice. *Neuropharmacology*, Apr. 2003, 44(5):653–661.

Koehne, G., Doubrovin, M., Doubrovina, E., Zanzonico, P., Gallardo, H. F., Ivanova, A., Balatoni, J., Teruya-Feldstein, J., Heller, G., May, C., Ponomarev, V., Ruan, S., Finn, R., Blasberg, R. G. Bornmann, W., Rivierre, I., Sadelain, M., O'Reilly, R. J., Larson, S. M., Tjuvajev, J. G. Serial in vivo imaging of the targeted migration of human HSV-TK-transduced antigen-specific lymphocytes *Nat. Biotechnol.*, Apr. 2003, 21(4):405–413.

Koike, C., Oku, N., Watanabe, M., Tsukada, H., Kakiuchi, T., Irimura, T., Okada, S. Real-time PET analysis of metastatic tumor cell trafficking in vivo and its relation to adhesion properties. *Biochim. Biophys. Acta*, Sept. 13, 1995, 1238(2):99–106.

Lammertsma, A. A. Measurement of tumor response using [^{18}F]-2-fluoro-2-deoxy-D-glucose and positron-emission tomography, *J. Clin. Pharmacol.*, 2001, Suppl., 104S–106S.

Lappin, G., Garner, R. C. Big physics, small doses: the use of AMS and PET in human microdosing of development drugs. *Nat. Rev. Drug Discov.*, Mar. 2003, 2(3):233–240.

Lawrence, S. Priming the pipeline, *Acumen* Vol. II, p. 20, 2004.

Lehmann Brothers and McKinsey & Co., *The Fruits of Genomics*, Jan. 2001.

Luker, G. D., Sharma, V., Pica, C. M., Dahlheimer, J. L., Li, W., Ochesky, J., Ryan, C. E., Piwnica-Worms, H., Piwnica-Worms, D. Noninvasive imaging of protein–protein interactions in living animals. *Proc. Natl. Acad. Sci. USA*, May 14, 2002, 99(10):6961–6966.

Luker, G. D., Sharma, V., Piwnica-Worms, D. Visualizing protein–interactions in living animals. *Methods*. Jan 2003, 29(1):110–122.

Marrosu, F., Serra, A., Maleci, A., Puligheddu, M., Biggio, G., Piga, M. Correlation between GABA(A) receptor density and vagus nerve stimulation in individuals with drug-resistant partial epilepsy. *Epilepsy Res.*, June–July, 2003, 55(1–2):59–70.

Mayer-Kuckuk, P., Doubrovin, M., Gusani, N. J., Gade, T., Balatoni, J., Akhurst, T., Finn, R., Fong, Y., Koutcher, J. A., Larson, S., Blasberg, R., Tjuvajev, J. G., Bertino, J. R., Banerjee, D. Imaging of dihydrofolate reductase fusion gene expression in xenografts of human liver metastases of colorectal cancer in living rats. *Eur. J. Nucl. Med. Mol. Imaging, Sept.*, 2003, 30(9):1281–1291.

Maziere, B., Coenen, H. H., Halldin, C., Nagren, K., Pike V. W. PET radioligands for dopamine receptors and re-uptake sites: chemistry and biochemistry. *Int. J. Radiat. Appl. Instrum. P. B Nucl. Med. Biol.*, 1992, 19(4):497–512.

Meguro, K., Meguro, M., Tanaka, Y., Akanuma, K., Yamaguchi, K., Itoh, M. Risperidone is effective for wandering and disturbed sleep/wake patterns in Alzheimer's disease. *J. Geriatr. Psychiatry Neurol.*, June 2004, 17(2):61–67.

Mitterhauser, M., Wadsak, W., Wabnegger L., Sieghart, W., Viernstein, H., Kletter, K., Dudczak, R. In vivo and in vitro evaluation of [^{18}F]FETO with respect to the adrenocortical and GABAergic system in rats. *Eur. J. Nucl. Med. Mol. Imaging*, Oct. 2003, 30(10)1398–1401.

Morris, E. D., Babich, J. W., Alpert, N. M., Bonab, A. A., Livni, E., Weise, S., et al. Quantification of dopamine transporter density in monkeys by dynamic PET imaging of multiple injections of 11C-CFT. *Synapse*, 1996, 24(3):262–272.

Myers, R., Hume, S. Small animal PET. *Eur. Neuropsychopharmacol.*, Dec. 2002, 12(6): 545–555.

Offord, S. J., Wong, D. F., Nyberg, S. The role of positron emission tomography in the drug development of M100907, a putative antipsychotic with a novel mechanism of action. *J. Clin. Pharmacol.*, 1999, Suppl., 17S–24S.

Osman, S., Rowlinson-Busza, G., Luthra, S. K., Aboagye, E. O., Brown, G. D., Brady, F., Myers, R., Gamage, S. A., Denny, W. A., Baguley, B. C., Price, P. M. Comparative biodistribution and metabolism of carbon-11-labeled *N*-[2-(dimethylamino)ethyl]acridine-4-carboxamide and DNA-intercalating analogues. *Cancer Res.* Apr. 1, 2001, 61(7):2935–2944.

Pantuck, A. J., Berger, F., Zisman, A., Nguyen, D., Tso, C. L., Matherly, J., Gambhir, S. S., Belldegrun, A. S. CL1-SR39: a noninvasive molecular imaging model of prostate cancer suicide gene therapy using positron emission tomography. *J. Urol.*, Sept. 2002, 168(3):1193–1198.

Piwnica-Worms, D., Schuster, D. P., Garbow, J. R., Molecular imaging of host–pathogen interactions in intact small animals. *Cell Microbiol.*, Apr. 2004, 6(4):319–331.

Ponomarev, V., Doubrovin, M., Lyddane, C., Beresten, T., Balatoni, J., Bornman, W., Finn, R., Akhurst, T., Larson, S., Blasberg, R., Sadelain, M., Tjuvajev, J. G. Imaging TCR-dependent NFAT-mediated T-cell activation with positron emission tomography in vivo. *Neoplasia*, Nov.–Dec. 2001, 3(6):480–488.

Qi, J., Leahy, R. M., Cherry, S. R., Chatziioannou, A., Farquhar, T. H. High-resolution 3D Bayesian image reconstruction using the microPET small-animal scanner. *Phys. Med. Biol.*, Apr. 1998, 43(4):1001–1013.

Salazar, D. E., Fischman, A. J. Central nervous system pharmacokinetics of psychiatric drugs. *J. Clin. Pharmacol.*, Aug. 1999, Suppl., 10S–12S.

Shinotoh, H., Fukushi, K., Nagatsuka, S., Irie, T. Acetylcholinesterase imaging: its use in therapy evaluation and drug design. *Curr. Pharm. Des.*, 2004, 10(13):1505–1517.

Smith, G. S., Koppel, J., Goldberg, S. Applications of neuroreceptor imaging to psychiatry research. *Psychopharmacol. Bull.*, Autumn, 2003, 37(4):26–65.

Smith-Jones, P. M., Solit, D. B., Akhurst, T., Afroze, F., Rosen, N., Larson, S. M. Imaging the pharmacodynamics of HER2 degradation in response to Hsp90 inhibitors. *Nat. Biotechnol.*, May 9, 2004.

Solomon, B., McArthur, G., Cullinane, C., Zalcberg, J., Hicks, R. Applications of positron emission tomography in the development of molecular targeted cancer therapeutics. *BioDrugs*, 2003, 17(5):339–354.

Tashiro, M., Yanai, K. A potential of positron emission tomography in the drug development of non-sedative antihistamines. *Nippon Yakurigaku Zasshi*, Nov. 2003, 122, Suppl., 78P–80P.

Tsukada, H., Harada, N., Ohba, H., Nishiyama, S., Kakiuchi, T. Facilitation of dopaminergic neural transmission doest not affect [(11)C]SCH23390 binding to the striatal D(1) dopamine receptors, but the facilitation enhances phosphodiesterase type-IV activity through D(1) receptors: PET studies in the conscious monkey brain. *Synapse*, Dec. 15, 2001, 42(4):258–265.

Tyler, J. L., Yamamoto, Y. L., Diksic, M., Theron, J., Villemure, J. G., Worthington, C., et al. Pharmacokinetics of superselective intra-arterial and intravenous [^{11}C]BCNU evaluated by PET. *J. Nucl. Med.*, 1986, 27:775–780.

Wells, P., Aboagye, E., Gunn, R. N., Osman, S., Boddy, A. V., Taylor, G. A., Rafi, I., Hughes, A. N., Calvert, A. H., Price, P. M., Newell, D. R. 2-[^{11}C]thymidine positron emission tomography as an indicator of thymidylate synthase inhibition in patients treated with AG337. *J. Natl. Cancer Inst.*, May 7, 2003, 95(9):675–682.

West, C. M., Price, P. Combretastatin A4 phosphate. *Anticancer Drugs*, Mar., 2004, 15(3): 179–187.

Wilms, H., Claasen, J., Rohl, C., Sievers, J., Deuschl, G., Lucius, R. Involvement of benzodiazepine receptors in neuroinflammatory and neurodegenerative diseases: evidence from activated microglial cell in vitro. *Neurobiol. Dis.*, Dec. 2003, 14(3):417–424.

Zheng, Q. H., Liu, X., Fei, X., Wang, J. Q., Ohannesian, D. W., Erickson, L. C., Stone, K. L., Hutchins, G. D. Synthesis and preliminary biological evaluation of radiolabeled O6-benzylguanine derivatives, new potential PET imaging agents for the DNA repair protein O6-alkylguanine-DNA alkyltransferase in breast cancer. *Nucl. Med. Biol.*, May 2003, 30(4):405–415.

15

POLYMERIC SEQUESTRANTS AS NONABSORBED HUMAN THERAPEUTICS

PRADEEP K. DHAL, CHAD C. HUVAL, AND S. RANDALL HOLMES-FARLEY
Genzyme Corporation
Waltham, Massachusetts

15.1 INTRODUCTION

The use of functional polymers in medicine has seen considerable growth during the past two decades.[1] Polymers as biomaterials have found applications in such areas as artificial organs, tissue engineering, components of medical devices, and dentistry. A growing aspect of the field is the recognition of polymers as useful therapeutic agents: polymers that either exhibit pharmacological properties themselves or that can be utilized as carriers for selective and sustained delivery vehicles for small molecule or macromolecular (e.g., proteins, genetic materials) pharmaceutical agents.

Over the past several years there has been a growing scientific interest in the use polymers as delivery agents for drug molecules. The goal of such research is to deliver drugs at a sustained rate, deliver drugs targeted at specific sites (to minimize toxicity and enhance selectivity for, e.g., certain antitumor agents), and to deliver macromolecular prodrugs with polymers acting as carrier molecules.[2–4] More recently, utility of polymers as nonviral vectors for the delivery of genetic materials for gene therapy has also been evaluated.[5] Significant advancements in the area of polymeric drug delivery systems (including commercial products) have taken place in recent years. Several research papers and review articles pertaining to the use of biomedical polymers for drug delivery have been published over the last 20 years.[6,7]

Although polymers are used extensively as drug delivery agents, *intrinsically* bioactive polymers (polymers as active pharmaceutical ingredients) are a relatively recent development.[8] Due in part to their high molecular weight, polymers would appear to offer several advantages over low-molecular-weight agents as potential therapeutic agents. These

Drug Discovery and Development, Volume 1: Drug Discovery, Edited by Mukund S. Chorghade

benefits may include lower toxicity, greater specificity of action, and enhanced activity due to multiple interactions (polyvalency). Nevertheless, the concept of polymeric drugs has not received widespread acceptance. Medicinal chemists have long considered synthetic polymers as an uninteresting class of compounds to be evaluated as potential pharmacophores. Some of the underlying concerns include the issue of undefined molecular weight, polydispersity, and compositional heterogeneity (copolymers in particular), since such properties could complicate drug development process. The high-molecular-weight characteristics of polymers would impede their systemic absorption through oral route as well as complicate subsequent ADME (absorption, distribution, metabolism, and elimination) profile. However, as we describe in this chapter, these potential shortcomings associated with pharmacological characteristics of polymers can be carefully exploited to design and develop therapeutic agents for disease conditions where low-molecular-weight drugs have either failed or produced inadequate therapeutic benefits. Although polymers do not exhibit properties that fit most druglike definitions and frequently violate Lipinski's rules, a reevaluation of the attributes of polymers highlights a number of benefits of polymeric drugs that are not achieved with traditional small molecule drugs.[9]

The fact that high-molecular-weight polymers are not generally absorbed from the gastrointestinal tract may be of particular advantage where it is desirable to prohibit a drug molecule from systemic exposure. For example, by combining this characteristic with the ability of polymers to selectively bind molecular and macromolecular species in gastrointestinal fluids, it has become possible to develop a new class of therapeutic agents that can selectively bind and remove detrimental species from the gastrointestinal (GI) tract. As with any pharmaceutical, sequestration of each target molecule or pathogen requires a unique strategy, and this strategy depends not only on the chemical nature of the target, but on the location, concentration, and quantity of the target that needs to be removed.

Since several books and review articles have been published covering the area of polymeric drug *delivery* systems,[2–7] this chapter is limited to case studies where polymers act as active drugs. We focus on some of the most recent efforts in this regard and concentrate on the development of polymeric drugs for the sequestration of low-molecular-weight species such as bile acids, phosphate, and iron as well as polyvalent interactions to bind toxins, viruses, and bacteria

15.2 POLYMERS AS SPECIFIC MOLECULAR SEQUESTRANTS

A number of potentially detrimental substances are present in the gastrointestinal (GI) tract. These molecules and pathogens are implicated for a number of disease conditions. These species can be either exogenic (i.e., they enter the body with food, drinks, and/or from the environment) or they may be produced endogenously as a result of body metabolism. Effective removal of these species in a selective manner offers a promising prophylactic and therapeutic approach for the treatment of a number of disease states. Biocompatible polymeric sequestrants represent an ideal class of agents for this purpose. Their nonabsorption through the intestinal wall should offer minimal toxicity, while the incorporation of appropriate functional groups and manipulation of polymer physicochemical characteristics provides opportunities to tailormake polymers with high selectivity and capacity. In the following sections we describe specific examples of polymeric drugs that have been developed during the last few years.

15.3 SEQUESTRATION OF INORGANIC IONS IN THE GI TRACT

Electrolytes play a key role in regulating the distribution of different inorganic anions and cations between the intracellular and extracellular fluid compartments (homeostasis) and are vital in maintaining myocardial and neurological functions, fluid balance, oxygen delivery, and acid–base balance. Electrolyte imbalance, caused by either excessive ingestion or impaired elimination of an electrolyte from the body, has important physiological effects. Although certain nonrenal tissues (e.g., muscle and liver) contribute to maintaining electrolyte balance, the kidney plays the predominant role in maintaining electrolyte balance.[10] Therefore, renal impairment is the most common reason for electrolyte imbalance and can lead to critical health conditions. Conceptually, nonabsorbed polymeric sequestrants could be used to bind these excessive ions (implicated in various pathologic conditions) selectively in the GI tract.

15.4 POLYMERIC POTASSIUM SEQUESTRANTS: A NONABSORBED THERAPY FOR HYPERKALEMIA

Since the transmembrane potential is a major regulator of cell function, the ratio of potassium between intra- and extracellular fluids is critically important to all living cells.[11] Hyperkalemia (elevated level of serum potassium, usually greater than 5.0 mEq/L) can result from burn and crush muscle injuries, acidosis, or through the use of antihypertension drugs (angiotensin-converting enzyme inhibitors). A rise in serum potassium can manifest moderate to serious health problems, including paresthesias, areflexia, respiratory failure, and bradycardia.[12] Since the kidney is responsible for elimination of most excess potassium, patients with impaired renal function are incapable of maintaining potassium homeostasis. Traditional approaches to treat hyperkalemia include use of insulin, glucose, sodium bicarbonate, and calcium chloride. However, these treatments have their own shortcomings. For example, excess calcium could lead to hypercalcemia, which in turn leads to myocardial infarction, kidney stones, and a variety of other conditions.

Use of an insoluble polyanionic resin to sequester excess potassium in the GI tract with elimination in the feces could enable patients to excrete potassium despite impaired kidneys. There are a number of low-molecular-weight ligands that complex potassium ions.[13] Utilizing this principle, a cation-exchange resin based on sodium polystyrene sulfonate (**1**) (Scheme 15.1) was developed to sequester potassium ion in the GI tract. This polymer,

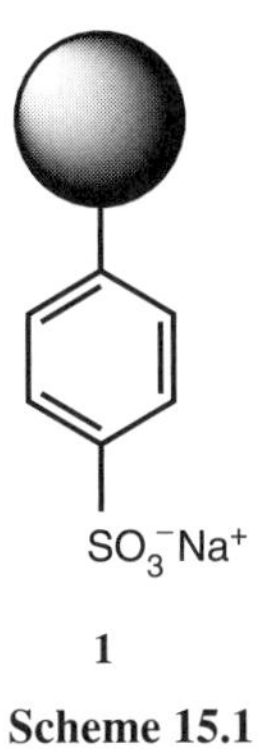

Scheme 15.1

marketed under the brand name Kayexalate[14] and approved in the United States for the treatment of hyperkalemia in 1975, is administered orally or rectally. A potential problem with its use is induction of hypernatremia (elevated serum sodium), since the material exchanges 1.0 eQ potassium for 1.5 eQ sodium as well as incidences of intestinal necrosis.[15] These adverse effects resulting from Kayexalate treatment may provide an opportunity to discover new generations of potassium sequestering polymers.

15.5 POLYMERIC DRUGS FOR CHRONIC RENAL FAILURE

Management and control of elevated-level serum phosphosphorus (hyperphosphatemia) is a severe issue for patients suffering from chronic and end-stage renal failure.[16,17] The consequences of inadequate control of serum phosphate level leads to various pathologies of clinical significance. They include soft tissue calcification (leading to cardiac calcification and cardiac-related complication), renal osteomalacia leading to reduced bone density, and secondary hyperparathyroidism. These conditions make hyperphosphataemia a major risk (including mortality) among patients suffering from end stage renal disease, such as dialysis patients.

The major route for the excretion of phosphate from healthy human body is through the kidney. Therefore, patients with an impaired renal system exhibit systemic accumulation of phosphate due to an ingestion–excretion imbalance. Phosphate binder therapy has been the mainstay for the treatment of hyperphosphatemia. The traditional phosphate binders have been calcium- and aluminum-based agents that remove phosphate through the formation of insoluble calcium or aluminum phosphate in the GI tract.[18,19] Unfortunately, since aluminum and calcium salts have the propensity for systemic absorption through the linings of the GI tract, they can lead to undesirable toxic and metabolic side effects (e.g., neurological disorders, cardiac calcifications) in patients with impaired renal function. As a result, the side effects associated with these inorganic salt–based phosphate sequestrants limit their long-term use.

Design and development of nonabsorbed cationic polymers as sequestrants for phosphate ions offer an ideal approach to treat hyperphosphatemia in renal failure patients. The known binding of phosphate by polycationic species has long been a well-studied phenomenon. A number of studies have been carried out over the last two decades in designing novel compounds such as macrocyclic oligomeric amines and oligomeric guanidinium compounds (e.g., **2** and **3**, Scheme 15.2) as receptors to study molecular recognition events

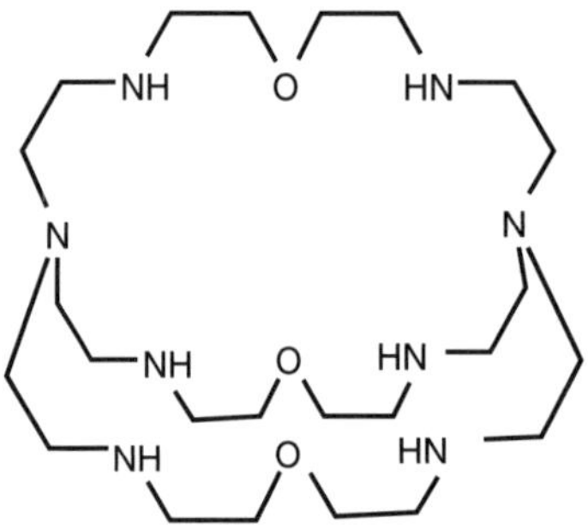

NH O HN N N NH O HN NH O HN

2

NH2 NH HN NH HN NH2

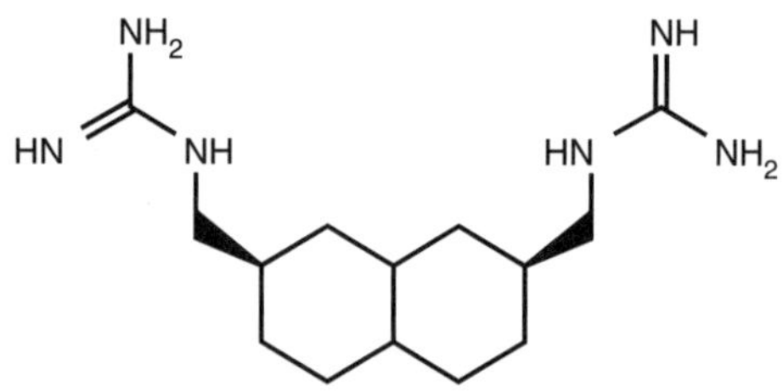

3

Scheme 15.2

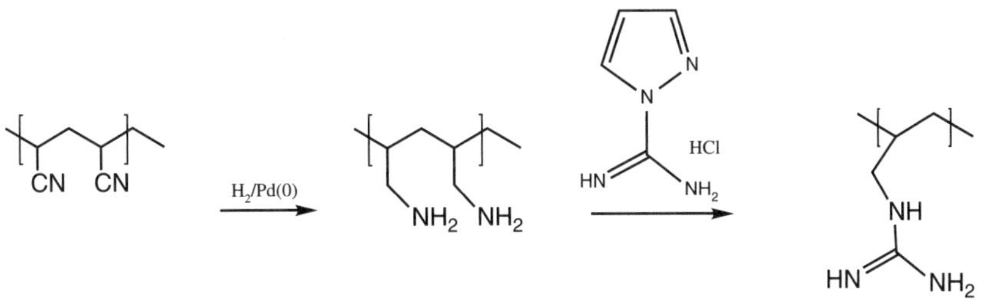

Scheme 15.3 Polymer-bound guanidinium salts obtained by chemical modification of poly(acrylonitrile) resins.

involving phosphate, pyrophosphate, and phosphonate anions as guests.[20,21] Electrostatic interaction is the primary force for this complexation process. Hydrogen bonding is considered to contribute to further binding strength. Utilization of this principle of physical organic chemistry has led to the discovery of a series of nonabsorbed polymeric amine compounds and polymeric guanidinium compounds that show affinity toward dietary phosphate. Being nonabsorbed, these polymeric sequestrants would be confined to the GI tract and thus would act as effective therapeutic agents to treat hyperphosphatemia that are free from the side effects arising from the use of calcium and related metal salts as phosphate binders.

Hider and Rodriguez have discovered a series of cross-linked polymer resins containing guanidine groups as specific sequestrants for phosphate anions.[22,23] Binding of phosphate ions to guanidinium groups of arginine residues of proteins involving two electrostatic bonds and two stereochemically favorable hydrogen bonds is well known in biological systems.[24] On the basis of biological principles, they developed the synthetic procedure to prepare insoluble polymer resins containing guanidinium groups. The chemical synthetic procedure to prepare these polymeric guanidinium salts is presented in Scheme 15.3. In vitro phosphate binding studies involving these polymeric guanidinium salts shows that the polymers bind phosphate more selectively in the presence of other biologically important anions, such as salts of bile acids, chloride, bicarbonate, and so on.

In our laboratories a series of amine containing polymers were evaluated as nonabsorbed phosphate sequestrants. Using the knowledge of phosphate binding properties of macrocyclic polyamines and related hosts (as molecular receptors for anions), a series of functional polymers bearing pendant and backbone amino groups was evaluated as phosphate sequestrants. These amine-functionalized polymer hydrogels were prepared either by cross-linking of amine polymers or by the cross-linking polymerization of amine-containing vinyl monomers.[25,26] Polymers bearing primary, secondary, and tertiary amine groups as well as quaternary ammonium groups were evaluated for this purpose. The general method for syntheses of polymeric amine gels by cross-linking of soluble polymers is shown in Scheme 15.4. The structural repeat units of a selection of these polymers are illustrated in Scheme 15.5.

Like their low-molecular-weight phosphate receptor counterparts, the binding strengths and capacities of these polymeric phosphate sequestrants have been found to depend on a number of parameters, including the pH of the medium. Since more ammonium groups are present at lower pH, the polymers exhibit higher binding capacity at lower pH. Due to higher concentrations of amine groups along polymer chains and the divalent (and possibly

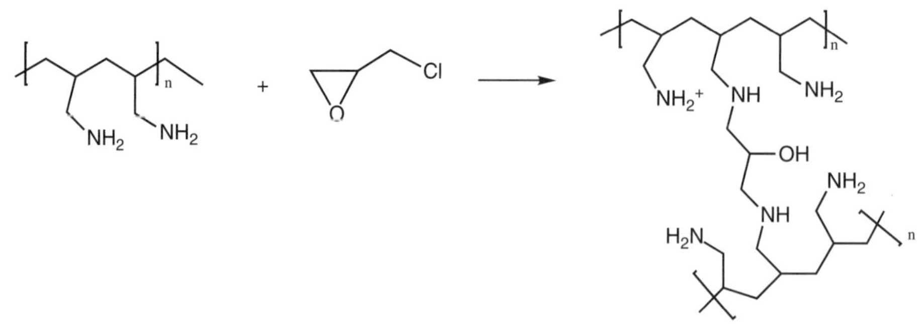

Scheme 15.4 Synthesis of polymeric hydrogels bearing amino groups by cross-linking of soluble polymeric amines.

trivalent) nature of phosphate anions, polymeric amines probably also exhibit stronger affinity toward phosphate anions than small molecule receptors, due to a polydentate chelate effect. This phenomenon has not, however, been examined in detail. Furthermore, since the degree of protonation of polymeric amines is dependent on polymer architectures (due to charge–charge repulsion), incorporation of appropriate spacing between the amine groups in the polymer chain and the backbone is key to the phosphate binding capacities of these polymers. Finally, polymers containing primary amine groups were found to be better phosphate sequestrants than polymers bearing secondary and tertiary amines, while polymers containing quaternary amines were found to exhibit the least phosphate binding property.[27,28]

A systematic structure–activity relationship (SAR) study was carried out by using different polymeric amine, different cross-linking agents, and the extent of cross-linking of these polymer gels. The study revealed that cross-linked polyallylamine gels are the most potent phosphate-binding polymers and possess properties suitable for pharmaceutical applications. This class of polymers exhibited maximum phosphate binding in the pH range encountered in the ssmall intestine, whereas the polymer gel was found to be nontoxic and is essentially nonabsorbed. From the systematic SAR study we identified

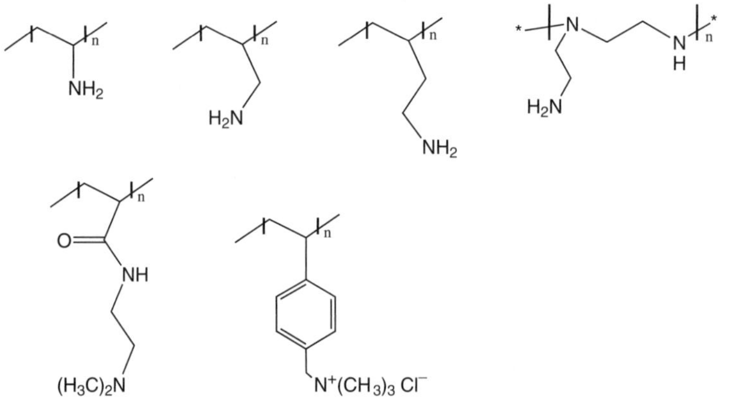

Scheme 15.5 Structural repeat units of representative polymers used as precursors for sequestrant synthesis.

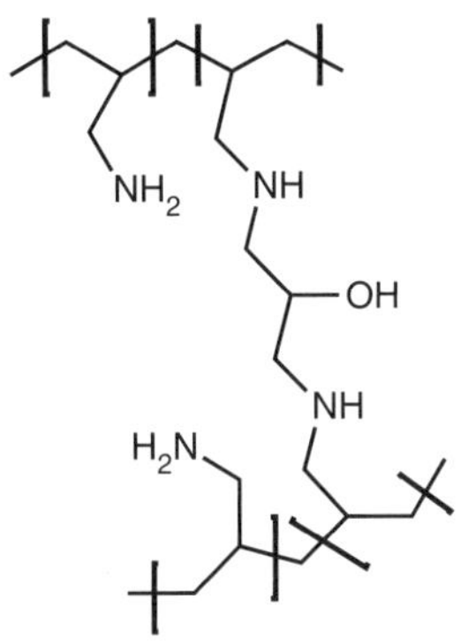

4

Scheme 15.6

epichlorohydrin crosslinked polyallylamine (**4**, Scheme 15.6) as the lead candidate for subsequent preclinical and clinical development. This polymer was found to exhibit maximum capacity for phosphate. It is believed that the polymer binds phosphate anion through electrostatic and possibly hydrogen-bonding interactions (see Scheme 15.7).

Thus, epichlorohydrin cross-linked polyallylamine hydrogel underwent clinical development as the first metal-free phosphate sequestrant for the treatment of hyperphosphatemia. The compound was approved in the United States by the Food and Drug Administration in 1998 under the generic name sevelamer hydrochloride, and it has been marketed since under the brand name Renagel by Genzyme Corporation. Subsequently, it was approved in the Europe, Japan, and a number of other countries. Since its approval, Renagel has demonstrated effective, long-term control of serum phosphate levels and has shown several advantages over and above traditional agents for the management of hyperphosphatemia in renal failure patients.[29,30] The ability for improved control of serum phosphate without increasing the exposure to toxic metal ions such as aluminum and eliminating the intake of additional calcium offers a number of clinical advantages. For example, without increasing the calcium load or promoting calcification, Renagel may help prevent cardiac complications in end-stage renal disease patients. Furthermore, Renagel has been found to reduce serum parathyroid hormone and total and low-density lipoprotein (LDL) cholesterol in hemodialysis patients. Since cardiovascular events are the most common causes of mortality of dialysis patients, Renagel thus offers a very promising treatment in the management of renal failure.[31] It represents one of the first tailormade polymeric drugs that exhibits its prophylactic and therapeutic properties through selective sequestration and removal of unwanted dietary components in the GI tract without presenting any systemic side effects. These documented benefits of Renagel provide an opportunity to affect patient survival and morbidity as well as to reduce health care expenses.

15.6 POLYMERIC IRON SEQUESTRANTS FOR THE TREATMENT OF IRON OVERLOAD DISORDERS

Although iron is essential for the proper functioning of all living cells, it is toxic if present in excess.[32] In the presence of molecular oxygen, excess iron can produce oxygen-derived

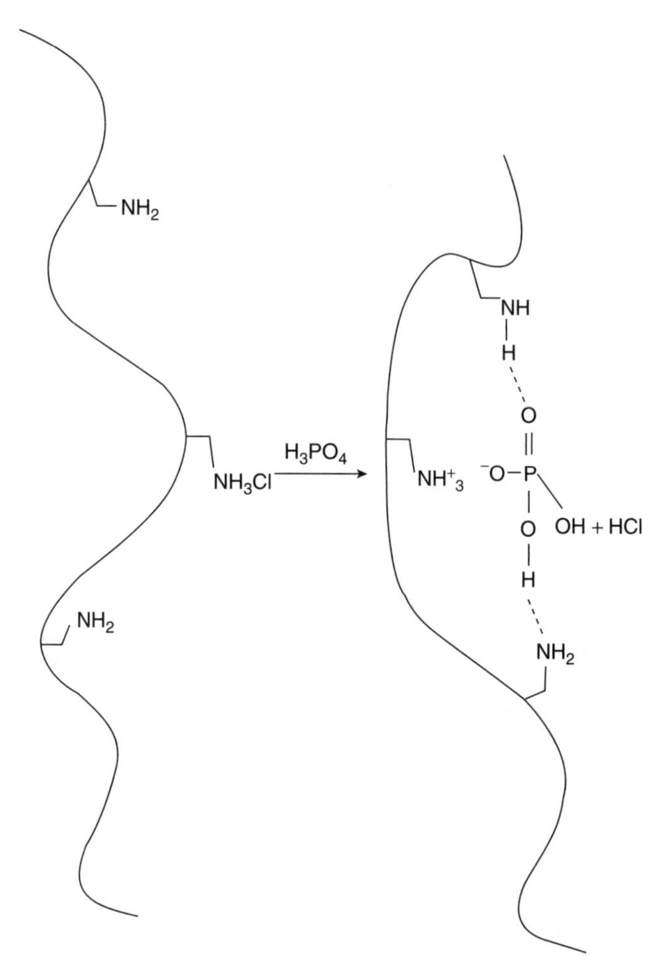

Scheme 15.7 Binding interaction between polymer-bound ammonium groups and phosphate anions.

free radicals such as hydroxyl radical. These reactive free radicals interact with many biological molecules, resulting in peroxidative tissue or organ damage. These adverse events form the basis for a number of pathological conditions. Under normal conditions, iron metabolism is highly conserved, with the majority of the iron being recycled within the body. There is no normal iron loss mechanism; iron is not normally present in urine, feces, or bile. Its removal from the body occurs only through bleeding or normal sloughing of epithelial cells. Certain genetic disorders can, however, lead to increased absorption of dietary iron (as in the case of hemochromatosis) or transfusion-induced iron over load (as in the case of β-thalassaemia and sickle cell anemia).[33,34]

Although excess iron can be removed by venesection (especially in hemochromatosis patients), removal of iron using an iron chelator is the only effective way to relieve iron overload in patients with β-thalasemia or sickle cell anemia.[35] The current standard of care is desferrioxamine (**5**, Scheme 15.8), the only approved iron chelator for this condition in the United States. This drug presents several shortcomings. It exhibits a narrow therapeutic window and is orally inactive. As a result, it requires administration by parenteral infusions

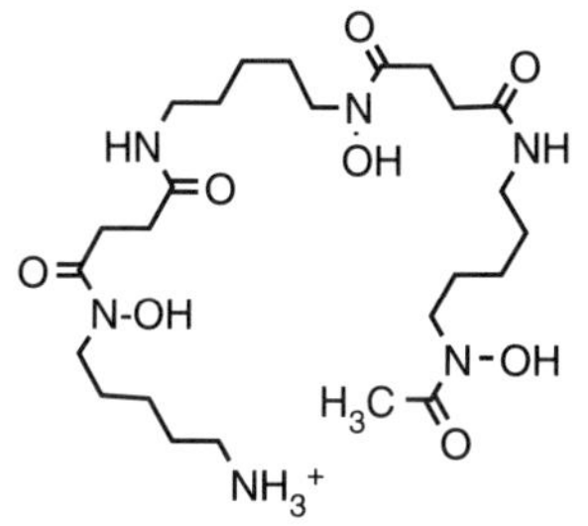

5

Scheme 15.8

for 8 to 12 hours per day.[36] Thus, there is a need for the discovery and development of orally active iron chelating agents for the treatment of iron overload.

Development of nonabsorbed polymeric ligands to sequester and remove dietary iron selectively from the GI tract appears to be an attractive method for the treatment of iron overload. To be clinically useful, polymeric iron chelators need to possess several important features. First, the polymeric ligand must possess high affinity, capacity, and selectivity toward iron. Furthermore, the chelator should be biocompatible and should not be absorbed from the GI tract. In general, design principles of polymeric chelators are based on the knowledge of low-molecular-weight iron chelators. For clinical applications, the properties of chelators in terms of metal ion selectivity and ligand–metal complex stability are important. Since iron exists in two oxidation states [ferrous (+2) and ferric (+3)], chelators can be designed for sequestering both forms of iron. Soft donor atoms (e.g., nitrogen-containing ligands such as bipyridine and phenanthroline) can be employed to sequester Fe(II). Although these ligands are selective for Fe(II), they also possess affinity for other biologically important divalent metal ions, such as Zn(II) and Cu(II). On the other hand, oxyanions such as hydroxamates and catecholates are selective toward Fe(III). These oxyanion ligands in general show higher selectivity toward trivalent metal ions than toward divalent metal ions. Natural iron chelators such as the siderophores desferrioxamine (**5**) and enterobactin (**6**, Scheme 15.9) contain hydroxamate and catechol groups, respectively, and are selective toward Fe(III).[37]

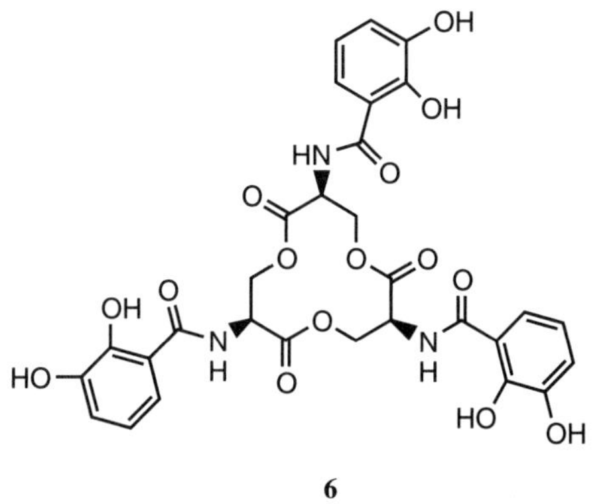

6

Scheme 15.9

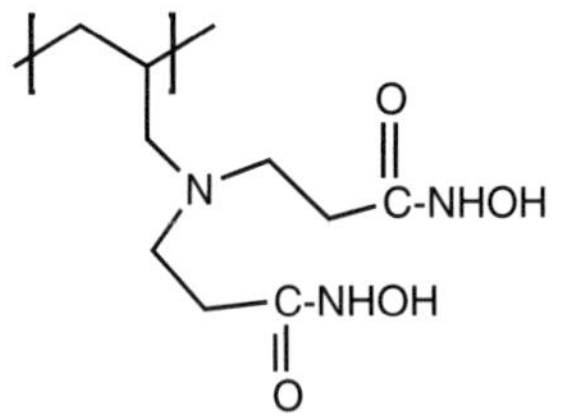

7 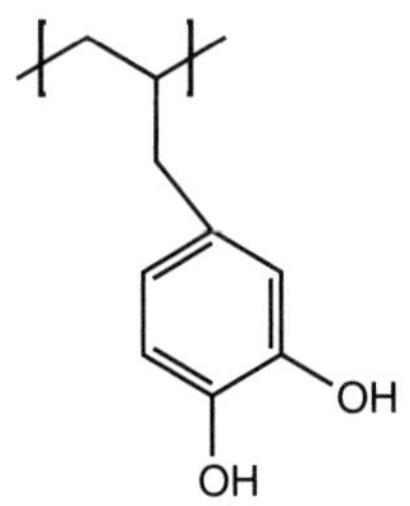 8

Scheme 15.10

By careful consideration of the foregoing facts, polymeric hydrogels containing hydroxamic acid and catechol moieties (**7** and **8**, Scheme 15.10) as well as cross-linked polymeric amines were prepared and were evaluated for their iron-binding properties.[38] Under in vitro conditions, all of these polymers sequester iron at higher pH. At lower pH, the polymers containing hydroxamic acids maintained their iron-binding properties, whereas other polymers substantially lost their iron-binding properties. In vivo studies using rodents have shown that treatment of the animals with the polyhydroxamic acid polymer lead to a decrease in absorption of dietary iron.[39] However, no human clinical data is available on the iron chelating polymers.

15.7 SEQUESTRATION OF BILE ACIDS: POLYMERS AS CHOLESTEROL-LOWERING AGENTS

Increased plasma total cholesterol and low-density lipoprotein cholesterol (LDLc) are established risk factors for antherosclerosis, which is the underlying cause of coronary heart disease and most strokes.[40] The reduction of elevated LDLc is one of the most common therapeutic treatments for cardiovascular disease. The majority of the people at risk require only a modest (20 to 30%) reduction in LDLc level to minimize the risk of this serious disease.[41] HMG-CoA reductase inhibitors (more commonly known as *statins*) are the most widely used drugs for reducing blood levels of LDLc and have been shown to reduce the risk of coronary events and strokes significantly. These findings have led to recent guidelines for expanding the use of cholesterol-lowering drug therapies to more patients.[42] In the United States alone there are an estimated 36 million people who are in need of cholesterol-lowering drug therapy.

Despite the spectacular success of statins in treating cardiovascular diseases, there is still a need for new therapies to reduce blood LDLc. Statins are, for example, not indicated for pregnant women, for pediatric use, or for patients with liver disease. Moreover, some patients do not achieve the LDLc goal with statin therapy alone. There are also long-term potential safety issues associated with statins, such as liver dysfunction and musculoskeletal symptoms. This is clearly evident from the recent withdrawal of a statin, cerevastatin (Baycol), from the market.[43]

The molecular mechanism underlying cholesterol metabolism was investigated systematically by Brown and Goldstein.[44] According to this metabolic pathway, cholesterol is synthesized in the liver by the enzyme HMG-CoA reductase. Subsequently, it is transformed into a bile acid in the liver and secreted to the gallbladder. The statins inhibit the function of HMG-CoA reductase, which is the rate-limiting enzyme in cholesterol biosynthesis. The presence of bile acid in the cholesterol metabolism pathway suggests that another approach to reduce plasma LDLc is through effective removal of the bile acid from the bile pool, resulting in upregulation of bile acid biosynthesis. Since the body attempts to maintain a steady state of bile acid pool, the process of sequestration and removal of bile acid leads to a corresponding drop in plasma cholesterol levels.[45]

Bile acid sequestrants (BASs) are cross-linked polymeric cationic gels that bind anionic bile acids in the GI tract and result in elimination of bile acid from the body.[46] The use of these polymeric gels for sequestering bile acids is an established approach for treating elevated cholesterol.[47] Being systemically nonabsorbed, these polymeric cholesterol-lowering drugs are free from the systemic side effects that are associated with statins. Moreover, the BAS have over 30 years of clinical experience with a good safety record. Until recently, two cationic polymers, cholestyramine (**9**) and colestipol (**10**, Scheme 15.11), have been the only approved bile acid sequestrants on the market. Despite the appeal of their safety profiles, these two first-generation bile acid sequestrants have, despite their high in vitro capacity, low clinical potency. This has led to reduced patient compliance and hence limited use. For example, the doses required for a 20% cholesterol reduction with cholestyramine and colestipol are typically 16 to 24 g/day.[48] This low clinical efficacy of BAS has been ascribed to competition for bile acids with the active bile acid transporter system of the GI tract.[49] It appears that for a cationic polymer to be a potent BAS, it must have high binding capacity, strong binding strength, and selectivity toward bile acids in the presence of competing desorbing forces of the GI tract. Thus, a potent BAS needs to

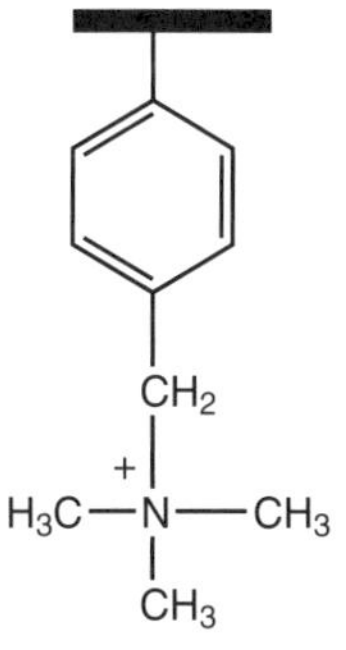

CH_2 $H_3C-N^+-CH_3$ CH_3

9

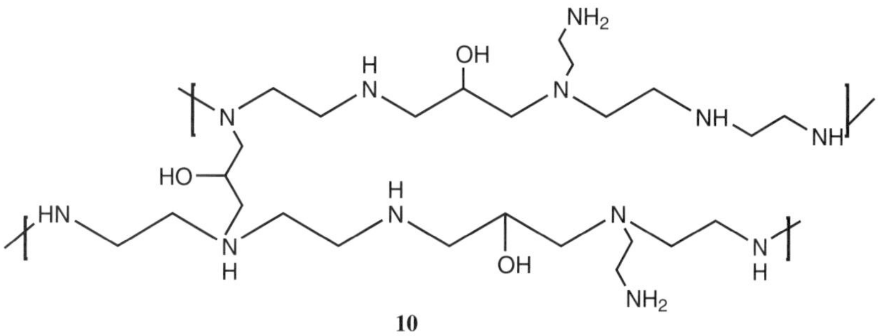

NH$_2$ OH H N N N NH NH HO H N N HN N H OH NH$_2$ N H

10

Scheme 15.11

11 a: $n = 1$, X = COOH; b: $n = 2$, X = SO_3H

Scheme 15.12

exhibit slow off-rates of bound bile acids from the polymer resin to effectively overcome the active transport of bile acids from the GI tract.[50] The critical design rules for discovering more potent BAS need to take into consideration the physicochemical features of bile acids. A typical bile acid (**11**, Scheme 15.12) possesses an anionic group and a hydrophobic core. Therefore, while electrostatic interaction is the primary force for bile acid binding by cationic polymers, a second attractive force that needs to be considered is the hydrophobic interaction between the sequestrant and bile acid. Furthermore, favorable swelling characteristics of these cationic hydrogels in physiological environments are required for attaining high capacity. Thus, a balanced combination of hydrophilicity (high capacity) and hydrophobicity (to slow down the rate of desorption), along with optimum density of cationic groups, would constitute key features of the most potent BAS.[51]

Consideration of these desired features has led to the discovery of a number of bile acid sequestrants over the last decade from our laboratories as well as from other groups.[52–56] Chemical structures of some of these representative bile acid sequestrants are summarized in Scheme 15.13. The common features of these hydrogels are the presence of amine and ammonium groups. Additional structural features of these polymers include the presence of hydrophobic chains. A variety of polymer backbones, including vinyl and allyl amine polymers, (meth)acrylates, (meth)acrylamide, styrene, carbohydrates, polyethers, and other condensation polymers, have been considered.[57–60] Although a large number of polymers have been synthesized and tested in preliminary in vitro and in vivo for their bile acid sequestration properties, very few polymers have proceeded to preclinical development and clinical trials. Some of these promising BASs that entered the clinic include DMP-504 (**12**), colestimide (**13**), SK&F 97426-A (**14**), and colesevelam hydrochloride (**15**, Scheme 15.14).

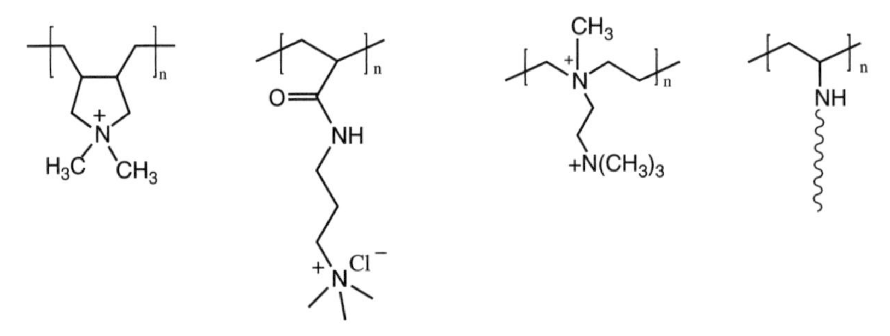

Scheme 15.13 Structural repeat units of representative ammonium salts as bile acid sequestrants.

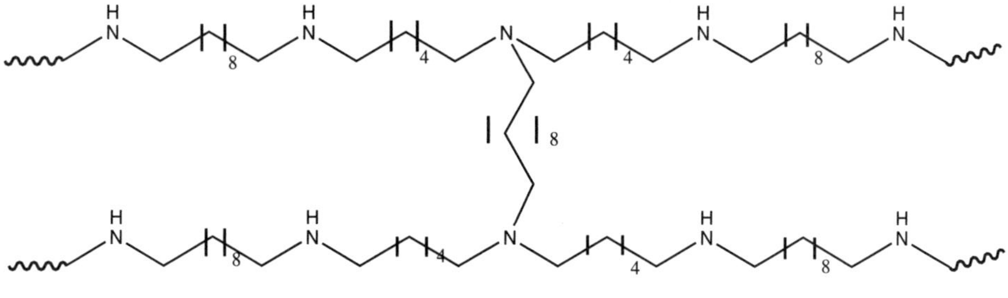

12

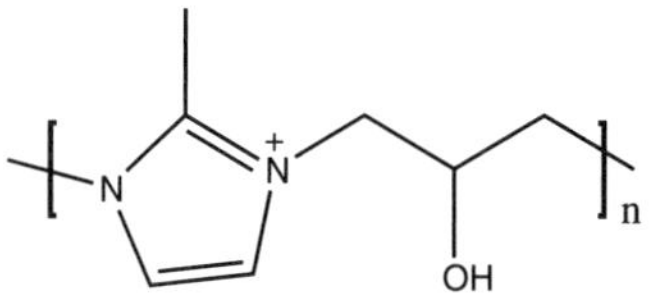

13

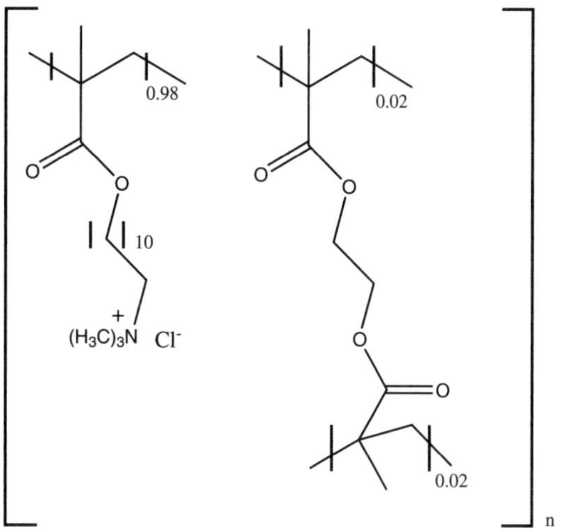

14

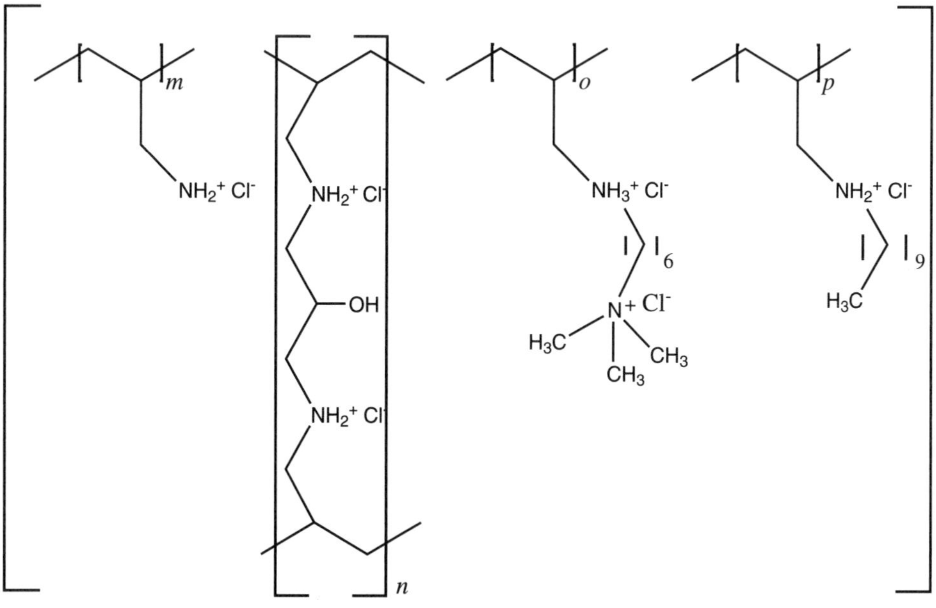

15

Scheme 15.14

Structural features of these polymers comprise an optimum combination of charge density, hydrophobic tails, and water swelling properties. From this new generation of bile acid sequestrants that entered clinical trials, only two have been approved for marketing. Colestimide has been approved for marketing in Japan and is sold under the trade name Cholebine; colesevelam hydrochloride has been approved for marketing in the United States and is being sold under the trade name WelChol.[61,62] Both polymers exhibit lower rates of side effects and are better tolerated than previously marketed BASs. These BASs can be used as monotherapy or in combination therapy by coadministration with statins.[63]

15.8 SEQUESTRATION OF PATHOGENS: POLYMERIC ANTI-INFECTIVE AGENTS

Polyvalent interactions are defined as simultaneous binding interactions between multiple ligands on one molecular entity and multiple receptors on another (cells, viruses, proteins, etc.). These phenomena are prevalent in biological systems. Polyvalent interactions can be collectively much stronger than the corresponding monovalent interactions.[64] These interactions are responsible for the early stages of a large variety of important biological processes, such as cell-surface or receptor–ligand recognition events. These processes can provide the basis for mechanisms of both agonizing and antagonizing biological interactions that are fundamentally different from those encountered in monovalent systems. A schematic illustration of the principle of polyvalency is presented in Scheme 15.15. The

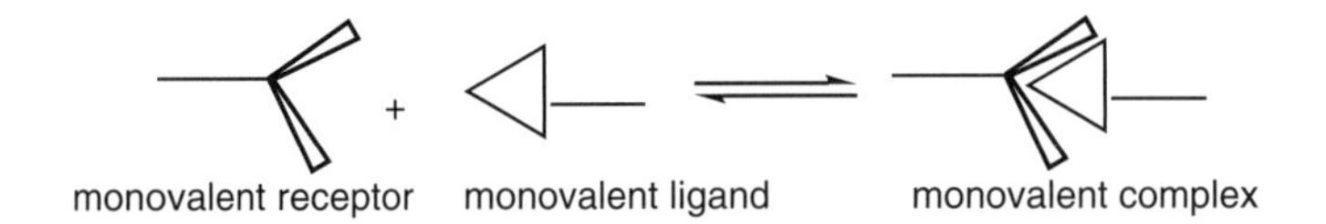

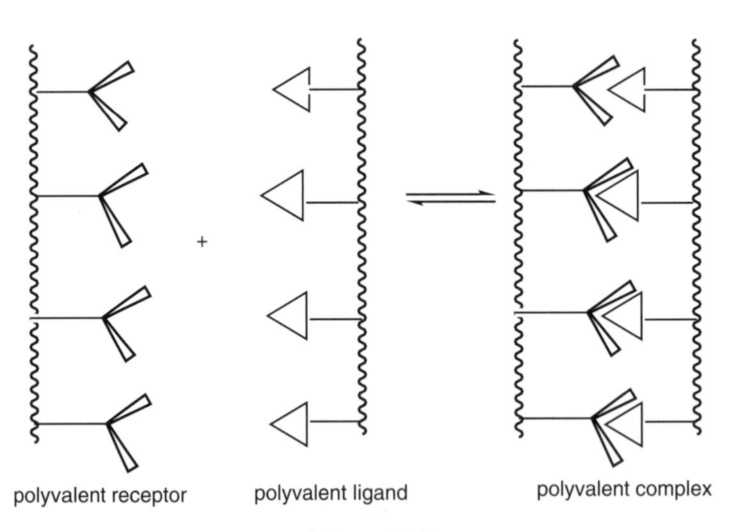

Scheme 15.15

underlying theory of polyvalency and its importance in biological processes have been reviewed in detail.[65]

The polyvalent interaction in biological systems has been considered as a new paradigm for the design and discovery of a new generation of therapeutic agents.[66] Since polyvalency can enhance the binding strength significantly, the approach is particularly appealing when the interaction between a monovalent ligand and a polyvalent receptor is weak.. For example, a polyvalent agent carrying two or more linked ligands can bind at two or more receptor sites on a target pathogen, resulting in enhanced binding strength. This would lead to enhanced inhibition and/or sequestration of the target agent, such as a pathogen. Polymeric systems provide an attractive platform to expand this approach for discovering novel drug molecules wherein a collection of similar or different ligands can be linked together covalently to a single polymer chain. The flexibility to choose different polymer architectures (e.g., block, alternate, random, comb, graft, branched, dendritic), monomer types, and the ease of polymer synthesis and modification enables one to synthesize a wide variety of well-defined polyvalent species with desired features such as controlled ligand density, optimum binding strength and capacity, hydrophilicity and hydrophobicity, and the incorporation of ancillary groups to enhance the recognition and binding of the target of interest. Interplay of these features in polyvalent ligands can, in principle, lead to long-lasting and more potent therapeutic agents. This new concept of ligand–substrate interaction has led in recent years to the discovery of a variety of polymeric drugs that have been found to sequester or inhibit toxins, viruses, and bacteria. Some of these polyvalent polymeric drugs have advanced into human clinical trials, which provide further support to the validity of this concept of drug discovery and development.

15.9 SEQUESTRATION OF TOXINS

Toxins are produced by pathogenic microorganisms in the host body. There are two kinds of toxins: exotoxins and endotoxins.[67,68] *Exotoxins* are generally proteinous materials released by pathogenic microorganisms; *endotoxins* are lipopolysaccharides and consist of polysaccharide segments and glycolipid segments that constitute the outer cell membranes of all gram-negative bacteria.

Upon secretion from microorganisms, these toxins can travel within the body of the host organism and can cause damage in regions of the body farther away from the site of infection. The pathogenic effects of these toxins—hemolysis, destruction of leucocytes, paralysis, diarrhea, and septic shock—could be fatal.[69] Some of the pathogens that produce life-threatening toxins in the GI tract are food poisoning organisms such as *Staphylococcus aureus*, *Clostridium perfringens*, and *Bacillus cereus* and intestinal pathogens such as *Vibrio cholerae*, *Escherichia coli*, and *Salmonella enteritidis*. Furthermore, anthrax toxin is produced by *Bacillus anthracis*. In almost all cases, the causative agents responsible for major symptoms of diseases associated with these pathogens are the exotoxins produced by the organisms.[70,71]

***Clostridium Difficile* Toxin** *Clostridium difficile* is responsible for large numbers of episodes of diarrhea that arise as a result of antibiotic treatment.[72] The outbreak of this disease has been attributed to the disruption of normal colonic flora by antibiotic treatment. As a result, *C. difficile* invades and colonizes in the gut. Therefore, *C. difficile* infection is prevalent in hospital settings. This bacterium releases two high-molecular-weight proteins (toxins A and B), which are implicated primarily in causing diarrhea.[73,74] The traditional way to treat *C. difficile*–associated diarrhea has been the use of one of two antibiotics: metronidazole or vancomycin. Although the use of antibiotics is effective in eliminating *C. difficile* infection initially, repeated treatment with antibiotics increases the chance of further infection by *C. difficile*, due to sterilization of the gut. Furthermore, there has been increasing body of documentation pointing to growing resistance of *C. difficile* to antibiotic treatment.[75]

Since the disease symptoms are attributed to the toxins produced by this bacterium, selective sequestration, neutralization, and elimination of these toxins by a polymeric sequestering agent appears to offer an attractive and safe alternative to antibiotic therapy. This method of treatment would treat the infection without disrupting the reestablishment of normal bacterial growth in the gut.

Since *C. difficile* toxins are proteins, they possess multiple binding sites that are derived from amino acid side chains. These side chain functional groups of the toxin can effectively interact with multifunctional polymers carrying complementary binding sites through simultaneous polyvalent interactions. Some of the early studies using polymeric ligands to sequester *C. difficile* toxins include the use of anion-exchange resins such as cholestryramine and colestipol (the bile acid sequestrants described above). However, the effectiveness of these polymeric ion exchangers as sequestrants to bind and remove *C. difficile* toxins was found to be quite modest.[76]

Systematic investigation in our laboratories has led to the discovery of some novel polymeric multivalent ligands as sequestrants that effectively bind and neutralize *C. difficile*–derived toxins. Among the various polymers evaluated, a series of high-molecular-weight water-soluble anionic polymers (**16–18**, Scheme 15.16) were found to be particularly effective in sequestering and neutralizing *C. difficile* toxins.[77,78] In general, these polymers contain sulfonic acid groups and have high molecular weight. Careful in vitro studies using pulsed ultrafiltration binding experiments and fluorescence polarization spectroscopy revealed that the binding constants for complexation between one of the polymers and toxins A and B were 133 nM and 8.7 μ*M*, respectively.[79] The ability to bind both the toxins is not a general feature of these anionic polymers. For example, while sodium salt poly(styrene sulfonic acid) effectively binds the polymers, poly(sodium 2-acrlamido-2-methyl-1-propanesulfonate), a high-molecular-weight polyanion of similar charge density does not bind either of the toxins to any measurable extent. Furthermore, the binding strength is dependent on molecular weight, with lower-molecular-weight polymers exhibiting very low binding strength. These results thus suggest that sequestration of *C. difficile* toxins by polyanions is not purely electrostatic in origin. On the other hand, it appears that the

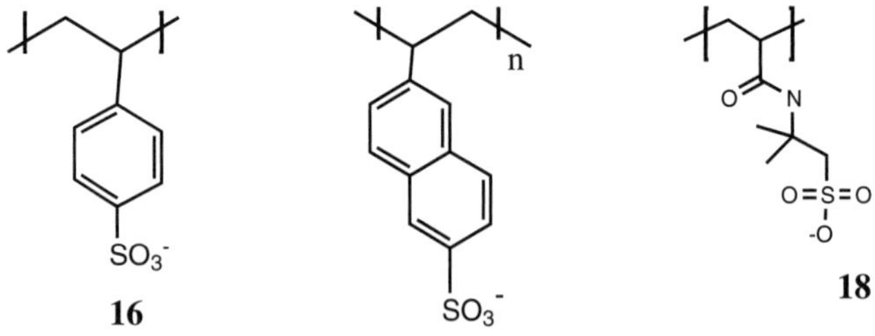

SO_3^- **16** SO_3^- **17** **18**

Scheme 15.16

sodium salt of poly(styrene sulfonic acid) and related anionic polymers interact with toxins A and B through multiple weak interactions that amplifies to a very high binding strength as a result of polyvalent interaction. From fluorescence polarization data, it was estimated that 1 molecule of toxin A interacts with about 800 monomer units on the polymer. Thus, a single polymer chain of molecular mass 300 kDa would wrap around a toxin molecule about three or four times. This suggests effective large polyvalent interactions between the polymer and protein surface.[79] The in vitro binding activity of these high-molecular-weight polyanions correlated well with the in vivo biological activity. The polymer GT prevented the mortality of 80% of hamsters with severe *C. difficile* colitis.[80] These polymers are non-antimicrobial and hence do not interfere with the activities of standard antibiotics. Therefore, they are likely to overcome the problems associated with antibiotic resistance.

High-molecular-weight sodium salt of a poly(styrene sulfonic acid) derivative was selected as the lead candidate for clinical development as a non-antibiotic-based polymer therapy for treating *C. difficile* infection. This compound, Tolevamer, has been successful in human clinical trials in both phases I and II. This polymer is currently undergoing phase III human clinical trials.

In addition to the polyanions cited above, another class of polymers bearing pendant oligosaccharide groups has been investigated as possible sequestrants for *C. difficile* toxins. Some of these compounds have progressed to human clinical trial.[81,82] The underlying principle behind these polymeric sugar agents for the treatment of *C. difficile* infection is that toxin A has shown lectinlike activity, which allows it to bind to an oligosaccharide receptor on epithelial cells. Furthermore, toxin B has been found to bind erythrocytes. These observations suggest that the cell invasion and binding of *C. difficile* may be mediated by cell-surface carbohydrate receptors. Therefore, polymers bearing pendant sugar residue may compete with human cells toward *C. difficile* toxin, After identifying oligosaccharide sequences, appropriate oligosaccharide molecules that are specific for both toxins were conjugated to different polymer backbone (**19**, Scheme 15.17). Lengths of tethering arms linking the polymer backbones with oligosaccharide moieties were optimized appropriately to maximize the binding strength of ligand–polymer (polymer–toxin) interaction. The oligosaccharide sequences that were found to improve toxin binding include maltose, cellobiose, isomaltotriose, and chitobiose. Polymeric carriers examined include substituted polystyrene and other olefin backbones. These polymeric toxin binders (under the generic

H OH H O HO H OH H O H H OH H OH HO O H O H n

19

Scheme 15.17

name of Synsorb) were found to be effective in neutralizing toxins and in controlling diarrhea in animal models. One of these polyvalent polymeric carbohydrate agents had also progressed to human clinical trials. However, due to lack of therapeutic efficacy, it was withdrawn after a phase II clinical trial.

15.10 POLYMERIC ANTIMICROBIAL AGENTS

The emergence of microbial pathogens that are resistant to multiple classes of available antimicrobial agents is becoming a global public health concern of significant magnitude. These multidrug resistant bacteria are prevalent in both hospital and community environments.[83] The majority of these strains have been found to carry multiple drug resistance factors. The only effective treatment for multiply resistant bacterial infections that is currently available is vancomycin. However, resistance to vancomycin is also a growing concern. For example, methicillin-resistant *Staphylococcus aureus* (MRSA) strains, vancomycin-resistant enterococci, and amikacin and β-lactam-resistant *Kleisiella pneumoniae* are some of the bacterial species that are resistant to vancomycin.[84,85] This rapid emergence of multidrug-resistant bacterial strains poses a potential threat to human life. As a result, there is an urgent need to discover and develop novel antibacterial agents that can circumvent the challenges posed by multidrug-resistant microorganisms.

Polyvalent ligands as antibacterial agents have been thought to exhibit potential advantages over monomeric antimicrobial agents. Cluster effects from polyvalent ligands would lead to amplification of weak nonbonding interaction between the bacterial surface receptors and the polymeric ligands. Aggregation and precipitation of bacteria by polyvalent ligands is potentially another favorable feature that could be achieved by using polyvalent antimicrobial agents. Finally, polyvalent ligands could enhance the rate of lysis of the bacterial cell membrane and wall more rapidly, due to multipoint attachment.

Most bacterial infections are initiated by adhesion of microorganisms to the mucosal surfaces of hosts, mediated in part by bacterial protein adhesins.[86] These adhesins interact with carbohydrate determinants of host cell glycolipids or glycoproteins. The mechanism for bacterial infection through this pathway suggests that development of appropriate polyvalent sugar derivatives could competitively block the attachment of microbial adhesin to host mucosal surface, resulting in protection against infection. This concept has been explored through the synthesis of a number of polymers bearing acid-functionalized glycoside moieties. Olefinic monomers containing glycoside moieties and acid functional groups such as *O*-sulfo and *O*-carboxymethyl groups were prepared and converted to various copolymers. These polymers were found to be effective in vitro against a number of bacterial targets.[87]

A second approach to design polyvalent ligands as antimicrobial agents based on cationic polymers has also been explored systematically in our laboratories. The mechanism of the antimicrobial action of these polymers has been attributed to their ability to enhance the rate of cell lysis. These polymers were designed as mimetics of certain cationic amphiphilic peptides containing multiple arginine and lysine residues. These cationic peptides, known as *antimicrobial peptides*, cause cell lysis through interaction of the positive charges of the peptides with negative phosphate head groups of cell membrane phospholipids[88]]. A series of amphiphilic cationic polymers were prepared bearing amine and quaternary ammonium groups as well as hydrophobic tails (see Scheme 15.18). These polymers were found to exhibit antimicrobial activity against a number of microbes.[89] Until the advent of

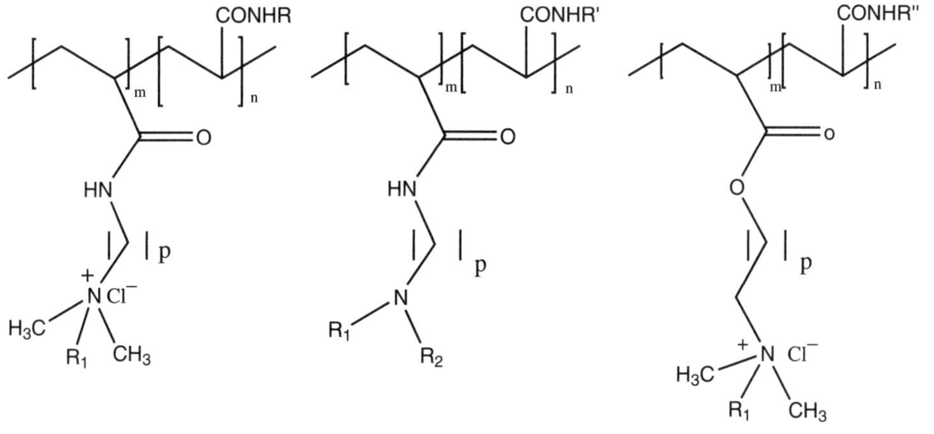

Scheme 15.18

protease inhibitor, *C. parvum* was a primary target for drug discovery to treat GI tract infections in persons with HIV infection.[90] Some of these polymers were found to be superior to the commonly prescribed antibiotic, such as paromomycin.

15.11 CONCLUSIONS AND OUTLOOK

Although polymeric materials have been developed to produce useful biomaterials and drug delivery systems, until recently intrinsically bioactive polymers as therapeutic agents have remained largely unappreciated. In the present chapter we have attempted to illustrate the potential of polymers in the discovery and development of novel therapeutic agents for treating a number of human diseases. By careful consideration of both disease targets and their mechanisms of action, a number of functional polymers have been discovered that exhibit promising pharmacological properties. These polymeric drugs capitalize on the unique physicochemical properties of polymer, and on many occasions, these polymers exhibit therapeutic properties that cannot be achieved by traditional small molecule drugs. The usefulness of these bioactive polymers for the treatment of different diseases that have been marketed or are being developed is quite impressive. Polymeric sequestrants that are confined to the GI tract and carry out their disease-modifying activities are important examples that attest to the validity of the concept of polymeric drugs. Recognition of the phenomenon of polyvalency as a tool in drug discovery strategy has been carefully utilized to discover polymeric drugs that provide a new paradigm for developing novel therapeutics. Although a number of these polymers have not yet met the goal of exhibiting in vivo biological activity, the development of toxin sequestrants suggests that it is not simply an academic curiosity. Finally, further understanding of cellular and molecular basis of various disease targets with the recent advancements of genomics and proteomics research would further enable pharmaceutical researchers in designing novel polymeric drugs with desired immunological and other related pharmacological properties for systemic applications of polymers against other disease targets. Once these design criteria are identified and fine tuned, potentially more selective polymer therapeutics will be discovered for medical needs unmet or met inadequately.

REFERENCES

1. R. Duncan, *Nature Rev. Drug Discov.*, 2, 347 (2003).
2. R. M. Ottenbrite, in R. L. Dunn and R. M. Ottenbrite (Eds.), *Polymeric Drugs and Drug Delivery Systems*, ACS Symposium Series 469, American Chemical Society, Washington, DC, 1991, 3 pp.
3. K. Ulbrich, M. Pechar, J. Strohalm, and V. Subr, *Macromol. Symp.*, 118, 577 (1997).
4. R. Duncan and J. Kopecek, *Adv. Polym. Sci.*, 57, 51 (1984).
5. M. C. Garnett, *Crit. Rev. Ther. Drug Carrier Syst.*, 16, 147 (1999).
6. R. Langer, *Nature*, 392, 5 (1998).
7. M. Ahlers, W. Muller, A. Reichert, H. Ringsdorf, and J. Venzmer, *Angew. Chem. Int. Ed. Engl.*, 29, 1269 (1990).
8. R. Duncan, in K. Park and R. J. Mrsny (Eds.), *Controlled Drug Delivery*, ACS Symposium Series 752, American Chemical Society, Washington, DC 2000, 350 pp.
9. S. R. Holmes-Farley, *Polym. Mater. Sci. Eng.*, 80, 246 (1999).
10. L. R. Johnson, *Physiology of the Gastrointestinal Tract*, Raven Press, New York, 1994.
11. R. M. Rosa, M. E. Williams, and F. H. Epstein, in D. W. Seldin and G. Giebisch (Eds.), *The Kidney: Physiology and Pathophysiology*, Raven Press, New York, 1992, 2165 pp.
12. G. O. Perez, J. R. Oster, R. Pelleya, P. V. Catalis, and D. C. Kem, *Nephron*, 36, 270 (1984).
13. T. J. Marrone and K. M. Terz, *J. Am. Chem. Soc.*, 114, 7542 (1992).
14. G. W. Gardiner, *Can. J. Gastroenterol.*, 11, 573 (1997).
15. B. B. Gerstman, R. Kirkman, and R. Platt, *Am. J. Kidney Dis.*, 20, 159 (1992).
16. J. Delmez and E. Slatopolsky, *Am. J. Kidney Dis.*, 4, 303 (1992).
17. J. W. Coburn and E. Slatopolsky, in B. M. Brenner and F. C. Rector, Jr. (Eds.), *The Kidney*, W.B. Saunders, Philadelphia, 1991, 2036 pp.
18. C. Hsu, *Am. J. Kidney Dis.*, 29, 641 (1997).
19. S. Ribiro, A. Ramos, and A. Brandao, *Nephrol. Dial. Transplant*, 13, 2037 (1998).
20. F. P. Schmidtchen and M. Berger, *Chem. Rev.*, 97, 1609 (1997).
21. M. W. Hosseim, and J. M. Lehn, *Helv. Chim. Acta,* 70, 1312 (1987).
22. R. C. Hider and A. Canas-Rodriguez (to British Technology Group Ltd.), U.S. patent 5,698,190 (December 1997).
23. R. C. Hider and A. Canas-Rodriguez (to British Technology Group Ltd.), U.S. patent 1,132,706 (March 2000).
24. B. Dietrich, D. L. Fyles, T. M. Fyles, and J. M. Lehn, *Helv. Chim. Acta.*, 62, 2763 (1979).
25. S. R. Holmes-Farley, W. H. Mandeville, and G. M. Whitesides (to Genzyme Corporation), U.S. patent 5,496,545 (January 1996).
26. S. R. Holmes-Farley, W. H. Mandeville, and G. M. Whitesides (to Genzyme Corporation), U.S. patent 5,667,775 (September 1997).
27. S. R. Holmes-Farley and W. H. Mandeville (to Genzyme Corporation), U.S. patent 6,726,905 (April 2004).
28. S. R. Holmes-Farley, W. H. Mandeville, J. Ward, and K. L. Miller, *J. Macromol. Sci. Pure. Appl. Chem.*, A36, 1085 (1999).
29. E. A. Slatopolosky, S. K. Burke, and M. A. Dillion, *Kidney Int.*, 55, 299 (1999).
30. N. Amin, *Nephrol. Dial. Transplant*, 17, 340 (2002).
31. A. J. Bleyer, S. K. Burke, and M. A. Dillon, *Am. J. Kidney Dis.*, 33, 694 (1999).
32. R. R. Crichton, in *Inorganic Biochemistry of Iron Metabolism*, Ellis Horwood, New York, 1991.
33. G. M. Brittenham, in R. Hoffman, E. Benz, S. Shattil, B. Furie, and H. Cohen (Eds.), *Hematology: Basic Principles and Practice*, Churchill Livingstone, New York, 1991, 327 pp.

34. D. J. Weatherall and J. B. Clegg, in *The Thalassaemia Syndromes*, 3rd ed., Blackwell Scientific, Oxford, 1981.
35. Z. D. Liu and R. C. Hider, *Med. Res. Rev.*, 22, 26 (2002).
36. J. B. Porter, M. C. Jawson, E. R. Huehn, C. A. East, and J. W. P. Hazell, *Br. J. Haematol.*, 73, 403 (1989).
37. K. N. Raymond, G. Muller, and B. F. Matzanke, *Top. Curr. Chem.*, 58, 49 (1984).
38. W. H. Mandeville and S. R. Holmes-Farley (to GelTex Pharmaceuticals, Inc.), U.S. patent 5,487,888 (January 1996).
39. P. K. Dhal, P. Cannon, S. R. Holmes-Farley, W. H. Mandeville, S. R. Polomoscanik, and T. X. Neenan, *Biomacromolecules,* 6, 2946 (2005).
40. S. M. Grundy, *Circulation*, 97, 1436 (1998).
41. S. M. Grundy, *Endocrinol. Metab. Clin. North Am.*, 27, 655 (1998).
42. Executive summary of the third report of the National Cholesterol Education (NCEP) Expert Panel on Detection, Evaluation, and Treatment of High Blood Cholesterol in Adults, *J. Am. Med. Assoc.*, 285, 2486 (2001).
43. D. J. Rader, *Nature Med.*, 7, 1282 (2001).
44. M. S. Brown and J. L. Goldstein, *Proc. Natl. Acad. Sci. USA*, 96, 11041 (1999).
45. S. M. Grundy, in R. Fears (Ed.), *Pharmacological Control of Hyperlipidaemia*, Prous Science Publishers, Barcelona, Spain, 1986, 3 pp.
46. E. R. Stedronsky, *Biochim. Biophys. Acta*, 1210, 255 (1994).
47. W. H. Mandeville and D. I. Goldberg, *Curr. Pharm. Design*, 3, 15 (1997).
48. G. M. Benson, C. Haynes, S. Blanchard, and D. Ellis, *J. Pharm. Sci.*, 82, 80 (1993).
49. M. Ast And W. H. Frishman, *J. Clin. Pharmacol.*, 30, 99 (1990).
50. W. H. Mandeville, W. Braunlin, P. Dhal, A. Guo, C. Huval, K. Miller, J. Petersen, S. Polomoscanik, D. Rosenbaum, R. Sacchiero, J. Ward, and S. R. Holmes-Farley, *Mater. Res. Soc. Symp. Proc.*, 550, 3 (1999).
51. P. Zarras and O. Vogl, *Prog. Polym. Sci.*, 24, 485 (1999).
52. S. R. Holmes-Farley and W. H. Mandeville (to GelTex Pharmaceuticals, Inc.), U.S. patent 5,607,669 (March 1997).
53. C. C. Huval, S. R. Holmes-Farley, J. S. Petersen, and P. K. Dhal (to Genzyme Corporation), U.S. patent 6,248,318 (1997).
54. S. R. Holmes-Farley, P. K. Dhal, and J. S. Petersen (to Genzyme Corporation), U.S. patent 6,203,785 (1998).
55. S. R. Holmes-Farley, W. H. Mandeville, S. K. Burke, and D.I. Goldberg (to Genzyme Corporation), U.S. patent 6,423,754 (2002).
56. G. M. Benson, D. R. Alston, D. M. B. Hickey, A. A. Jaxa-Chamiec, C. M. Whittaker, C. Haynes, A. Glen, S. Blanchard, S. R. Cresswell, and K. E. Suckling, *J. Pharm. Sci.*, 86, 76 (1997).
57. C. C. Huval, M. J. Bailey, S. R. Holmes-Farley, W. H. Mandeville, K. Miller-Gilmore, R. J. Sacchiero, and P. K. Dhal, *J. Macromol. Sci. Pure Appl. Chem.*, A38, 1559 (2001).
58. C. C. Huval, S. R. Holmes-Farley, W. H. Mandeville, R. J. Sacchiero, and P. K. Dhal, *Eur. Polym. J.*, 40, 693 (2004).
59. L. Zhang, V. Janout, J. L. Renner, M. Uragami, and S. L. Regen, *Bioconjug. Chem.*, 11, 397 (2000).
60. N. S. Cameron, A. Eisenberg, and G. R. Brown, *Biomacromolecules*, 3, 116 (2002).
61. B. Gaudilliere and P. Berna, *Annu. Rep. Med. Chem.*, 35, 331 (2000).
62. B. Gaudilliere, P. Bernardellli, and P. Berna, *Annu. Rep. Med. Chem.*, 36, 293 (2001).
63. W. H. Mandeville and C. Arbeeny, *Idrugs*, 2, 237 (1999).
64. S.-K. Choi, *Synthetic Multivalent Molecules: Concepts and Biomedical Applications*, Wiley-Interscience, New York, 2004.

65. M. Mammen, S.-K. Choi, and G. M. Whitesides, *Angew. Chem. Int. Ed. Engl.*, 37, 2754 (1998).
66. S. Borman, *Chem. Eng. News*, 78, 48 (2000).
67. B. Henderson, M. Wilson, and B. Wren, *Trends Microbiol.*, 5, 454 (1997).
68. M. Rauchhaus, A. J. Coats, and S. D. Anker, *Lancet*, 356, 930 (2000).
69. B. Beutler and A. Poltorak, *Crit. Care Med.*, 29, 52 (2001).
70. Y. Gasche, D. Pittet, and P. Sutter, in W. J. Sibbald and J. L. Vincent (Eds.), *Clinical Trials for Treatment of Sepsis*, Springer-Verleg, Berlin, 1995, 35 pp.
71. T. C. Dixon, M. Meselson, J. Guillemin, and P. C. Hanna, *N. Engl. J. Med.*, 341, 815 (1999).
72. J. G. Bartlett, *Clin. Infec. Dis.*, 18, (Suppl. 4), S265 (1994).
73. C. P. Kelly and T. LaMont, *Annu. Rev. Med.*, 19, 375 (1998).
74. R. C. Spencer, *J. Antimicrob. Chemother.*, 41 (Suppl. C), 5 (1998).
75. S. Lebel, S. Bouttier, and T. Lambert, *FEMS Microbiol. Lett.*, 238, 93 (2004).
76. N. S. Taylor and J. G. Bartlett, *J. Infect. Dis.*, 141, 92 (1980).
77. C. Bacon-Kurtz and R. Fitzpartick (to GelTex Pharmaceuticals, Inc.), U.S. patent 6,290,946 (September 2001).
78. C. Bacon-Kurtz and R. Fitzpartick (to GelTex Pharmaceuticals, Inc.), U.S. patent 6,270,755 (August 2001).
79. W. Braunlin, Q. Xu, R. Fitzpatrick, C. B. Kurtz, R. Burrier, and J. D. Klinger, *Biophysics*, 87, 534 (2004).
80. C. B. Kurtz et al., *Antimicrob. Agents Chemother.*, 45, 2340 (2001).
81. L. D. Heerze and G. D. Armstrong (to Synsorb Biotech, Inc.), U.S. patent 6,013,635 (January 2000).
82. U. J. Nilsson, L. D. Heerze, Y.-C. Liu, G. D. Armstrong, M. M. Palcic, O. Hindsgaul, *Bioconj. Chem.*, 8, 466 (1997).
83. H. Pearson, *Nature*, 418, 469 (2002).
84. C. Walsh, *Nature*, 406, 775 (2000).
85. L. P. Kotra, D. Golemi, S. Vakulenko, and S. Mosashery, *Chem. Ind.*, 10, 341 (2000).
86. G. Reid and J. D. Sobel, *Rev. Infect. Dis.*, 9, 470 (1987).
87. W. H. Mandeville and V. R. Garigapati (to Genzyme Corporation), U.S. patent 5,821,312 (October 1998).
88. W. L. Maloy and U. P. Kari, *Biopolymers*, 37, 105 (1995).
89. W. H. Mandeville, T. X. Neenan, and S. R. Holmes-Farley (to Genzyme Corporation), U.S. patent 6,034,129 (March 1998).
90. J. M. Matukaitis, *J. Community Health Nurs.*, 14, 135 (1997).

16

BOTANICAL IMMUNOMODULATORS AND CHEMOPROTECTANTS IN CANCER THERAPY

BHUSHAN PATWARDHAN AND MANISH GAUTAM

University of Pune
Pune, India

SHAM DIWANAY

Abasaheb Garware College
Pune, India

16.1 INTRODUCTION

Botanicals being chemically complex and diverse may provide combinations of synergistic moieties useful in cancer therapy. With suitable examples, in this chapter we highlight the importance of traditional medicine in natural product drug discovery related to cancer. Most cancer chemotherapeutic agents are associated with toxicity toward normal cells and tissues that share many characteristics with tumor cells, particularly high cell turnover. Optimal dosing of cancer chemotherapeutic agents is often limited because of severe nonmyelosuppressive and myelosuppressive toxicities. Therefore, it is a continuing challenge to design therapy that is effective and also efficiently targeted to tumor cells. Cytoprotective agents are expected to control or prevent these toxicities and include use of synthetic and natural products. Specific chemoprotectants are emerging for cisplatin and anthracyclin antibiotics. None of the available agents satisfy criteria for an ideal chemoprotection. This has stimulated research for discovering natural resources with immunomodulatory and cytoprotective activities. Various botanicals and ethnopharmacological agents used in traditional medicine have been investigated by various workers for their chemoprotective, immunomodulating, adaptogenic, and antitumor activities and have revealed promises toward developing into a potential drug for cancer treatment per se or as adjuvant.

Drug Discovery and Development, Volume 1: Drug Discovery, Edited by Mukund S. Chorghade

Cancer chemotherapy is a major treatment modality used for the control of advanced stages of malignancies and also as a prophylactic against possible metastases in combination with the radiotherapy.[1] The chemotherapeutic agents available today have mainly immunosuppressent activity. Most of them are cytotoxic and exert a variety of side effects. The metabolism and clinical safety of these agents has not been clearly established. This has given rise to stimulation in research for locating natural resources showing immunological activity. In past, the variety of naturally occurring agents, including living and attenuated microorganisms, autologous and heterologous proteins, and injections of animal organ preparations, were used to restore and repair defense mechanism. A number of plant extracts have been shown to be immunomodulators.[2] Chemical agents preventing site-specific toxicity of cytotoxic drugs are in current practice in cancer chemotherapy. These agents exhibiting chemoprotective and immunomodulatory activities are reviewed here.

16.2 IMMUNOMODULATION

The control of disease by immunologic means has two objectives: the development of immunity and the avoidance of undesired immune reactions. Immunostimulation in a drug-induced immunosuppression model and immunosuppression in an experimental hyperreactivity model by the same preparation can be said to be true immunomodulation. Immunomodulators are biological response modifiers (BRMs), used to treat cancer, which exert their antitumor effects by improving host defense mechanisms against the tumor. They have a direct antiproliferative effect on tumor cells and also enhance the ability of the host to tolerate damage by toxic chemicals that may be used to destroy the cancer. Modulation of immunity was previously attempted with glucocorticoids and cytotoxic drugs such as cyclophosphamide. It is now recognized that immunomodulatory therapy could provide an alternative to conventional chemotherapy for a variety of diseased conditions, especially when the host's defense mechanisms have to be activated under the conditions of impaired immune responsiveness or when a selective immunosuppression has to be induced in situations such as inflammatory diseases, autoimmune disorders, and organ/bone marrow transplantation.[3] All three classes of immunomodulators—biologicals, chemical, and cytokines—will continue to play a major role in advancing and improving the quality of treatment of several human as well as animal diseases.[4] There is need for further research to better understand the biochemical mechanisms involved in immunoregulation to maximize the benefits of chemical immunomodulators as single agents or adjuvants in cancer therapy.[5]

16.3 ETHNOPHARMACOLOGY AND BOTANICAL IMMUNOMODULATORS

There are two major ways of bioprospecting natural products for investigation. The first is the classical method, which relies on phytochemical factors, serendipity, and random screening approaches. The second method uses traditional knowledge and practices as the drug discovery engine. Known as the *ethnopharmacology approach*, this method is time- and cost-effective and may lead to better success than routine random screening. Various ethnopharmacological agents are under investigation as immunomodulators. Traditional Chinese medicine, Japanese Kampo, Indian Ayurveda, and such are becoming important bioprospecting tools. Ayurveda gives a separate class of immunomodulating botanicals named *Rasayanas*. Ayurveda, one of the most ancient and yet living traditions practiced widely in India, Sri Lanka, and other countries, has a sound philosophical and experiential basis. India has about 45,000 plant species; medicinal properties have been assigned to many to several thousands. Ayurveda has detailed descriptions of over 700 herbs and 400,000-registered Ayurvedic practitioners routinely prescribe them, particularly for treatment of chronic disease conditions. A considerable research on pharmacognosy, chemistry, pharmacology, and clinical therapeutics has been carried out and the Ayurvedic database has detailed descriptions of over 700 medicinal plants.[6,7] Rasayanas are nontoxic herbal preparations or individual herbs used to rejuvenate or attain the complete potential of a healthy or diseased person in order to prevent diseases and degenerative changes that lead to disease. Pharmacodynamic studies on Rasayana botanicals have suggested many possible mechanisms, such as nonspecific and specific immunostimulation, free-radical quenching, cellular detoxification, cell proliferation, and cell repair. Ayurveda (with particular reference to botanicals) may play an important role in modern health care, particularly where satisfactory treatment is not available. There is a need to evaluate the potential of Ayurvedic remedies as adjuvant to counteract side effects of modern therapy and compare the cost-effectiveness of certain therapies vis-à-vis modern therapeutic schedules.[8]

16.4 ADAPTOGENS OR ADJUSTIVE MEDICINE

Most of the synthetic chemotherapeutic agents available today are immunosuppressants, are cytotoxic, and exert a variety of side effects. N. V. Lazarev, who developed the concept of a state of nonspecifically increased resistance of an organism (SNIR), laid down the theoretical basis for separation of a new group of medicinal substances. The medicinal substances causing SNIR were named *adaptogens*.[9] Generally, adaptogens are those drugs that enable one to withstand the stress and strain of life, impart immunity to give protection against diseases, postpone aging, and improve vigor, vitality, and longevity. The concept is also referred to as *adjustive medicine*. The concept of adjustive remedies has been difficult to prove experimentally. A bifunctional information exchange network between the nervous and immune systems is established by specific receptors for humoral substances on cells of nervous and immune systems. In particular, neuroregulators (neurotransmitters and neuromodulators) can modulate specific immune system function(s), and immunoregulators (immunomodulators) can modulate specific nervous system function(s). Acute and chronic inflammatory processes, malignancy, and immunological reactions stimulate the synthesis and release of immunomodulators in various cell systems. These immunomodulators have pivotal roles in the coordination of the host defense mechanisms and repair and induce a series of endocrine, metabolic, and neurologic responses.[10] However, with recent insight into the neuroendocrine immune system regulation, such adjustive effects on the homeostatic system of the body seem very likely.

16.4.1 Botanicals with Adaptogenic Activity

Mistletoe Lectin Defined nontoxic doses of the galactoside-specific mistletoe lectin (mistletoe lectin-I, a constituent of clinically approved plant extract) have immunomodulatory potencies. The obvious ability of certain lectins to activate nonspecific mechanisms supports the assumption that lectin–carbohydrate interactions may induce clinically beneficial immunomodulation. Randomized multicenter trials are being performed to evaluate the ability of complementary mistletoe lectin-I treatment to reduce the rate of tumor recurrences and metastases, to improve overall survival and the quality of life, and to exert immunoprotection in cancer patients under tumor-destructive therapy.[11]

Panax Ginseng *Panax ginseng* (Araliaceae), a Korean and Chinese medicine employed for its putative medicinal properties in South Asia, stimulated basal natural killer (NK) cell activity following subchronic exposure and helped stimulate the recovery of NK function in cyclophosphamide-immunosuppressed mice but did not further stimulate NK activity in poly I:C-treated mice; T- and B-cell responses were not affected. *P. ginseng* provided a degree of protection against infection with *Listeria monocytogenes* but did not inhibit the growth of transplanted syngenetic tumor cells. Increased resistance to *L. monocytogenes* was not detected in challenged mice previously given immunosuppressive doses of cyclophosphamide. These data suggest that *P. ginseng* have some immunomodulatory properties, associated primarily with NK cell activity.[12] Ginseng alone, or in combination with vitamins and minerals, is promoted primarily as general tonics which increase nonspecific resistance and sometimes even as an aphrodisiac. On prolonged use, ginseng shows a few adverse effects, most notable of which is the Ginseng abuse syndrome.[13]

Achyranthes Bidentata *Achyranthes bidentata* polysaccharide (ABP) root extract (25 to 100 mg/kg, day −1 to 7) could inhibit tumor growth (S-180) by 31 to 40%. A combination of cyclophosphamide and ABP increased the rate of tumor growth inhibition by 58%. ABP could potentiate LAK cell activity and increase the Con A–induced production of tumor necrosis factor (TNF-β) from murine spleenocytes. The S-180 cell membrane content of sialic acid was increased, and phospholipid decreased after ABP had acted on cells for 24 hours. Data suggest that the antitumor mechanism of ABP may be related to potentiation of host immunosurveillance mechanism and the changes in cell membrane features.[14]

Viscum Album and Echinacea Purpurea Extracts of *Viscum album* (Plenosol) and *Echinacea purpurea* (Echinacin) are used clinically for their nonspecific action on cell-mediated immunity. These two were shown to possess a stimulating effect on the production of lymphokines by lymphocytes and in the transformation test. A toxic effect on cells was produced only with very high, clinically irrelevant concentrations. Clinical application of these extracts can produce a stimulation of cell-mediated immunity (one therapeutic administration followed by a free interval of one week) or can have a depressive action (daily administration of higher doses). These observations were confirmed by lymphokine production and assay, 3H for at least three months, thymidine incorporation, and a skin test with recall antigens.[15]

16.4.2 Rasayana Botanicals as Adaptogens

Tinospora Cordifolia Treatment with aqueous, alcohol, acetone, and petroleum ether extracts of stem of *T. cordifolia* resulted in significant improvement in mice swimming time and body weights, and petroleum ether extract showed significant protective effect against cyclophosphamide-induced immunosuppression. Prevention of cyclophosphamide-induced anemia was also reported.[16]

Withania Somnifera Pretreatment with *W. somnifera, T. cordifolia,* and *Asparagus racemosus* induced a significant leucocytosis in cyclophosphamide-induced myelosuppresed animals. In terms of phagocytosis and intracellular killing, PMN functions, were stimulated, and reticuloendothelial system functions were greatly activated in treated animals. The phagocytic functions of peritoneal and alveolar macrophages were also stimulated with *T. cordifolia, A. racemosus*, and *Emblica officinalis. T. cordifolia, A. racemosus, E. officinalis, Terminalia chebula, Bacopa monira*, and *W. somnifera* improved the carbon clearance, indicating stimulation of the reticuloendothelial system. A significant increase in the proliferative fraction in the bone marrow was observed in mice treated with *T. cordifolia*, as revealed in flow cytometry analysis.[17] A comparative pharmacological investigation of *W. somnifera* (Ashwagandha) and ginseng showed a significant difference in antistress activity. Ginseng exhibited higher antistress activity; however, gastric ulcers due to swimming stress were notably less in ashwagandha. The anabolic study revealed that the ashwagandha-treated group had a greater gain in body weight than did the ginseng group.[18]

16.5 CHEMOPROTECTION

Modern cancer therapy produces substantial acute and chronic toxicity, which impairs quality of life and limits the effectiveness of treatment. Recent clinical and laboratory data suggest that repair of treatment-related injury is a multiphase and continuous process providing multiple opportunities for pharmacologic intervention. A host of agents (toxicity antagonists) are under development that modulate normal tissue response or interfere with mechanisms of toxicity. Although significant challenges remain, the routine application of such agents promises to reduce treatment-related morbidity substantially and potentially to allow treatment intensification in high-risk disease.[19] The concept of site-specific inactivation of cytotoxic anticancer agents has been explored with numerous modalities. The goal of such chemoprotection is to improve the therapeutic ratio of an agent by selectively reducing its toxicity in non-tumor-bearing tissue, which is target for dose-limiting toxicity. Furthermore, a chemoprotectant cannot add new toxicities that might otherwise limit the administration of maximally tolerated doses of chemotherapeutic agent.[20,21]

16.5.1 Drug Targets and Current Trends

Chemoprotection and cytoprotection are studied under preventive oncology and are interchangeable terms in cancer chemotherapy. Preventive oncology applies pharmacological agents to reverse, retard, or halt progression of neoplastic cells to invasive malignancy.[22] Cancer chemoprevention is one of the newer approaches in the management of cancer. Epidemiological observations, preclinical animal pharmacology, knockout models, cancer cell lines, and clinical trials have shown the efficacy of this approach. Many drug targets are under clinical development; prostaglandin pathway, estrogen receptor modulation, gluthathione peroxidase inhibition, and immunomodulation appear promising. Celecoxib, tamoxifen, retinoids, rexinoids, selenium, tocopherols, and mofarotene are some of the promising leads and are in clinical development.[23] New opportunities in clinical chemoprevention research include investigating chemopreventive effects of phytochemicals.[24] Safer immunomodulating agents suitable for long-term therapy remain an unmet therapeutic need. Figure 16.1 gives a schematic overview of cytoprotection and immunomodulation.

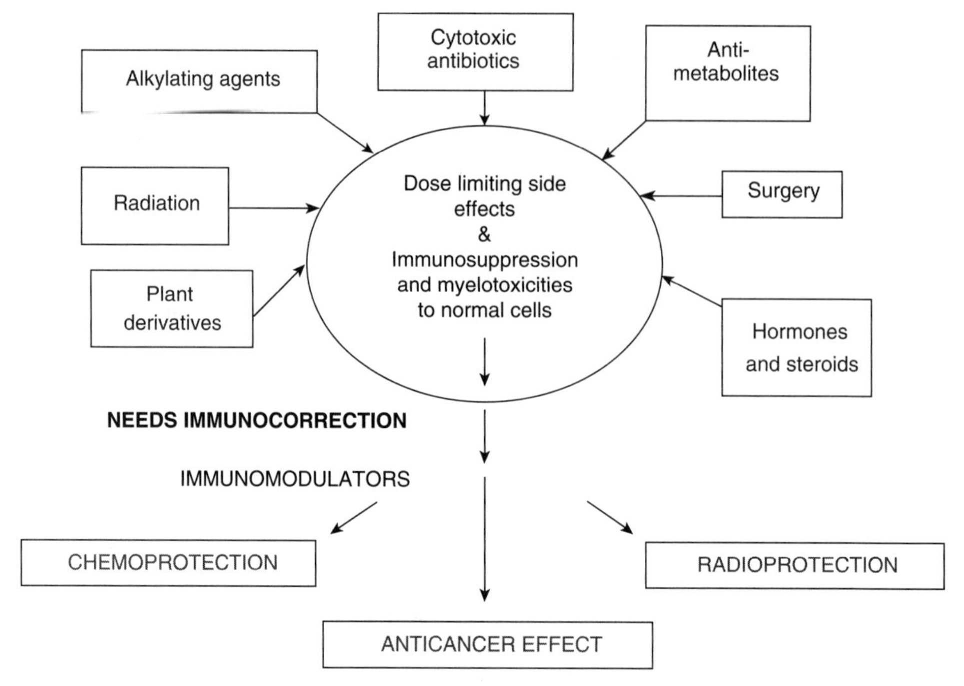

Figure 16.1 Chemoprotection and immunomodulation.

16.5.2 Chemoprotectants for Antimetabolites

Methotrexate The best-studied chemoprotectant regimen involves rescue from high-dose methotrexate bone marrow toxicity using reduced folate derivative leucovorin (folinic acid). This agent is able to substitute for the endogenous reduced folate cofactor, which methotrexate diminishes. Leucovorin can thereby replenish intracellular reduced folate pools and prevent methotrexate-induced cytotoxicity via blockade of thymidine synthesis. However, leucovorin rescue is not truly site-specific and will principally depend on differences in tumor and normal cell growth rates. Schedule-dependent chemoprotection is described for asparaginase and methotrexate. Preclinical studies show that asparaginase administered before methotrexate can attenuate its toxicity by inhibiting cellular protein synthesis. Reduced intracellular polyglutamation of methotrexate by asparaginase reduces drug retention and thereby causes more rapid methotrexate efflux, with reduced toxicity.[25,26]

Fluorouracil (5-FU) Chemoprotectants for 5-FU have been directed primarily at the formation of 5-fulorouridylate triphosphate (5-FUTP), which is closely associated with normal tissue toxicity in the gut and bone marrow. The best-studied clinical interaction involves the modulation of 5-FU myelotoxicity with concomitant allopurinol. Selective chemoprotection involves inhibition of orotidine monophosphate decarboxylase by allopurinol metabolite oxipurinol. This enzymatic blockade leads to a buildup of orotidine and orotic acid, which ultimately blocks pathways leading to FUTP formation and halts RNA inhibition by 5-FU. Selectivity for allopurinol depends on different primary pathways or ratios of 5-FU activation in normal cells and tumor cells.[27] The RNA base uridine has also been reported to block 5-FUlethal toxicity in mice and humans. Modulation of

5-FU associated gastrointestinal toxicity leading to protection was afforded by uridine diphosphoglucose (URDP), a precursor of uridine. It was found that URDP reduces the toxic side effects and increases the therapeutic index. The incorporation of 5-FU metabolites into RNA is blocked primarily as a result of increased intracellular uridine pools. Uridine rescue from 5-FU toxicity depends on prolonged exposure to uridine rather than attaining high peak plasma levels of uridine.[28–30]

Arotinoid Ro 40-8757 The arotinoid Ro 40-8757 (mofarotene) exhibits a high degree of activity against established chemically induced mammary tumors in rats. Treatment of animals with high doses of arotinoid leads to reductions in tumor numbers, with many animals becoming free of palpable tumors. The toxicities associated with these therapeutic effects are relatively mild compared to those of all-*trans* retinoic acid or 13-*cis* retinoic acids given at doses with little or no antitumor efficacy. However, long-term treatment with Ro 40-8757 results in new growth of tumors. In a rat mammary tumor model, chronic administration of cyclophosphamide (5 days/week at 10 mg/kg) plus daily administration of arotinoid at a relatively low dose (75 mg/kg per day) indicated an additive antitumor effect. However, the therapeutic effects were synergistic because all of the animals treated with cyclophosphamide as a single agent died after 6 weeks of treatment, whereas all of the animals given the combination survived the full 10 weeks of the experiment. The results of detailed studies on the hemopoietic progenitor cells in mice treated with this combination demonstrated that the protective effect of Ro 40-8757 occurred at the level of the bone marrow progenitors.[31] Combination therapy of arotinoid Ro 40-8757 and 5-FU of established chemically induced mammary tumors in rats significantly enhanced the reduction in tumor burden and tumor number. Ro 40-8757 did not have an effect on tumor burden. This protective effect of arotinoid makes it a useful potential partner for combination therapy with 5-FU.[32]

Nucleophilic Sulfur Thiols as Alkylating Agent Chemoprotectants Administration of supplemental thiols or other compounds with an available reduced sulfur atom against DNA-alkylating or DNA-binding drugs forms a biochemical rationale of chemoprotection. Commercially available thiols, or compounds containing a free sulfhydryl residue, include the amino acid cysteine, its N-acetylated derivative mucomyst, and the sodium sulfonate salt of 2-mercaptoethane sulfonate mesna. The specificity for chemoprotection by sulfur mucleophiles lies in their physical and pharmacokinetic properties. For example, sodium thiosulfate is a small molecule that is highly charged and hydrophilic at physiologic pH. Therefore, it distributes in an active state in the bloodstream but achieves poor uptake into lipophilic tissue compartments, including the central nervous system and the bone marrow. Sodium thiosulfate concentrates in the renal tubules during its rapid urinary elimination, indicating that thiosulfate will directly (and nonspecifically) inactivate any electrophilic (alkylating) species in the urine or in the bloodstream. For this reason, sodium thiosulfate is useful only as a local chemoprotectant for alkylating agents such as mechlorethamine. The poor distribution of sodium thiosulfate into bone marrow further lessens its chemoprotectant utility for agents such as mechlorethamine.[33]

16.5.3 Thiol-Based Chemoprotectants for Cisplatin and Oxazophosphorine-Based Alkylating Agents

Reduction of associated nephrotoxicity and bone marrow suppression was studied by allowing dose escalation and designing pharmacokinetically based dosing schedules.

However, new dose-limiting toxicities consisting of peripheral neuropathy and ototoxicity have emerged, which continue to restrict the potential use of high-dose cisplatin therapy. Chemoprotective agents, including sodium thiosulfate, WR2721, and diethyldithiocarbamate (DDTC) are being examined extensively as "rescue agents" for either regional or systemic administration of cisplatin. Sodium thiosulfate (STS) is given intravenously concurrently or following intraperitoneal (i.p.) cisplatin administration. This can produce a 12-fold greater exposure to cisplatin in the i.p. space and has been associated with tumor regression. However, this approach is limited to locoregional drug treatment since active form of sulfur nucleophile is distributed in the bloodstream and could complex with cisplatin to abrogate its intended systemic therapeutic effects.[34] Many studies suggest local application of STS to prevent cisplatin toxicity and maintain the systemic antitumoral effectiveness of cisplatin. Cocheal admistration resulted in complete protection against CDDP-induced hearing loss, with no change in compound action potential (CAP).[35] However, alleviation of cisplatin-induced side effects by administration of sodium thiosulfate was not observed in a mice xenograft tumor model, suggesting that cisplatin–STS treatment would provide no benefit in patients treated with cisplatin.[36] Several other thiol-based compounds have shown activity in preventing cisplatin-induced toxicities. These include the experimental aminothiol WR-2721, the disulfide metal chelator diethyldithiocarbamate (DDTC), mesna, *N*-acetylcysteine (NAC), and thiourea.

The oxazophosphorine-based antitumor agents include cyclophosphamide and ifosfamide. These alkylating agents require metabolic activation by hepatic microsomal (P450) enzymes to active species, including the bis functional alkylator phosphoramide mustard and the protein-reactive aldehyde acrolein. A difference between cyclophosphamide and its chloroethyl isomer ifosfamide involves primarily a slower rate of metabolic activation of ifosfamide to 4-hydroxyifosfamide, the precursor of the active species of the drug. Large amounts of toxin acrolein are produced and accumulate in the urinary bladder. In the absence of a chemoprotectant, active doses of ifosfamide can produce dose-limiting urotoxicity manifested by hemorrhagic cystitis, bladder fibrosis, and heightened long-term risk of bladder cancer. Similar toxicities are noted with chronic or acute high-dose cyclophosphamide dosing. Thus, both ifosfamide and high-dose cyclophosphamide have limited clinical efficiency in the absence of the effective chemoprotectant of the urinary bladder.[20]

Mesna Mesna is the sodium salt of 2-mercaptoethanesulfonic acid, a selective urinary tract protectant for oxazophosphorine-type alkylating agents. Mesna prevents bladder damage from major toxic metabolites of ifosfamide and cyclophosphamide. Mesna can bind specifically to cisplatin or alkylating agent–generated free radicals or alkylating agent metabolites to reduce the incidence of cisplatin-associated neurotoxicity and nephrotoxicity or alkylating agents associated with myelosuppression and urothelial toxicity.[37] Mesna does not block the antitumor activity of oxazophosphorines or of other classes of antitumor agents. Mesna is superior to previous urinary prophylaxis regimens among a large series of SH-based uroprotectants with the least toxicity of any of the agents tested.[38] Mesna also attenuates the lethal effect and hematological toxicity of vespeside and taxol but does not reduce specific activities in mice with transplanted tumors.[39]

Amifostine Amifostine is organic thiophosphate compound able to protect normal tissues selectively against cytotoxic agents in cellular and animal models without protecting tumor tissues. Amifostine is a prodrug that is dephosphorylated into its active metabolite, a free thiol derivative, by the membrane alkaline phosphatase of target tissue. This unique metabolism supports its cellular selectivity and its preferential uptake by normal tissues. Preclinical animal studies have demonstrated that administration of amistofine protects against irradiation and a variety of chemotherapy-related toxicities, including cisplatin-induced nephrotoxicity, neurotoxicity, and cyclophosphamide- and bleomycin-induced pulmonary toxicity and cardiotoxicity induced by doxorubicin and related chemotherapeutic agents. In nonrandomized and randomized trials in malignant melanoma, colorectal cancer, head and neck cancer, non-small cell lung cancer, and epithelial ovarian carcinoma, amifostine significantly reduced the hematological and nonhematological toxicity of DNA-damaging agents such as alkylators, platinum compounds, and mitomycin C. In more recent studies, amistofine also protected patients from side effects produced by taxanes or topoisomerase I inhibitors.[40] Currently, there is no evidence that amifostine compromises the antineoplastic effect of the drugs studied. Moreover, amifostine appears to produce growth factor–like properties, resulting in growth-promoting effects on primitive blood progenitor cells ex vivo. In a randomized phase III study conducted in patients with ovarian carcinoma receiving a combination of cisplatin and cyclophosphamide, a significant decrease in hematological, renal, and neurologic toxicity was observed in amifostine-treated patients compared with the control group.[41] The protective effect of amifostine has been demonstrated for cisplatin-induced toxicity in lung and ovarian cancer, with particular regard to nephrotoxicity, neurotoxicity, and neutropenia. No protective effect has been seen for tumor cells, owing to a selective action of amifostine on healthy tissues. A frequent side effect of amifostine is a transient decrease in blood pressure; it is usually asymptomatic if an easily handled premedication is given. Cytoprotection by amifostine is also well known for alkalyting drugs and radiation therapy, whereas it is still the object of study for new drugs, especially taxanes.[42] Amifostine also protects bone marrow from cumulative toxicity arising from chronic exposure to therapeutic agents such as alkylating agents. Well-controlled clinical trials have shown that amifostine can ameliorate cumulative bone marrow toxicity and the acute and chronic neutropenic and/or thrombocytopenic effects of cyclophosphamide. In a pivotal phase III study of cisplatin–cyclophosphamide with or without amifostine, amifostine reduced course-by-course cumulative bone marrow damage compared with the course-by-course cumulative myelosuppression experienced by those treated with cisplatin or cyclophosphamide alone.[43]

Disulfiram Disulfiram (tetraethylthioperoxidicarbonic diamide), an aldehyde dehydrogenase inhibitor, prevented cyclophosphamide-induced bladder damage in a dose-dependent manner when administered simultaneously with cyclophosphamide [100 to 400 mg/kg, intraperitoneally (i.p.)] but failed to diminish the acute toxicity, leucotoxicity, and immunotoxicity of cyclophosphamide. Diethyldithiocarbamate (DDTC), a metabolite of disulfiram, did not interfere with cyclophosphamide antitumor activity when administered 3 hours after cyclophosphamide. The protective effect of disulfiram on the bladder was critically dependent on administration timing. Disulfiram slightly potentiated the antitumor activity of cyclophosphamide against Sarcoma-180 or EL-4 leukemia in vivo when administered simultaneously with cyclophosphamide.[44]

Disulfiram is an effective protective agent against bladder damage caused by ifosfamide treatment.[45] Disulfiram (DSF), in combination with ifosfamide (IFX), prevented IFX-induced bladder damage but failed to diminish the acute lethal toxicity or leukocytotoxicity of IFX. Diethyldithiocarbamate (DDTC) prevented IFX-induced bladder damage when administered simultaneously with IFX or 1 to 5 hours afterward. The antitumor activity of IFX in ddY-mice inoculated with Sarcoma-180 or in C57BL/6J mice inoculated with

EL-4 leukemia was not impaired when it was given simultaneously with DSF or 3 hours before DDTC. Thus, neither DSF nor DDTC impaired the antitumor effect of IFX, and both diminished its adverse effects. The bladder protection of DSF and DDTC appeared to be resulting from adduct formation with acrolein and not from inhibition of the metabolic activation of IFX.[46] Disulfiram (DSF) blocks the urotoxicity of cyclophosphamide in mice and increases the oncolytic effect of cyclophosphamide in the L1210 murine leukemia. However, mice treated with cyclophosphamide and DSF appeared to have longer-lasting neutropenia than did animals treated with cyclophosphamide alone. Bone marrow granulocyte/macrophage progenitor cells (GM-CFCs) were relatively well preserved and the recovery of the GM-CFCs was not prolonged by DSF, indicating that the acute cytotoxic effect of cyclophosphamide on the granulocyte/macrophage progenitor cells is not enhanced by DSF.[47]

16.5.4 Chemoprotectants for Anthracyclines

Anthracyclins such as doxorubicin and daunomycin comprise one of the most important classes of DNA-binding antitumor agents. Short- and long-term cardiotoxicity can occur at lower doses, and the use of these drugs is limited by a characteristic clinical cardiomyopathy that develops in approximately 5 to 15% of patients after cumulative doxorubicin doses greater than 450 mg/m^2. Children and adolescents appear to be particularly sensitive to the cardiotoxic effects of doxorubicin. Although cumulative doxorubicin doses are usually limited to $\leq$450 mg/m^2, up to 70% of long-term survivors of childhood cancer have evidence of cardiac dysfunction, including overt congestive heart failure.[48] Anthracyclines complexed with metals can sustain lipid peroxidation, and scavengers of reduced oxygen free radicals, including superoxide dismutase, catalase, and mannitol, do not block its action.[49] A variety of putative free-radical scavengers have been shown to protect against doxorubicin cardiotoxicity in experimental animals. These include the lipid-soluble antioxidant vitamin E[43] and *N*-acetylcysteine (NAC). Despite positive preclinical results, NAC was not effective in cancer patients given high cumulative doxorubicin doses.[50]

ICRF-187 The piperazine derivative of ethylenediaminetetraacetate razoxane (ICRF-187) is a prodrug that is converted intracellularly to an iron-chelating agent that removes iron from doxorubicin-iron complexes in vitro. Dexrazoxane is a cardioprotective antioxidant that is used clinically to reduce the cardiotoxicity of the chemotherapeutic drugs doxorubicin, paclitaxel, and other anthracyclins.[51] Although the cardioprotective effect of dexrazoxane in cancer patients undergoing chemotherapy with anthracyclines is well documented, the potential of this drug to modulate topoisomerase II activity and cellular iron metabolism may hold the key for future applications of dexrazoxane in cancer therapy, immunology, or infectious diseases.[52]

16.5.5 Botanical Immunomodulators as Chemoprotectants

Withania Somnifera *Withania somnifera* is an official drug mentioned in the *Indian Herbal Pharmacopoeia*[48] and *Ayurvedic Pharmacopoeia*.[53] Studies indicate that *W. somnifera* (ashwagandha) (WS) possesses anti-inflammatory, antitumor, antistress, antioxidant, immunomodulatory, hemopoietic, and rejuvenating properties. The chemistry of WS has been studied extensively, and over 35 chemical constituents have been identified, extracted, and isolated. The biologically active chemical constituents are alkaloids (isopelletierine, anaferine), steroidal lactones (withanolides, withaferins), saponins containing an additional acyl group (sitoindoside VII and VIII), and withanolides with a glucose at carbon 27 (sitoindoside IX and X).[54]

The suppressive effect of cyclophosphamide-induced toxicity by WS extracts was observed in mice. Administration of WS extracts significantly reduced leucopenia induced by cyclophosphamide treatment, resulting in an increase in bone marrow cellularity. Administration of *W. somnifera* extract for 5 days along with cyclophosphamide (CTX) (1.5 mmol/kg body weight, i.p.) reduced the CTX-induced urotoxicity.[31] Treatment of *W. somnifera* resulted in the enhancement of interferon gamma (IFN-gamma), interleukin-2 (IL-2), and granulocyte macrophage colony stimulating factor (GM-CSF), which were lowered by cyclophosphamide administration. Pharmacodyanmic studies reveal that the major activity of *W. somnifera* may be due to the enhancement of cytokine production and stem cell proliferation and its differentiation.[55] These studies indicate that *W. somnifera* could reduce the cyclophosphamide toxicity and its usefulness in cancer chemotherapy.

The major activity of WS may be due to stimulation of stem cell proliferation, indicating the fact that WS could reduce cyclophosphamide toxicity and its usefulness in cancer chemotherapy.[56] WS was also shown to prevent lipid peroxidation (LPO) in stress-induced animals, indicating its adjuvant as well as chemoprotectant activity.[57] Glycowithanolides, consisting of equimolar concentrations of sitoindosides VII to X and withaferin A, isolated from the roots of WS were evaluated for protection in iron-induced hepatoxicity in rats. Ten days of oral administration of these active principles, in graded doses (10, 20, and 50 mg/kg), resulted in attenuation of hepatic lipid peroxidation (LPO) and the serum enzymes alanine aminotransferase, aspartate aminotransferase, and lactate dehydrogenase during iron-induced hepatoxicity.[58] Antistress activity observed with *W. somnifera* will be an additional benefit, along with chemoprotectant activity.

Tinospora Cordifolia *Tinospora cordifolia* is widely used in ayurvedic medicines and is known for its immunomodulatory, antihepatotoxic, antistress, and antioxidant properties. It has been used in combination with other plant products to prepare a number of ayurvedic preparations. The chemistry has been studied extensively, and its chemical constituents can be broadly divided into alkaloids, diterpenoids, steroids, flavanoids, and lignans. Reviews have appeared on quaternary alkaloids and biotherapeutic diterpene glucosides of *Tinospora* species. Much of the work has been carried out on berberine, jatrorrhizine, tinosporaside, and columbin. Extracts of *T. cordifolia* (TC) have been shown to inhibit lipid peroxidation and superoxide and hydroxyl radicals in vitro. The extract was also found to reduce the toxic side effects of cyclophosphamide (25 mg/kg, 10 days) in the mice hematological system by free-radical formation as seen from total white cell count, bone marrow cellularity, and α-esterase-positive cells.[59] The active principles of TC were found to possess anticomplementary and immunomodulatory activities.[60] TC is reported for its various immunopharmacological activities (e.g., inhibition of C3-convertase of the classical complement pathway). Humoral and cell-mediated immunity were reported for cardioside, cardifolioside A, and cardiol and their activation was more pronounced with increasing incubation time.[61] Extracts of *T. cordifolia* has been shown to inhibit lipid peroxidation and superoxide and hydroxyl radicals in vitro. The extract was also found to reduce the toxic side effects of cyclophosphamide (25 mg/kg, 10 days) in the mice hematological system by free-radical formation as seen from total white cell count, bone marrow cellularity, and α-esterase-positive cells.

Tinospora Bakis Dose-dependent cytoprotection by *Tinospora bakis*, a plant from the Senegalese pharmacopoeia, was observed in an in vitro model. A lyophilized aqueous extract of plant roots decreased intracellular enzyme release (LDH and ASAT) from CCl_4-intoxicated hepatocytes isolated from rats. The cytoprotective effect was more effective for long-course treatment.[62]

Asparagus Racemosus Chemically, *Asparagus racemosus* (AR) contains steroidal saponins, known as shatavarins, isoflavaones, and isoflavones, including 8-methoxy-5, 6,4′-trihydroxyisoflavone 7-*O*-β-D-glucopyranoside, asparagamine, a polycyclic alkaloid, racemosol, a cyclic hydrocarbon (9,10-dihydrophenanthrene), polysaccharides, and mucilage. AR has been shown to stimulate macrophages and influence long-term adaptation favorably. Possible links between immunomodulatory and neuropharmacological activity have been suggested. Extracts of *A. racemosus* were evaluated for its neuroendocrine immune modulating effect. It prevents stress-induced increase in plasma cortisol along with activation of peritoneal macrophages and inhibition of gastric vascular damage.[63] A comparative study between *A. racemosus, T. cordifolia*, glucan, and lithium carbonate against the myelosuppressive effects of single does (200 mg/kg, subcutaneously) and multiple doses (three doses, 30 mg/kg, i.p.) of cyclophosphamide in mice revealed that all four drugs prevented, to varying degrees, leucopenia produced by cyclophosphamide.[64] Treatment of *A. racemosus* significantly inhibited ochratoxin A–induced suppression of chemotactic activity and production of IL-1 and TNF-α by macrophages.[65]

Our studies on these plants (i.e., WS, TC, and AR) revealed that they have significant immunomodulatory activity. This activity can be useful in a variety of conditions, such as myeloprotection in cancer chemotherapy and immunoprotection during infection. We observed that treatment of ascitic sarcoma–bearing mice with formulation of total extracts of WS and TC and alkaloid-free polar extract of WS resulted in protection toward cyclophosphamide-induced myelo- and immunosuppression. In another situation, these plants were evaluated for their immunoadjuvant potential in a pertussis model, where aqueous extracts of these plants reduced the dose of vaccine required to confer protection against pertussis intracerebral challenge, increasing survival percentage and significantly increasing in pertussis antibody titers. These observations are of major importance in the immunochemical industry and in vaccination strategies.[66–68]

Crocetin A natural carotenoid, crocetin, at a dose of 50 mg/kg, modulated the release of chloroaceteldehyde, a urotoxic metabolite of cyclophosphamide in the urine of mice given a combined treatment. Crocetin at the same dose significantly elevated glutathione-*S*-transferase enzyme activity in both the bladder and liver of mice treated with cyclophosphamide. In Sarcoma-180 tumor–bearing mice, crocetin has the ability to protect against cyclophosphamide-induced bladder toxicity without altering its antitumor activity.[69] Crocetin also inhibited benzo[*a*]pyrene–induced genotoxicity and neoplastic transformation in C3H10T1/2 cells. Crocetin was found to increase the activity of GST and decreases the formation of a benzo[*a*]pyrene–DNA adduct.[70]

UL-409 Oral administration of UL-409, a herbal formulation, at a dose of 600 mg/kg significantly prevented the occurrence of cold-resistant stress-induced ulcerations in Wistar rats, alcohol- and aspirin-induced gastric ulceration, and cysteamine- and histamine-induced duodenal ulcers in rats and guinea pigs, respectively. The volume and acidity of gastric juice in pylorius-ligated rats was also reduced by UL-409. It also significantly, and dose dependently, promoted gastric mucus secretion in normal as well as in stress-, drug-, and alcohol-induced ulceration in animals.[71]

Mikania Cordata Induction of phase 2 enzymes is an effective and sufficient strategy for achieving protection against the toxic and neoplastic effects of many carcinogens. Literature reports suggest that the chemopreventive action of *Mikania chordata* is based on its effect on phase 2 enzymes. *M. cordata* oral administration resulted in increased activity of microsomal uridine diphosphoglucose dehydrogenase and reduced nicotinamide adenine dinucleotide (phosphate): quinine reductase and cytosolic glutathione *S*-transferases, with a concomitant elevation in the contents of reduced glutathione. *M. chordata* was also found to increase the total protein mass, fractional rate of protein synthesis, ribosomal capacity and efficiency (rate/ribosome), and high turnover rate of protein (protein/DNA) on pretreatment in CCl_4-treated hepatic tissue. This indicated tissue repair leading to a functional improvement in CCl_4 disorganized hepatocytes.[72] Oral administration of a methanolic fraction of *M. cordata* (Burm., B. L. Robinson) significantly prevented the occurrence of water-immersion stress-induced gastric ulcers in a dose-responsive manner. The extract also dose-dependently inhibited gastric ulcers induced by ethanol, aspirin, and phenylbutazone. The volume, acidity, and peptic activity of the gastric juice in pylorus-ligated rates were not altered upon administration of the extract but significantly and dose-dependently promoted gastric mucus secretion in normal as well as stress- and ethanol-induced ulcerated animals. It was claimed that the activity observed might be due to the modulation of defensive factors through an improvement in gastric cytoprotection.[73]

Indigenous Herbal Drug Formulations Brahma Rasayana and Ashwagandha Rasayana were found to protect mice from cyclophosphamide-induced (50 mg/kg daily for 14 days) myelosuppression and subsequent leucopenia.[74] Treatment with *A. racemosus, T. cordifolia, W. somnifera*, and *Picrorhiza kurrooa* significantly inhibited carcinogen ochratoxin A (OTA)–induced suppression of chemotactic activity and production of interleukin-1 (IL-1) and tumor necrosis factor–alpha (TNF-α) by macrophages. Immu-21, a polyherbal formulation that contains extracts of *Ocimum sanctum, W. somnifera, E. officinalis*, and *T. cordifolia*, at 100 mg/kg daily over 7 days and 30 mg/kg daily over 14 days prevented cyclophosphamide-induced genotoxicity in mice.[75]

16.6 RADIOPROTECTION

Radiotherapeutics is still one of the major treatment modalities practiced for control of localized solid tumors. The major goal of therapy is the achievement of total tumor control with limited toxicity and complications. Radiation doses that can be delivered without causing severe damage to surrounding normal tissues can be insufficient to eradicate a tumor. New strategies for the prevention of radiation injuries are currently being explored with the ultimate aim of developing globally radioprotective nontoxic pharmacologics. These include the development of agents as radiosensitizers, apoptosis inducers in tumor cells, and radioprotectants, which can increase the radiosensitivity of such resistant tumors, reduce the radiation dose required, and protect the normal tissue morbidity associated with tumoricidal radiation doses. Radioprotective agents, although widely studied in the past four decades and including several thousand agents, have not reached the level of providing the agent that conforms to criteria required of an optimal radioprotective, including effectiveness, specificity, availability, toxicity, and tolerance. The prophylactic treatments under review encompass diverse pharmacological classes as novel immunomodulators, nutritional antioxidants, and cytokines.[76]

With the advancement in understanding of tumor cell biology, many molecular mechanisms or targets have been identified for modulation of radiation response. Some important ones are COX-II, MMPs, TRAIL (TNF-α-related apoptosis-inducing ligand)/Apo2L, and epidermal growth factor receptor (EGFR).[77–79] Substances reducing the radiation-induced toxicity by modulating the biological response to radiation injury may represent an alternative concept in radioprotection, and hence there is interest in and the need for new compounds that can protect tissues from radiation injury. Natural compounds may have an advantage, being more structurally diverse and safer, hence more acceptable for human application.

16.6.1 Radioprotectants from Botanicals

Withaferin A is a steroidal lactone from *W. somnifera* inhibited growth of Ehrlich ascites carcinoma in Swiss mice with increased survival and life span. Antitumor and radiosensitizing effects were observed with a combination treatment of abdominal gamma irradiation with withaferin A, resulting in increased tumor cure and tumor-free survival.[80] Withaferin A showed growth inhibitory effect in vitro on both Chinese hamster V79 and HeLa cells. It reduced the survival of V79 cells in a dose-dependent manner.[81] Combination treatment of an alcoholic extract of *Withania somnifera* (500 mg/kg, i.p., 10 days) with one local exposure to gamma radiation (10 Gy) followed by hyperthermia (43°C for 30 minutes) significantly increased the tumor cure (Sarcoma-180 grown on the dorsum of adult BALB/c mouse), growth delay, and animal survival. This combination also depleted the tumor GSH level significantly and synergistically. Thus, in addition to having a tumor inhibitory effect, ashwagandha acts as a radiosensitizer, and heat enhances these effects. The severe depletion in the tumor GSH content by the combination treatment must have enhanced the tumor response, as the inherent protection of the thiol will be highly reduced.[82] The presence of a wide variety of effects, such as hypotensive, antispasmodic, antitumor, antiarthritic, antipyretic, analgesic, anti-inflammatory, and hepatoprotective activities and antistress properties, show that *W. somnifera* may be acting by nonspecifically increasing the resistance of the animals to various stressful conditions.[83]

16.6.2 Botanical Immunomodulators as Antitumor Agents

Plant products have contributed several novel compounds that possess promising antitumor activity. Crude extract of *W. somnifera* root has a strong tumoricidal and tumor growth inhibitory activity. Withaferin A, an alkaloid isolated from the leaves, has been reported to show marked tumor inhibitory activity in vitro against cells derived from human carcinoma of nasopharynx and experimental mouse tumors. A single i.p. dose of withaferin A injection 24 or 48 hours after Ehrlich's ascites tumor transplantation produced an immediate growth reduction in 3 to 80% of mice, followed by complete disappearance of tumor cells in the peritoneal cavity of the surviving mice with no signs of tumor development. However, withaferin A is toxic in mice, with an LD_{50} of 54 mg/kg body weight after an i.p. injection. An effective dose of crude extract was much higher (a cumulative dose of more than 10 g—750 mg/kg daily for 15 days) with less toxicity than reported doses of purified withaferin A and withanolide D, which exhibited toxic effects. Crude extract included a range of chemicals (e.g., a few flavanoids, several alkaloids, and other withanolides) in addition to withaferin A. In the ayurvedic system of treatment, dry powders or crude extract is used, and hence the effects observed may not be attributed to a single component. The rationale for this type of treatment is that the toxicity of an active component may be counteracted by another component, which may not have the desired therapeutic property.

Resinous material from a methanol extract and orange-colored oil from a petroleum ether extract of *Semecarpus anacardium* Linn. F. have been found to possess antitumor properties in a P388 lymphocytic leukemia model.[84] Acetylated oil of *Semicarpus anacardium*, which itself does not possess antitumor activity against experimental transplantable tumors, enhances the antitumor effect of anticancer drugs such as mitomycin-C, 6-mercaptopurine, and methotrexate when used in combination against P388 and S180 (ascites) tumor systems.[85]

Extracts of *T. cordifolia* have been shown to inhibit lipid peroxidation and superoxide and hydroxyl radicals in vitro. The extracts were also found to reduce the toxic side effects of cyclophosphamide administration in mice. Moreover, their administration partially reduced elevated lipid peroxides in serum and liver as well as alkaline phosphatase and glutamine pyruvate transaminase thus indicating the value of *Tinospora* extracts in reducing the chemotoxicity induced by free radical–forming chemicals.[86]

Crude saponins obtained from shoots of *Asparagus racemosus* [asparagus crude saponins (ACSs,] were found to have antitumor activity. They inhibited the growth of human leukemia HL-60 cells in culture and macromolecular synthesis in a dose- and time-dependent manner. The ACS in the range 75 to 100 μg/mL was cytostatic; at concentrations greater than 200 μ/mL it was cytocidal to HL-60 cells. ACSs at 6 and 50 μg/mL inhibited the synthesis of DNA, RNA, and protein in HL-60 cells by 41, 5, and 4% or by 84, 68, and 59%, respectively. The inhibitory effect of ACSs on DNA synthesis was irreversible.[87]

16.7 CONCLUSIONS

Many chemical agents are used as chemoprotectants for conventional cancer chemotherapy and/or radiation therapy. However, their effect is locoregional and is dependent on dose and time of administration, in contrast to anticancer drugs. The limitations and inconvenience of their use has stimulated research to discover natural resources with immunological activity. Various botanicals and ethnopharmacological agents of traditional medicine are under investigation for chemoprotective and immunomodulatory activities. The results are encouraging. Chemoprofiling of these botanicals have been reported, but most activity reports are for crude semiprocessed extracts or fractions. Several reports suggest cytostatic and cytotoxic properties, along with enhanced immune function in extract and or polyherbal formulations. Nevertheless, any single component isolated from an extract or formulation may not retain all three desired properties. There have been attempts to isolate and characterize active moieties, with limited success. Many authors have hypothesized the presence of synergism and buffering in extracts; however, systematic scientific investigations on pharmacodyanmics, kinetics, dosing, and interactions need to be undertaken to study these principles. Furthermore, studies are required to better understand the molecular and biochemical mechanisms involved in immunoregulation and its role in cytoprotection and radioprotection. Such efforts might lead to effective integration of botanical medicine in cancer therapy. This review will be useful in bioprospecting exercises toward the development of newer, safer, and effective agents for therapeutic management of cancer.

REFERENCES

1. Devi, P. U., Sharada, A. C., Solomon, E. F., and Kamath, M. S., In vivo growth inhibitory effect of *Withania somnifera* (ashwagandha) on a transplantable mouse tumor, Sarcoma 180, *Indian J. Exp. Biol.*, 30; 169–172, 1992.
2. Patwardhan, B., Kalbag, D., Patki, P. S., and Nagasampagi, B. A., Search of immunoregulatory agents: a review, *Indian Drugs*, 28(2); 56–63, 1990.
3. Upadhyay, S. N., Plant products as immune response modulators, *Proceedings of the International Ayurveda Conference' 97*, Sanjay Gandhi Post Graduate Institute of Medical Sciences, Lucknow, Indian 10, pp. 1997.
4. Chirigos, M. A., Immunomodulators: current and future development and applications, *Thymus*, 19(1); S7–S20, 1992.
5. Turowski, R. C., and Triozzi, P. L., Application of chemical immunomodulators to the treatment of cancer and AIDS, *Cancer Invest.*, 12(6); 620–643, 1994.
6. Patwardhan, B., Ayurveda: the designer medicine—review of ethnopharmacology and bioprospecting research, *Indian Drugs*, 37; 213–227, 2000.
7. Patwardhan, B., Vaidya, A., and Chorghade, M., Ayurveda and natural products drug discovery, *Curr. Sci.*, 86; 789–799, 2004.
8. Rege, N. N. Adaptogenic properties of six rasayana herbs in ayurvedic medicine, *Phytother. Res.*, 13; 275–291,1999.
9. Brekhman, I. L., and Dardimov, I. V., New substances of plant origin which increase nonspecific resistance, *Annu. Rev. Pharmacol.*, 21; 419–426, 1969.
10. Plata-Salaman, C. R., Immunomodulators and feeding regulation: a humoral link between the immune and nervous systems, *Brain Behav. Immunol.*, 3(3); 193–213, 1989.
11. Beuth, J., Clinical relevance of immunoactive mistletoe lectin-I, *Anticancer Drugs*, 8(1); S53–S55, 1997.
12. Kim, J. Y., Germolee, D. R., and Luster, M. I., Panax ginseng as a potential immunomodulator: studies in mice, *Immunopharmacol. Immunotoxicol.*, 12(2); 257–276, 1990.
13. Siegel, R. K., Ginseng abuse syndrome: problems with the panacea, *J. Am. Med. Assoc.*, 241(15); 1614–1615, 1979.
14. Yu, S., and Zhang, Y., Effect of *Achyranthes bidenta* polysaccharide (ABP) on antitumor activity and immune function of S-180 bearing mice, *Chung Hua Chung Liu Tsa Chih* (China), 17(4); 275–278, 1995.
15. Coeugniet, E. G., and Elek, E., Immunomodulation with *Viscum album* and *Echinacea purpurea* extracts, *Onkologie*, 10(3); 27–33, 1987.
16. Patil, M, Patki, P., Kamath, H. V., and Patwardhan, B., Antistress activity of *Tinospora cordifolia* (Wild) Miers, *Indian Drugs*, 34(4); 211–215, 1997.
17. Dahanukar, S. A., and Thatte, U. M., Current status of ayurveda in phytomedicine, *Phytomedicine*, 4(4); 359–368, 1997.
18. Grandhi, A., Mujumdar, A. M., and Patwardhan, B., A comparative pharmacological investigation of ashwagandha and ginseng, *J. Ethnopharmacol.*, 44; 131–135, 1994.
19. Trotti, A., Toxicity antagonists in cancer therapy, *Curr. Opin. Oncol.*, 9(6); 569–578, 1997.
20. Dorr, R. T., Chemoprotectants for cancer chemotherapy, *Semi. Oncol.*, 18(1); 48–58, 1991.
21. Matthew, L., and Craig, L., Chemoprotectants: a review of their clinical pharmacology and therapeutic efficacy, *Drugs*, 57; 293–308, 1999.
22. Krzystyniak K. L. Current strategies for anticancer chemoprevention and chemoprotection, *Acta Pol. Pharm.*, 59(6); 473–478, 2002.

23. Krishnan, K., Campbell, S., Abdel-Rahman, F., Whaley, S., and Stone, W. L., Cancer chemoprevention drug targets, *Curr. Drug Targets*, 4(1); 45–54, 2003.

24. Kucuk, O., New opportunities in chemoprevention research, *Cancer Invest.*, 20(2); 237–245, 2002.

25. Bertino, J. R., Rescue techniques in cancer chemotherapy: use of leucovorin and other rescue agents after methotrexate treatment, *Semin. Oncol.*, 4; 203–216, 1977.

26. Jolivet, J., Cole, D. E., and Holcenberg, J. S., Prevention of methotrexate cytotoxicity by asparaginase inhibition of methotrexate polyglutamate formation, *Cancer Res.*, 45; 217–220, 1985.

27. Howell, S. B., Wung, W., and Taetle, R., Modulation of 5-fluorouracil toxicity by allopurinol in man, *Cancer*, 48; 1281–1289, 1981.

28. Klubes, P., and Cerna, I., Use of uridine rescue to enhance the antitumor selectivity of 5-fluorouracil, *Cancer Res.*, 43; 3182–3186, 1983.

29. Stolfi, R. L., Martin, D. S., and Sawyer, R. C., Modulation of 5-fluorouracil-induced toxicity in mice with interferon or with the interferon inducer, polyinosinic–polycytidylic acid, *Cancer Res.*, 43; 561–566, 1983.

30. Van Groeningen, C. J., Petes, G. J., and Leyva, A., Reversal of 5-fluorouracil-induced myelosuppression by prolonged administration of high-dose uridine, *J. Natl. Cancer Inst.*, 81; 157–162, 1989.

31. Eliason, J. F., Inoue, T, Kubota, A., Teelmann, K., Hurh, I., and Hartmann, D., The antitumor arotinoid Ro 40-8757 protects bone marrow from the toxic effects of cyclophosphamide, *Int. J. Cancer*, 55; 492–497, 1993.

32. Eliason, J. F., Inoue, T., Kubota, A., Hurh, I., and Hartmann, D., The anti-tumor arotinoid Ro 40-8757 protects bone marrow from the toxic effects of 5-fluorouracil, *Int. J. Cancer*, 57; 192–197, 1994.

33. Shea., M., Koziol, J. A., and Howell, S. B., Kinetics of sodium thiosulfate, a cisplatin neutralizer, *Clin. Pharmacol. Ther.*, 35; 419–425, 1984.

34. Pfeifle, C. E., Howell, S. B., Felthouse, R. D., et al., High-dose cisplatin with sodium thiosulfate protection, *J. Clin. Oncol.*, 3; 237–244, 1985.

35. Wang, J., Lloyd Faulconbridge, R. V., Fetoni, A., Guitton, M, J., Pujol, R., and Puel, J. L., Local application of sodium thiosulfate prevents cisplatin-induced hearing loss in the guinea pig, *Neuropharmacology*, 45(3); 380–393, 2003.

36. Masaki, I., Chikako, S., Hiromu, S., and Osamu, T., Neutralizing effect of sodium thiosulfate on antitumor efficacy of cisplatin for human carcinoma xenografts in nude mice, *Gynecol. Oncol.*, 40; 34–37, 1991.

37. Lewis, C. M., Chemoprotectants: a review of their clinical pharmacology and therapeutic efficacy, *Drugs*, 57(3); 293–308,1999.

38. Dorr, R. T., Daniel, D., and Hoff, V., *Cancer Chemotherapy Handbook*, 2nd ed., Appleton & Lange, East Norwalk, CT, 1994, pp. 685–688.

39. Hensley, M. L., Schuchter, L. M., Lindley, C., Meropol, N. J., Cohen, G. I., and Broder, J., *J. Clin. Oncol.*, 17(10); 3333–3355, 1999.

40. Santini, V., Amifostine: chemotherapeutic and radiotherapeutic protective effects, *Expert Opin. Pharmacother.*, 2(3); 479–489, 2001.

41. Lenoble, M., Amifostine: current and future applications in cytoprotection, *Bull. Cancer*, 83(9); 773–787, 1996.

42. Castiglione, F., Dalla, M., and Porcile, G., Protection of normal tissues from radiation and cytotoxic therapy: the development of amifostine, *Tumori*, 85(2); 85–91,1999.

43. Alberts, D. S., Protection by amifostine of cyclophosphamide-induced myelosuppression, *Semin. Oncol.*, 26(2); 37–40, 1999.

44. Ishikawa, M., Aoki, T., Yomogida, S., Takayanagi, Y., and Sasaki, K., Disulfiram as protective agent against cyclophosphamide-induced urotoxicity without compromising antitumor activity in mice, *Pharmacol. Toxicol.*, 74(4–5); 255–261,1994.

45. Ishikawa, M., and Takayanagi, Y., Inhibition of ifosfamide-induced urotoxicity by disulfiram in mice, *Jpn. J. Pharmacol.*, 49(1); 147–150, 1989.

46. Ishikawa, M.,Takayanagi, Y., and Sasaki, K., Drug interaction effects on antitumor drugs: VIII. Prevention of ifosfamide-induced urotoxicity by disulfiram and its effect on antitumor activity and acute toxicity of alkylating agents in mice, *Pharmacol. Toxicol.*, 68(1); 21–25, 1991.

47. Gamelli, R. L., Ershler, W. B., Kacker, M. P., and Foster, P. S., The effect of disulfiram on cyclophosphamide-mediated myeloid toxicity, *Cancer Chemother. Pharmacol.*, 16(2); 153–155, 1986.

48. Wexler, L. H., Andrich, M. P., and Venzon, D., Randomized trial of the cardioprotective agent ICRF-187 in pediatric sarcoma patients treated with doxorubicin, *J. Clin. Oncol.*, 14(2); 362–372, 1996.

49. Gianni, I., Vigano, L., and Lanzi, C., Role of daunosamine and hydroxyacetyl side chain in reaction with iron and lipid peroxidation by anthracyclins, *J. Natl. Cancer Inst.*, 80; 1104–1111, 1988.

50. Myers, C. E., McGuire, W. P., and Liss, R. H., Adriamycin: the role of lipid peroxidation in cardiac toxicity and tumor response, *Science*, 197; 165–167, 1977.

51. Weiss, G., Loyevsky, M., and Gordeuk, V.R., Dexrazoxane (ICRF-187), *Gen. Pharmacol.*, 32(1); 155–158, 1999.

52. Della Torre P., Imondi A, R., Bernardi, C., Podesta, A., Moneta, D., Riflettuto, M., and Mazue, G., Cardioprotection by dexrazoxane in rats treated with doxorubicin and paclitaxel, *Cancer Chemother. Pharmacol.*, 44(2); 138–142, 1999.

53. *Indian Herbal Pharmacopoeia*, joint publication of Indian Drugs Manufacturers' Association and Regional Research Laboratory, Jammu-Tawi India, 1998, pp. 165–173.

54. Mishra, L. C., Singh, B. B., and Dagenais, S., Scientific basis for the therapeutic use of *Withania somnifera* (ashwagandha): a review, *Altern. Med. Rev.*, 5(4); 334–346, 2000.

55. Davis, L., and Kuttan, G., Effect of *Withania somnifera* on cytokine production in normal and cyclophosphamide treated mice, *Immunopharmcol. Immunotoxicol.*, 21(4); 695–703, 1999.

56. Davis, L., and Kuttan, G., Suppressive effect of cyclophosphamide-induced toxicity by *Withania somnifera* extract in mice, *J. Ethnopharmcol.*, 62(3); 209–214, 1998.

57. Dhuley, J. N., Effect of ashwagandha on lipid peroxidation in stress-induced animals. *J. Ethnopharmacol.*, 60(2); 173–178, 1998.

58. Bhattacharya, A., Ramanathan, M., Ghosal, S., and Bhattacharya, S. K., Anti-stress activity of sitoindosides VII and VIII, new acylsterylglucosides from *Withania somnifera, Phytother. Res.*, 1; 32–37, 1987.

59. Mathew, S., and Kuttan, G., Antioxidant activity of *Tinospora cordifolia* and its usefulness in the amelioration of cyclophosphamide activity, *J. Exp. Clin. Cancer Res.*, 16(4); 407–411,1997.

60. Kapil, A., and Sharma, S., Immunopotentiating compounds from *Tinospora cordifolia, J. Ethnopharmacol.*, 58(2); 89–95, 1997.

61. Patil, M., Patki, P., Kamath, H. V., and Patwardhan, B., Antistress activity of *Tinospora cordifolia* (Wild) Miers, *Indian Drugs*, 34(4), 211, 1997.

62. Diallo-Sall, A., Niang-Ndiaye, M., Ndiaye, A. K., and Dieng, C. F., Hepato-protective effect of a plant from the Senegalese pharmacopoeia: *Tinospora bakis* (Menispermaceae) using an in vitro model, *Dakar Med.*, 42(1); 15–18, 1997.

63. Dahanukar, S., and Thatte, U., Rasayana concept of ayurveda myth or reality: an experimental study, *Indian Pract.*, 245–252, 1988.

64. Thatte, U. M., and Dahanukar, S. A., Comparative study of immunomodulating activity of Indian medicinal plants, lithium carbonate and glucan, *Methods Find. Exp. Clin. Pharmacol.*, 10; 639–644, 1988.
65. Dhuley, J. N., Effect of some Indian herbs on macrophage functions in ochratoxin A treated mice, *J. Ethnopharmacol.*, 58(1); 15–20, 1997.
66. Gautam, M., Diwanay, S. S., Gairola, S., Shinde, Y. S., Patki, P. S., and Patwardhan, B., Immunoadjuvant potential of *Asparagus racemosus* in experimental system, *J. Ethnopharmacol.*, 91; 251–255, 2004.
67. Gautam, M., Diwanay, S. S., Gairola, S., Shinde, Y. S., Jadhav, S. S., and Patwardhan, B., Immune response modulation to DPT vaccine by aqueous extract of *Withania somnifera* in experimental system, *Int. Immunopharmacol.*, 4, 841–849, 2004.
68. Diwanay, S., Chitre, D., and Patwardhan, B., Immunoprotection by botanical drugs in cancer chemotherapy, *J. Ethnopharmacol.*, 90, 49–55; 2004.
69. Chang, W. C., Lin, Y. L., Lee, M. J., Shiow, S. J., and Wang, C. J., Inhibitory effect of crocetin on benzo[*a*]pyrene genotoxicity and neoplastic transformation in C3H10T1/2 cells, *Anticancer Res.*, 16(6B); 3603–3608, 1996.
70. Nair, S. C., Panikar, K. R., and Parthod, R. K., Protective effect of crocetin on bladder toxicity induced by cyclophosphamide, *Cancer Biother.*, 8(4); 339–343, 1993.
71. Mitra, S. K., Gopumadhavan, S., Hemavathi, T. S., Muralidhar, T. S., and Venkataranganna, M. V., Protective effect of UL-409, a herbal formulation against physical and chemical factor induced gastric and duodenal ulcers in experimental animals, *J. Ethnopharmacol.*, 52(3); 165–169, 1996.
72. Bishayee, A., and Chatterjee, M., Anticarcinogenic biological response of *Mikania cordata*: reflections in hepatic biotransformation systems, *Cancer Lett.*, 30; 81(2); 193–200, 1994.
73. Mandal, P. K., Bishayee, A., and Chatterjee, M., Stimulation of hepatic protein synthesis in response to *Mikania cordata* root extract in carbon tetrachloride–induced hepatotoxicity in mice, *Ital. J. Biochem.*, 41(6); 345–351,1992.
74. Praveenkumar, V., Kuttan, R., and Kuttan, G., Chemoprotective action of rasayanas against cyclophosphamide toxicity, *Tumori*, 80; 306–308, 1994.
75. Jena G. B., Nemmani, K. V., Kaul, C. L., and Ramarao, P., Protective effect of a polyherbal formulation (Immu-21) against cyclophosphamide-induced mutagenicity in mice, *Phytother. Res.*, 17(4); 306–310, 2003.
76. Weiss, J. F., and Landauer, M. R., Protection against ionizing radiation by antioxidant nutrients and phytochemicals, *Toxicology*, 189(1–2); 1–20, 2003.
77. Lammering, G., Anti-epidermal growth factor receptor strategies to enhance radiation action, *Curr. Med. Chem. Anti-cancer Agents*, 3(5); 327–333, 2003.
78. Marini, P., and Belka, C., New strategies for combined treatment with ionizing radiation, *Curr. Med. Chem. Anti-cancer Agents*, 3(5); 334–342, 2003.
79. Petersen, C., Baumann, M., and Petersen, S., New targets for the modulation of radiation response: selective inhibition of the enzyme cyclooxygenase 2, *Curr. Med. Chem. Anti-cancer Agents*, 3(5); 354–359, 2003.
80. Devi, P. U., *Withania somnifera* Dunal (ashwagandha): potential plant source of a promising drug for cancer chemotherapy and radiosensitization, *Indian J. Exp. Biol.*, 34(10); 927–932, 1996.
81. Sharada, A. C., Soloman, F. E., Devi, P. U., Udupa, N., and Srinivasan, K. K., Antitumor and radiosensitizing effects of withaferin A on mouse Ehrlich ascites carcinoma in vivo, *Acta Oncol.*, 35(1); 95–100, 1996.
82. Devi, P. U., Sharada, A. C., and Soloman, F. E., Antitumor and radiosensitizing effects of *Withania somnifera* (ashwagandha) on a transplantable mouse tumor, Sarcoma-180, *Indian J. Exp. Biol.*, 31(7); 607–611, 1993.

83. Asthana, R., and Raina, M. K., Pharmacology of *Withania somnifera* (Linn) Dunal: a review, *Indian Drugs*, 26(5); 199–205, 1989.

84. Indap, M. A., Ambaye, R. Y., and Gokhale, S. V., Anti-tumor and pharmacological effects of the oil from *Semicarpus anacardium* Linn. F., *Indian J. Physiol. Pharmacol.*, 27(2); 83–91, 1983.

85. Indap, M. A., Ambaye, R. Y., and Gokhale, S. V., Potentiation of activity of anti-cancer drugs by acetylated oil of *Semicarpus anacardium* Linn. F. in experimental tumors, *Indian Drugs*, 23(8); 447–451, 1986.

86. Mathew, S., and Kuttan, G., Antioxidant activity of *Tinospora cordifolia* and its usefulness in the amelioration of cyclophosphamide activity, *J. Exp. Clin. Cancer Res.*, 16(4); 407–411, 1997.

87. Shao, Y., Chin, C. K., Ho, C. T., Ma, W., Garrison, S. A., and Huang, M. T., Anti-tumor activity of the crude saponins obtained from asparagus, *Cancer Lett.*, 104(1); 31–36, 1996.

INDEX

Drug Discovery and Development, Volume 1: Drug Discovery, Edited by Mukund S. Chorghade